Aquaculture: Principles and Practices

Aquaculture: Principles and Practices

Edited by Geoffrey Gilbert

SYRAWOOD PUBLISHING HOUSE

New York

Published by Syrawood Publishing House,
750 Third Avenue, 9th Floor,
New York, NY 10017, USA
www.syrawoodpublishinghouse.com

Aquaculture: Principles and Practices
Edited by Geoffrey Gilbert

© 2019 Syrawood Publishing House

International Standard Book Number: 978-1-68286-668-9 (Hardback)

Cataloging-in-Publication Data

Aquaculture : principles and practices / edited by Geoffrey Gilbert.
    p. cm.
Includes bibliographical references and index.
ISBN 978-1-68286-668-9
1. Aquaculture. 2. Agriculture. I. Gilbert, Geoffrey.
SH135 .A68 2019
639.8--dc23

TABLE OF CONTENTS

Preface..IX

Chapter 1   **An Analysis of Genetic Variability in Wild and Hatchery Populations of Swimming Crab *(Portunus trituberculatus)* Using AFLP Markers** 1
Ling-Xiao Liu, Yun-Guo Liu and Shi Chao Xing

Chapter 2   **Growth, Mortality and Exploitation of *Sardinella maderensis* (Lowe, 1838) in the Liberian coastal waters** 6
Wehye AS, Amponsah SKK and Jueseah AS

Chapter 3   **Different Heat Shock Application Effect on Gynogenetic Production of Zebrafish *(Danio rerio)*** 11
Rahmi Can Ozdemir and Aygül Ekici

Chapter 4   **Effects of Seasonal Variation on Fish Catching in Jebel Aulia Reservoir on the White Nile, Sudan** 17
Ahmed Mohammed Musa Ahmed

Chapter 5   **Features of Sperm Motility and Circadian Rhythm in Japanese Anchovy *(Engraulis japonicus)*** 22
Dipak Pandey, Yong-Woon Ryu and Takahiro Matsubara

Chapter 6   **Classical Fisheries Theory and Inland (Floodplain) Fisheries Management; Is there Need for a Paradigm Shift? Lessons from the Okavango Delta, Botswana** 28
Mosepele K

Chapter 7   **Growth Patterns and Condition Factor of *Hepsetus odoe* (Bloch, 1794) Captured in Eleyele Lake, Southwest Nigeria** 36
Kareem OK, Olanrewaju AN, Osho EF, Orisasona O and Akintunde MA

Chapter 8   **Farmed Salmon and Farmed Rainbow Trout - Excellent Sources of Vitamin D?** 40
Jette Jakobsen and Cat Smith

Chapter 9   **Growth Performance, Feed Utilization and Body Composition of *Clarias gariepinus* (Burchell 1822) Fed Marine Fish Viscera-based-diet in Earthen Ponds** 45
Vincent Oké, Youssouf Abou, Alphonse Adité and Jean-André T Kabré

Chapter 10   **Effect of Natural and Hydroponic Barley Plant and Sprout on the Common Carp *(Cyprinus Carpio)* Growth Performances** 52
Hazem S Abedalhammed, Nasreen M Abdulrahman and Haitham L Sadik

Chapter 11   **Benthic Fish Fauna and Physicochemical Parameters of Otamiri River, Imo State, Nigeria** 56
Ikenna Kelvin Obiyor, Christopher Didigwu Nwani, Gregory Ejikeme Odo, Josephine Chinenye Madu, Doris Ulumma Ndudim and Ifeanyi Oscar Ndimkaoha Aguzie

Chapter 12   **Habitat Suitability Index Relationships for the Northern Clearwater Crayfish,** *Orconectes Propinquus* **(Decapoda: Cambaridae)**......64
Thomas P Simon and Nicholas J Cooper

Chapter 13   **Effects of Dietary Protein Levels on the Growth, Feed Utilization and Haemato-Biochemical Parameters of Freshwater Fish,** *Cyprinus Carpio Var. Specularis*......71
Imtiaz Ahmed and Amir Maqbool

Chapter 14   **De Novo Assembly and Analysis of the Testes Transcriptome from the Menhaden,** *Bervoortia tyrannus*......83
Frank J Zadlock IV, Satshil B Rana, Zain A Alvi, Ziping Zhang, Wyatt Murphy and Carolyn S Bentivegna

Chapter 15   **Fisheries of Jemma and Wonchit Rivers: As a Means of Livelihood Diversification and its Challenges in North Shewa Zone, Ethiopia**......91
Erkie Asmare, Sewmehon Demissie and Dereje Tewabe

Chapter 16   **Dietary Encapsulated Butyric Acid (Butipearl™) and Microemulsified Carotenoids (Quantum GLO™ Y) on the Growth, Immune Parameters and their Synergistic Effect on Pigmentation of Hybrid Catfish (***Clarias macrocephalus* × *Clarias gariepinus***)**......97
Edwin Pei Yong Chow, Kah Heng Liong and Elke Schoeters

Chapter 17   **Have Centuries of Inefficient Fishing Sustained a Wild Oyster Fishery**......103
Stephen Long, Richard Ffrench-Constant, Kristian Metcalfe and Matthew J Witt

Chapter 18   **Profitability of Selected Ventures in Catfish Aquaculture in Ondo State, Nigeria**......110
Thompson OA and Mafimisebi TE

Chapter 19   **Isolation and Identification of** *Edwardsiella tarda* **from Lake Zeway and Langano, Southern Oromia, Ethiopia**......117
Kebede B and Habtamu T

Chapter 20   **Myeloperoxidase Inactivation Affects Neutrophil Recruitment in Zebrafish Injury-Induced Model**......123
Yajuan Li, Ren DL, Chen M Ge SC and Bing Hu

Chapter 21   **Population Dynamics of** *Pseudotolithus Senegalensis* **and** *Pseudotolithus Typus* **and Their Implications for Management and Conservation within the Coastal Waters of Liberia**......131
Austin Saye Wehye, Patrick K Ofori-Danson and Angela Manekuor Lamptey

Chapter 22   **Impact Assessment on By-catch Artisanal Fisheries: Sea Turtles and Mammals in Cameroon, West Africa**......140
Ayissi I and Jiofack TJE

Chapter 23   **Stock Assessment of Indian Scad, Decapterus Russelli in Pakistani Marine Waters and Its Impact on the National Economy**......145
Muhammad Talib Kalhoro, Mu Yongtong, Kalhoro Muhsan Ali, Shah Syed Babar Hussain, Memon Aamir Mahmood, Mohsin Muhammad and Pavase Tushar Ramesh

Chapter 24   **Pathogenic Bacteria in *Oreochromis Niloticus* Var. Stirling Tilapia Culture**...**155**
Huicab-Pech ZG, Castaneda-Chavez MR and Lango-Reynoso F

Chapter 25   **Survey on Penaeidae Shrimp Diversity and Exploitation in South East Coast of
India**..**162**
Perumal Rajakumaran and Baskralingam Vaseeharan

Chapter 26   **Marketing and Livelihood Contribution of Fishermen in Lake Tana, North
Western Part of Ethiopia**..**170**
Kidanie Misganaw and Addis Getu

Chapter 27   **Histopathological Study in Stomach and Intestine of *Anabas testudineus*
(Bloch, 1792) under Almix Exposure**..**175**
Palas Samanta, Sandipan Pal, Aloke Kumar Mukherjee, Debraj Kole
and Apurba Ratan Ghosh

Chapter 28   **The Production of Catfish and Vegetables in an Aquaponic System**........................**181**
Nawwar Zawani Mamat, Mohd Idrus Shaari and Nur Amirul Anas Abdul Wahab

Chapter 29   **Microbial Evaluation of Selected Post Harvest Processing Techniques for
Quality Fish Product at Bahir Dar Town, Ethiopia**...**184**
Adamu Yimer, Minwyelet Mingist and Behailu Bekele

Chapter 30   **Investigations on Mass Mortalities among *Oreochromis Niloticus* at Mariotteya
Stream, Egypt: Parasitic Infestation and Environmental Pollution Impacts**...........**189**
Nisreen E Mahmoud, MM Fahmy and Mohga FM Badawy

Chapter 31   **Price Modulation Policy of Federal Government of Nigeria: Effects on Fish
Production**..**196**
Ayeloja AA, George F, Sodeeq E and Adebisi GL

Chapter 32   **Iranian Fisheries Status: An Update (2004-2014)**...**200**
Harlioglu MM and Farhadi A

Chapter 33   **To Reduce Mortality of Fry Fish (*Oncorhynchus mykiss*) Caused with Viral
Infection (IPNV and VHSV) by Water Treatment with Chloramin-T as
Disinfectant**..**208**
Saeed Ganjoor M

Permissions

List of Contributors

Index

PREFACE

Aquaculture is the science that deals with the sustainable breeding, rearing and harvesting of aquatic organisms. It is an emerging practice to meet the global demands of food across the world. Fisheries worldwide are suffering due to intensive fishing, a lack of adequate scientific methods to support the marine ecosystem and excessive pollution and toxicity. The field of aquaculture has witnessed consistent research and study in recent years, aimed at improving breeding practices, production statistics and environmental sustainability. This book explores the principles and practices of aquaculture. It elucidates the concepts and innovative models around prospective developments with respect to this field. The book is meant for students and all professionals who are looking for an elaborate reference text in this area of study.

The information shared in this book is based on empirical researches made by veterans in this field of study. The elaborative information provided in this book will help the readers further their scope of knowledge leading to advancements in this field.

Finally, I would like to thank my fellow researchers who gave constructive feedback and my family members who supported me at every step of my research.

Editor

An Analysis of Genetic Variability in Wild and Hatchery Populations of Swimming Crab *(Portunus trituberculatus)* Using AFLP Markers

Ling-Xiao Liu[1,2], Yun-Guo Liu[1*] and Shi-Chao Xing[3*]

[1]College of Life Sciences, Yantai University, Yantai 264005, China

[2]Linyi Academy of Agricultural Sciences, Linyi 276012, China

[3]Gout laboratory, The Affiliated Hospital of Qingdao University, Qingdao 266003, China

*Corresponding authors: Yun-Guo Liu, College of Life Sciences, Yantai University, Yantai 264005, China, E-mail: yguoliu@163.com

Shi-Chao Xing, Gout laboratory, The Affiliated Hospital of Qingdao University, Qingdao 266003, China, E-mail: ahmcqdu@gmail.com

Abstract

The genetic variation of wild and hatchery populations of swimming crab *Portunus trituberculatus* based on observation of amplified fragment length polymorphism (AFLP) was described. A group of 180 genotypes belonging to five wild samples, Dongying (DY), Weifang (WF), Weihai (WH), Qingdao (QD), Rizhao (RZ) and one hatchery population, Yantai (YT) were screened using eight different AFLP primer combinations. A total of 396 loci were screened in the six studied populations. 49.9%, 48.5%, 52.3%, 51.2%, 50.3% and 44.5% of these loci were polymorphic among the individuals tested in the DY, WF, WH, QD, RZ and YT populations, respectively. The number of polymorphic loci detected by single primer combinations ranged from 22 to 37. The average heterozygosity of the DY, WF, WH, QD, RZ and YT populations were 0.087, 0.085, 0.096, 0.092, 0.090 and 0.068, respectively. The WH population showed the highest genetic diversity in terms of total number of AFLP bands, total number of polymorphic bands, average heterozygosity and percentage of low frequency (0-0.2) polymorphic loci among all the populations, while the WF population was the lowest among the wild populations. Compared with the wild populations, the hatchery population showed a low genetic viability.

Keywords: Swimming crab; *Portunus trituberculatus*; AFLP marker; Genetic variability

Introduction

The swimming crab *Portunus trituberculatus*, is distributed mainly on sandy and muddy bottoms in the coastal waters of Japan, Korea, and China. It is one of the most common edible crabs in China and Korea and supports a large crab fishery and aquaculture in China [1]. It is now being cultured in North China, especially in Shandong Peninsula, because of its high commercial interest. A few reports are available on molecular phylogeny and population structure in this species using different molecular marker techniques [1-4]. Long-term conservation of genetic diversity is important for any species [5]. Swimming crab resource management and enhancement are a recent practice to maintain long-term resource sustainability. A basic understanding of stock structure among geographical swimming crab samples is thus required.

In general, the effective sizes of founder populations are restrained by farming conditions, in which only a few individuals as broodstock are used. This practice may lead to the erosion of genetic diversity of stocks, thereby compromising industrial performance. Intentional and accidental release of cultured swimming crabs into wild environment could have major ecological consequences. If a large number of cultured swimming crabs escape or are released from aquaculture facilities, they could significantly alter the genetic composition of wild populations by either displacing them or interbreeding with them [6]. Most hatchery stocks typically show a reduced genetic variability, which may possibly result in the reduction of the population's capability to adapt to new environments [7]. Therefore, it is important to establish baseline information on genetic background of the aquaculture population both for genetic enhancement programs as well as protection of the genetic integrity of natural populations. Unfortunately, to date little is known about the population structure of swimming crab in China. Molecular markers provide useful tool for the assessment of genetic variations. Among the several marker systems, amplified fragment length polymorphism (AFLP) is highly reliable for the assessment of genetic variation among and within populations [8-10]. AFLP does not require previously known genetic information, a feature especially useful with species for which there are no established polymorphic markers, or for which there is limited sequence information [11,12]. AFLP is a PCR-based, multi-locus fingerprinting technique that has tremendous power for revealing polymorphism [13].

The aim of this study is to assess the genetic diversity of among samples of swimming crab from Shandong peninsula in China, and that genetic differences were observed among five wild and one cultured samples of swimming crab.

Materials and Methods

Swimming crab sampling

A total of 180 individuals of swimming crab specimens, based on six sample sets, 30 individuals each, were collected in 2009 and genetically screened in the present study. The Weights of the samples are from 80 to 120 g. Geographic locations, sample sizes are given in Figure 1. Wild swimming crab were collected at five sites, the coast of Dongying (DY), the coast of Weifang (WF), the coast of Weihai (WH), the coast of Qingdao (QD) and the coast of Rizhao (RZ). Hatchery swimming crabs were from a hatchery station in Yantai (YT). The hatchery sample was founded using wild caught individuals from Bohai sea. Samples were stored frozen (-20ºC) until genetic analysis was performed.

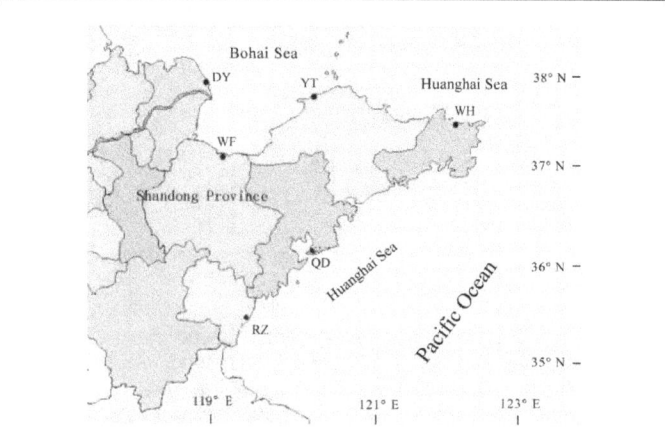

Figure 1: Sampling localities of swimming crab *Portunus trituberculatus* samples. DY, Dongying (n=30); WF, Weifang (n=30); YT, Yantai (n=30); WH, Weihai (n=30); QD, Qingdao (n=30); RZ, Rizhao (n=30).

Genomic DNA extraction

Genomic DNA was extracted from swimming crab muscle. About 150 mg muscle tissue was digested overnight at 37°C in 0.85 ml of lysis buffer (6 M urea, 10 mM Tris–HCl, 125 mM NaCl, 1% SDS, 10mM EDTA, pH 7.5) and 50µl of proteinase K (20 mg/ml). DNA was extracted twice with phenol and once with chloroform. DNA was precipitated by adding 200µl of 7.5 M ammonium acetate and 500µl of ethanol. DNA was collected by brief centrifugation and washed twice with 75% ethanol, air-dried, and dissolved in TE buffer.

AFLP reactions

Procedures of AFLP analysis were essentially based on Vos et al. [13]. Genomic DNA was processed using the "AFLP Analysis System I" (Invitrogen Corp.) according to the manufacturer's instructions with some modifications. The EcoR I primers used were not radioactively labeled. Instead, a modified silver staining method detailed below was used. After partial digestion and ligation to adaptors, samples were pre-amplified and amplified through two sequential steps. PCR pre-amplification, with primers carrying one selective nucleotide, was performed with 30 cycles at 94º C for 30s, 56º C for 1 min and 72º C for 1 min. Samples were processed immediately for the following step or stored at -20º C. After dilution up to 20 times in water, PCR products from the pre-amplification reaction were used as templates for selective amplification. The amplification step consisted of 12 cycles at 94º C for 30 s, 65º C for 30 s (with a decreasing ramp of 0.7º C each cycle) and 72º C for 1 min, followed by 30 cycles at 94º C for 30 s, 56º C for 30 s and 72º C for 1 min. Altogether eight primer combinations that produced clear and reproducible fragments were selected for further analysis (Table 1).

Gel electrophoresis and silver staining

The PCR products were mixed with an equal volume of formamide dye (99% formamide, 10 mM EDTA, 0.05% bromophenol blue and 0.05% xylene cyanol). The samples were heated to 95º C to denature for 5 min and immediately placed on ice. The gel was pre-electrophoresed at 60 W for 30 min, then 5.0 µl of the amplified DNA was loaded and run through a 5% denaturing polyacrylamide gel (4.75% acrylamide, 0.25% bisacrylamide, 7.5 M urea and 1×TBE buffer) with 1×TBE buffer on a DNA sequencing system (Liuyi Corporation, China) at 80 W for 120 min.

Silver staining was conducted using the procedures of Merril and Liu with modifications [14,15]. After electrophoresis, the gel was fixed in 10% ethanoic acid for at least 30 min. The gel was rinsed in distilled water three times and stained with a mixture of 0.1% silver nitrate and 0.15% formaldehyde for 30 min. The stained gel was rinsed again with distilled water and immersed in a developing solution (3% sodium carbonate, 0.15% formaldehyde, 0.02% sodium thiosulphate). The development was subsequently stopped with 10% ethanoic acid when the bands became visualized and reached desirable intensity. Band sizes were estimated by a standard AFLP DNA ladder (Takara Corporation, China).

Primer combination	EcoRI primers	MseI primers
P1	EcoR I primer + ACG	Mse I primer + CTG
P2	EcoR I primer + ACT	Mse I primer + CTC
P3	EcoR I primer + AGC	Mse I primer + CTT
P4	EcoR I primer + AAC	Mse I primer + CTC
P5	EcoR I primer + AGC	Mse I primer + CTA
P6	EcoR I primer + AAC	Mse I primer + CAT
P7	EcoR I primer + AAC	Mse I primer + CAG
P8	EcoR I primer + AGG	Mse I primer + CTG

Table 1: AFLP primer combinations used in the study. The sequences for the EcoRI primer were 5'-GACTGCGTACCAATTC-3', and for MseI primer were 5'-GATGAGTCCTGAGTAA-3'.

AFLP reproducibility test

The reproducibility of AFLP fingerprints was evaluated to increase the consistency of the results by comparing the fingerprints of the same swimming crab individual. Two selected individuals were independently processed from the beginning of the AFLP analysis for three replicates with two different primer combinations (No. 1 and 2 from Table 1). Reproducibility was calculated as the percentage of bands that showed consistent results over the three replicates analyzed per swimming crab sample.

Data analysis

AFLP bands were scored for presence (1) or absence (0) using the Crosscheck freeware 8 [16], and transformed into a 0/1 binary character matrix. Fragments that could not be scored unambiguously were not included in the analysis. The data matrix was analyzed for population genetic diversity using POPGENE software package 1.3.1[17]. Population genetic relationships were estimated by constructing a UPGMA tree based on Nei's standard genetic distance [18]. Analysis of Molecular Variance (AMOVA) was performed to analyze genetic distance among samples using ARLEQUIN 3.1 [19]. Average heterozygosities and percent polymorphic loci were estimated using the TFPGA program 1.3 [20]. Average heterozygosity estimates were calculated for each locus and then averaged over loci according to Nei's unbiased heterozygosity formula [18]. The percentages of polymorphic loci were estimated based on the percent of loci not fixed for one allele. Confidence intervals were generated by bootstrapping analysis at the 99% confidence level with 1000 replications. All the above significances were tested using t-test ($P<0.05$ and $P<0.01$). Estimation of pairwise Fst values for all sample combinations were also performed using ARLEQUIN program and were evaluated by a test analyogous to the Fisher's exact test using the Markov-Chain method. Significance value was adjusted for multiple comparisons using the sequential Bonferroni correction [21].

Results

AFLP polymorphism of six populations of swimming crab

Three-time repeated analysis of two randomly selected DNAs with primer combinations 1 and 2 showed highly reproducible results (reproducibility score = 99%), although some quantitative variances of the given amplicons were observed (data not shown). AFLP analysis of 180 swimming crab individuals using eight primer combinations produced a total of 396 scoreable bands, of which 49.9%, 48.5%, 52.3%, 51.2%, 50.3% and 44.5% were polymorphic over all the individuals tested in the Dongying, Weifang, Weihai, Qingdao, Rizhao and Yantai populations, respectively (Table 2). The total number of polymorphic bands over all populations varied from 22 (for primer combination 6) to 37 (for primer combination 3) per primer combination. The average heterozygosity of the DY, WF, WH, QD, RZ and YT populations were 0.087, 0.085, 0.096, 0.092, 0.090 and 0.068, respectively.

Population structure and genetic differences between wild and hatchery populations

The WH population showed the largest number of total AFLP bands, total polymorphic bands and average heterozygosity among all the populations, while the YT population had the smallest number. The WF population displayed the smallest number of total AFLP bands, total polymorphic bands and average heterozygosity in wild populations. A greater number of total AFLP bands were observed from the wild populations than from the hatchery population. There were 375, 373, 396, 383 and 378 AFLP loci detected in the DY, WF, WH, QD and RZ populations, while it was 366 in the YT population, respectively (Table 2). The total number of polymorphic bands was also higher ($P<0.05$) in the wild populations than in the hatchery population. The total polymorphic loci were 187 in the DY population, 181 in the WF population, 207 in the WH population, 196 in the QD population, 190 in the RZ population and 163 in the YT population. However, no significant difference was found in the proportion of polymorphic bands among the five populations.

WH	23	24	37	23	26	22	27	25	207	396	0.1	52.3
QD	22	26	34	21	23	21	25	24	196	383	0.09	51.2
RZ	19	25	35	20	24	18	26	23	190	378	0.09	50.3
YT	17	18	30	19	22	16	21	20	163	366	0.07	44.5

Table 2: A summary of AFLP analysis of genetic variations in swimming crab using the eight primer combinations as listed in Table 1. P1 through P8 indicate primer combinations 1 to 8 as listed in Table 1. Total number of AFLP bands (N), average heterozygosity (H), percentage of polymorphic loci (P), Dongying (DY), Weifang (WF), Yantai (YT), Weihai (WH), Qingdao (QD) and Rizhao (RZ) are given.

Differences in the percentage of polymorphic loci were found among populations as shown in Figure 2. The WH population showed the highest percentage of low frequency (0-0.2) polymorphic bands among all the studied populations, while the YT population had the lowest one. Among the wild populations, the WF population displayed the smallest percentage of low frequency polymorphic bands. In total, the percentage of low frequency polymorphic bands in the hatchery population was lower ($P<0.05$) than that in the wild populations, while the percentage of high frequency (0.6-1.0) polymorphic bands had a tendency to be higher in the hatchery population as compared with the wild populations.

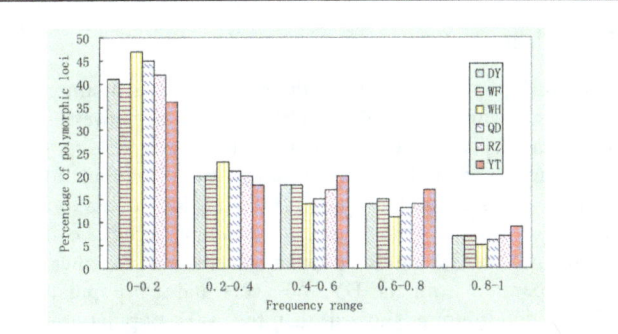

Figure 2: Distribution of percentage of polymorphic loci at each frequency range in the six populations of swimming crab *Portunus trituberculatus*.

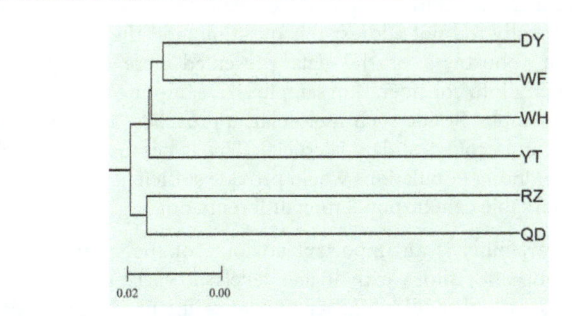

Figure 3: UPGMA dendrogram showing the phylogenetic relationship among six samples of swimming crab *Portunus trituberculatus*.

The UPGMA dendrogram constructed on the basis of the inter sample genetic similarity is shown in Figure 3. Genetic distances

Population	Number of polymorphic loci									N	H	P
	P1	P2	P3	P4	P5	P6	P7	P8	Total			
DY	18	22	34	23	23	19	25	23	187	375	0.09	49.9
WF	20	21	33	21	24	17	23	22	181	373	0.09	48.5

between samples are summarized in Table 3. Significant genetic differentiation was detected using AMOVA among samples (P<0.05). The pairwise Fst values in Table 3 also indicated significant differentiation among the six samples except between DY and WF, WF and YT, WH and YT, and QD and RZ. These results suggested that genetic divergence between these samples may have arisen.

Pop ID	DY	WF	WH	QD	RZ	YT
DY	***	0.017	0.022*	0.038*	0.037*	0.029*
WF	0.14	***	0.021*	0.036*	0.033*	0.019
WH	0.146	0.144	***	0.031*	0.029*	0.018
QD	0.158	0.156	0.15	***	0.015	0.028*
RZ	0.151	0.16	0.152	0.147	***	0.028*
YT	0.149	0.142	0.147	0.15	0.153	***

Table 3: Genetic distance and Fst values for pairwise comparison among different samples of *Portunus trituberculatus*. Genetic distance and Fst values are below and above the diagonal respectively. Dongying (DY), Weifang (WF), Yantai (YT), Weihai (WH), Qingdao (QD) and Rizhao (RZ) are given. Asterisk indicates significant genetic differentiation test by Fisher's technique after a sequential Bonferroni correction (P<0.0125).

Discussion

The WH population showed the highest diversity among all the populations while the YT population was the lowest. The WF population displayed the lowest diversity among wild populations. This phenomenon could be explained that the WF population lives in the Bohai sea, a semi-closed water. QD and RZ population are near geographically and have more chance for gene flow. In the dendrogram, the QD and RZ populations clustered together and were clearly separated from the DY, WF, WH and YT populations. The UPGMA dendrogram showed that the WH population clustered closer to the WF population than to the QD population. The reason is presumably due to the absence of physical barriers to migration between the WH and QD populations. Some differences in the percentage of polymorphic loci among populations were also found in the present study. The question is whether the polymorphic loci distribution maintained in each population temporally is stable or not. The present study did not address this question, and the only way to improve the robustness of the data presented here is repetitive sampling taking into consideration sample size, sampling time and the life stage of sample. Beside such molecular approaches, accumulation of biological and ecological data is crucial. This is because high levels of migration among populations would provide sufficient gene flow to prevent remarkable genetic population differentiation.

Genetic variability is an important attribute of the species under domestication, since those with higher levels of variation are most likely to present high additive genetic variance for productive traits. Wild populations represent the primary source of genetic variability for aquacultured stocks. Genetic variability of the hatchery strain seems likely to have been substantially reduced. This would be caused by losses of many low-frequency alleles due most likely to the small effective number of parents when the strain was founded, suggesting that the hatchery strain was bottlenecked. The YT population was founded using about 36 individuals, but the number of effective

parents may be smaller than that. This suggests that the genetic diversity in the wild swimming crab is still not being fully exploited. In view of this, there will be a higher chance of enhance diversity viability by frequent outcrossing of the cultivated swimming crab population with the wild population rather than by selective breeding followed by inbreeding among cultivated varieties. In this way, the genetic variability among the cultivated varieties would also be enhanced, thus preventing the occurrence of further different inbreeding that will probably take place if the present situation of unplanned breeding persists. Therefore, for high sustainability of the culture of swimming crab, proper breeding programs must be implemented with careful management and monitoring such that there is frequent outcrossing with the wild forms as well as maintenance of any newly emerged traits by inbreeding.

Many cultured aquatic stocks represent genetically exogenous populations, thus, the intra-specific hybridization with wild stocks may result in reduction of fitness in wild populations [22-24]. Even cultured populations that originated from the same local population may threaten the fitness of the local population through the reduction of its effective population size [25], especially when the absolute size of the wild population is small [26]. Hence, caution should be exercised to avoid significant release of hatchery stocks of swimming crab into the wild, either intentionally, or accidentally.

Acknowledgements

This work was supported by a grant from Yantai University (SM13B22).

References

1. Liu Y, Liu R, Ye L, Liang J, Xuan F, et al. (2009) Genetic differentiation between populations of swimming crab Portunus trituberculatus along the coastal waters of the East China Sea. Hydrobiologia, 618: 125-137.

2. Imai H, Fujii Y, Karakawa J (1999) Analysis of the population structure of the swimming crab, Portunus trituberculatus in the coastal waters of Okayama Prefecture, by RFLPs in the whole region of mitochondrial DNA. Fisheries science 65: 655-656.

3. Dai YJ, Liu P, Gao BQ, Li J, Wang QY (2010) Sequence analysis of mitochondrial 16s rRNA and CO I Gene Fragments of four wild populations of Portunus trituberculatus. Periodical of Ocean University of China 40: 54-60.

4. Liu YG, Guo YH, Hao J, Liu LX (2012) Genetic diversity of swimming crab (Portunus trituberculatus) populations from Shandong peninsula as assessed by microsatellite markers. Biochemical Systematics and Ecology 41: 91-97.

5. Hamrick JL, Godt MJW, Murawski DA, Loveless MD (1991) Correlations between species traits and allozyme diversity: implications for conservation biology. Genetics and Conservation of Rare Plants. Oxford Univ Press Oxford 75–86.

6. Waples RS (1999) Dispelling some myths about hatcheries. Fisheries 24: 12-21.

7. Allendorf FW, Phelps SR (1980) Loss of genetic variation in a hatchery stock of cutthroat trout. Trans Am Fis Soc 109: 537–543.

8. Travis SE, Maschinski J, Keim P (1996) An analysis of genetic variation in Astragalus cremnophylax var. cremnophylax, a critically endangered plant, using AFLP markers. Mol Ecol 5: 735-745.

9. Keim P, Kalif A, Schupp J, Travis SE, Richmond K, et al. (1997) Molecular evolution and diversity in Bacillus anthracis as detected by amplified fragment length polymorphism markers. J Bacterio 179: 818–824.

10. Keiper FJ, McConchie R (2000) An analysis of genetic variation in natural populations of Sticherus flabellatus [R. Br. (St John)] using

amplified fragment length polymorphism (AFLP) markers. Mol Ecol 9: 571-581.

11. Han TH, de Jeu M, van Eck H, Jacobsen E (2000) Genetic diversity of Chilean and Brazilian alstroemeria species assessed by AFLP analysis. Heredity (Edinb) 84: 564-569.

12. Ajmone-Marsan P1, Negrini R, Crepaldi P, Milanesi E, Gorni C, et al. (2001) Assessing genetic diversity in Italian goat populations using AFLP markers. Anim Genet 32: 281-288.

13. Vos P, Hogers R, Bleeker M, Reijans M, van de Lee T, et al. (1995) AFLP: a new technique for DNA fingerprinting. Nucleic Acids Res 23: 4407-4414.

14. Merril CR, Switzer RC, Van Keuren ML (1979) Trace polypeptides in cellular extracts and human body fluids detected by two-dimensional electrophoresis and a highly sensitive silver stain. Proc Natl Acad Sci U S A 76: 4335-4339.

15. Liu Y, Wang X, Liu L (2004) Analysis of genetic variation in surviving apple shoots following cryopreservation by vitrification. Plant Sci 166: 677-685.

16. Buntjer BJ (1999) Software Crosscheck .Developed in Wageningen University and Research Centre.

17. Yeh FC, Yang RC, Boyle T (1999) POPGENE version 1.3.1. Microsoft Window-bases Freeware for Population Genetic Analysis. University of Alberta and the Centre for International Forestry Research.

18. Schneider S, Roessli D, Excoffier L (2000) Arlequin: A software for population genetics data analysis. Genetics and Biometry Laboratory. Switzerland.

19. Miller MP (1997) Tools for Population Genetic Analysis (TFPGA) Version 1.3. A Windows Program for the Analysis of Allozyme and Molecular Population Genetic Data. AZ.

20. Nei M (1978) Estimation of average heterozygosity and genetic distance from a small number of individuals. Genetics 89: 583-590.

21. Rice WR (1989) Analyzing tables of statistical tests. Evolution 43: 223-225.

22. Hindar K, Ryman N, Utter F (1991) Genetic effects of cultured fish on natural fish populations. Can J Fish Aquat Sci 48: 945-957.

23. Lester LJ (1992) Marine species introductions and native species vitality: genetic consequences of marine introductions. Introduction and Transfers of Marine Species. South Carolina Sea Grant Consortium 79-89.

24. Ferguson A, McGinnity P, Stone C, Clifford S, Taggart J, et al.(1995) The genetic impact of escaped farm Atlantic salmon on natural populations. Aquaculture 137: 55-56.

25. Ryman N, Utter F, Laikre L (1995) Protection of intraspecific biodiversity of exploited fishes. Rev Fish Biol Fish 5: 417-446.

26. Ryman N, Laikre L (1991) Effects of supportive breeding on the genetically effective population size. Conserv Biol 5: 325-329.

Growth, Mortality and Exploitation of *Sardinella maderensis* (Lowe, 1838) in the Liberian coastal waters

Wehye AS[1*], Amponsah SKK[2], and Jueseah AS[1]

[1]*Bureau of National Fisheries, Ministry of Agriculture, Liberia*

[2]*Food Research Institute, Box M20, Accra, Ghana*

*Corresponding author: Wehye AS, Bureau of National Fisheries, Ministry of Agriculture, Monrovia, Liberia, E-mail: austinwehye@yahoo.com

Abstract

This study examined some aspects of population dynamics of 1776 specimen of *S. maderensis* (Lowe, 1988) from Liberian coastal waters, from April 2013 to September 2013 (total of six months) using the FiSAT II for analysis. From the results, the growth was assumed to follow the von Bertalanffy growth function with asymptotic length ($L\infty$) and the growth coefficient (K) estimated at 44.63 cm total length and 0.38 year^{-1} respectively. The growth performance index, longevity and the theoretical age at birth (t_0) were estimated as 2.88, 7.51 years and -0.387 year^{-1} respectively. The length at first capture (L_{c50}=13.99 cm) was lower than the length at first maturity (L_{m50}=29.75 cm), an indication that most of the harvested stock were juveniles. Instantaneous rate of total mortality, natural mortality and fishing mortality were estimated as 1.24 year^{-1}, 0.81 year^{-1} and 0.43 year^{-1} respectively. The current exploitation rate (E) and maximum exploitation rate (E_{max}) were calculated as 0.34 and 0.36 respectively. Results from the study indicated that the exploitation of *S. maderensis* is at the maximum sustainable yield coupled with the presence of growth overfishing and intense fishing pressure. Therefore, urgent management actions including increasing fishing gears mesh size and regulating fishing effort is needed to protect the *S. maderensis* stock.

Keywords: Liberia; *Sardinella maderensis*; Growth; Mortality; Exploitation rate

Introduction

Sardinella maderensis also known as flat 'sardinella' forms part of the commercially important fish species of Liberia which prefers areas of lower salinities close to the mouth of river. Though *S. maderensis* appears throughout the year in Ivory Coast coastal waters with a strong reduction from May to July due to the transition from warm season to cold season, adults *S. maderensis* are more sedentary [1,2]. *S. maderensis* fishery is of great importance to fishing households within most coastal communities in Liberia, both economically and food security wise. As a result, *S. maderensis* fishery like other commercially important fish species in Liberia is currently subjected to intense fishing pressure. Intensive fishing pressure on marine biodiversity by location and depth has led to decline of marine capture fisheries [3,4].

In Liberia, factors such as poor fisheries data collection, limited resources, conflicts and illegal, unregulated and unreported (IUU) do not only make it difficult to estimate the status of almost all of the marine biodiversity but also presents a great challenge to fisheries managers [5,6]. However, Togba [7] reported that *Sardinella, Barracudas*, Croakers, Sharks and *Ilisha africana* constituted 83% and 59.06% of local fish supply in 2004 and 2005 respectively; indicating that there has been a declined in fish catches.

Furthermore, the paucity of information on population parameters and biology pertaining to commercially important fish species within Liberian coastal waters cripples any management interventions geared towards sustainable fisheries in Liberia. It is against this backdrop that the present study sought to estimate some population parameters of *S.*

maderensis residing in Liberian coastal waters to enhance already existing management interventions.

Materials and Methods

Study area

The coastline of Liberia is 579 kilometres in length and consist of nine (9) coastal counties with an exclusive economic zone (EEZ) that extends 200 nautical miles off-shore, characterised by relatively warm waters with low nutrient content.

Figure 1: Map showing the study area.

The study focussed on eight counties namely Grand Cape Mount, Montserrado, Grand Bassa, River Cess, Sinoe, Maryland and GrandKru (Figure 1). A two-stage sampling strategy was applied in the selection of the study areas, namely the intense level of fishing activity and geographical location. The main source of livelihood for majority of inhabitants within the study areas was fishing and its related activities. However, a few are engaged in alternative source of livelihoods.

Data collection

The length frequency data was collected by Fisheries Enumerators of the Bureau of National Fisheries (BNF) from artisanal fishers from selected landing sites in eight (8) of the nine coastal counties for six months from April, 2013 to September, 2013 (6 months). Data was collected from fishers who operated mostly with multifilament fishing gears with morphometric measurement recorded on-site. For instance, total length was measured to the nearest 0.1 cm using the 100 cm measuring board while the weight was weighed using electric weighing scale. Identification of specimen was done to the species level using identification keys by Fischer et al. [8,9]. In all, a total of 1,776 specimens of S. maderensis were sampled.

Growth parameters

The growth rate (K), asymptotic length (L∞) and the growth performance index (Φ) of the fish was assumed to follow Von Bertanlaffy Growth Function (VBGF). These growth parameters were obtained using the VBGF fitted in FISAT II [10]. According to VBGF as expressed below, individual fishes grow on average towards the asymptotic length at an instantaneous growth rate (K) with length at time (t) following the expression: $L_t = L\infty\left(1-e^{-K(t-to)}\right)$ [11]. The theoretical age at birth (to) was calculated using the empirical formula: $log10\left(-t_0\right) = -0.3922 - 0.275*log10L\infty - 1.038*log10K$ [11]. The longevity (tmax) was estimated as: tmax=3/K + to Pauly [12]. The growth performance index was calculated from the below expressed equation: $(\Phi) = 2logL\infty + logK$ [13].

Mortality parameters

The total instantaneous mortality rate (Z) was estimated using length converted catch curve method as implemented in FiSAT II. Natural mortality rate (M) was estimated using Pauly's empirical relationship, using a mean surface temperature (T) of 25.5°C:

Log M = -0.0066 - 0.279 log L∞ + 0.6543log K + 0.4634log T [14],

Where M is the instantaneous natural mortality, L∞ is the asymptotic length, T is the mean surface temperature and K refers to the growth rate coefficient of the VBGF. Fishing mortality (F) was calculated using the relationship: F=Z – M [15], where Z is the total mortality, F the fishing mortality and M is the natural mortality. The exploitation level (E) was obtained using the relationship: E=F/Z [15].

Length at first capture (L_{c50}) and maturity (L_{m50})

The ascending left arm of the length converted catch curve incorporated in FiSAT II tool was used to estimate the probability of length at first capture (Lc50) in addition to the length at both 25 and 75 captures which corresponded to the cumulative probability at 25% and 75% respectively. The probability of capture gives clear idea about the estimate of the real size of the fish in the fishing area that is being caught by specific gear. It is an important tool for fisheries managers in sustainably managing a target fishery, because it helps would be managers determining the minimum mesh size of a fishing fleet. The length at first maturity was estimated using the expression: Length at first maturity (L_{m50}) = $(2*L\infty)/3$ [16].

Relative yield per recruit (Y'/R)

The relative biomass per recruit (B'/R) was estimated as B'/R=(Y'/R)/F. Emax which depicts exploitation rate producing maximum yield, $E_{0.1}$ highlighting exploitation rate at which the marginal increase of Y'/R is 10% of its virgin stock with $E_{0.5}$ implying exploitation rate under which the stock is reduced to half its virgin biomass were computed using the procedure incorporated using the Knife-edge option fitted in the FiSAT II Tool.

Data Analysis

The length frequency data were pooled into groups with 1cm length intervals. Then the data was analyzed using the FiSAT II (FAO-ICLARM Stock Assessment Tools) software [17].

Result

Length frequency distribution

The monthly pooled length frequency data from the 1776 specimen of S. maderensis were group into one-centimeter interval. Figure 2 shows the length frequency distribution for the assessed fish species. From Figure 2, midlength of 19.5 cm showed the highest frequency distribution, follow by length 22.5 cm. The highest length recorded throughout the study period was 42 cm.

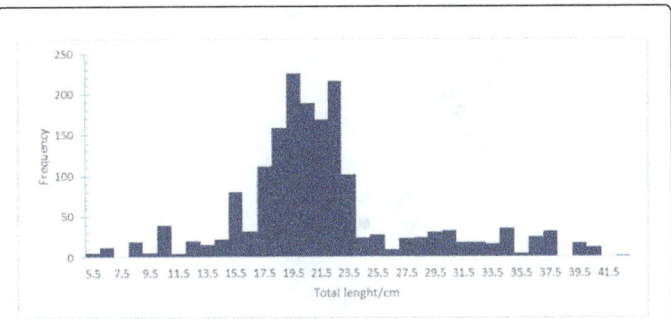

Figure 2: Length frequency distribution of *Sardinella maderensis* from the fisheries waters of Liberia.

Growth Parameters

The Length at infinitive (L∞) and growth constant (K) were estimated at 44.63 cm and 0.38 year-1 respectively with its longevity as 7.51 years. The growth performance index (Φ) and theoretical age at birth (t0) were estimated at 2.88 and -0.387 year respectively. Using the growth parameters (L∞, K and t0), the VBGF for length at time (t) was expressed as:

$$L_t = 44.63 (1-e^{-0.38 (t-(-0.387))}).$$

Figure 3 below showed the restructured length frequency with superimposed growth curves with bimodal population structure, indicating probably the existence of six cohorts within the population.

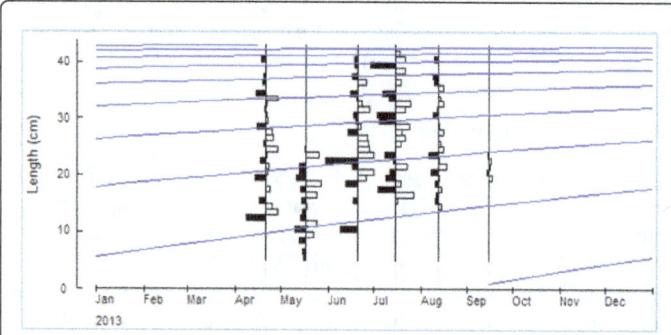

Figure 3: Restructured Length frequency distribution output from FiSAT II with superimposed growth curves (Dark bars=actual frequency bars & White bars=reconstructed bars).

Mortality coefficients and current exploitation rate

Figure 4 showed the calculated motilies from FiSAT II output of the length converted catch curve. The instantaneous total mortality coefficient (Z) was estimated as 1.24 year^{-1}. The natural mortality (M) and fishing mortality (F) were estimated to be 0.81 year^{-1} and 0.43 year^{-1} respectively. The current exploitation rate was estimated as E=0.34.

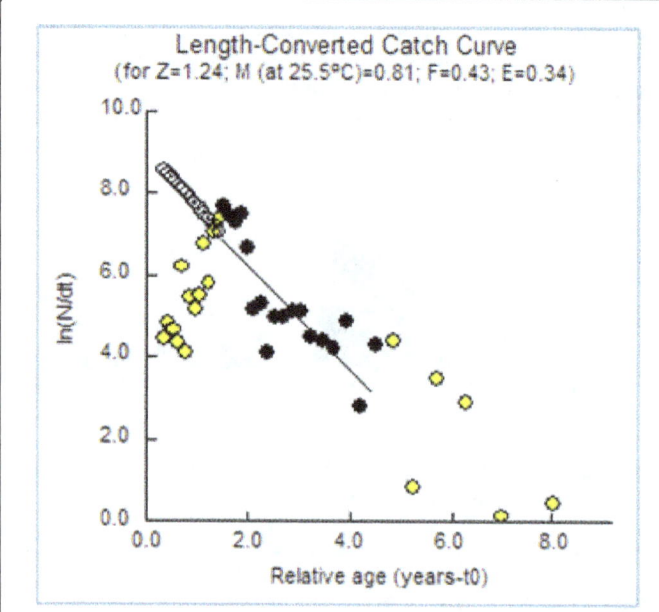

Figure 4: FISAT II output of linearized length-converted catch curve for *S. maderensis* (Yellow dots are dots used in calculation and White dots are dot not used in calculations).

Length at first capture (L_{c50}) and length at first maturity (L_{m50})

The probability of capture of *S. maderensis* at 25%, 50% and 75% which provides a clear indication of the estimated real size of fish in the fishing area that are being caught by specific gear were estimated as: L_{25}=12.41 cm, L_{50}=13.99 cm and L_{75}=15.58 cm (Figure 5). Therefore, the length at first capture (L_{c50}) was 13.99 cm. The length at first maturity (L_{m50}) was estimated at 29.75 cm.

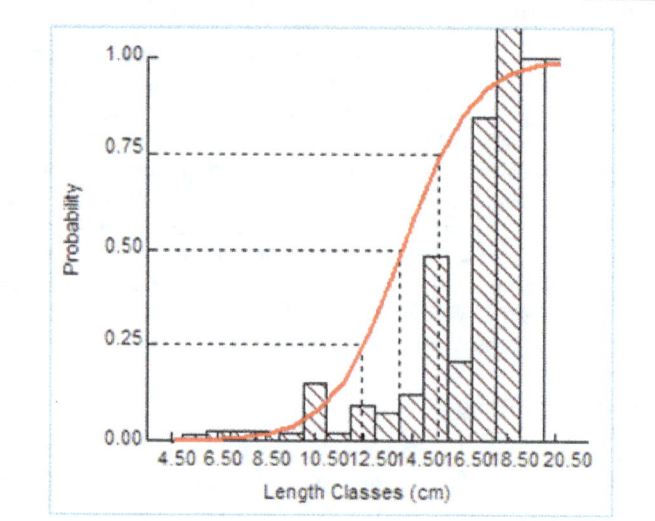

Figure 5: FiSAT II output of the probability of capture of *S. maderensis* in the fisheries waters of Liberia (0.2, 0.50 and 0.75 relates to 25%, 50% and 75% respectively).

Relative yield per recruit (Y'/R)

The Beverton and Holt relative yield per recruit model in figure 6 showed that the indices for sustainable yield were 0.236 for optimum sustainable yield ($E_{0.5}$), 0.363 for the maximum sustainable yield (E_{max}) and 0.254 for economic yield target ($E_{0.1}$).

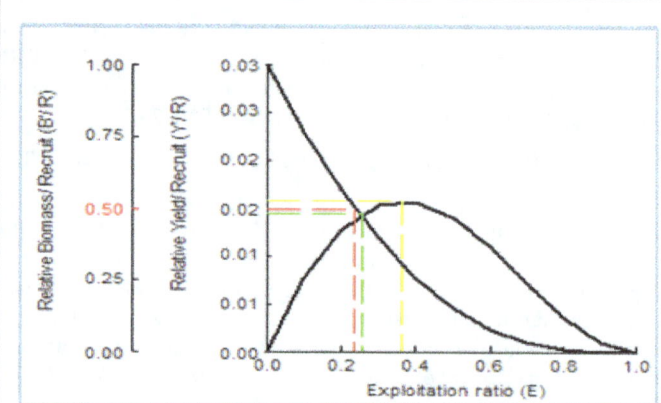

Figure 6: Beverton and Holt's relative yield per recruit and average biomass per recruit models, showing levels of yield indices for *S. maderensis* in the Coastal waters of Liberia (Red dashes=$E_{0.1}$, Green dashes=$E_{0.5}$ and Yellow dashes=E_{max}).

Discussion

Arguably, the present study appears to be the maiden work done on *S. maderensis* stock resident in Liberian coastal waters, therefore information gained will serve as a springboard for further research pertaining to this commercially important species. The asymptotic length in the present study is greater than results reported by other researchers (Table 1). This differences could be attached to factors such as the selectivity of the gears, the sampling methods and geographical locations. The estimated growth rate (K=0.38 year^{-1}) from this study was favourable with estimates by Gabche et al. [18,19]. However, it was relatively lower than estimates from studies done elsewhere (Table 1), possibly as a result of variation in geographical locations, the data analysis method used and the size classes obtained [20]. Further, the growth rate (K) of this study was within the range: 0.34 per year and 0.67 per year, suggesting that *S. maderensis* is an intermediate growing fish species, evinced by its lifespan of 7.51 years [21]. The growth performance index (=2.88) appeared to be in line with estimates from other studies (Table 1). This finding demonstrates that they are of similar taxonomic family. Further, the growth performance index indicates the important availability of food and other favorable environmental conditions [1].

TL∞	K	Φ	t0	Countries	Authors
44.63	0.38	2.88	-0.387	Liberia	Current study
33.6	0.65	2.86	0.24	Benin	Sossoukpe et al., 2016 [1]
27.24	0.48	1.76	-0.06	Cameroon	Gabche & Hockey, 1995 [2]
35	0.6	2.88		Senegal	Postel, 1955 [22]
37.5	0.34	2.68	-0.25	Nigeria	Marcus, 1989 [19]
32.5	0.59	2.79		Cameroon	Djama et al., 1989 [23]
24.93	0.98	2.79	0.024	Congo	Gheno & Le Guen, 1968 [24]

Table 1: Estimated growth parameters of *S. maderensis* of the fisheries waters of Liberia compared to those off other regions.

The length at first maturity (L$_{m50}$=29.75 cm) was relatively higher than the length at first capture (L$_{50}$=13.99 cm), signifying that the *S. maderensis* stock are harvested before they could reach the matured stage, a characteristic feature of growth overfishing [19]. Furthermore, the ratio L$_{c50}$/L∞ from the study was estimated as 0.31, relatively lower than 0.5, which implied that the harvested catch is mostly made up of small sized *S. maderensis* [25]. This observation affirmed the earlier assertion that growth overfishing exists within the fishery of *S. maderensis* resident in Liberian coastal waters.

The natural mortality (M=0.81 year^{-1}) was greater than the fishing mortality (F=0.43 year^{-1}), contrary to estimates reported by Sossoukpe et al. [1] from Benin (M=1.30 year^{-1}; F=2.62 year^{-1}). This observation could be due to the fact that *S. maderensis* stock in Liberian coastal waters is more susceptible to natural mortality conditions than to fishing gears. The exploitation rate (E$_{current}$=0.34) from the present study was lower than 0.5, depicting that the *S. maderensis* stock is currently underexploited. Further, the current rate of exploitation (Ecurrent=0.34) of *S. maderensis* was slightly lower than maximum exploitation rate (E$_{max}$=0.36). Such observation indicates that the maximum sustainable yield for *S. maderensis* could be reached earlier than expected amidst continuous and intensive fishing as well as the use of the small mesh size fishing gears within the coastal waters of Liberia.

Conclusion

This study has shown that the stock of *S. maderensis* within the Liberian coastal waters is experiencing exploitation rate close to the maximum sustainable yield amidst the presence of heavy fishing pressure. Further, growth overfishing is currently present within the *S. maderensis* stock. Therefore, to ensure sustainable exploitation of *S. maderensis* stock, fishing effort should be regulated along with increase in mesh size.

Acknowledgement

Thanks are due to Bureau of National Fisheries (BNF), WARFP-Liberia and field technicians for assisting in data collection.

Conflict of Interest

No conflicts of interests

References

1. Sossoukpe E, Djidohokpin G, Fiogbe ED (2016) Demographic parameters and exploitation rate of Sardinella maderensis (Pisces: Lowe 1838) in the nearshore waters of Benin (West Africa) and their implication for management and conservation. Fish Res 4: 165-171.

2. Marchal E (1993) Biologie et écologie des poissons pélagiques côtiers du littoral ivoirien pp: 269.

3. Christensen V, Guenette S, Heymans JJ, Walters CJ, Watson R, et al. (2003) Hundred-year decline of North Atlantic predatory fishes. Fish and Fisheries 4: 1-24.

4. Swartz W, Sala E, Tracey S, Watson R, Pauly D (2010) The spatial expansion and ecological footprint of fisheries 5: 1-6.

5. MRAG (2014) Fisheries Stock Assessment. Report produced under WARFP/BNF Contract 11/001. Republic of Liberia, West Africa.

6. Sherif SA (2014) The development of fisheries management in Liberia: vessel monitoring system (vms) as enforcement and surveillance tools: national and regional perspectives. World Maritime University Dissertations pp: 462.

7. Togba GB (2008) Analysis of profitability of trawl fleet investment in Liberia, University of Akureyri.

8. Fischer W, Bianchi G, Scott W (1981) FAO species identification sheets for fishery purposes. Eastern Central Atlantic; Fishing area 34, 47 (in part). Canada Funds n-Trust. Ottawa, Department of Fisheries and Oceans Canada, by arrangement with Food and Agriculture Organization of the United Nations.

9. Schneider W (1990) Field guide to commercial marine resources of the Gulf of Guinea. Food and Agricultural organization of the United Nations. Rome pp: 268.

10. Sparre P, Venema SC (1992) Introduction to tropical fish stock assessment. Part I. Manual. FAO Fish Tech pp: 376.

11. Pauly D (1979) Theory and management of tropical multispecies stocks: a review with emphasis on the Southeast Asian demersal fisheries stud.

12. Pauly D (1983) Length converted catch curves. A powerful tool for fisheries research in tropics. ICLARM Fishbyte 1: 9-13.

13. Munro JL, Pauly DA (1983) Simple method for comparing the growth of fishes and invertebrates. ICLARM Fish byte pp: 5-6.

14. Pauly DA (1980) selection of simple methods for the assessment of tropical fish stocks. FAO Fisheries Circular 729, FAO, Rome.

15. Gulland J (1971) The Fish Resources of the Oceans. FAO/Fishing News Books, Surrey pp: 255.

16. Hoggarth DD, Abeyasekera S, Arthur R, Beddington JR, Burn R.W, et al. (2006) Stock Assessment for fishery management-A framework guide to the stock assessment tools of the Fisheries Management Science Programme (FMSP). Fisheries Technical Rome. FAO pp: 261.

17. Gayanilo F, Sparre P, Pauly D (2005) FAO-ICLARM Stock Assessment Tools II (FiSAT II). Revised. User's guide. Computerized Information Series (Fisheries). No. 8. Revised version. FAO, Rome pp: 168.

18. Gabche CE, HUP, Hockey (1995) Growth, mortality and reproduction of Sardinella maderensis (Lowe, 1841) in the artisanal fisheries off Kribi, Cameroon. Fisheries Research 24: 331-344.

19. Marcus O (1989) Breeding, age and growth in Sardinella maderensis (Lowe 1839) Pisces: Clupeidae from coastal waters around Lagos, Nigeria. Niger J Sci 5: 1.

20. Amponsah SKK, Patrick DO, Nunoo FKE (2016) Fishing regime, growth, mortality and exploitation status of Scomber japonicus from catches landed along the eastern coastline of Ghana. Int J Fish Aqua Res 1: 5-10.

21. Kienzle MO (2005) Estimation of the population parameters of the Von Bertalanffy Growth Function for the main commercial species of the north sea. Fisheries Research Services Internal Report pp: 34.

22. Postel E (1955) Les faciès bionomiques des côtes de Guinée française. Rapp Cons Int Expl Mer 137: 10-13.

23. Djama T, Gabche C, Njifonjou 0 (1989) Growth of Sardinella maderensis in the Lobe estuary, Cameroon. ICLARM Fishbyte 7: 8-10.

24. Gheno Y, Le Guen JC (1968) Détermination de l'âge et croissance de Sardinella eba (Val.) Dans la région de Pointe-noire. Cah ORSTOM, Sér Océanogr 2: 6982.

25. Pauly D, Soriano ML (1986) Some practical extensions to Beverton and Holt's relative yield-per-recruit model (1986). In: JL Maclean, LB Dizon, LV Hosillos (Eds.) The First Asian Fisheries Forum, Asian Fisheries Society. Manila, Philippines pp: 491-496.

3

Different Heat Shock Application Effect on Gynogenetic Production of Zebrafish (*Danio rerio*)

Rahmi Can Ozdemir[1*] **and Aygül Ekici**[2]

[1]*Department of Fisheries, Kastamonu University, Kastamonu, Turkey*

[2]*Department of Fisheries, Istanbul University, Turkey*

Corresponding author: Rahmi Can Ozdemir, Department of Fisheries, Kastamonu University, Kastamonu, Turkey, E-mail: rozdemir@kastamonu.edu.tr

Abstract

The present study aimed to obtain gynogenetic zebrafish. For this purpose, zebrafish spermatozoa exposed to UV irradiation to make it haploid gynogenetic fish and heat shock was applied to haploid zygote in order to obtain gynogenetic diploid fish.

Temperature, which is an important factor of the production, is taken into consideration in this study. In this respect, this study compared the results of 41.4°C and 41°C heat-shock applications and found that 12nd-24th-48th-72nd hour survival rate was maximum at 41,4°C ($P<0.05$). When considered from hatching rate (72th-78th hour) view at 41.4°C was 17.3 ± 3% in gynogenetic diploid group and heat-shock application at 41°C survival rate was 14 ± 2% in gynogenetic diploid group and there is no survivor in haploid group, was observed ($P<0.05$).

The result of the karyotype analysis in haploid gynogenetic embryos, ruptured chromosome fragments was identified. Also in karyotype analysis of diploid gynogenetic embryos, 2n=50 chromosomes was identified. On 3rd day after fertilization, total body length of the haploid gynogenetic fish was 39.6% shorter and body thickness was 33% thicker than the diploid gynogenetic group. Our gynogenetic fish producing method and the different heat-shock applications improved survival rate of gynogenetic fish.

Keywords: Zebrafish*; Danio rerio;* Gynogenesis; Heat-shock

Introduction

The spermatozoa, of which genetic material has been destroyed by using Gamma (γ), X or Ultraviolet (UV) rays in order to eliminate the male genetic material, is used in fertilization of the egg cell during the gynogenesis; one of the chromosome manipulation techniques [1-5]. Various shock applications including heat / cold shock, hydrostatic pressure and chemical agents (Colchicine, Cytochalasin B and N_2O, etc.) have been used in the study in order to convert these haploid embryos, fertilized with spermatozoa and untreated, to diploid embryos [4,6-10].

The first gynogenesis study was conducted on trout (*Onchorhynchus mykiss*) by Opperman [11]. After this study, gynogenesis studies were conducted on weatherfish (*Misgurnus fossilis* L.) by Neifach [12], on *Gadus morhua* by Van Eenennaam et al. [13], on goldfish (*Carasius auratus*) by Paschos et al. [14] and on *Acipenser transmontanus* by Ottera et al. [15], respectively with various shock applications.

The zebrafish (*Danio rerio*) mentioned in this study is used as a live model in aquaculture and human diseases due to the facts that its embryo has the characteristics of in vitro development, its eggs have the transparent quality, its mutants can be examined morphologically and also it has high fecundity (100-300 eggs) and it can be manipulated easily [16-23]. Its embryonic development is highly rapid

at 26.5°C-28°C and the hatching of the eggs will begin 48-72 hours after the fertilization [24].

The method of gynogenesis is used for production of the generations with female genetic materials, easily conduction of genetic analysis of characters in living creatures and information gathering relevant to the issues such as nutrition and disease resistance when accurate clones are produced [25-29]. This study was conducted for biotechnology and molecular genetic studies. The problems encountered in here have been created in order to attain more comfortable solutions as mentioned above.

Materials and Methods

Care & feeding of fish

A male (7-8 months old) and a female (15-16 months old) zebrafish (Danio rerio), supplied from an importing company and brought to biotechnology laboratory of Istanbul University Faculty of Fisheries Department of Aquaculture, have been used in the study. The fish have been fed with *Artemia sp.,* frozen tubifex, dried Daphnia sp and fishmeal powder and the photoperiod has been applied to them (14-10 h, light-dark).

Mature fish selection and anesthesia treatment

A slim line, male zebrafish with yellowish colour and also a female fish with bloated appearance and silver colour on its abdomen have been used as the mature fish. Than an anesthesia procedure has been applied by taking the mature zebrafish in a solution (Tricaine 0.2 mg/ml) for sucking eggs and sperm.

Gamete collection

The testicular tissues (two-piece) of male zebrafish obtained by dissection method have been transferred to 50 µl Hank's solution (+4°C) and the testicles have been smashed with the help of forceps [30]. They have been protected at +4°C in this solution in order to immobilize the spermatozoa without losing viability till the phase of in vitro fertilization [31,32].The sperm motility has been identified by light microscopy and then the sperm has been diluted with an activation solution (9% NaCl) in 1:10 ratio at 22°C-23°C. In the gynogenesis study, motility of spermatozoa over 90% has been used in fertilization.

The eggs have been collected from female zebrafish by stripping to a dried Petri dish by means of massage method. During the fertilization period, at least 100 eggs have been used in each of haploid gynogen, diploid gynogen and control groups and the sperm/egg ratio to be performed has been 0.5 ml/number. The care and incubation of the fertilized eggs and larvae of the zebrafish in the oven have been performed as reported by Westerfield et al. [24,33] in order to prevent the risk of contamination.

Gynogenesis application

The sperm taken from 4 male zebrafish (sperm+Hank's solution) has been taken on ice by performing pooling procedure and transferred to a watch glass for the gynogenesis application. The watch glass with ice has been placed in UV cross-linker device (UVP TL-2000) so as to be 28 cms away from the UV lamb and (254 nm) UV has been applied for 2 minutes. Thirteen minutes later from using the UV-treated sperm in the fertilization process, the heat shock treatment (41°C or 41.4°C) has been conducted to the fertilized egg for two minutes.

Morphological analysis of gynogenetic zebrafish

The morphological developments of control, haploid and diploid gynogen groups from fertilization day to fifth day have been examined by means of shooting their photos with a stereo (Olympous SZ-PT) and invert microscope (Olympus CK40-F200) that has a camera attachment. Shock results at 41.4°C have been used in monitoring the embryo development of each group.

Karyotype analysis

Dechorionate: The chorion has been removed by the help of a fine-tipped forceps in 24 hours after the fertilization and the embryo of zebrafish has been transferred to Petri box by using a sterile Pasteur pipette.

Karyotype analysis: The embryo of zebrafish has been transferred into (4 mg/ml) freshly made colchicines solution after the dechorionation process and incubated at 28.5°C for 90 minutes in the dark. After that, the embryos has been rinsed at room temperature (~21°C) and transferred into 1.1% Sodium citrate ($C_6H_5Na_3O_7$). The yolk of the embryo has been punctured and a timer has been started. After the dissection of the whole yolk, the Petri dish has been taken on the ice and hold there for 8 minutes. At the end of this time, the embryos have been transferred in 3:1 mixture of methanol/ acetic acid and hold

there for 20 minutes then, after this process methanol/acetic acid fixative has been added to the same degree in the solution and the embryos have been stored in a freezer overnight [24].

Spreading of chromosome: The embryos have been picked up one by one with forceps and blotted until dried. The dried embryos have been transferred on the watch glass and 2-3 drops of 50% acetic acid have been added on it and the pressing process with forceps has lasted for one minute. The suspended cells have been liquidized for two times in a 50 µl wiretrol micro capillary and droplets of the cell suspension have been dropped on a slide, pre warmed to 50°C, and the liquid has been pulled back to the wiretrol quickly. ~6 droplets have been dropped per slide. The slides have been left at 50°C for about 10 minutes to completely dry, then stained in Giemsa for 20 minutes and after that rinsed twice with H_2O. The slides have been left at room temperature to dry and the samples covered with entellan have been observed by light microscopy [24].

Osmotic pressure measurement

20 µl samples taken from Hank's solution are used in this study and measured in a micro osmometer with F FISKE (Fiske-210 brand name). 20 µl supernatant has been taken by centrifuging the sperm conserved in Hank's solution for 5 minutes at 12,000 rpm and its osmotic pressure has been measured in micro osmometer device.

Results

Results of osmotic pressure measurement

The osmotic pressures of Hank's solution and sperm cells measured in Fiske brand micro osmometer device are mOsm/kg and 306 mOsm/kg, respectively.

Results of embryo survival rates

Time dependent death rates of haploid gynogen, diploid gynogen and control group embryos are shown at Table 1. According to these data, the death rates in diploid gynogen group were observed as 80 ± 3% at 41.4°C and 85 ± 2% at 41°C (P<0.05).

Shock treatment temperature (°C)	Control		Diploid Gynogen		Haploid Gynogen	
	Survive	Death	Survive	Death	Survive	Death
41.4°C	75 ± 5	20 ± 5	17.3 ± 3	80 ± 3	0.5 ± 0.5	99 ± 1
41.0°C	80 ± 6	15 ± 5	14 ± 2	85 ± 2	0	100

Table 1: Survival and death ratio (%) of shock treatment applied embryo.

Time dependent death rates in heat shock implementation at 41.4°C are shown at Table 2. Accordingly, it is observed that the death rates of diploid gynogen group are higher than those of control group at time between 12th-24th hours and 24th-48th hours after the fertilization. In haploid gynogen group, the deaths occurred at the rate of 60.32% ± 13.46% and 53.03% ± 15.8% at time between 12th-24th hours and, 24th-48th hours, respectively. Moreover, no survival is observed in haploid group at 75th hour.

Experimet group	Control	Diploid gynogen	Haploid gynogen
1. hour	5.1 ± 6.8	18.89 ± 14.44	36.24 ± 21.02
Between 1-12 hours	7.3 ± 5.95	27.54 ± 8.05	39.16 ± 12.76
Between12-24 hours	8.7 ± 6.35	41.28 ± 5.12	60.32 ± 13.46
Between 24-48 hours	8.75 ± 7.03	48.89 ± 21.22	53.03 ± 15.80
Between 48-72 hours	0 ± 0	28,35 ± 20.83	90 ± 8.6

Table 2: Percentage of deaths result of shock treatment at 41.4°C for each experimental groups (%).

Time dependent death rates in heat shock implementation at 41°C are shown at Table 3. Accordingly, it is observed that the death rates of haploid gynogen are much more than those of diploid gynogen of which death rates are much more than those of control group. And furthermore, no survival is observed in haploid group.

Experimet group	Control	Diploid gynogen	Haploid gynogen
1. hour	4.9 ± 4.6	20.49 ± 16.24	31.84 ± 22.72
Between 1-12 hours	6.2 ± 4.75	27.54 ± 10.25	43.46 ± 14.36
Between12-24 hours	8.1 ± 4.15	43.78 ± 5.82	71.52 ± 15.16
Between 24-48 hours	8.25 ± 7.03	54.89 ± 21.72	73.03 ± 15.80
Between 48-72 hours	0 ± 0	30.35 ± 27.83	100

Table 3: Percentage of deaths result of shock treatment at 41°C for each experimental groups (%).

Morphology screening of embryo development

At 12th hour after the fertilization; normal embryo development was observed in diploid gynogen (41.4°C shock applied) and control group (bud-stage), however, haploid embryos showed slower development than them, that is; in 8-hour stage (75% epiboly). At 24th hour after the fertilization; haploit gynogen embryo was observed at 18th hour stage (18-somit), on the other hand, diploid gynogen and control group was observed at 24th hour stage (prim-6). At 48th hour after the fertilization; haploit gynogen embryo was observed at 36th hour stage (prim-22), diploid gynogen embryo was observed at 42nd hour stage (beginning of pectoral fin) and control group was observed at 48th hour stage (long tail stage). At 72nd hour after the fertilization; haploit gynogen embryo was observed at 60th hour stage (pectoral fin), and diploid gynogen and control group were observed at 72nd hour stage (mouth development) (Table 4). On the basis of these data, the morphological development of haploid embryos is the latest and the morphological developments of diploid gynogen group and control group are found as similar. It is observed that, embryogenesis in haploid gynogen group comes up several phases from diploid gynogen group (Table 4). When embryo development of haploid gynogen group is compared with the other groups (control and diploid gynogen), it is seen that the phases of embryo development do not have normal development stage and the development stages differ from each other (Table 4). Furthermore, it is determined that the Haploid gynogen fry has body disorder which means it has shorter body length and also curvature of spine while the development of diploid gynogen fry has normal shape (Figure 1).

Experiment Groups	Morphological Development Stages			
	12nd. hour	24th hour	48th hour	72nd hour
Control	Bud stage	Prim-6	Long tail structure	Mouth development
Haploid Gynogen	%75 epiboli	18 somit	Prim-22	Pectoral fin stage
Diploid Gynogen	Bud stage	Prim-6	Beginning of Pectoral fin stage	Mouth development

Table 4: Comparison of embryonic development of control and gynogen groups.

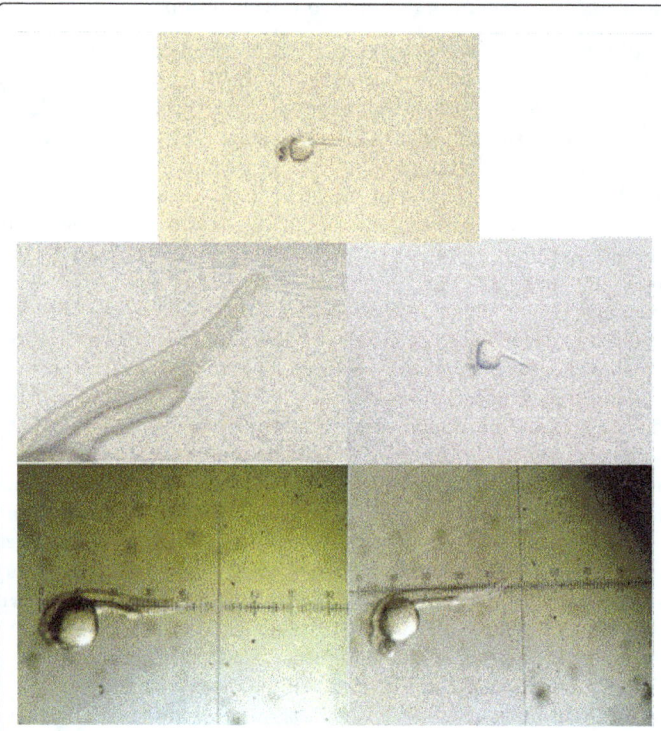

Figure 1: Embryonic developmental stage of Zebra fish. A: Diploid gynogen embryo body shape (72nd hour), B: Haploid embryo tail deformation structure (4x), C: Haploid gynogen embryo body shape (72nd hour), (x1.8), D: Haploid larvae tall stature (x1.8), E: Diploid gynogen larvae tall stature (x1.8).

Data of karyotype analysis

The results of the karyotype analysis of embryos of the experimental groups (control, haploid and diploid gynogen) at 24th hour after the fertilization, is shown at Figure 2. In haploid gynogenetic embryos, the ruptured chromosome fragments were identified at Figure 2C. Moreover, the number of the chromosomes was determined as 2n=50 in the karyotype analysis of diploid gynogenetic and control group embryos (Figure 2A and 2B).

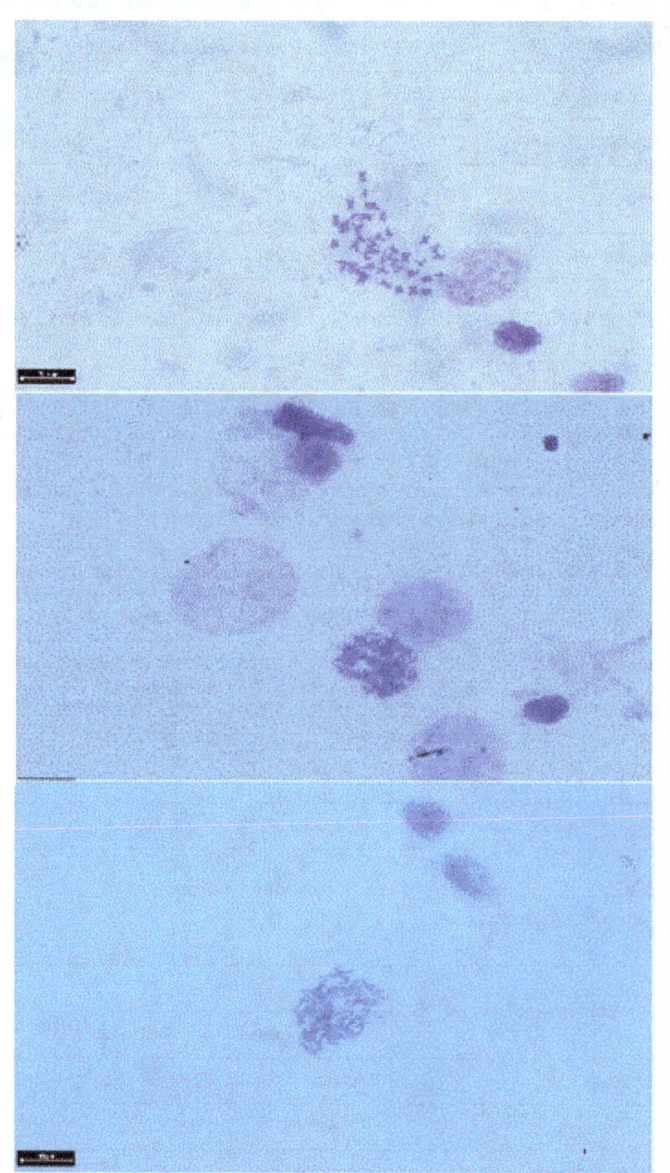

Figure 2: A: Control group (x100); B: Diploid gynogen group (x100) and C: Haploid gynogen group chromosome spreads (x100).

Discussion

The results of the heat shock implementation at 41.4°C used in this study show similarities with those of concerning the zebrafish [34,35]. It is determined that the survival rates of diploid gynogen embryos are higher at 41.4°C is used in this study- than at 41°C (Table 3).

The shocks implemented to get the diploit gynogen embryos have been done 13 minutes after the fertilization for 2 minutes, and this is the same implementation with that conducted by Walker and Streisinger. So; the results were similar.

10% haploid larvae were obtained as a result of the gynogenesis implementation on *Carasius auratus*, however all larvae were dead after this stage. The heat-shock implementation at 39.5°C for 2.5

minutes was conducted at 15, 20, 30, 35 and 45 minutes later than the fertilization in order to get diploid gynogen. It is determined that mito-gynogenetic and mayo-gynogenetic larvae can be obtained with the hot shock implementations conducted at different intervals in the study [14]. The shock implementation was conducted 13 minutes after the fertilization and mayo-gynogenetic larvae were obtained in the study.

254 nm UV implementation from 4 cms away was used for 3-4 minutes to corrode spermatozoa genome in a study on *Cyprinus carpio* [36]. The spermatozoa of *Oreochromis niloticus*, of which genetic material was neutralized with an exposure to radiation, was used for the fertilization in a study carried out on Betta splendens and after the fertilization, gynogen fish was obtained by implementing pressure shock.

It is informed that 50% heterozygote gynogenetics fish can be obtained when the pressure shock is implemented 2.5 minutes after the fertilization for 6 minutes. However, the rate of success achieved has been 21% when the pressure shock is implemented 34 minutes after the fertilization for 5 minutes. According to the results; under these conditions, the leave time of 2nd polar body of Betta splendens eggs actualizes 2.5-8.5 minutes after the fertilization [26].

In a study of Schwark [35] performed on *Danio rerio*, the sperm was exposed to UV implementation from 27.5 cm distance for 2 minutes. All haploid larvae exposed to heat shock implementation (41.4°C) 13 minutes after the fertilization for 2 minutes were dead 2-3 days after running out of vitellus and their caudal length is 1/3 shorter when compared to those of diploid group [35].

Conclusions

The gynogenetic fish producing method and the different heat-shock applications mentioned in this study enhanced the survival rate of gynogenetic fish. Some important stages are in gynogenetic production which is UV irradiation, short-term storage of sperm, heat shock time and duration, karyotype analysis and morphological screening were applied.

In our study, time dependent death rates of diploit gynogen embryos in heat shock implementation at 41.4°C at times between 12th-24th hours and 24th-48th hours were observed as 41.28% ± 5.12% and 48.89% ± 21.22% respectively, such as the same those in the studies of Walker et al. [34,35] on *Danio rerio*, In haploid gynogen group, deaths occurred at the rate of 60.32% ± 13.46% at times between 12th-24th hours and 24th-48th hours were observed as 60.32% ± 13,46% and 53.03% ± 15%, respectively. Furthermore, survived haploit larvae were observed on the second day after the hatch of larvae from the eggs. Time dependent death rates of gynogen diploid embryos in heat shock implementation at 41°C used in our study were observed as 43.78% ± 5.82% and 54.89% ± 21.72% at times between 12th-24th hours and 24th-48th, respectively. In haploid gynogen group, the rate of death occurred was observed as 71.52% ± 15.16% and 73.03% ± 15.80% at times between 12th-24th hours and 24th-48th , respectively. Moreover, there is no larvae were survived in between 48th-72nd hours (Table 2 and 3).

In the study, the sperm was exposed to UV (254 nm) from 28 cms away for 2 minutes and the survival rate of diploid embryos was higher (17% ± 3%) in 41.4°C hot shock implementation when compared to that in 41°C (14% ± 2%).

In screening the morphological development of haploid and diploid embryos; assembly of cell cycle, axis, epiboly at 12nd hour; general body shape, eyes, notocord and death cells at time between 24th-30th hour; hearth beat, blood cells, pectoral fins and newly death cells at time between 48th-54th and swimming movement and embryo death at time between72nd -78th hour were observed [2]. In this study; bud stages were observed in control and diploid gynogen groups at 12nd hour after the fertilization but, this stage is observed in haploid group at 12-18th hour after the fertilization. Delay in the development stages, shortness in body length and death cells were observed more in haploid embryos than gynogen embryos at time between 24th -30th hour. In haploid embryos, heart beats are slower than those of gynogen embryos (heart beats observed at 36-42nd) at 48-54th hour. In this case, deformation in the embryonic development was caused by the fact that circulatory system was mot developed enough at time between 48th-54th hour as it is pointed out in study [37,38].

In the diploid embryos; the developments of eye, pigment cells, notochord and otic placod were observed as normal. Hatching from the egg, normal swimming movement and body shape with normal evolution were observed in the investigation on diploid gynogen and control group at time between 72nd-78th hours. Deformation of tail, body shape and, slow blood flow were observed in haploid gynogen. In conjunction with these deformations, they died 1 or 2 days after hatching from the eggs.

On the 3rd day after the fertilization, total body length of the haploid gynogenetic fish (1277 µm) was 39.6% shorter and its body thickness was 33% thicker than those of the diploid gynogenetic group (2553 µm).

In this study, ruptured chromosome fragments were identified as a result of the karyotype analysis in haploid gynogenetic embryos. Furthermore, the number chromosomes were detected as 2n=50 in the karyotype analysis of diploid gynogenetic embryos. All these results share similarities with those of found in study [35].

This study forms the basis of gynogenesis which will be held later and also the method of further biotechnological studies. The study also puts forward the points to be considered in the gynogenesis application. The data obtained in this study will contribute to the gynogenesis applications to be conducted on the fish with high commercial value.

Authors Contributions

AE was consulted in the study. RCO carried out the animal experiments and data analysis, and drafted the manuscript. RCO and AE designed the study and revised the manuscript together. RCO and AE participated in the animal trial. RCO made the data collection and analyses. All authors read and confirmed the final manuscript.

Acknowledgment

The research is a part of master thesis with 10850 project number which was supported by Istanbul University Scientific Research Project Agency.

References

1. Amsterdam A, Hopkins N (2006) Mutagenesis strategies in zebrafish for identifying genes involved in development and disease, Trends Genet 22: 473-478.

2. Moens C, Walker CS, Walsh G (2009) Making gynogenetic diploid Zebrafish by early pressure. J Vis Exp 28: 1396.

3. Cherfas NB, Kozinsky O, Rothbard S, Hulata G (1990) Induced Diploid Gynogenesis and Triploidy in Ornamental (Koi) Carp, Cyprinus carpio L. Theor Appl Genet 42: 3-9.

4. Dahm R, Geisler R, Nusslein C (2005) Zebrafish (Danio rerio) Genome and Genetic, Robert A. Meyers, encyclopedia of molecular cell biology and molecular medicine, Volhard Max Planck Institute for Development Biology.

5. Dieter A, Quillet E, Chourrout D (1993) Suppression of first egg mitosis induced by heat shocks in the rainbow trout, J Fish Biol 42: 777-786.

6. Dunham RA (2004) Gynogenesis, Androgenesis, Cloned population and nuclear transplantation, aquaculture and fisheries biotechnology: Genetic Approaches, Section 4: 54-64.

7. Ekici A (2007) Döllenmiş zebra balığı (Danio rerio) yumurtalarına gen (gfp) transferi üzerinde bir araştırma, Doktora Tezi, İstanbul Üniversitesi, Fen Bilimleri Enstitüsü.

8. Ekici A (2011) Sazan Balıklarında Islah Çalışmaları, Sazan Balığı Üretim Tekniği, Edt. TİMUR, M., İstanbul Üniversitesi Yayını.

9. Grunwald DJ, Eisen JS (2002) Headwaters of the zebrafish emergence of a new model vertebrate, Nature Rewievs Genetics, 3: 717-724.

10. Jiang Y (1993) Transgenic fish-gene transfer to increase disease and cold resistance. Aquaculture 111: 31-40.

11. Jing R, Huang C, Bai C, Tanguay R, Dong Q (2009) Optimization of activation, collection, dilution, and storage methods for zebrafish sperm, Aquaculture 290: 165-171.

12. Jungalwalla PJ (1991) Production of nonmaturing Atlantic Salmon in Tasmania. In: Pepper VA (ed.), Proceedings of the atlantic canada workshop on methods for the production of non-maturing salmonids.

13. Karayucel İ, Karayucel S, Penman D, Mcandrew B (2002) Production of androgenetic Nile tilapia, Oreochromis niloticus L.: Optimisation of heat shock duration and its application time to induce diploid androgenesis. Israel Journal of Aquaculture 54: 145-156.

14. Karayucel İ, Karayucel S, Bircan R (2001) Tek eşeyli veya steril balık üretim metotları. Türkiye Su Ürünleri Dergisi 2: 17-19.

15. Komen H, Thorgaard GH (2007) Androgenesis, gynogenesis and the production of clones in fishes: A review, Aquaculture 269: 150-173.

16. Lutz CG (2001) Practical genetics for aquaculture.

17. Malison JA, Kayes TB, Held JA, Barry TB, Amundson CH (1993) Manipulation of ploidy in yellow perch (Perca flavescens) by heat shock, hydrostatic pressure shock, and spermatozoa inactivation, Aquaculture 110: 229-242.

18. Manickam P (1991) Triploidy induced by cold shock in the Asian catfish, Clarias batrachus (L) Aquaculture 94: 377-379.

19. Neifach AA (1959) Radiative ionization effect on early development of fish. Trud Inst Morf Zhiv Akad Nauk SSSR 24: 135-59.

20. Opperman K (1953) Die Entwicklung von Forelleneiern nach Befruchtung mit radiumbestrahlten Samenfäden. Arch mikroskop Anat 83: 141-89.

21. Paschos I, Natsis L, Nathanailides C, Kagalou I, Kolettas E (2001) Induction of gynogenesis and androgenesis in Goldfish. Reprod Dom Anim 36: 195-198.

22. Purdom CE (1993) Genetics and Fish Breeding. Chapman & Hall, Fish and Fisheries Series 8.

23. Rubinstein AL (2003) Zebrafish: from disease modelling to drug discovery. Curr Opin Drug Discov Devel 6: 218-223.

24. Streisinger G, Walker C, Dower N, Krauber D, Singer F (1981) Production of clones of homozygous diploid zebrafish. Nature 291: 293-296.

25. Tanalp R, Uzalp B (1975) Fizyoloji Pratik Ders Kitabı, Ankara Üniversitesi Eczacılık Fakültesi Tıp Bilimleri Kürsüsü sayfa: 25-27.

26. Teskeredzic E, Donaldson EM (1993) Comparison of hydrostatic pressure and thermal shocks to induce triploidy in Coho salmon (Onchorhynchus kisutch) Aquaculture 117: 47-55.

27. Thorgaard GH (1983) 8 Chromosome set manipulation and sex control in fish, Fish physiology, vol. 9 B. Academic Press, New York pp: 405-434.

28. Turgut T (2011) İnsan pnömonisinin hayvan modelleri, Göğüs hastalıklarında in vivo ve in vitro araştırma yöntemleri, AVES Yayıncılık, ADA Ofset Matbaacılık Tic Ltd.

29. Westerfield M (1995) The zebra fish book: A guide for the laboratory use of zebrafish (Danio rerio), University of Oregon Pres Eugene.

30. Westerfield M, Doerry E, Douglas S (1999) Zebrafish in the Net. Trends Genet 6: 248-249.

31. Varadaraj K, (1990) Dominant red colour morphology used to detect paternal contamination in batches of Oreochromis mossambicus (Peters) gynogens. Aquaculture Res 21: 163-172.

32. Vascotta SG, Beckham Y, Kelly GM (1997) The zebrafish's swim to fame as an experimental model in biology. Biochem Cell Biol 75: 479-485.

33. Schwark GH (1993) Production of homozygous diploid zebrafish (Brachydanio rerio). Aquaculture 112: 25-37.

34. Eenennaam JPV, Medrano JF, Vedoroshov SI (1995) Induction of meiotic gynogenesis and polyploidy in White Sturgeon (Acipenser transmontanus), Department of Animal Science University of California.

35. Walker C (1999) Haploid Screens and Gamma-Ray Mutagenesis The Zebrafish: genetics and genomics, Methods in Cell Biology Series. Academic Press, San Diego CA.

36. Ottera H, Thorsen A, Peruzzi S, Karlsen O (2011) Induction of meiotic gynogenesis in Atlantic cod (Gadus morhua). J Appl Ichthyol 27: 1298-1302.

37. Walker C, Streisinger G (1995a) Embryo production by in vitro fertilization. In: The Zebrafish Book-A Guide for Laboratory Use of Zebrafish (Danio rerio). Westerfield M (editor). University of Oregon, Eugene, OR.

38. Hollebecq MG, Chourrout D, Wohlfarth G, Billard R (1985) Diploid Gynogenesis induced by heat shocks after activation with UV-irradiated sperm in Common carp Aquaculture 54: 69-76.

4

Effects of Seasonal Variation on Fish Catching in Jebel Aulia Reservoir on the White Nile, Sudan

Ahmed Mohammed Musa Ahmed*

Department of Fish Sciences, Neelain University, Khartoum, Sudan

***Corresponding author:** Ahmed Mohammed Musa Ahmed, Assistant Professor of Fish Science, Department of Fish Sciences, Neelain University, Khartoum, Sudan, E-mail: nifidy@yahoo.com

Abstract

This study was conducted to see the effects of seasons on fish production, in Jebel aulia dam south of Khartoum 45 km during the period January to December 2014, (containing 12 months) includes three seasons, summer, autumn and winter. 23 species belonging to 14 families were recorded during the period of investigation. Distribution production of fish in seasons as follow: in summer the high production is *Tilapia* in March 61.2%, April 53.3%, May 40%, finally June 32%. Bagrus bayad in March 9.9%, April 5.6%, May 12.6%, finally 4.9% in June. The fish which is rare is *Disticodus niloticus* and *Citharinus citharus*. High production months in the summer are June 36%, April 23%, March 21% and May 21%. In autumn the fish species which very high production is *tilapia* in July 25.9%, August 31.6%, September 33.5% and October 9.9% followed by *Schall fish* and *Labeo niloticus*. In winter the study found the high production of species is *Tilapia, labeo niloticus and Hydrocon Forskalli*. The months which is high production in winter containing November, December, February and January. The study showed that the fish production seasons are summer 37.15%, autumn 35.95% and finally winter 26.90%.

Keywords: Jebel Aulia Dam; Reservoir; Seasons; Fish species; Investigation production; Variation

Introduction

Jebel Aulia Dam was constructed in 1937 across the White Nile some 45 kilometres south of Khartoum. It resulted in the formation of larger shallow lake and covered estimated area of about 12,000 hectares. The Dam stores about 3.5 Millard cubic meters of water. Maximum depth of the reservoir is about 15 meters during the time of high flood (late August to mid-September) while a minimum depth of 5 meters is attained in May, when the reservoir is nearly emptied to a normal river level. Fish and fisheries of White Nile have been investigated by several workers. The taxonomy and characteristic of fish were compiled by Boulenger [1] in his treaties or fish fauna of the Nile. Girgis [2] recorded 18 families and 62 species from the swamps and the southern tributaries of the White Nile. The feeding and breeding habits of some common Nile fish were studied by Pekkola [3] and his investigations were further extended by Sandon [4]. Sandon [5] investigated the fishes of northern Bahr Elgazal and stated that fishing follows the seasonal regime of flooding and fallowing water. One hundred and eight species were recorded by Sandon [6] from the Sudan waters of the White Nile system. They belonged to 51 genera and 23 families. More recently [7] reported that the fish fauna of the Nile basin is rich and diversified and includes at least 54 genera and well over 300 species. Gidiri [8] carried out a detailed study on the biology of genus *Synodontis* at Khartoum which its establishment of *Synodontis khartoumensis* as a new species. The present paper however is an attempt to consider the distribution and abundance of fish of the White Nile in the area affected by Jebel Aulia Dam.

Materials and Methods

The area investigated was Jebel Aulia Dam two sets of gill nets were used to catch the fish the first set with mesh size ranging between 40 mm-120 mm and 1.5 meters-2 meters in depth. The second set of gill nets had a mesh size ranging from 70 mm-90 mm and 1.55 meters-1.80 meters in depth. Gill nets were set overnight. The catch was sorted out immediately after collection and fish identified down to species level. Total weight of fish was recorded in kilogram (kg). The period of study was (January-December 2014).

Results

Fish population in the study area obtained 23 species belonging to 14 families were recorded during the period of investigation. These are listed as follows in Table 1. In summer *Tilapia fish*, and *Bagrus bayad* is high production in month March, April, May and June. The rare fish were *Disticodus niloticus* and *Citharinus citharus* in Table 2. High production of months in summer is June 36% (Table 3).

Family (14)	Species (23)
(1) Mormyridae	1-*Mormyrus caschive*
	2-*Mormyrus bebe*
	3-*Mormyrus cyprinoids*
(2) Mochokidae	1-*Synodontis schall*
(3) Bagridae	1-*Bagrus bayad*
	2-*Bagrus domac*
	3-*Chrysichthys auratus*
	4-*Auchenoglanis occidentals*

(4) characidae	1-Alestes dentex
	2-Hydrocynus forskalli
	3-Alestes nurse
(5) Citharinidae	1-Disticodus niloticus
(6) Schilbeidae	1-Schilbe mystus
(7) Cyprinidae	1-labeo niloticus
	2-Labeo horii
	3-Barbus bynii
(8) Cichlidae	1-Oreochromis niloticus

(9) Clariidae	1-Clarias lazera
(10) Protopteridae	1-Protopterus aethiopicus
(11) Centropomidae	1-Lates niloticus
(12) Tetraodontidae	1-Tetraodon lineatus
(13)Malapteroidae Malapteroidae	1-Malapterurus electricus
(14) Osteoglossidae	1-Heterotis niloticus

Table 1: Showed the families and species of the study.

Species	March		April		May		June	
	%	Wt.(kg)	%	Wt.(kg)	%	Wt.(kg)	%	Wt.(kg)
Lates niloticus	0.7	385.28	1.2	677.57	1.6	868	1.1	961
Bagrus domac	0.2	93	0.3	150.57	0.1	48.71	0.3	318.86
Bagrus bayad	9.9	5194.71	5.6	3255	12.6	6642.86	4.9	4455.14
Tilabia niloticus	61.2	32200.14	53.3	31093	40	21040.1	32	29255.1
Labeo niloticas	7.6	4003.43	18.8	10978.4	17.1	9007.71	18	16443.3
Barbus bynni	0.1	75.26	0.4	252.43	0.5	279	0.9	828.4
Labeo horii	0	0	0	0	0	0	0	0
Kanumm sp	0.2	93	0.4	212.57	1.5	788.29	0.4	366.57
Distichodus niloticus	0	0	0	0	0.1	75.29	0.1	53.14
Cithrius cithrius	0	0	0	0	0	0	0	0
Synodontis schall	2.1	1098.29	2.2	1266.57	2.9	1501.29	23.1	21124.3
Alestes dentex	2.4	1240	2.2	1266.57	2.5	1302	1.5	1377.29
Clarias lazera	5.5	2896.28	7.6	4424.14	8.7	4561.43	5.3	4867
Hydrocon Forskalli	2.3	1217.85	1.8	1049.57	4.4	2338.29	1.8	1603.14
Shelbe mystes	0.5	261.28	0.5	314.43	1.5	766.14	0.8	744
Other	7.3	3835.14	5.7	3343.57	6.5	3432.14	10	9114
Total	100	52593.66	100	58284.39	100	52651.25	100	91511.24

Table 2: Showed fish production in months of summer 2014.

Months of summer	Production(kg)	%Percentage
March	52594	21%
April	58284	23%
May	52651	21%
June	91511	36%

Total	254740	100%

Table 3: Showed months of summer and production (kg).

In autumn also *tilapia* and *Labeo niloticus* is high production (Table 4). Months of autumn which is high are July, October, August and September, in Winter also *Tilapia, Labeo niloticus* and *Hydrocon Forskalli* is high production but the fish was very rarely is *Bynii, Citharus, Disticodus* and *Labeo horii* (Tables 5 and 6).

Species	July		August		September		October	
	%	Wt.(kg)	%	Wt.(kg)	%	Wt.(kg)	%	Wt.(kg)
Lates niloticus	1.6	1169.11	1.1	635.71	1.5	827.14	0.1	681.43
Bagrus domac	0.2	172.71	0.5	270	0.4	205	0.1	381.43
Bagrus bayad	5.6	4034.72	3.5	2044.29	5.9	3325.71	1.2	6930
Tilabia.niloticus	25.9	18639.9	31.6	18475.71	33.5	18831.43	9.9	59258.57
Labeo niloticas	19	13666.6	15.3	8961.43	13.9	7787.14	2.9	17288.57
Barbus bynii	0.1	66.43	0	7.57	0	0	0	0
Labeo horii	0	0	0	12.86	0	0	0.1	488.57
Kannume sp	0.8	571.29	0.4	244.29	0.3	180	0	0
Disticodus niloticus	0.1	97.43	0	0	0.1	30	0	0
Cithrius cithrius	0	0	0	0	0	0	1.9	11472.86
Synodontis schal	25.3	18205.9	22.1	12882.86	15	8400	0.2	1015.71
Alestes dentex	1.1	819.29	15.6	9092.8	1.2	668.57	1.3	7920
Clarias lazera	6.2	4464	4.3	2502.86	7.6	4285.71	0.2	1345.71
Hydrocon Forskalli	1.6	1169.14	1.7	1002.86	1.4	801.43	0.1	660
Shelbe mystes	0.9	624.43	1.5	890.71	1.4	775.71	9.3	55774.29
Other	11.6	8356.71	2.4	1398	17.9	10050	72.7	435623
Total	100	72057.66	100	58421.95	100	56167.84	100	59884.14

Table 4: Showed the fish production in months of autumn 2014.

Months of autumn	Production (kg)	%Percentage
June	72058	29.20%
August	58422	23.70%
September	56188	22.80%

October	59884	24.30%
Total	246532	100%

Table 5: Showed months of autumn and production kg.

Species	November		December		January		February	
	%	Wt.(kg)	%	Wt.(kg)	%	Wt.(kg)	%	Wt.(kg)
Lates niloticus	1	801.43	0.2	681.43	2.2	540.14	2.9	1085
Bagrus domac	1	827.14	9.9	38143	2.1	513.71	1.3	496.71
Bagrus bayad	5	4165.71	1.8	6930	6	1483.57	8.1	3038
Tilabia niloticus	39.9	33407.14	15.4	59258.57	61.3	15256.43	58.5	21908.14
Labeo niloticas	18.9	15810	4.5	17288.57	5.5	1377.29	6.2	2329.43
Barbus bynii	0	0	0	0	1.4	354.29	1.1	403
Labeo horii	0	0	0	0	0	0	0	0
Kanumm sp	0.5	420	12.7	48857	0.3	84.14	0.6	234.71
Distichodus niloticus	0	0	0	0	0	0	0	0

Cithrius cithrius	0	0	0	0	0	0	0	0
Synodontis schall	3.3	2790	3	11472.8	4	987.57	1.8	669.43
Alestes dentex	1.6	1307.14	0.3	1015.71	3	752.86	3.2	1204.5
Clarias lazera	5.1	4264.29	2.5	9720	2.2	552.71	5.3	1970.71
Hydrocon Forskalli	11.1	9272.86	35	134571	4.6	1155.86	3.1	1160.29
Shelbe mystes	0.7	604.29	0.2	660	2.1	513.17	1.1	394.14
Other	11.9	9972.86	14.5	55774.29	5.3	1324.14	6.9	2568.57
Total	100	83642.9	100	38437.24	100	24895.9	100	37462.6

Table 6: Showed fish production in months of winter 2014.

High months in production in winter is November, December and February the study show the high season in production is summer 37.15%, autumn 35.95% and winter 26.90% (Tables 7 and 8). The boats were made of woods, metal, and fiberglass. Gill net and cast net are very famous nets in the Reservoirs. The boats made of locally woods (Table 9).

Months of winter	Production (kg)	%Percentage
November	83643	45.40%
December	38437	20.80%
January	24896	13.50%
February	37463	20.30%
Total	184439	100%

Table 7: Showed months of winter and production kg.

Seasons	Fish production	%Percentage
Summer	254.741	37.15%
Autumn	246.532	35.95%
Winter	184.439	26.90%
total	685.712	100%

Table 8: Show seasons variation and fish production 2014.

Boat kinds	Nets kinds	Woods kinds
Wood boat	Gill net	haraz
Metal boat	Cast net	sonut
sharoug	crabs	neem
flouka		mahogani
Fiber glass		sayal

Table 9: Show the boat, net and kind of wood in fish catching.

Discussion

The study showed that high production was recorded in summer season 37.145%, autumn 35.95% and finally winter 26.95%. 23 species and 14 families were found in the study. The important species in this study were *Oreochromis niloticus*, *Labeo niloticus* and *Synodontis schall*. The rare fish in the dam were *Disticodus niloticus*, *Labeo niloticus* and *Citharinus citharus*. The present study agrees [9] had found 14 families and 21 species. The study agrees [10] that the boats in Jebel aulia were made from woods, metal, shroug and fiberglass. Otherwise the local names of the nets were gill nets and caste net. The kinds of wood by local name are Haraz, Sonut, Neem and Sayal.

Girgis [2] recorded 18 families and 62 species from the swamps and the southern tributaries of the White Nile but in this study found 14 families only. 108 species were recorded by Sandon [6] from the Sudan waters of the White Nile system. They belonged to 51 genera and 23 families but in this study 14 families may be return to periods between the two studies. More recently [7] reported that the fish fauna of the Nile basin is rich and diversified and includes at least 54 genera and well over 300 species [8].

1967 studied all fishes in Sudan but this study is included only Jebel aulia dam (Figure 1).

Figure 1: Show the Jebel aulia dam (location study).

Acknowledgement

We express deep sense of gratitude to the head of department of fish sciences, the technical support of Mr. Elnuman Babikir.

References

1. Boulenger GA (1907) Zoology of Egypt. The fished of the Nile Hugh ross Ltd, London.

2. Girgis S (1948) A list of Common fish from the Upper Nile with their shilluk, Dinka and Nuer names. Sudan Notes and Records pp: 120-125.

3. Pekkola W (1919) Notes on habits, breeding and food of some white Nile fish. Sudan Notes and Records pp: 112-121.

4. Sandon E (1953) Problems of fisheries. Sudan Notes and Records 32: 5-36.

5. Sandon H, Tayib AA (1953) The food of some common Nile fish, Sudan Notes and Records pp: 205-229.

6. Sandon E (1950) An illustrated guide to the freshwater fishes of the Sudan. Sudan Notes and Records, Khartoum.

7. Hammerton D (1972) The Nile River-Symposium on river ecology and the impact of man. Hydrobiol Res unit Khartoum.

8. Gidiri BA (1967) Fishes of the Blue Nile between Khartoum and Roseires. Rev Zool Bot Afri 3-4.

9. Ahmed J (2010) Distribution of the fish in Jebel Aulia dam reservoir on White Nile.

10. Mohammed OM (2012) A short Review on: fishing boat used in Sudan freshwater fisheries. Bull Environ Pharmacol Life Sci 1: 93-99.

5

Features of Sperm Motility and Circadian Rhythm in Japanese Anchovy (*Engraulis japonicus*)

Dipak Pandey[1,2], Yong-Woon Ryu[2] and Takahiro Matsubara[2*]

[1]*The United Graduate School of Agricultural Sciences, Bioresource Production Science, Ehime University, Japan*

[2]*South Ehime Fisheries Research Center, Ehime University, Japan*

*Corresponding author: Dr. Takahiro Matsubara, South Ehime Fisheries Research Center, Faculty of Agriculture, Ehime University, 25-1 Uchidomari, Minamiuwa-gun, Ainan, Ehime 7984206, Japan, E-mail: matsu@agr.ehime-u.ac.jp, b741012u@mails.cc.ehime-u.ac.jp

Abstract

Japanese anchovy (*Engraulis japonicus*) is a commercially and ecologically important fish that exhibits group synchronous and multiple spawning. However, the reproductive characteristics of the male in this species, especially sperm features and activation, are still largely unknown. In this study, we confirmed that features of the sperm and characteristics of the activations, regarding sperm motility and moving velocity. The average size of the sperm was 51 ± 1.3 μm in total length and possessed a normal structure with clockwise, anticlockwise, and linear motion. The initial motility at one minute after activation in seawater was 75 ± 12% during spawning time in this species (21:00–22:00), and the initial moving velocity (196 ± 26 μm/sec) remained constant for fifteen minutes post activation. While, comparatively low motility (30 ± 10%) was found until 17:00, and the sperm was almost immotile in the morning (08:00–09:00). Swimming ability was also confirmed with sperm that swam for more than one hour in seawater without an exogenous energy supply derived from the ovary in females, suggesting the trigger for sperm activation in multiple spawning fish is possibly species dependent. This report is the first to demonstrate time specific activation, that is, circadian rhythm, in teleost males.

Keywords: *Engraulis japonicus*; Sperm motility; Synchronous spawning; Circadian rhythm

Introduction

Japanese anchovy is a small pelagic fish belonging to the order Clupeiformes and is widely distributed around Japan. This species is commercially important to fisheries in Japan [1], China [2], Korea [3], and Taiwan [4], and furthermore, it plays an important role as a key member of marine ecosystems [5]. Japanese anchovy is known as a multiple spawning fish having a long spawning season [6,7] and spawn oval-shaped pelagic eggs. Females spawn periodically at intervals of two or more days during the spawning period [8], and the spawning rhythm is regulated by ambient water temperature [9]. However, the reproductive characteristics of the male of this species, especially sperm activity, have yet to be clarified.

To date, research on sperm in fish has centered on initiation mechanisms of motility [10] and sperm quality for artificial fertilization [11]. However, the acquisition mechanisms of sperm motility have not yet been demonstrated. In Japanese eel, it has been shown that spermatogenesis completes in an *in vitro* culture with 17α, 20β-dihydroxy-4-pregnen-3-one (DHP), and the sperm is morphologically the same as functional sperm naturally matured from males, but the sperm does not move in seawater (SW) [12]. In addition, *in vitro* incubation of sperm, which was from artificially matured males of Japanese eel with bicarbonate medium, showed increased sperm motility [13]. These results indicate that spermatogenesis and acquisition of sperm activation mechanism are controlled by different mechanisms.

In oviparous teleosts, spermatogenesis is generally known to take a long time (several weeks to months), in which spermatozoa are released at the same time and place with female ovulation [14]. Therefore, the intrinsic quality and quantity of both gametes affect the success of fertilization. Motility is one important function of the male gametes (sperm), which allows them to actively reach and penetrate the female gametes (eggs). A teleost egg is covered with a thick envelope called a chorion, which has a narrow pore designated as a micropyle that helps to avoid polyspermy [15]. Sperm can only enter and reach the ooplasmic membrane through the micropyle [16,17]. Thus, sperm motility might directly influence fertilization.

In most external fertilization types of teleost, sperm remains quiescent in the seminal plasma and becomes transiently motile when released into hypotonic fresh water or hypertonic seawater depending on the spawning environment [18]. In most freshwater species, sperm usually moves for less than 2 min and high activation is observed for less than 30 s [19]. Meanwhile, longer durations of sperm movement have been reported in various marine fish, including *Anarhichas minor* [20] and *Abramis brama orientalis* [21], and the time might be related to their reproductive strategies. It is known that several factors affect sperm motility, such as pH, temperature, ions, osmolality, and ovarian fluid (OF) [22]. Ovarian fluid is slightly viscous fluid found in the gonad cavity of oviparous fishes after ovulation. Ovarian fluid has been studied for its role in fish reproduction, its chemical composition, novel proteins and utility in testing for the presence of fish diseases [23]. There are also some reports on sperm being activated by ovarian fluid, which is extruded with ovulated eggs in Rainbow trout, *Oncorhynchus mykiss* [24] and Arctic charr, *Salvelinus alpinus* (L.) [25,26]. However, such information about the sperm activation of male Japanese anchovy is not well known.

Although understanding of sperm activation in teleost species is important, especially for commercial seed production, it is still a largely unknown area due to the technical difficulties of analyzing sperm motility. A microscope connected to a computer analyzing system (CASA: - is commonly used to evaluate sperm motility in fish, such as common carp [27], African catfish [28], and zebrafish [29]. The aim of the present study is to clarify the features of anchovy sperm and time course changes in sperm motility using such a CASA system, which should provide basic knowledge about male reproduction of a multiple spawning teleost. Moreover, the effect of OF on motility was also examined to verify the activation system of the sperm.

Materials and Methods

Experimental animals

Juvenile Japanese anchovy captured by a commercial fishing boat were transferred to a sea surface aquaculture cage (5 × 5 × 5 m) located in Mishou Bay in Southern Ehime in December 2014. Captive reared anchovy were transferred to a 30-ton aquarium at the South Ehime Fisheries Research Center, Ehime University, and reared under natural photoperiod and water temperature conditions (19.5–20.9°C). About two hundred adults (body weight range: approximately 10–12 g) were placed into two one-ton tanks. Fish were fed 40 g of commercial feed (Otohime S2; Nissin-Marubeni, Japan) per day, which corresponded to about 2% of body weight, under a controlled photoperiod of 14L:10D and reared from May to June 2015.

Sampling

Sampling was carried out at four different time points (Table 1). All males from each time sampling point were examined. Gonads were excised and weighed to calculate the gonadosomatic index (GSI; [gonad weight/body weight] × 100). Intra-testicular semen was collected by syringe after making an incision in the posterior part of the excised testes and stored in a closed test tube (1.5 mL) on ice until use (up to 1 h). Sperm features were individually assessed from all sampled males. Stripped semen was diluted with Hank's solution, which is an inhibiting solution of sperm motility (Nacalai Tesque, Kyoto, Japan) containing calcium and magnesium, at a ratio of 1:9 (semen:solution) and used within 1 h. Ovarian fluid was collected from 5 ovulated females sampled at 21:00, the spawning time in this species. After expelling the eggs onto a plastic plate by gentle abdominal pressure, transfer in a sieve (mesh size 1 mm^2) and the OF was poured off, collected and store at 4°C for analysis or -30°C for lateral use. The obtained OF was then diluted with Hank's solution at a ratio of 1:1 before use.

	Sampling Point			
	08:00 - 09:00	12:00 - 13:00	16:00 - 17:00	20:00 - 21:00
Total sample fish	19	15	20	21
Total male	7	6	12	9
Physiological status of female (Pandey et al., In press)		Final oocyte maturation		Ovulation and spawning
All experimental fishes were reared under 14L-10D photoperiod (light, from 05:00 to 20:00) with water natural temperature (19.5°C ~ 20.9°C).				

Table 1: Sampling information of Japanese anchovy in this study.

Sperm motility analysis

Each 5 µL of semen diluent was added to 500 µL of filtered SW and mixed gently. Then, 6 µL of the sperm suspension was pipetted into a Standard Count 2 Chamber slide (Leja products B.V., GN Nieuw Vennep, Netherlands) and observed under a microscope (VANOX-T; Olympus, Japan) connected to a digital high speed camera (HAS-L1 Ver. 2.14; Detect Inc., Japan). Video tracking was carried out at 1, 5, 10, 15, 45, and 60 min post activation. Each video was recorded at 100 frame per second (FPS) at a resolution of 640 × 680 CFG. Sperm motility was analyzed using capture video under sperm tracking software DIPP-Motion Pro (Ditect Inc., Japan).

Statistical analysis

All data are presented as mean ± standard error (SEM) and were subjected to one-way ANOVA followed by the Tukey and Kramer HSD test. Statistical significance was set at $P \leq 0.05$. All statistical analyses were performed using SAS 10.0 packed in Jump (SAS; Cary, NC, USA).

Results

Morphology of Japanese anchovy sperm

The sperm of the Japanese anchovy was quite small and moved in a linear, semi linear, and circular direction. The total length of the sperm was approximately 51 ± 1.3 µm including the oval head (3.0 ± 0.2 µm), mid piece (1.7 ± 0.1 µm), and tail (48.3 ± 1.7 µm) (Figure 1). There were no changes in morphological features at the different sampling times (data not shown).

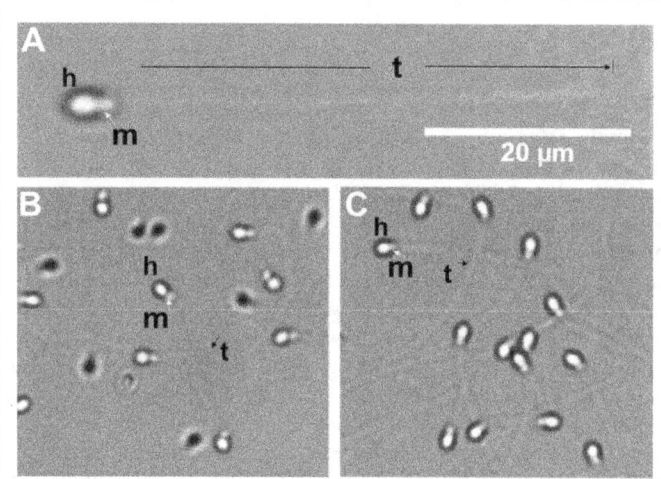

Figure 1: Japanese anchovy sperm under a light microscope. A; single sperm including head, mid piece and tail, B; motile sperm at 1 min post activation, C; Motionless sperm after 60 min post activation. h: head, m: mid piece, t: tail.

Gonadosomatic index

Our previous study showed that the GSI in female Japanese anchovy is approximately 6 to 8 when they are undergoing final maturation (FOM), and reaches 20 at the spawning time around 21:00 under natural environmental conditions [30]. In this experiment, there was no significant (p>0.05) difference in the GSI of the males between the sampling times (Figure 2). However, the values were slightly higher and reached 11 ± 2 at 17:00 and 21:00 rather than 09:00 and 13:00, indicating the GSI was higher before spawning.

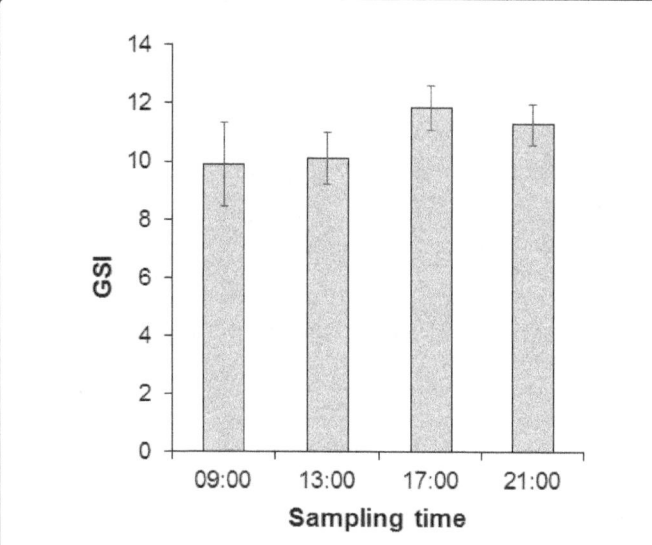

Figure 2: Changes in male gonadosomatic index (GSI) at different sampling times.

Sperm motility and moving velocity

The percentage of motile sperm significantly differed among the four groups: highest motility was confirmed at 21:00 (P<0.0001, Figure 3). Initial motility, at 1 min after SW dilution, at the sampling times of 9:00, 13:00, 17:00, and 21:00 were 4 ± 5%, 12 ± 9%, 28 ± 7%, and 75 ± 12%, respectively. Motility decreased in accordance with time (minutes post activation) and became almost inactive after 60 min post activation. On the other hand, the moving velocity of the motile sperm was not significantly different between the time sampling points, even though the values were slightly higher close to the spawning time (Figure 4). The average moving velocity at 1 min post activation was 123 ± 17, 135 ± 17, 127 ± 25, and 196 ± 26 µm/s, respectively. The moving velocity remained constant for 15 min post activation, and reached zero after 60 min post activation.

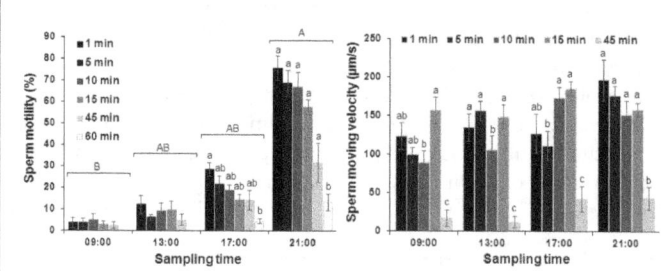

Figure 3: Variations in sperm motility and moving velocity under seawater activation by sampling time and duration. Data are expressed as mean ± standard error. Means sharing a letter superscript are not significantly different (P<0.05).

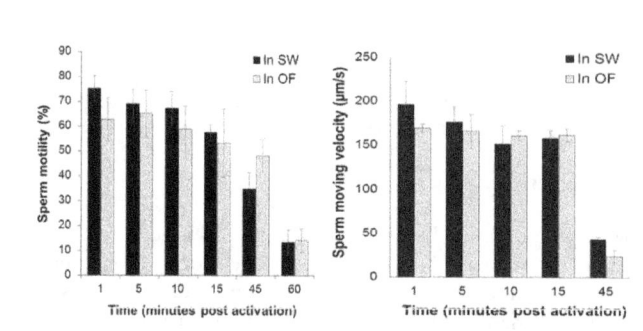

Figure 4: Effect of ovarian fluid on motility and moving velocity in different tracking point post activation. Data are expressed as mean ± standard error.

Effect of ovarian fluid on sperm motility and moving velocity

Comparisons of two activation mediums, SW and SW + OF, on sperm motility and moving velocity showed no significant differences (Figure 4). Initial motility at 1 min post activation for SW and SW + OF was 75 ± 12% and 62 ± 20%, respectively. The levels continuously decreased in accordance with time and reached 13 ± 12% and 14 ± 11%, respectively, at 60 min post activation. On the other hand, the initial moving velocity was 196 ± 26 µm/s in SW and 169 ± 10 µm/s in SW + OF, remained constant until 15 min post activation, and then decreased.

Discussion

Sperm motility in teleosts can be evaluated using a number of methods as follows: ratio of motile sperm, moving speed, and motile duration. In this study, we accurately analyzed the ratio of moving spermatozoa, moving speed, moving style whether straight or circular or both, and moving duration using a CASA system to clarify the features and characteristics of activation in sperm of Japanese anchovy.

Japanese anchovy is known as a multiple spawning fish [6] and the females spawn at intervals of two days or more [8]. Thus, spawning occurs in group stocking tanks every day under appropriate rearing conditions. In our previous report about final oocyte maturation in Japanese anchovy, spawning occurred from 21:00 to 22:00 every day under captive conditions as follows: 14L:10D photoperiod in water of 19.5 to 20.9°C. Moreover, final oocyte maturation started from 13:00 to 15:00 and progressed to reach ovulation by around 21:00 [30]. Group synchronized spawning of Japanese anchovy is also reported under wild [6] and captive conditions [30]. It is noteworthy that circadian rhythm in sperm motility was also observed in each sampled male under the same captive condition in this study, revealing that male anchovy has a daily rhythm in acquisition of sperm motility. Moreover, this finding suggests that all mature males can synchronously participate in fertilization at the same time of spawning. As shown in Table 2, time duration of sperm movements in Japanese anchovy is quite longer, at approximately 60 min, compared with other fish species. The characteristics of the spermatozoa may ensure successful fertilization in independently spawned unfertilized eggs without pairing behavior. In addition, this characteristic may enable a high genetic variability because a clutch of eggs from one female can be fertilized by spermatozoa, suspended in the surrounding seawater, from many males.

	Duration of motility (s or min)	Initial velocity (µm/s)	References
Fresh-water fish species			
Esox lucius	60-80 s	160-170	[37]
Oreochromis mossambicus	>30 min	70-80	[38]
Anguilla anguilla	<30 min	120-160	[39]
Carassius auratus	3 min at 18-21°C	ND	[40]
Cyprinus carpio	200 s	140	[41]
Salmo irideus	60-105 s	ND	[42]
Sea-water fish species			
Thunnus thynnus	140 s	215-230	[43]
Merluccius merluccius	400-500 s	65-130	[43]
Gadus morhua	700-800 s	130	[43, 44]
Dicentrarchus labrax	50-60 s	120	[45]
Acipenser persicus	1.5-5 min at 15-20°C	ND	[46]
Anarhichas minor	>2 days	40-50	[20]
Scophthalmus maximus	600 s	220	[47]
Takifugu niphobles	50 s	160	[48]
Oncorhynchus mykiss	30 s	220	[49]
Abramis brama orientalis	20 min	ND	[21]
Engraulis Japonicus	**60 ± 11 min at RT**	**196 ± 26**	**Present result**

* Duration of motility refers to the period of time from activation to complete inactivation where no single sperm is seen to be active. Temperature of experimental test is indicated in °C; RT refers to 'room temperature'.

**Initial velocity refers to the earliest value of sperm moving velocity at activation. ND: not defined.

Table 2: Experimental values for duration of motility and initial velocity for spermatozoa of various fish species.

In Pacific herring, which belongs to the same order (Clupeiformes) as Japanese anchovy, intensive studies related to the initiation of sperm motility have been conducted [10,31]. Although the Pacific herring sperm did not move in seawater when released for spawning, they started moving near unfertilized eggs through the action of herring sperm activating protein (HSAP) contained in OF [32], and entered an egg micropyle through the action of sperm motility initiating factor (SMIF), which is present near the micropyle [31,33]. Also, the functional relationship of these two physiological activation factors in fertilization is discussed in Pacific herring [10]. Such activation is also observed in the nest-building marine sculpin *Hemilepidotus gilberti* in the presence of OF, resulting in an increased period of sperm motility of up to 90 min, six times longer than in SW alone [34]. Sperm of freshwater species, such as bullhead *Cottus gobio* L. [35] and the wolf fish *Anarhichas minor* Olafsen [20], also showed an extended period of sperm movement in the presence of OF, with motility lasting for 2 h in the bullhead and 48 h in the wolf fish. In contrast, ovarian fluid did not influence sperm motility in fifteen-spined stickleback *Spinachia spinachia* [36]. In this study, the moving ability of the sperm also not change in the presence of OF, indicating that OF may not influence activation of sperm in Japanese anchovy. This finding may be related to species-specific egg characteristics, for example, the eggs of Japanese anchovy are pelagic and non-adhesive, whereas herring eggs are demersal and adhesive. However, the mechanism is not clear, and thus should be verified with further research.

In this study, we have clarified the process in which spermatozoa acquire motility during a time period of approximately half a day. It is suggested that the process occurs concomitant with female final oocyte maturation. Synchronous functionalization of both female and male gametes is suggested to ensure efficient fertilization in Japanese anchovy. Moreover, long duration sperm moving seems likely to guarantee a group synchronous non-pairing spawning style. To show the rhythm is possible among teleost species having multiple spawning characteristics, further research is necessary to determine such a rhythm of sperm motility in other species. Finally, this is the first report showing the time course of acquisition of sperm motility in a multiple spawning teleost.

Acknowledgement

We would like to thank the entire team at SEFRC for their kind co-operation and help during this research work and we are grateful to Dr.

Rie Goto (Ehime University), Dr. Haruhisa Fukada (Kochi University), Dr. Kazuhiro Mochida (National Research Institute of Fisheries and Environment of Inland Sea, Hiroshima), and Dr. Sayumi Sawaguchi (Japan Fisheries Research and Education Agency, Yokohama) for their kind suggestions. We also thank Dennis Murphy (The United Graduate School of Agriculture Sciences, Ehime University) for his critical reading of this manuscript. This work is included in the doctoral dissertation for The United Graduate School of Agricultural Sciences, Ehime University.

References

1. Aoki I, Miyashita K (2000) Dispersal of larvae and juvenile of Japanese anchovy Engraulis japonicus in the Kuroshio Extension and Kuroshio-Oyashio transition regions, western North Pacific Ocean. Fish Res 49: 155-164.

2. Chiu TS, Young SS, Chen CS (1997) Monthly variation of larval anchovy fishery in I-lan Bay NE Taiwan with an evaluation for optimal fishing season. J Fish Soc Taiwan 24: 273-282.

3. Kim JY, Kang YS, Oh HJ, Suh YS, Hwang JD (2005) Spatial distribution of early life stages of anchovy (Engraulis Japonicus) and hairtail (Trichiurus lepturus) and their relationship with oceanographic features of the East China Sea during the 1997–1998 El Nino event. Estuar Coast Shelf Sci 63: 13-21.

4. Zhao X, Wang Y, Dai F (2008) Depth-dependent target strength of anchovy (Engraulis japonicus) measured in situ. ICES J Mar Sci 65: 882-888.

5. Takasuka A, Oozeki Y, Kimura R, Kubota H, Aoki I (2004) Growth-selective predation hypothesis revisited for larval anchovy in offshore waters: cannibalism by juveniles versus predation by skipjack tunas. Mar Ecol Prog Ser 278: 297-302.

6. Funamoto T (2001) Maturation and spawning of Japanese anchovy. Nippon Suisan Gakkaishi 67: 1129-1130.

7. Tsuruta Y (1993) Geographical variation in the reproductive traits of Japanese anchovy as related to offshore distribution. Bull Jpn Soc Fish Oceanogr 57: 368.

8. Yoneda M, Kitano H, Selvaraj S, Matsuyama M, Shimazu A (2013) Dynamics of gonadosomatic index of fish with indeterminate fecundity between subsequent egg batches: application to Japanese anchovy Engraulis japonicus under captive conditions. Mar Biol 160: 2733-2741.

9. Yoneda M, Kitano H, Tanaka H, Kawamura K, Selvaraj S, et al. (2014) Temperature- and income resource availability-mediated variation in reproductive investment in a multiple-batch-spawning Japanese anchovy. Mar Ecol Prog Ser 516: 251-262.

10. Cherr GN, Morisawa M, Vines CA, Yoshida K, Smith EH, et al. (2008) Two egg-derived molecules in sperm motility initiation and fertilization in the Pacific herring (Clupea pallasi). Int J Dev Biol 52: 743-752.

11. Ochokwu IJ, Apollos TG, Oshoke JO (2015) Effect of egg and sperm quality in successful fish breeding. IOSR-JAVS 8: 48-57.

12. Miura T, Ando N, Miura C, Yamauchi K (2002) Comparative studies between in vivo and in vitro spermatogenesis of Japanese eel (Anguilla japonica). Zool Sci 19: 321-329.

13. Ohta H, Ikeda K, Kagawa H, Tanaka H, Unuma T (1999) Acquisition and loss of potential for motility of spermatozoa of the Japanese eel Anguilla japonica. UJNR Technical Report 28: 77-81.

14. Babin PJ, Cerda J, Lubzens E (2007) The fish oocyte: from studies to biotechnological applications. Springer.

15. Murata K (2003) Blocks to polyspermy in fish: a brief review. Aquaculture and pathobiology of crustacean and other species. Proceedings of the Thirty-second US Japan Symposium on Aquaculture Panel Symposium, Davis and Santa Barbara, California USA.

16. Yanagimachi R (1957) Some properties of the sperm-activating factor in the micropyle area of the herring egg. Annot Zool Japon 30: 114-119.

17. Yanagimachi R, Cherr G, Matsubara T, Andoh T, Harumi T, et al. (2013) Sperm attractant in the micropylar region of fish and insect eggs. Biol Reprod 88: 1-11.

18. Morisawa M (1994) Cell signaling mechanism for sperm motility. Zool Sci 11: 647-662.

19. Kime DE, Van Look KJW, McAllister BG, Huyakens G, Rurangwa E, et al. (2001) Computer assisted sperm analysis (CASA) as a tool for monitoring sperm quality in fish. Comp Biochem Physiol 130: 425-433.

20. Kime DE, Tveiten H (2002) Unusual motility characteristics of sperm of the spotted wolf fish. J Fish Biol 61: 1549-1559.

21. Gosteeva MN (1957) Ecological and morphological characteristics of development of Aral Bream, Abramis brama orientalis (Berg). Trudy Institute Morfologia Zhivotnykh AN USSR 20: 121-147.

22. Alavi SMH, Cosson J (2006) Sperm motility in fishes, (II) Effects of ions and osmolality: a review. Cell Biol Int 30: 1-14.

23. Sawyer ES, Sawyer PJ, Janmey PA (2003) Method of using fish ovarian fluid for culture and preservation of mammalian cells. United States Patent: US006562621 B1 (Invention).

24. Dietrich GJ, Wojtczak M, Slowinska M, Dobosz S, Kuzminski H, et al. (2008) Effects of ovarian fluid on motility characteristics of rainbow trout (Oncorhynchus mykiss Walbaum) spermatozoa. J Applied Ichthyology 24: 503-507.

25. Turner E, Montgomerie R (2002) Ovarian fluid enhances sperm movement in Arctic charr. J Fish Biol 60: 1570-1579.

26. Urbach D, Folstad I, Rudolfsen G (2005) Effect of ovarian fluid on sperm velocity in Artic charr (Sylvanius alpinus). Behavior Ecology and Sociobiology 57: 438-444.

27. Christ SA, Toth GP, McCarthy HW, Torsella JA, Smith MK (1996) Monthly variation in sperm motility in common carp assessed using computer-assisted sperm analysis (CASA). J Fish Biol 48: 1210-1222.

28. Rurangwa EF, Volckaert AM, Huyskens G, Kime DE, Ollevier F (2001) Quality control of refrigerated and cryopreserved semen using computer-assisted sperm analysis (CASA), viable staining and standardized fertilization in African catfish (Clarias gariepinus). Theriogenology 55: 751-769.

29. Wilson-Leedy JG, Ingermann RL (2007) Development of a novel CASA system based on open source software for characterization of zebrafish sperm motility parameters. Theriogenology 67: 661-672.

30. Pandey D, Ryu YW, Goto R, Matsubara T (2017) Morphological and biochemical changes of oocytes during final oocyte maturation in Japanese anchovy (Engraulis japonicus). Aquacult Sci 65: 29-40.

31. Cherr G, Vines CA, Smith HE, Pillai M, Griffin F, et al. (2015) Sperm motility initiation in pacific herring. Flagellar Mechanics and Sperm Guidance 4: 208-224.

32. Morisawa M, Tanimoto S, Ohtake H (1992) Characterization and partial purification of sperm- activating substance from eggs of the herring, Clupea palasii. J Exp Zool 264: 225-230.

33. Griffin FJ, Vines CA, Pillai MC, Yanagimachi R, Cherr GN (1996) Sperm motility initiation factor is a minor component of the pacific herring egg chorion. Dev Growth Diff 38: 193-202.

34. Hayakawa Y, Munehara H (1998) Fertilization environment of the non-copulating marine sculpin, Hemilepidotus gilberti. Env Biol of Fishes 52: 181-186.

35. Lahnsteiner F, Berger B, Weismann T, Patzner RA (1997) Sperm structure and motility of the freshwater teleost Cottus gobio. J Fish Biol 50: 564-574.

36. Elofsson H, Look KV, Borg B, Mayer I (2002) Influence of salinity and ovarian fluid on sperm motility in the fifteen-spined stickleback. J Fish Biol 63: 1429-1438.

37. Hulak M, Rodina M, Alavi SMH, Linhart O (2008) Evaluation of semen and urine of pike (Esox lucius L.): ionic composition and osmolality of the seminal plasma and sperm volume, density and motility. Cybium 32: 189-190.

38. Morita M, Takemura A, Okuno M (2004) Requirement of Ca2+ on activation of sperm motility in euryhaline tilapia (Oreochromis mossambicus). J Exp Biol 207: 337-345.

39. Woolley DM (1998) Studies on the eel sperm flagellum. The kinematic of normal motility. Cell Motility and the Cytoskeleton 39: 233-245.

40. Suzuki R (1959) Sperm activation and aggregation during fertilization in some fishes: III Non species specificity of stimulating factor. Annotation Zoology Japan 32: 105-111.

41. Musselius VA (1951) How to store carp milt and to determine its quality. Rybnoe Khozyaistov 27: 51-53.

42. Dorier A (1951) Conservation of the vitality and fertile power of the spermatozoids of rainbow trout. Trav Labor Hydrob Piscic Grenoble Annees 75-85.

43. Cosson J (2008a) Methods to analyze the movements of fish spermatozoa and their flagella in fish spermatology. Alavi SMH, Cosson JJ, Coward K and Rafiee G (eds.) Oxford: Alpha Science 63-101.

44. Cosson J (2008b) The motility apparatus of fish spermatozoa. In Fish Spermatology. Alavi SMH, Cosson JJ, Coward K and Rafiee G (eds.) Oxford: Alpha Science 281-316.

45. Abascal FJ, Cosson J, Fauvel C (2007) Characterization of sperm motility in European sea bass. The effect of heavy metals and physicochemical variables on sperm motility. J Fish Biol 70: 509-522.

46. Alavi SMH, Cosson J, Karami M, Abdolhay H, Amiri BM (2004) Chemical composition and osmolality of seminal fluid of Acipenser persicus; their physiological relationship with sperm motility. Reproduction 35: 1238-1243.

47. Dreanno C, Seguin F, Cosson J, Suquet M, Billard R (1999) Metabolism of turbot (Scophthalmus maximus) spermatozoa: relationship between motility, intracellular nucleotide content and mitochondrial respiration. Mol Rep and Dev 53: 230-243.

48. Oda A, Morisawa M (1993) Rises of intracellular Ca2+ and pH mediate the initiation of sperm motility by hyper osmolality in marine teleosts. Cell Motility and the Cytoskeleton 25: 171-178.

49. Christen R, Gatti JL, Billard R (1987) Trout sperm motility: the transient movement of trout sperm is related to changes in the concentration of ATP following the activation of the flagellar movement. European Journal of Biochemistry 166: 667-671.

6

Classical Fisheries Theory and Inland (Floodplain) Fisheries Management; Is there Need for a Paradigm Shift? Lessons from the Okavango Delta, Botswana

Mosepele K*

Senior Research Scholar- Fisheries Biologist, Research Services and Training, Botswana

***Corresponding author:** Senior Research Scholar - Fisheries Biologist, Research Services and Training, Botswana, E-mail: mosepelek@gmail.com

Abstract

This paper reviews the fisheries management question of inland (floodplain) systems in the developing world and proposes a paradigm shift in approach. Inland fisheries management is largely based on classical fisheries formulations derived on temperate freshwater and marine single-stock fisheries. The basic models to manage inland fisheries are based on steady state equilibrium models. However, inland, flood-pulsed fisheries are dynamic and driven by external factors which are incongruent with the classical approach. Therefore, adopting this management approach in inland, flood-pulsed fisheries has created a management conundrum because of the obvious fundamental differences that exist between these two systems. Marine fisheries contribute to the macroeconomic growth of fishing countries, inland fisheries from developing countries are largely focused on recreational activities, while inland (floodplain) fisheries are key sources of food and nutrition security for marginalized riparian communities in the developing world. This review also uses lessons from the Okavango Delta fishery to illustrate the uniqueness of floodplain fisheries and the management questions therein. One key debate highlighted in this review is that inland fisheries are a livelihood of the last resort for poor (and sometimes malnourished) communities. Management should therefore mainstream this value into management interventions, especially since a sustainable utilization of this resource can assist developing countries to achieve some of the MDG's. The paper concludes with an argument of the need for a paradigm shift in inland fisheries management, where key factors such as enhanced data collection, co-management regimes based on "real" democratic principles constitute some of the germane attributes of fisheries management plans.

Keywords: Flood pulse; Classical fisheries management; Okavango Delta; Floodplain fisheries

Introduction

Fisheries management theory, based on single-species exploitation [1]) is replete with "theoretical" equilibrium models that have been used to manage global (marine) fisheries (e.g. [2-4]). These models used the concept of maximum sustainable yield (MSY) as the foundation of the fisheries management paradigm. The MSY assumes a state steady, constant parameter system [1,5-8]. One of the basic tenets of this concept was the development of mesh size regulations, as an attempt to exploit a fishery at its maximum sustainable level [9]. The main management philosophical construct has always focused on internal drivers (e.g. fishing effort) as the main agents of change that need to be managed to attain sustainable fish utilization [3,8,10]. The classical management paradigm has always neglected external drivers (e.g. flooding, nutrients, etc.), primarily because they can't be managed [10]. Fisheries theories were developed in relatively static temperate, single species marine [11] and freshwater [12] fisheries but were tropicalized [13] into comparatively dynamic inland, flood-pulsed multi-species fisheries. Several classical overfishing scenarios were also developed based on fisheries theory [14]. These include growth overfishing [15], recruitment overfishing [16], biological overfishing [17], economic overfishing [18] and ecosystem overfishing [19]. These concepts also contributed to, and entrenched the classical fisheries management paradigm in inland (floodplain fisheries). The dichotomies created by these conflicting paradigms (i.e. marine vs. inland freshwater fisheries) have created a fisheries management dilemma, in floodplain fisheries.

Tropical floodplain fisheries are characterized by diverse species assemblages, diverse fishing gears, and diverse threats [20-22] driven primarily by the seasonal flood pulse in floodplain systems [23-25]. While these fisheries are significantly different from marine fisheries in form and structure [10], management interventions, premised on "over-exploitation" [26,27] have adopted classical fisheries management paradigms, ostensibly to ensure sustainable utilization [10]. These interventions were borne on the back of development projects that were implemented in these fisheries, to secure livelihoods of the socio-economically marginalized riparian communities. While marine fisheries contribute significantly to national GDP growth of fishing countries (FAO, 1951[28]), floodplain fisheries are a major source of food and nutrition security for the rural poor [29], though full-time commercial food fisheries have disappeared from most "developed" countries [30]. Floodplain fisheries from developing countries are also significantly different from those of the developed world [30]. Floodplain fisheries from developed countries place more emphasis on recreational value, while those from developing countries are valued for their food value. This is the fundamental distinction between these two fisheries, which should be reflected in their management approaches. If managed sustainably, these fisheries can contribute towards African countries' achievement of the MDG's [31], including those of developing countries in Asia and South America.

Globally, floodplain fisheries are a key livelihood resource, particularly in Africa, Asia and South America. They are a major source of employment and food for the poor communities [21,26] and contribute to the economic development of many African rural economies [32]. These fisheries are also the mainstay of livelihoods in Asia and South America. The fisheries of the Yala swamp (in Kenya's coastal area of Lake Victoria) are a major source of income to its riparian community, and income derived from fishing is estimated to be four times the agricultural income [33]. Generally, African inland fisheries provide employment to several people [20]. Bangladeshi floodplain fisheries provide an essential source of protein and income seasonally, which underpin the livelihoods of the riparian communities [34]. The Mekong River, which covers several countries in South-East Asia [35], contains the world's largest inland fishery [20]. The Mekong contributes directly to food security of riparian communities in its system [35], is a major source of protein and micro-nutrients to 22 million people in Cambodia and Laos [20] and is a key livelihood resource for the nearly 70 million people in the Lower Mekong River Basin [36]. Subsistence fishing in the Amazon is an important source of animal protein for the riparian community of this river system [37], where this fishery and related activities provides employment for a substantial number of people in the central Amazon floodplains [38]. Inland fisheries in the developing world are therefore a key livelihood resource whose management determines where the next meal is coming from for the socio-economically marginalized riparian communities.

Understanding the biology and ecology of freshwater systems is critical towards a comprehensive management of floodplain fisheries. Most classical fisheries models are theoretical formulations (e.g. [9, 15]), without any firm basis on empirical observations. Therefore, there is need for a paradigmatic shift, at least in floodplain fisheries, to ensure sustainable utilization of the fish resources. Because of their dynamicity and heterogeneity at both spatial and temporal scales, floodplain fisheries management needs to be approached from a basic understanding of ecosystem ecology of freshwater systems. The major concept underpinning (inland) floodplain fisheries management is the flood pulse concept [23,25,36], which suggests that floodplains are the major areas of production in flood-pulsed systems [23]. The food-web dynamic created by the seasonal flood pulse regulates fish species interactions with their prey in floodplains [39]. Ultimately, this creates a dynamic aquatic-terrestrial interface that makes floodplains some of the most productive systems globally. This characteristic flood pulse regulates energy storage in fish [40], due to the food-web dynamic created by seasonal flooding. Moreover, seasonal flooding also regulates spawning and recruitment of floodplain fish species [41]. Therefore, Junk [36] argues that sustainable development should include the maintenance of the natural flood regime in floodplain systems to ensure sustained fish production. This suggests that external drivers (e.g. flooding), are major drivers of change, which then nullifies assumptions for classical management approaches in floodplain fisheries.

Diverse management approaches have been implemented to manage inland fisheries, ranging from top-down approaches (e.g. [30]) to co-management regimes (e.g. [42,43]). Top-down strategies are based on classical approaches which involve input and output controls [44], which are essentially effort and mesh regulations. According to Pauly [45], classical management approaches are premised on the assumption that fishers are in a "social and financial" position to either comply or implement these measures. However, this is not the case in the developing world, where the challenges of putting food on the table daily are sometimes insurmountable. The fundamental philosophy behind community based fisheries management is that a

holistic management of fisheries enhances benefits gained from the resource and encourages sustainability [46]. Chuenpagdee and Jentoft [47], however, highlight that the success of co-management initiatives also depends on pre-implementation stages of the co-management process. Ultimately, co-management is seen as a sustainable response to (classical) fisheries management failure, which has been predominantly based on top down control measures [42]. According to Pauly [48,49], the major fisheries management objective in small scale inland fisheries is to stem the tide of Malthusian over-fishing. This Malthusian overfishing occurs when poor fishers, without any alternative livelihood strategies, continue fishing even when the resources are severely depleted [48]. In this scenario, women, who would have migrated to urban areas in search of employment, subsidise men fishers through remittances. It can be argued, therefore, that Malthusian overfishing, premised on Hardin's [50], "tragedy of the commons" thesis is one of the drivers of fisheries management paradigms in inland fisheries.

This review analyses the Okavango Delta fishery using the framework developed by the preceding literature, to highlight the nature and character of this fishery and its fish community. It is envisaged that this will highlight the dynamicity of floodplain fisheries, and illustrate that dynamic fisheries theory modeling is incompatible with floodplain fisheries management.

Classical Fisheries Management Theory

Classical fisheries management is premised on a single stock paradigm which essentially argues that the productivity of a stock is a function of its size and its reproductive potential [51]. Subsequently, it is argued that the basic objective of fisheries management is to exploit this stock at a level where its reproductive ability is equal to its natural mortality [9,51], through mesh regulations/ selective fishing [9]. Therefore, the classical approach to fisheries management necessitates the need to estimate growth and mortality parameters from exploited populations [51,52], which are then used as input parameters to estimate MSY [53], which is the key objective of fisheries management [54]. This fisheries management philosophy is codified in Beverton and Holt's [55] yield-per-recruit model which is a major seminal work in fisheries literature [13]. Over time, regulations were gradually introduced to manage fisheries resources to achieve optimum utilization (i.e. maximum sustainable yield) of the fish resources. According to Pauly [45]), some of these include reducing fishing effort, mesh regulations, closed fishing seasons, and fishing gear restrictions. Welcomme [56]) defines these as technical measures (e.g. mesh and gear limitations, closed seasons, etc), input controls (e.g. licensing to control effort and access, ownership, etc), and output controls (e.g. quotas, size limits on fish landed, etc). These classical regulations have subsequently been assiduously implemented in floodplain fisheries [57,58].

Several management approaches, based on the classical paradigm, have been developed to manage fisheries resources globally. Because one of the premises of classical fisheries management is Hardin's [50] Tragedy of the Common's scenario, the basic approach to mitigate against this has been to privatize fisheries resources. Subsequently, Pauly [59]) proposes individual transferable quotas (ITQ) as an alternative approach to privatize the commons, ostensibly to inculcate a conservation ethic in exploitation regimes. Other classical management approaches to safeguard fish resources include the delineation of fish refuges, known as marine protected areas [59-61] which are essentially meant to act as game reserves or national parks used in wildlife management. Other approaches include the ecosystem approach [62], multi-species models [63-65], dynamic system models [60]. The fundamental question that this paper highlights then, is whether these approaches are relevant towards management of floodplain fisheries, when the very premise of classical approaches, that of a "constant parameter system [52]", are nullified?

Lessons of Fisheries Management; the Okavango Delta

Hydrology and limnology

The Okavango Delta (Figure 1) is a flood-pulsed inland floodplain system that receives water annually from the Angolan highlands [65]. Generally, seasonal floods peak in the panhandle in April (see Figure 2) and are lowest towards the end of the year. There are also strong inter-annual variations in flooding in the delta (Figure 1), and generally no two years have similar hydrographs. This flood pulse is out of phase with the rainy season [66] and peaks during the cold winter months when biological production is low [67]. Maximum flooded area is observed between August and October (Figure 2), several months after peak discharge. The floods flow the delta in a pulse that reaches Mohembo (Figure 1) in December/ January and only reaches the distal ends of the delta in July, several months later [68]. Different floodplain classes exist in the Delta due to different hydrological characteristics [69]. Floodplains that are permanently flooded are classified as permanent floodplains and are characterized by permanent marsh vegetation communities; regularly flooded seasonal floodplains (4-8 months of the year) are characterized by seasonal marsh vegetation communities; occasionally flooded seasonal floodplains (1-4 months of the year) are characterized by flooded grasslands vegetation communities; while floodplains that only receive water at high floods (and flood for less than 2 months) are characterized by forbland/ grassland vegetation communities. Similarly, there are strong intra-annual variations in water chemistry in the panhandle. Mladenov et al. [70-71], observed that DOC in the Delta is flood pulse driven, originating from the terrestrial environment. This terrestrial origin of DOC is a classic illustration of the existence of the flood pulse in the Delta's hydrology and limnology.

The main channel waters of the Delta are oligotrophic, while those of floodplains range from oligotrophic and mesotrophic [72]. Conversely, lagoons in the Delta's seasonal floodplains are highly enriched, primarily from cattle dung [73] or large herbivore dung in the seasonally flooded floodplains [74], and hence support high fish biomass [73]. This high production in the seasonal floodplains agrees with Junk et al. [23] flood pulse concept (FPC) that seasonal flooding drives primary production in floodplains due to a dynamic interchange between terrestrial and aquatic habitats. A similar dynamic exists in the Delta where primary production is driven by the flood pulse [75] and zooplankton diversity in the seasonal floodplains is driven by the seasonal flood regime [74]. Seasonal floods also drive zooplankton production in temporary floodplains of the Okavango Delta [76]. Other studies have revealed that flooding frequency in the delta, especially in rarely flooded floodplains, is the major factor determining the diversity and viability of zooplankton [77], which is key food for juvenile fish.

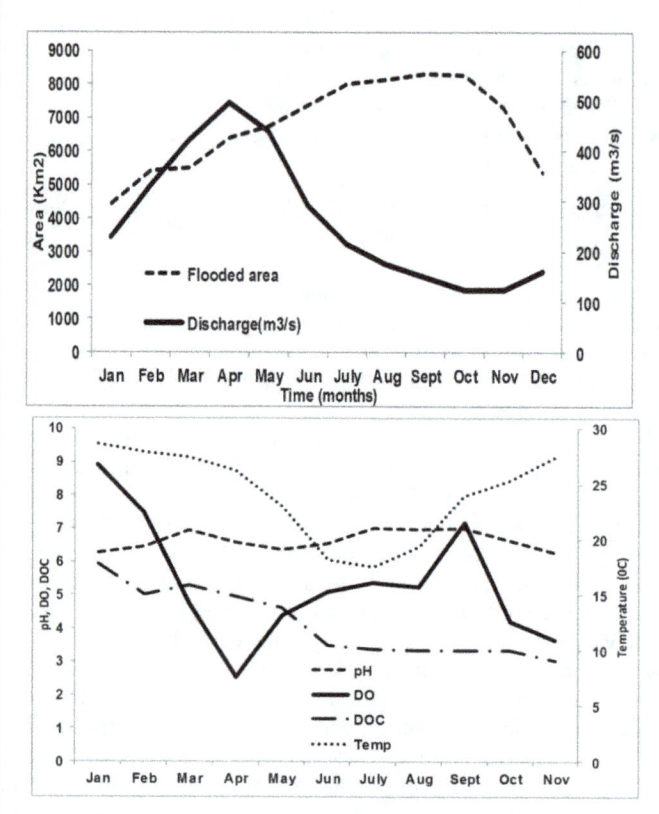

Figure 2: Hydro-chemical variations in the panhandle of the Okavango Delta

Biology and ecology of fisheries:

There are 71 different fish species in the Okavango Delta [68] and range in size from small species like *Barbus haasianus* (maximum size is 32 mm SL) to large species like *Clarias gariepinus*, whose maximum size is 1.4 m SL [78]. The majority of fish species are insectivores (Table 1), which highlights the importance of insects/ aquatic macro-invertebrates in fish biomass production in the delta. This notwithstanding, the delta's fish species have a relatively plastic feeding behavior [79,80], because of the dynamic nature of the system.

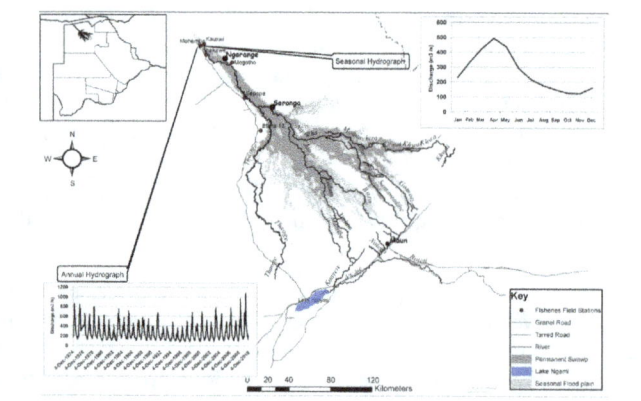

Figure 1: Map of the Okavango Delta showing some of the fishing villages around the panhandle. The map also illustrates a seasonal and annual hydrograph

Schilbe intermedius is one of the dominant species in the delta's fish community [79,81], possibly due to its opportunistic feeding ecology [79,81], where it also feeds on terrestrial food items, driven by the seasonal flood pulse [80]. Cichlids are the dominant family in the delta's fish community [81], and Mosepele [80] also found out that they have a plastic feeding behavior, where detritivorous species like *Oreochromis andersonii* were found to prey on fish under stressed conditions. Currently, there are no endemic and or introduced species in the Okavango delta [68], making the delta a relatively pristine environment. While the permanence and flow rate of water are the key variables regulating fish distribution in the system [68], habitat partitioning has resulted in variable life history strategies of similar species in the delta. Similar species have different growth rates from different parts of the delta [83]. Furthermore, Mosepele et al. [81] also observed habitat partitioning in fish species assemblages of the delta, where upper delta habitats have relatively higher species diversity than lower delta habitats [79].

Feeding guild	Relative proportion (%)
Insectivores	44
Mixed predators	29
Omnivores	8
Piscivores	7
Molluscivores	6
Detritivores	4
Planktivores	1
Herbivores	1

Table 1: Fish feeding guilds from the Okavango delta based on Skelton [78]

Several studies have shown that the flood pulse is a key driver in the delta's fish community [75,80, 84-86]. Similar fish species from different habitats in the Delta have different life history strategies [82, 87], which is driven primarily by differences in environmental conditions [85]. Furthermore, Mosepele et al. [80] showed that the seasonal flood pulse is the major driver of the feeding ecology of selected fish species from the Okavango Delta. Their study revealed that most fish species from the delta are opportunistic feeders whose diet is changed seasonally regulated by the seasonal flood pulse. In fact, the feeding rate of some fish species is driven by changes in either mean depth or discharge [80]. This study [80] revealed that fish diet varied between terrestrial and aquatic sources, depending on the seasonal hydrograph. At a larger time scale, years of low floods resulted in low fish biomass while years of high floods resulted in high fish biomass [75]. High flood years inundate rarely flooded floodplains which are massive engines of zooplankton production which is then grazed heavily by juvenile fish, especially cichlids [88]. Therefore, fish dynamics in the Delta are driven by the length of time water is present in the system and by the nature of its flow [84]. Subsequently, Linhoss et al. [86] developed a statistical model to show that the seasonal flood pulse is a major driver of fish population dynamics in the Okavango Delta.

Exploitation regime of fisheries:

Fishers use different fishing gears to exploit the delta's diverse fish species assemblage [89-92], where fishers use different fishing gears to adapt to variations in fish catch ability [91]. Fishers also use indigenous knowledge to target their preferred species [93]. The three major exploitation regimes in the delta's fishery, which use different fishing gears, and exploit different fish species are subsistence [90], commercial [92] and recreational [94] fishing. The main species exploited by the small-scale commercial fishery are *Oreochromis andersonii, O. macrochir* and *Tilapia rendalli* [84,89], and these three species are also exploited by the artisanal fishery (i.e. using homemade fishing gear). However, different artisanal fishing gears exploit different fish species as shown in Table 2. Moreover, fishers of different ages using similar fishing gears exploit different strata of the delta's fish community, from small sized species to large sized individuals [89].

C.p.u.e (nos/set)	Species richness	Mean fish size (mm)	Shannon's index (H')	Evenness (J')	Most important species
3.8	22	38	2.26	0.73	*A. johnstoni*
4.8	43	259	2.29	0.61	*Clarias spp*
19	46	57	2.62	0.68	*T. sparrmanii*
21.1	24	261	2.31	0.73	*O. andersonii*
58.5	17	228	2.36	0.83	*C. gariepinus*

Table 2: Summary of catching efficiency (c.p.u.e), species diversity and the most important fish species per fishing gear used by subsistence fishers in the Delta where the most important species in the catch is determined by using the Index of Relative Importance (IRI) (Source: Mmopelwa et al. [91]) Note: 1 net set is defined as a 25 m long gill net set in the water overnight for 12 hours.

Socio-economics and management of fisheries:

Fishing is a major livelihood activity in the Okavango Delta [87,90, 92,95-96], though characterized by conflict due to competing uses [87, 96]. Subsistence fisher households in the delta are mostly headed by female-headed households who are relatively poor and tend to depend on fish as a major s ource of food [90,95] and nutrition security [97]. Generally, subsistence fishers are women, have very little formal education, with an average household size of 7 people [94]. Apart from being a major source of food for poor households [91,95], fish is also a natural safety net and buffers those who are chronically ill, especially those who suffer from HIV/AIDS (mostly from HIV/AIDS illnesses) people against deterioration into chronic poverty [98] and for children (under the age of 5 years) during times of food scarcity [95]. Conversely, commercial fishing in the delta is a major source of rural income [99] which (i.e. income) at times exceeds agricultural (e.g. cattle) income [92]. Research has also shown that commercial fishers utilize indigenous traditional knowledge to target their preferred species in the delta, and have developed efficient fishing techniques compared to subsistence fishers [93].

Fisheries management in the delta is based on a classical fisheries management approach, where fishing effort, mesh regulations, fishing season and fishing methods are the main elements of the management regime [100]. Due mostly to competing use value between (small-scale) commercial and recreational fishing in the delta, allegations of fish over-exploitation by commercial fishing created serious conflict in the fishery [87]. However, a fish stock assessment revealed that the fish stocks were still healthy [87,101,102] and that perceptions of over-exploitation were driven by a motive for exclusive access to the fish resource by recreational fishers [87,101]. Despite the obvious reliance on indigenous knowledge by commercial fishers, current government policy has not integrated this into natural resource management in the delta [103]. Moreover, the lack of a national fisheries policy [83,100] has always made it difficult to implement comprehensive management strategies which include gender equity in the fisheries sector [104]. This issue was also highlighted by Mosepele [96] who developed a co-management model for the Okavango delta fishery, with special emphasis on the role of women in the fishery. The co-management model developed for the delta does not have any legal foundation, and is based purely on people's willingness to comply [96]. The lack of a national fisheries policy, has compounded management concerns in this fishery, where critical elements like indigenous knowledge, gender equity and contribution of this fishery towards achievement of the MDG's have not been integrated into the management regime. Despite these drawbacks, a code of conduct is currently operational in the fishery, and was developed through consensus among the stakeholders, facilitated through an Okavango Management Fisheries Committee (OFMC). This committee was formed as part of the co-management regime that was developed in the fishery [96].

Among others, classical management approaches in the delta prohibit use of mosquito nets, drive fishing, and encourage use of larger meshed nets [105]. According to Mosepele [89], the major fish species exploited by mosquito nets are topminnows (Aplocheilichthys spp) which are consumed at the household level. These contribute a major source of high quality nutrients like zinc, calcium and amino acids, similar to some fisheries in Asia [106,107] which are critical micro-nutrients to women and children under the age of 5. Furthermore, Mosepele [93] has shown that drive fishing is a highly effective indigenous fishing method that small-scale commercial fishers use to target their preferred species. This fishing method exploits species like Tilapia rendalli, which are otherwise not exploited by other more conventional fishing methods [42,93]. Prohibiting some fishing methods, restricting certain fishing gears, and encouraging large meshed gill nets encourages selective fishing in a multi-species complex. Walsh [107] cautions that selective fishing drives the exploited fish population towards more r life history strategists, that it results in the loss of highly fecund (i.e. larger/ older) females, and this according to Harvey [108], may reduce the reproductive potential of exploited populations. Ultimately, selective fishing (by targeting larger sized individuals mostly) results in loss of resilience of exploited populations which may invariably result in their crash [109]. It is on this basis that Jul-Larsen [110] argues for a non-selective exploitation pattern in multi-species inland fisheries, primarily to protect the biodiversity of these fisheries. This is the non-selective fishing pattern that traditionally existed in the delta [88], before the advent of classical management approaches.

Synthesis

Because of the multiple threats facing freshwater systems [20], inland fisheries management cannot be sector specific because they transcend the fisheries sector [96]. It was on this basis that a fisheries Code of Conduct was developed to alleviate conflicts that prevailed in the fishery. This code of conduct essentially integrated different management issues that affect the fisheries of the Okavango Delta.

Another important management issue is what Pauly [45] terms the marginality of the inland fisheries, which he defines as their "geographic, socio-economic and political remoteness from decision makers in the major population centers". In his opinion, the socio-economic remoteness of these inland fisheries, which are small-scale in nature, is related to the relatively low incomes of their fishers and the fact that they belong to ethnic groups of low status. This is particularly relevant in systems such as the Okavango Delta where women are a major subsistence group that is generally marginalized in the prevailing management regime as observed by Ngwenya [103]. Therefore, there is need for a comprehensive management of floodplain fisheries which should be sensitive to issues of (ethnic and gender) equity. This marginalization is also manifest in lack of political power by these low socio-economic status groups [45]. Pauly [45,48] argues that marginalization creates Malthusian over-fishing, which is seen my some managers as ground for strict management regulations. Conversely, lessons from the Delta have shown that total fishing effort (defined as number of total fishers) is in constant flux, driven by changes in the macro-economic situation in the country [86]. According to Mosepele [86], more people enter the fishery when their livelihood opportunities are limited, but immediately move out when more opportunities are available elsewhere in the national economy. This preceding argument then calls for a paradigm shift in inland fisheries management. It is evident that inland fisheries management issues are much more complex than simply implementing either "input" or "output" controls in the fishery (i.e. based on the classical management paradigm), ostensibly to exploit the fishery optimally at the MSY.

Small scale fishers in the developing world are generally poor with few employment possibilities, whose families are also usually malnourished [111]. Therefore, inland fisheries management in the developing world should focus more on securing food in these fisheries, compared to the developed world where the focus is more on recreational values [30]. The basic argument here is that there are fewer alternatives for food in the developing world, especially high protein diet that fish provides in riparian communities. Essentially, top down management regimes cannot provide food and nutrition security for subsistence fishing households to whom fish is a key life and death resource [112]. Based on lessons learnt from the Okavango Delta, a co-management regime is recommended based on truly democratic principles [96]. This argument is based on the observation that fisheries are "non-excludable" but "subtractable" resources [43]. They [43] highlight that non-excludability means it is costly to either legally or physically exclude users from accessing the resource while "subtractability" means that users cannot jointly consume the resource. This is particularly critical because classical approaches to management of fisheries focus on control and command interventions which are invariably associated with high operational costs. Enforcing a closed fishing season in the Okavango Delta essentially diverts resources (human and financial) from more productive pursuits, to stop women basket fishers from harvesting a mere 90g of fish/ day (the efficiency of basket fishing as calculated by Mosepele et al. [88]. This is certainly an over-kill and not cost effective. Therefore, co-management in this management scenario is seen as a way of minimizing management costs [42]. This not to suggest however, that a co-management approach is a panacea to inland fisheries issues. What is important nonetheless, is that the people should have some element of participation in the decision making process of how the fisheries resource needs to be managed.

Inland fisheries management, especially in flood pulsed systems, determines access to food and nutrition of the last resort to socio-economically marginalized households. The value of these fisheries is much more critical in Africa, Asia and South America where the majority of the riparian communities exist under conditions of chronic poverty. These fisheries are also a critical entry point for state intervention into achievement of the MDG's in the developing world. As discussed above, inland fisheries management is a political exercise because it defines access to fisheries by excluding some people and including others [42]. It is therefore incumbent upon fisheries managers, policy makers and fisheries scientists to acknowledge this fact, that they hold the keys to political power that should be used effectively and efficiently for the benefit of the politically voiceless. Moreover, Welcomme et al. [21] highlights that the key management questions in inland (flood-pulsed) fisheries are predicated on environmental flows because of the multifaceted nature of these fisheries. Therefore, inadequate data collection systems that exist in these fisheries are a major hindrance to inland fisheries development [21]. Hence, there is need to expand monitoring efforts and management capacity in the developing world [113]. Obtaining catch and effort data in African (inland) fisheries is problematic [42], which provides a compelling argument to institute proper catch and effort monitoring strategies as part of a management regime. These data are needed to illustrate the "real" intrinsic value of these fisheries to policy makers. Lack of knowledge from these fisheries, compounds the management dilemma, where managers resort to classical management approaches, because these are easy to implement, but not necessarily to enforce. A shift in the paradigmatic approach to the inland fisheries management question, will allow managers and scientists to appreciate the intrinsic values of these fisheries to marginalized riparian communities.

Conclusion

This review has shown that inland (floodplain) fisheries are diametrically opposite from marine fisheries. While marine fisheries are more focused on large industrial fisheries, the value of inland (floodplain) is more about putting food on the table for marginalized riparian communities. Furthermore, while marine fisheries management is focused more on theoretical formulations aimed at maximizing fish yield, inland (floodplain) fisheries management should focus more on the ecosystem approach because of the multiple threats they face. Classical management approaches, based on input and output controls, do not have any philosophical foundation in inland (floodplain) fisheries. A key management objective in inland (floodplain) fisheries should be to secure the food and nutrition security value of these resources for riparian communities. Another key management objective should be to enhance fishery dependent and fishery independent data collection systems because these are currently lacking in inland fisheries. Another major objective should be to enhance knowledge about the socio-economic value of these fisheries so that their values are mainstreamed into management. A key approach to inland (floodplain) fisheries management is co-management, as facilitated within the context of the Okavango Delta. These management objectives illustrated the need to shift paradigms in inland fisheries management, where a more holistic approach is required, so that the food value of these fisheries is maintained for posterity.

References:

1. Matsuda H, Abrams PA (2006) Maximal yields from multispecies fisheries systems: rules for systems with multiple trophic levels. Ecological Applications 16: 225-237.

2. Schaefer MB (1943) The theoretical relationship between fishing effort and mortality. Copeia 2: 79-82.

3. Ricker WE (1944) Further Notes on Fishing Mortality and Effort. Copeia 1: 23-44.

4. Beverton RJH (1998) 'Fish, fact and fantasy: A long view', Reviews in Fish Biology and Fisheries. 8: 229-249.

5. Ricker WE (1946) Production and utilization of fish populations. Ecological Monographs 16: 373-391.

6. Sparre P, Venema SC (1998) Introduction to tropical fish stock assessment. Part1: Manual. FAO Technical Paper No 306.

7. Hillborn R, Walters CJ (1992) Quantitative Fisheries Stock Assessment: Choice, Dynamics and Uncertainty. New York: Chapman and Hall.

8. Niw H-S (2007) Random-walk dynamics of exploited fish populations. ICES Journal of Marine Science. 64: 496-502.

9. Ricker WE (1945) A Method of Estimating Minimum Size Limits for Obtaining Maximum Yield. Copeia 2: 84-94.

10. Kolding J, Van Zwieten PAM (2011) The Tragedy of Our Legacy: How do Global Management Discourses Affect Small Scale Fisheries in the South? Forum for Development Studies 38: 267-297.

11. Graham M (1935) Modern theory of exploiting a fishery, and application to North Sea trawling. Journal du Conseil Interational por l'Exploration de la Mer 10: 264-274.

12. Ricker WE (1958) Maximum sustained yields from fluctuating environments and mixed stocks. Journal Fisheries Research Board of Canada 15: 991-1006.

13. Pauly D (1998) Beyond our original horizons: The tropicalization of Beverton and Holt. Reviews in Fish Biology and Fisheries 8: 307-334.

14. Pauly D (1994) From growth to Malthusian overfishing: Stages of fisheries resources misuse. SPC Traditional Marine Resource Management and Knowledge Information Bulletin.

15. Beverton RJH, Holt SJ (1957) On the Dynamics of Exploited Fish Populations. Chapman and Hall 533.

16. Ricker WE (1954) Stock and recruitment. Journal Fisheries Research Board of Canada 11: 559-623.

17. Ricker WE (1975) Computation and interpretation of biological statistics of fish populations. Bulletin for the Fisheries Research Board of Canada 191.

18. Gordon HS (1953) An economic approach to the optimum utilization of fisheries resources. Journal of the Fisheries Research Board of Canada 10: 442-457.

19. Pauly D (1979) Theory and management of tropical multispecies stocks: A review with emphasis on the Southeast Asian demersal fisheries. ICLARM Stud Rev.

20. Dugan P, Delaporte A, Andrew N, O'Keefe M, Welcomme R (2010) Blue Harvest: Inland Fisheries as an Ecosystem Service. World Fish Center Penang Malaysia.

21. Welcomme RL1, Cowx IG, Coates D, Béné C, Funge-Smith S, et al. (2010) Inland capture fisheries. Philos Trans R Soc Lond B Biol Sci 365: 2881-2896.

22. Welcomme R (2011) Review of the state of the world fishery resources: Inland fisheries. FAO, Rome.

23. Junk WJ, Bayley PB, Sparks RE (1989) The flood pulse concept in river-floodplain systems. Can Spec Publ Fish Aquat Sci 106: 110-127.

24. De Graaf G (2003) The flood pulse and growth of floodplain fish in Bangladesh. Fisheries Management and Ecology 10: 241-247.

25. Junk WJ, Wantzen KM (2004) The Flood Pulse Concept: New Aspects Approaches Applications. Cambodia 117-140.

26. Allan JD, Abell R, Hogan Z, Revenga C, Taylor BW, et al. (2005) Overfishing of inland waters. BioScience 55: 1041-1051.

27. Dudgeon D1, Arthington AH, Gessner MO, Kawabata Z, Knowler DJ, et al. (2006) Freshwater biodiversity: importance, threats, status and conservation challenges. Biol Rev Camb Philos Soc 81: 163-182.

28. Kolding J, Béné C and Bavinck M (2013) Small-scale fisheries - importance vulnerability deficient knowledge.

29. Richter BD, Postel S, Revenga C, Scudder T, Lehner B, et al. (2010) Lost in development's shadow: The downstream human consequences of dams. Water Alternatives 3: 14-42.

30. Arlinghaus R, Mehner, T, Cowx IG (2002) Reconciling traditional inland fisheries management and sustainability in industrialized countries with emphasis on Europe. Fish and Fisheries 3: 261-316.

31. Heck S, Bene C, Reyes-Gaskin R (2007) Investing in African fisheries: building links to the Millennium Development Goals. Fish and Fisheries 8: 211-226.

32. Béné C, Neiland AE (2003) Valuing Africa's inland fisheries: Overview of current methodologies with an emphasis on livelihood analysis. NAGA 26: 18-21.

33. Mwakubo SM, Ikiara MM, Abila R (2007) Socio-economic and ecological determinants in wetland fisheries in the Yala Swamp. Wetlands Ecology and Management.

34. Craig JF, Halls AS, Barr JJF, Bean CW (2004) The Bangladesh floodplain fisheries. Fisheries Research 66: 271-286.

35. Baran E, Jantunen T, Chong CK (2007) Values of inland fisheries in the Mekong River Basin. World Fish Center, Phnom Penh, Cambodia. 76.

36. Mainuddin M, Kirby M, Chen Y (2011) Fishery productivity and its contribution to overall agricultural production in the Lower Mekong River Basin. Colombo Sri Lanka: CGIAR Challenge Program for Water and Food (CPWF Research for Development Series 03).

37. Junk WJ (2001) Sustainable use of the Amazon River flood plain: Problems and possibilities. Aquatic Ecosystem Health & Management 3: 225-233.

38. Junk WJ and Piedade MTF (2000) Concepts for the sustainable management of natural resources of the Middle Amazon floodplain: A summary. German-Brazilian Workshop on Neotropical Ecosystems–Achievements and Prospects of Cooperative Research.

39. Arrington DA, Winemiller KO (2006) Habitat affinity the seasonal flood pulse and community assembly in the littoral zone of a Neo tropical floodplain river JN Am. Benthol Soc 25: 126-141.

40. Arrington DA, Davidson BK, Wine miller KO, Layman CA (2006) Influence of life history and seasonal hydrology on lipid storage in three neo tropical fish species. Journal of Fish Biology 68: 1347-1361.

41. Humphries P, King AJ, Koehn JD (1999) Fish flows and flood plains: links between freshwater fishes and their environment in the Murray-Darling River system Australia. Environmental Biology of Fishes 56: 129-151.

42. Jul-Larsen E, Van Zwieten P (2002) African Freshwater Fisheries: What needs to be Managed? Naga 25: 35-40.

43. Pinho PF, Orlove B, Lubell M (2012) Overcoming Barriers to Collective Action in

44. Welcomme RL (2001) Inland Fisheries: Conservation and management Black wells Oxford. 350.

45. Pauly D (1997) Small-scale fisheries in the tropics: Marginality, marginalization and some implication for fisheries management. 40-49.

46. Charles A (2008) Turning the tide: Toward community-based fishery management in Canada's Maritimes. American Fisheries Society Symposium 49: 569-573.

47. Chuenpagdee R, Jentoft S (2007) Step zero for fisheries co-management: What precedes implementation. Marine Policy 31: 657-668.

48. Pauly D (1994) On Malthusian overfishing. In on the sex of fish and gender of scientists: Essays in Fisheries Science Chapman and Hall London UK. 112-117.

49. Pauly D (1990) On Malthusian overfishing. NAGA 13: 3-4.

50. Hardin G (1968) The Tragedy of the Commons. Science 162: 1243-1248.

51. Hoggarth DD, Abeyasekera S, Arthur RI, Beddington JR, Burn RW, et al (2006) Stock assessment for fishery management- A framework guide to the stock assessment tools of the Fisheries Management Science Programme (FMSP). FAO: 261.

52. Sparre P and Venema SC (1998) Introduction to tropical fish stock assessment. Part1: Manual FAO Technical Paper No: 306.

53. Gayanilo FC, Pauly D (1997) The FAO-ICLARM stock assessment tools (FiSAT).

54. Rounsefell GA, Everhart WH (1953) Fishery science: Its methods and applications: 444.

55. Beverton RJH, Holt SJ (1957) On the Dynamics of Exploited Fish Populations: 533.

56. Welcomme RL (2007) Conservation of fish and fisheries in large river systems: American Fisheries Society Symposium 49: 587-599.

57. Malasha I (2003) Colonial and postcolonial fisheries regulations: The case of Zambia and Zimbabwe Isaac Malasha 253-266.

58. Pauly D (1999) Fisheries management: putting our future in places: 355-362.

59. Ngwenya B, Mosepele K (2008) Socio-economic survey of subsistence fishing in the Okavango Delta, Botswana.

60. Christensen V(1996) Managing fisheries involving predator species: Reviews in Fish Biology and Fisheries 6: 417-442.

61. Conover DO, Munch SB (2002) Sustaining Fisheries Yields Over Evolutionary Time Scales: Science 297: 94-96.

62. Hall SJ, Mainprize B (2004) Towards ecosystem-based fisheries management. Fish and Fisheries 5: 1-20.

63. James MK, Stark KP (1982) Application of the three bays ecosystem model to fisheries management: 99-121.

64. Kirkwood GP (1982) Simple models for multi-species fisheries.

65. Nnyepi M, Ngwenya B, Mosepele K (2007) Food in security and child nutrition in Ngamiland. 281 - 291.

66. Wolski P, Murray-Hudson M (2005) Flooding dynamics in a large low-gradient alluvial fan, the Okavango Delta, Botswana, from analysis and interpretation of a 30 year hydrometric record. Hydrology and Earth System Sciences J1 10: 127-137.

67. Mendelsohn JM, vander Post C, Ramberg L, Murray-Hudson M, Wolski P, et al. (2010) Okavango Delta: Floods of life.

68. Merron GS (1991) The ecology and management of the fishes of the Okavango Delta Botswana with particular reference to the role of the seasonal floods. Rhodes University, South Africa.

69. Ramberg L, Hancock P, Lindholm M, Meyer T, Ringrose S, et al. (2006) Species diversity of the Okavango Delta, Botswana. Aquatic Science 68: 310–337.

70. Murray-Hudson M, Wolski P, Ringrose S (2006) Scenarios of the impact of local and upstream changes in climate and water use on hydro-ecology in the Okavango Delta, Botswana. Journal of Hydrology 331: 73-84.

71. Mladenov N, McKnight DM, Wolski P, Ramberg L (2005) Effects of annual flooding on dissolved organic carbon dynamics within a pristine wetland, the Okavango Delta, Botswana. Wetlands 25: 622-638.

72. Mladenov N, McKnight DM, Wolski P, Murray-Hudson M (2007) Simulation of DOM fluxes in a seasonal floodplain of the Okavango Delta, Botswana. Ecological Modelling 205: 181-195.

73. Gronberg G, Gieske A, Martins E, Prince-Nengu J, Stenstrom IM (1995) Hydrobiological studies of the Okavango Delta and Kwando/Linyanti/Chobe River, Botswana. Surface water quality analysis: Botswana Notes and Records 27: 151-226.

74. Fox PJ (1976) Preliminary observations on fish communities of the Okavango Delta. 125-130.

75. Lindholm M, Hessen DO, Mosepele K, Wolski P (2007) Flooding size and energy pathways on a floodplain of the Okavango Delta. Wetlands 27: 775–784.

76. Lindholm M, Hessen DO, Ramberg L (2009) Diversity, dispersal and disturbance: cladoceran species composition in the Okavango Delta. African Zoology 44: 24–35.

77. Siziba N, Chimbari MJ, Masundire H, Mosepele K (2011) Spatial and temporal variations of microinvertebrates across temporary floodplains of the lower Okavango Delta, Botswana. Physics and Chemistry of the Earth 36: 939-948.

78. Siziba N, Chimbari MJ, Mosepele K, Masundire H, Ramberg L (2012) Inundation frequency and viability of microcrustacean propagules in soils of temporary aquatic habitats of lower Okavango Delta, Botswana. Ecohydrology.Skelton PH (2001) A complete guide to freshwater fishes of southern Africa. Struik Publishers. Cape Town.

79. Merron GS, Bruton MN (1991) The ecology and management of the fishes of the Okavango delta, Botswana, with special reference to the role of the seasonal floods. Rhodes University, South Africa.

80. Mosepele K, Mosepele B, Wolski P, Kolding J (2013) Dynamics of the feeding ecology of selected fish species from the Okavango delta, Botswana. Acta Ichthyologica et Piscatoria 42: 271-289.

81. Mosepele K, Mosepele B, Bokhutlo T, Amutenya K (2011) Spatial variability in fish species assemblage and community structure in four subtropical lagoons of the Okavango delta, Botswana. Physics and Chemistry of the Earth 36: 910-917.

82. Mosepele K, Basimane O, Mosepele B, Thethela B (2005) Using population parameters to separate fish stocks in the Okavango Delta fishery: a preliminary assessment. Botswana Notes and Records, 37: 292 – 305.

83. Mosepele K, Mosepele B (2005) Spatial and Temporal Variability in Fishery and Fish Community Structure in the Okavango Delta, Botswana: Implications towards fisheries management. Botswana Notes and Records 37: 280 – 291.

84. Mosepele K, Moyle PB, Merron G, Purkey D, Mosepele B (2009) Fish, floods, and ecosystem engineers: Aquatic conservation in the Okavango Delta. Botswana Bioscience 59: 53 – 61.

85. Mosepele K, Murray-Hudson M, Mosie I, Sethebe K (2013) Lagoons Fish Communities in Flood-Pulsed Floodplains: Heterogeneity in a Highly Dynamic System? The Case of the Okavango Delta. University of Botswana. 149-147.

86. Linhoss AC, Muñoz-Carpena R, Allen M, Kiker G, Mosepele K (2012) A flood pulse driven fish population model for the Okavango Delta, Botswana. Ecological Modeling 228: 27 – 38.

87. Mosepele K (2000) Preliminary Length Based Stock Assessment of the Main Exploited Stocks of the Okavango Delta Fishery. University of Bergen, Norway.

88. Siziba N, Chimbari MJ, Masundire H, Mosepele K, Ramberg L (2013) Variation in Assemblages of Small Fishes and Microcrustaceans After Inundation of Rarely Flooded Wetlands of the Lower Okavango Delta, Botswana. Environmental Management 52: 1386-1399.

89. Mosepele K, Mmopelwa TG, Mosepele B (2003) Characterization and monitoring of the Okavango delta artisanal fishery. 391 – 413.

90. Mmopelwa G, Mosepele K, Mosepele B, Moleele N, Ngwenya B (2009) Environmental variability and the fishery dynamics of the Okavango Delta, Botswana: The case of subsistence fishing. African Journal of Ecology 47: 1–9.

91. Mosepele K, Ngwenya B (2010) Socio-economic survey of commercial fishing in the Okavango Delta, Botswana.

92. Mosepele K, Mmopelwa G, Mosepele B, Kgathi DL (2007) Indigenous knowledge and fish utilization in the Okavango Delta, Botswana: Implications for food security: 292 - 302.

93. Kolding J (1996) Feasibility study and appraisal of fish stock management plan in Okavango. University of Bergen, Norway.

94. Mosepele K, Ngwenya BN, Bernard T (2006) Artisanal fishing and food security in the Okavango Delta, Botswana. 159 - 168.

95. Mosepele K (2014) Fish, floods and livelihoods in the Boteti River. University of Botswana, Botswana: 153-190.

96. Mosepele B, Mosepele K, Mogotsi S, Thamage D (2014) Fisheries co-management in the Okavango Delta's panhandle: The Okavango Fisheries Management Committee (OFMC) case study.

97. Ngwenya B, Mosepele K (2007) HIV/ AIDS, artisanal fishing and food security in the Okavango Delta. Physics and Chemistry of the Earth 32: 1339–1349.

98. Mmopelwa G, Raletsatsi S, Mosepele K (2005) Cost Benefit Analysis of Commercial Fishing in Shakawe, Ngamiland, Botswana. Botswana Notes and Records 37: 11 – 21.

99. Mosepele K (2008) Flood pulse in a subtropical floodplain fishery and the consequences for steady state management. 56 – 62.

100. Mosepele K, Kolding J (2003) Fish Stock Assessment in the Okavango delta, Botswana. Preliminary results from a length based analysis: 363 – 390.

101. Kgathi DL, Mmopelwa G, Mosepele K (2005) Natural Resource Assessment in the Okavango delta, Botswana; Case Studies of Some Key Resources. Natural Resource Forum 29: 70 - 81.

102. Cassidy L, Wilk J, Kgathi DL, Bendsen H, Ngwenya BN, et al. (2011) Indigenous Knowledge, Livelihoods and Government Policy. 75-98.

103. Ngwenya BN, Mosepele K, Magole L (2012) A case for gender equity in governance of the Okavango Delta fisheries in Botswana. Natural Resources Forum 36: 109 – 122.

104. Botswana Government (2008) Fish protection regulations. Government Printing and Publishing Services Gaborone, Botswana.

105. Roos N, Islam MM, Thilsted SH (2003) Small indigenous fish species in bangladesh: contribution to vitamin A, calcium and iron intakes. J Nutr 133: 4021S-4026S.

106. Roos N, Wahab MA, Hossain MA, Thilsted SH (2007) Linking human nutrition and fisheries: incorporating micronutrient-dense, small indigenous fish species in carp polyculture production in Bangladesh. Food Nutr Bull 28: S280-293.

107. Walsh MR, Munch SB, Chiba S, Conover DO (2006) Maladaptive changes in multiple traits caused by fishing: impediments to population recovery. Ecol Lett 9: 142-148.

108. Harvey CJ, Tolimieri N, Levin PS (2006) Changes in body size, abundance, and energy allocation in rockfish assemblages of the northeast Pacific. Ecol Appl 16: 1502-1515.

109. Law L, Plank MJ, Kolding J (2012) On balanced exploitation of marine ecosystems: results from dynamic size spectra. ICES Journal of Marine Science 69: 602-614.

110. Jul Larsen E, Kolding J, Overa R, Nielsen J R, Zwieten PAM (2003) Management, co-management or no management? Major dilemmas in southern African freshwater fisheries. FAO Fisheries Technical Paper 426/1.

111. Pauly D, Silvestre G, Smith IR (1989) On development, fisheries and dynamite: A brief review of tropical fisheries management. Natural Resource Modeling 3: 307 – 329.

112. Viner K, Ahmed M, Bjørndal T, Lorenzen K (2006) Development of Fisheries Co-management in Cambodia: A case study and its implications. World Fish Center Discussion Series No. 2.

113. Branch TA, Austin JD, Acevedo-Whitehouse K, Gordon IJ, Gompper ME, et al. (2012) Fisheries conservation and management: finding consensus in the midst of competing paradigms. Animal Conservation 15: 1-3.

7

Growth Patterns and Condition Factor of *Hepsetus odoe* (Bloch, 1794) Captured in Eleyele Lake, Southwest Nigeria

Kareem OK[1*], Olanrewaju AN[2], Osho EF[1], Orisasona O[3] and Akintunde MA[4]

[1]*Department of Aquaculture and Fisheries Management, University of Ibadan, Nigeria*

[2]*Federal College of Freshwater Fisheries Technology, P.M.B 1060, Maiduguri, Nigeria*

[3]*Department of Wildlife and Fisheries Management, Osun State University, Nigeria*

[4]*National University of Lesotho, Department of Agriculture, Roma 120, Kingdom of Lesotho, Southern Africa*

*Corresponding author: Kareem OK, Department of Aquaculture and Fisheries Management, University of Ibadan, Nigeria, E-mail: kaykaz2007@yahoo.co.uk

Abstract

Hepsetus odoe is a commercially valuable fish and is considered as endemic to Nigeria. The growth patterns and condition factor of *Hepsetus odoe* from Lake Eleyele, Oyo State were investigated as an aspect of its biology essential for bringing it to culture. A Total of 205 specimens (55 and 150, males and females, respectively) were collected between June, 2012 and August, 2012. The morphometric indices such as Total Length (TL), Standard Length (SL), Body Weight (BW) and Stomach Weight (SW) were assessed using standard methods. Also, the length-weight relationship ($W=aL^b$) and Condition factor ($100W/L^3$) were calculated. Sex ratio of 1:3 (Males and Females) was obtained which shows a female dominated population. The Standard length (SL) and body weight (BW) ranged from 16.60-30.50 cm and 51.0-250.0 g respectively. Length-weight relationship equations were calculated as:

Log BW=2.051+3.105log SL (r=0.93)

This analysis showed significant relationship between the standard length, body weight and stomach weight. The relative condition factor (Kn) calculated ranged from 0.99-2.14 while the mean K value was 1.24. The condition factor fall within the range recommended for freshwater fish species in the tropics. The growth pattern indicates that the fish follows cube law and exhibited positive allometry growth. This information provides important tool in fishery management and guide for future culture trials.

Keywords: *Hepsetus odoe*; Length-weight relationship; Allometry; Condition factor; Eleyele Lake

Introduction

Eleyele dam, an artificial lake constructed in 1942 supports fish consumption and conservation in South-west, Nigeria. Although, the quest to create a modern water supply system to meet the challenge of water scarcity for the emerging Ibadan metropolis led to the construction of Eleyele Dam on the main River Ona with a reservoir storage capacity of 29.5 million litres [1]. The lake experiences both dry and rainy seasons typical of tropical environment. Adebisi documented families like Cichlidae, Channidae, Gymnarchidae, Latidae, Hepsetidae, Clariidae, Osteoglossidae, e.t.c., as fauna resources of the lake [2]. Among these resources, *Hepsetus odoe* form an important fish species of commercial fish resources whose bionomic has not been adequately studied.

H. odoe (Bloch, 1974) commonly known as African pike remain the sole representative of the family Hepsetidae. The species is widely distributed around Western and Central Africa [3]. It inhabits slow and shallow waters of rivers in the plains as well as estuaries and a variety of other freshwater habitats. *H. odoe* is piscivorous, feeding on several species of smaller fish by laying ambush in dense vegetation, and they feed primarily on cichlids and mormyrids [4]. The African pike is a highly priced freshwater food fish species in Nigeria especially in the riparian community Nigeria, principally because of its availability (all year round), affordability, tasteful flesh, economic and nutritional value [5]. A medium size African pike contain 26.2% protein, 18.2% fat, 7.5% carbohydrate, 7.7% ash, 1.3% fibre and 128.5 µg/g [6]. This species prefers quiet and deep water, can reach up to about 70 cm in length and 4 kg in weight [7]. *H. odoe* form a significant part of commercial catches in Eleyele Lake but unfortunately little or no work has been reported on this specie from the Lake.

The study of length-weight relationships (LWRs) and condition factor of fishes has manifold importance in fisheries and fish biology. As much as LWRs provide valuable information on the habitat where the fish lives, it can also provide important clues on climatic and environmental changes and the change in human subsistence practice [8]. For effective fishery management and successful fish farming, knowledge of the growth patterns and condition factor is necessary. Hence, this study investigates the length-weight relationships and condition factor of *H. odoe* for effective management, sustainable exploitation and as a prelude to make it an aquaculture candidate.

Materials and Methods

Site description

Lake Eleyele is situated in North-west of Ibadan, Oyo State, Nigeria (Figure 1). The elevation is relatively low ranging between 100-150 m above sea level and surrounded by quartz-ridge hills toward the downstream section where the dam barrage is located. A number of stream channels serve as feeding/recharge streams to the Eleyele wetland basin. The lake has a surface area of 546 km² with a mean depth of 6.0 m. The widest and narrowest arm of the lake is about 250 m and 20 m respectively, having forest reserve with much vegetation stretch on each side. Eleyele Lake is usually flooded with the water-level rising during early period of rainy season. It covers some parts of Ijokodo, Apete, Awotan, Ologun-eru, Agbaje, Idi-osan, Polytechnic of Ibadan and Eleyele area with the rural fishing communities mostly dominated by Ilaje and Yorubas. Human activities in the area include fishing, farming, agro processing and boat traffic.

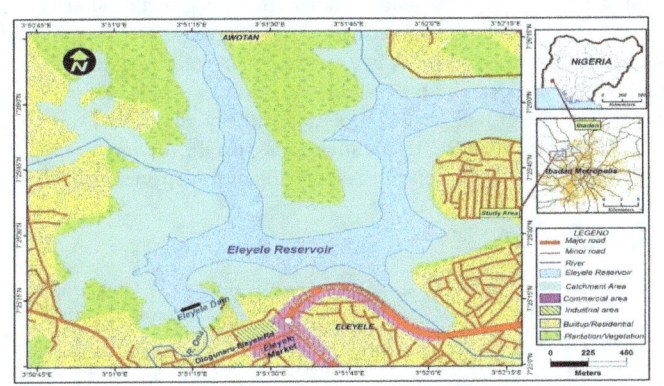

Figure 1: Map of Eleyele Lake.

Collection of sample

Samples of *H. odoe* for the study were collected fortnightly during June, 2012 to August, 2012 from catches landed by artisanal fishermen using baited longlines and gill nets of different mesh sizes. Two hundred and five (205) specimens of H. odoe were sampled, with length and weight ranging from 20.3 to 35.8 cm and 51.0 and 250.0 g respectively. Of the total number of specimens, 55 were males and 150 were females.

Morphometric measurements include Total Length (TL) and Standard Length (SL) (to the nearest cm) and Body Weight (BW) (to the nearest g) of each specimen were measured after blotting off water from their body. The TL was taken from the tip of the snout (mouth closed) to the extended tip of the caudal fin, and SL from mouth tip to the mid-point of caudal fin origin using a measuring board while BW was measured using a top loading Metler balance [9,10].

Length-weight relationships were estimated using the equation $W=aL^b$, where W is the total wet weight (g), L is the standard length (SL, cm), and a and b are the equation parameters calculated by the least squares method. To determine significant differences from the isometric value of b=3, and another t-test was to applied. The condition factor which shows the degree of well-being of the fish in their habitat was determined by the formula: Condition Factor $(K) = \dfrac{100W}{L^3}$ [11].

where W=weight in g, and

L=length in cm

Results

The length-weight frequency distribution of H. odoe sampled from Lake Eleyele is shown in Table 1. Similarly, the total length frequency of specimen was illustrated in Figure 2. The males were found to range from 20.3 to 31.9 cm in total length and total weight was ranged between 51.0 to 219.0 g.

Sex	n	Standard length (cm)			Body weight (g)		
		Min	Max	Mean ± SD	Min	Max	Mean ± SD
Males	55	16.6	26.2	19.7 ± 5.43	51	219	109.5 ± 50.8
Females	150	17.8	30.5	23.2 ± 7.64	95	250	127.4 ± 45.3
Combined sex	205	16.6	30.5	22.1 ± 6.53	51	250	119.8 ± 36.8

Table 1: Length-weight relationship of Hepsetus odoe in Lake Eleyele. where n=Number of fish sampled, SD = Standard Deviation.

In case of females, the TL and BW were ranged from 21.8 to 35.8 cm and 95.0 to 250 g respectively. The mean TL for male was calculated as 22.3 ± 7.14 cm and the mean BW calculated 109.5 ± 50.8 g (N=55). For female, the mean TL and BW were calculated as 23.6 ± 6.3 cm and 127.4 ± 45.3 g (N=155) respectively. Also, the mean SL calculated for the species are, Males 19.7 ± 5.43 cm, Females 23.2 ± 7.64 cm and Combined sex 22.1 ± 6.53 cm. The length-weight relationship for the sampled fish is expressed by the regression equation:

Log BW = -2.051 + 3.105 Log SL (r = 0.93).

Log length against log weight revealed a linear relationship hence there is a direct proportionality between the log length and log weight.

Sex	n	Standard length (cm)			Stomach weight (g)		
		Min	Max	Mean ± SD	Min	Max	Mean ± SD
Males	55	16.6	26.2	19.7 ± 5.43	0.5	2.5	1.2 ± 3.56
Females	150	17.8	30.5	23.2 ± 7.64	0.5	4.5	2.1 ± 5.33
Combined sex	205	16.6	30.5	22.1 ± 6.53	0.5	4.5	1.7 ± 6.0

Table 2: Length-stomach weight relationship of Hepsetus odoe in Lake Eleyele. n=Number of fish sampled, SD=Standard Deviation.

The SW of *H. odoe* ranged from 0.5 g to 4.5 g (Table 2). The length-stomach weight relationship was calculated as:

Log STWT = -3.487 ± 2.787 Log SL (r = 0.24).

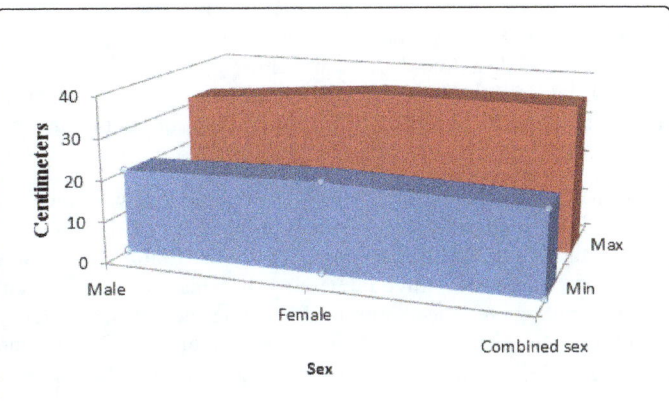

Figure 2: Total length frequency of *Hepsetus odoe* used for the study.

The Condition Factor (K) ranged between 0.99 to 1.24 for all the sexes (Table 3). The result showed that the females were significantly (p<0.05) larger than males.

Sex	Condition factor (K)	Mean value
Males	$\dfrac{\sum K}{n} = \dfrac{67.52}{55}$	1.23
Females	$\dfrac{\sum K}{n} = \dfrac{185.65}{150}$	1.24
Combined sex	$\dfrac{\sum K}{n} = \dfrac{253.17}{205}$	1.24

Table 3: Condition factor of Hepsetus odoe in Lake Eleyele, Nigeria. where ΣK=Summation of condition factors, n=Number of fish samples used.

Discussion

Length-Weight relationship is an effective tool for proper exploitation and management of the population of fish stock. According to Nagesh, LWRs have a significant importance in studying the growth, gonadal development and general well-being of fish population [12]. In the present study, the relationship between the standard length (SL) and the body weight (BW) of H. odoe shows a significant difference of 0.97 (p< 0.05). Also the significant difference of 0.49 was obtained as a relationship between the body weight (BW) and the stomach weight (STWT) of the species. This indicates that the standard length of the H. odoe increases as the body weight increases. Likewise, stomach weight increases as the body weight of the species increases. This shows that the growth pattern of the fish is allometric as indicated by the correlation coefficient (3.105). This result conformed to the findings of Adedokun who reported positive allometric growth in Hepsetus odoe in Ogbomoso reservoir [13]. Similar findings in H. odoe were reported by Idowu and Oso from Ado-Ekiti reservoir, Nigeria, with positive allometric, b value greater than 3 [14,15]. The Length-weight relationship of H. odoe in Eleyele Lake is also similar to that of other species in Nigeria water bodies. Nigeria. Ekelemu and Samuel recorded 3.03 for *Heterotis niloticus* in Ona lake, Bernard documented 3.04 for *Oreochromis niloticus* in Egah River while Ayoade and Ikulala found 3.34 for Chromidotilapia guentheri in Eleyele Lake [16-18]. However, Ayoade and Ikulala reported negative allometric growth patterns for Sarotherodon melanotheron and Hemichromis bimaculatus in Eleyele Lake [18].

Bagenal opined that a fish living in a favourable environment in term of food availability and good environmental conditions grow faster with "K" ≥ 1 [19]. The mean condition factor of 1.24 was obtained for H. odoe captured in Eleyele Lake. This result is consistent with the findings of Adedokun and Winemiller who recorded mean condition factor of 1.58 and 1.25 for H. odoe captured in Ogbomosho reservoir and River Zambezi, respectively [4,13]. Balogun also reported a mean condition factor of 1.23 for H. odoe in Asejire reservoir [20]. Comparing K values in this study with other species, it was observed that this value varies from >1 to <1. Abowei, reported 0.99 for Ilisha africana in Nkoro River, Kalu obtained 0.76 for *Clarias gariepinus* in

Lake Alau, Ikongben found 1.62 for Bagrus docmac in Lake Akata while Ayoade documented 1.11 for *Labeo oguensis* in Asejire Lake [21-24]. The condition factor value for male H. odoe (1.23) is lowered than that of the female (1.24). This agrees with the results obtained by Ugwumba and Idowu for male (1.23) and female (1.24) H. odoe in Ado Ekiti reservoir [9]. This indicates that females were in better condition than males during study period. This could be as a result of fatness and egg development in the females and hence increase in body weight.

Conclusion

This study has made available information on growth patterns and condition factor of H. odoe from Lake Eleyele. Also, significant relationships were established between the fish body weight, standard length and stomach weight. Based on the findings of this research, it is evident that the growth of H. odoe is positively allometric, therefore efforts should be directed to maintain and sustain the environmental condition of the lake for continuous thriving of this species and others.

References

1. Tijani MN, Olaleye AO, Olubanjo OO (2011) Impact of Urbanization on Wetland Degradation: A Case Study of Eleyele Wetland, Ibadan, South West, Nigeria.

2. Adebisi AA (1981) Analysis of stomach contents of the piscivorous fishes of the Upper Ogun River, Nigeria. Hydrobiologia 79: 167-177.

3. Idodo-Umeh G (2003) Fresh water fishes of Nigeria, Taxonomy, Ecological Notes, Diet and Utilization p: 232.

4. Winemiller KO (1993) Comparative Ecology of Serranochromis Species (Teleostei: Cichlidae) in the upper zimbezi River floodplain. Journal of Fish Biology 39: 617-639.

5. Reed W, Burchard J, Hopson AJ, Jenness J, Yaro B (1967) Fish and Fisheries of Northern Nigeria. Ministry of Agriculture, Northern Nigeria p: 226.

6. Fawole OO, Yekeen TA, Adewoye SO, Ogundiran MA, Ajayi OE (2013) Nutritional qualities and trace metals concentration of six fish species from Oba reservoir, Ogbomoso, Nigeria. Afr J Food Sci 7: 246-252.

7. Kareem OK, Ajani EK, Orisasona O, Olanrewaju AN (2015) The Sex ratio, Gonadosomatic index, Diet composition and Fecundity of African pike, Hepsetus odoe (Bloch, 1794) in Eleyele Lake, Nigeria. J Fisheries Livest Prod 3: 139.

8. Bolarinwa JB, Popoola Q (2013) Length-Weight Relationships of Some Economic Fishes of Ibeshe Waterside, Lagos Lagoon, Nigeria. J Aquac Res Development 5: 203.

9. Ugwumba AA (2007) The food and feeding ecology of fishes in Nigeria pp: 13-74.

10. Fafioye OO, Oluajo OA (2005) Length-weight relationships of five fish species in Epe Lagoon, Nigeria. Afr J Biotechnol 4: 749-751.

11. Pauly D (1993) Fish byte. Naga ICLARM Quarterly 16: 16-26.

12. Nagesh TS, Jana D, Khan I, Khongngain O (2004) Length-weight relationship and relative condition of Indian major carps from Kulia beel, Nadia, West Bengal, Aquacult 5: 85-88.

13. Adedokun MA, Fawole OO, Ayandiran TA (2013) Allometry and condition factors of African pike "Hepsetus odoe" actinopterygii in a lake. Afr J Agric Res 8: 3281-3284.

14. Idowu EO (2007) Aspects of the Biology of Hepsetus odoe in Ado-Ekiti Reservoir Ekiti, Nigeria Ph.D thesis, University of Ibadan, Ibadan, Nigeria.

15. Oso JA, Idowu EO, Fagbuaro O, Olaniran TS, Ayorinde BE (2011) Fecundity, Condition Factor and Gonado-Somatic Index of Hepsetus odoe (African Pike) in a Tropical Reservoir, Southwest Nigeria. World J Fish & Marine Sci 3: 112-116.

16. Ekelemu KJ, Samuel AAZ (2006) Growth Patterns and Condition Factors of Four Dominant Fish Species in Lake Ona, Southern Nigeria. J Fish Intl 1: 157-162.

17. Bernard E, Bankole NO, Akande GR, Adeyemi S, Ayo-Olalusi CI (2010) Organoleptic characteristics, Length-weight relationship and condition factor of Oreochromis niloticus in Egah River at Idah L.G.A of Kogi State, Nigeria. Internet Journal of Food Safety 12: 62-70.

18. Ayoade AA, Ikulala AOO (2007) Length Weight Relationship, Condition Factor and Stomach Contents of Hemichromis bimaculatus, Sarotherodon melanotheron and Chromidotilapia guentheri (Perciformes: Cichlidae) in Eleiyele Lake, Southwestern Nigeria. Rev Biol Trop 55: 969-977.

19. Bagenal TB (1978) Methods for assessment of fish production in freshwater (ed Bagenal) Black Well Scientific Publications pp: 219-255.

20. Balogun K (1980) The biological survey of fishes in Epe lagoon M.sc. Thesis, University of Lagos, Lagos.

21. Abowei JFN (2010) The Condition Factor, Length-Weight Relationship and Abundance of Ilisha africana (Block, 1795) from Nkoro River Niger Delta, Nigeria. Adv J Food Sci Technol 2: 6-11.

22. Kalu KM, Umeham SN, Okereke F (2007) Length-weight relationship and condition factor of Clarias gariepinus and Tilapia zillii in Lake Alau and Monguno hatchery, Borno State, Nigeria. Animal Research International 4: 635-638.

23. Ikongbeh OA, Ogbe FG, Solomon SG (2012) Length-Weight Relationship and Condition Factor of Bagrus docmac from Lake Akata, Benue state, Nigeria. Journal of Animal and Plant Sciences 15: 2267-2274.

24. Ayoade AA (2011) Length-weight relationship and Diet of African Carp Labeo ogunensis (Boulenger, 1910) in Asejire Lake Southwestern Nigeria. J Fish Aqua Sci 6: 472-478.

8

Farmed Salmon and Farmed Rainbow Trout - Excellent Sources of Vitamin D?

Jette Jakobsen[1*] and Cat Smith[2]

[1]*National Food Institute, Technical University of Denmark, Kemitorvet, DK-2800, Lyngby, Denmark*

[2]*Bantry Marine Research Station, Gearhies, Bantry, Co. Cork, Ireland*

[*]**Corresponding author:** Jette Jakobsen, National Food Institute, Technical University of Denmark, Kemitorvet DK-2800, Lyngby, Denmark, E-mail: jeja@food.dtu.dk

Abstract

Fatty fish are generally stated as having high vitamin D content and among these are salmon and trout. In the aquaculture industry of salmonids the two main species produced are *Salmo salar* (Atlantic salmon) and *Onchorhynchus mykiss* (rainbow trout). Published data have shown lower content of vitamin D in farmed than in wild species, but generally data on vitamin D in farmed salmon and rainbow trout are scarce. In commercial production facilities we aimed to study the variation of vitamin D in farmed salmon and rainbow trout prepared for sale to consumer. Thirteen organically produced salmon and 18 rainbow trout were sampled within the range 0.7-4.0 kg of gutted weight. All fish were ready for consumption, and analysed for content of vitamin D3, 25-hydroxyvitamin D3, and fat.

Mean vitamin D3 content in salmon and rainbow trout was 1.6 ± 0.5, and 5.0 ± 2.3 µg/100 g, respectively. Compared to vitamin D3, the content of 25-hydroxyvitamin D3 was 11% and 3%, respectively. In farmed salmon a linear relationship with vitamin D3 being dependent on weight ($P<0.05$) as well as to fat content ($P<0.05$), while no similar relationship was found for farmed rainbow trout. Despite this, both species exhibit a linear correlation between fat and gutted weight ($P<0.001$).

The results indicate that there is a difference in the storage of vitamin D between the two salmonids, as 25-hydroxyvitamin D3 amounted more to the vitamin D activity than in rainbow trout. The low content of vitamin D in i.e. found in the salmonids is challenging farmed salmon and farmed trout as an essential vitamin D source.

Keywords: Vitamin D; Salmon; Trout; Aquaculture

Introduction

Vitamin D is a fat-soluble vitamin that humans produce in skin when exposed to UV light (290-315 nm), and obtain through dietary intake. The vitamin D content of food is composed of the parent vitamin D, cholecalciferol (vitD3) and ergocalciferol (vitD2), and the metabolites, 25-hydroxyvitamin D3 (25OHD3) and 25-hydroxyvitamin D2 (25OHD2). There are several high risk groups associated with vitamin D deficiency: individuals who avoid sun exposure, people with dark-pigmented skin, older people and those in residential care or nursing homes [1,2]. Dietary intake is essential all year round for people in high risk groups and during winter for people living at latitudes above 35°. A recent review of dietary intake of vitamin D in adults and children in Europe shows that intake of vitamin D is generally inadequate compared to recommendations [3]. Fatty fish e.g. salmon is generally stated as having high vitamin D content, which for wild salmon has been reported to 8-55 µg vitD/100 g [4], but it has also been reported that the content of vitamin D in farmed salmon is lower than in wild salmon [5]. In fish, vitamin D is essential to preserve calcium and phosphate homoeostasis and the level rely on the intake in feed [6].

In the aqacultural production salmon and trout is the largest single commodity by value [7]. The dominant farmed salmon and farmed trout are the Atlantic salmon *Salmo salar* (*S. salar*) and the Pacific Ocean species of wild salmon *Oncorhynchus mykiss* (*O. Mykiss*), respectively; *O. Mykiss* is also named rainbow trout [8,9].In terms of volume, *S. salar* is the most important aquacultured salmon species, as it covers >90% of the farmed salmon market and >50% of the total global salmon market [8]. The main producers of farmed salmon are Chile, Norway, Scotland and Canada, [8], while the main producers of farmed rainbow trout are Iran, Chile, Turkey and Norway [10].

The dietary requirement of vitD3 for rainbow trout is 40-60 µg/kg feed [11], while for *S. salar* the estimated required amount is 60 µg vitD3/kg feed [12]. In Europe there are maximum limits for the addition of vitamin D to feed, namely 75 µg/kg feed [13]. Fish are very tolerant of high contents of vitD3 in feed [6], e.g. no effect of excessive amount of vitD3, up to 25000 µg/kg feed, for *O. mykiss* has been demonstrated [14].

In nine different species of fresh and salt water fish fattiness had no significant effect on content of vitamin D [15]. Similar results were obtained for herring (*Clupea harengus*) and mackerel (*Scomber scombrus*), as content of fat through the year varied from 2-32% and 11-35%, respectively, while content of vitamin D showed no association to fattiness [16]. Data on vitamin D in farmed salmon and trout are scarce. In commercial production facilities, we aimed to study the variation of vitamin D in farmed salmon and rainbow trout prepared for sale to consumer. The hypothesis was that the content of vitamin D depends on fat content and weight of the farmed fish.

Materials and Methods

Fish samples

Salmo salar-farmed salmon: Thirteen farmed salmon (0.7 kg to 3.9 kg gutted weight) were included in the study. The salmon fry hatched between Month 1-6, were put to sea in Month 12-16, and were harvested in Month 29, at an age of between two years and two years five months. The salmon were farmed by Murphy's Irish Seafood (Bantry, Ireland) using the organic feeding regiment Emerald Organic Feed (Skretting, Wincham, UK). One whole fillet from each of the 13 salmon was homogenized.

Oncorhynchus mykiss-farmed rainbow trout: Eighteen rainbow trout (0.9 kg to 3.9 kg gutted weight) were included in the study. The fry hatched in Month 1, were put to sea in Month 25, and were harvested in Month 32 at an age of between two years six months and two years seven months. The trout were farmed by Musholm A/S, Reersø, Denmark. During the period at sea (from week 15 to harvest), Aller Superior (Aller Aqua A/S, Christiansfeld, DK) and Efico Enviro 939 (BioMar A/S, Brande, DK), were used for feeding. For practical reasons the Norwegian Qualtiy Cut (NQC) [17] was sampled for each of the 18 rainbow trout and homogenised.

Quality control materials for the analyses of vitamin D: Certified reference materials, Milk powder (CRM421, IRMM, Geel, B), and proficiency testing materials, Milk Powder (FAPAS 2184, FAPAS, York, UK). In-house reference material, Control- salmon, is homogenised whole fillet of wild Atlantic salmon, divided into 50 individual 10 g portions and stored at max -20°C.

Homogenisation: All individual fish sampled were homogenised in a blender (Osterizer, Hohn Oster, WI, USA or 1094 Tecator, Höganäs. Sweden) for 30-60 seconds until homogenous. As a precaution against oxidation, the homogeniser instrument was purged with nitrogen before the homgenisation process was initiated. Homogenised samples were stored in plastic bags at maximum -20°C until analysis, which took place within 6 months.

Analytical methods

Vitamin D: The amounts of vitD3 and 25OHD3 in fish were detected and quantified by use of HPLC-UV/PDA [18,19] or liquid chromatography combined with triple quadrupole mass spectrometry and electrospray ionisation (LC-MS/MS-ESI) [20]. The equipment used being HPLC-UV/PDA from Waters (Alliance 2695; UV-detector 2487; PDA-detector 2996, Milford, MA) and LC-MS/MS from Agilent (LC 1200 and MS 6460, Agilent Technologies, Santa Clara, CA). The trueness of the methods was secured by acceptable results for the analyses of a certified reference material and participation in proficiency test scheme. The consistency of the methods during the studies was checked by analysis of Control-salmon. The HPLC-UV/DAD method (n=8) found 15.4 µg vitD3/100g ± 1.5%, and 0.268 µg 25OHD3/100 g ± 11%, which were similar values to the LC-MS/MS- method (n=18) being 15.2 µg vitD3/100 g ± 5.3% and 0.266 µg 25OHD3/100 g ± 5.8%. All analyses were conducted in a laboratory accredited to performing the analyses according to [21].

Fat content: The content of fat in the fish was determined gravimetrically by a modified Schmid-Bondzynski-Ratslaff (SBR) procedure [22].

Statistical tests: Linear regression and t-test with the probability level of P=0.05 was used. Excel 2010 was used for all calculations and all results are given as average ± standard deviation (x ± SD).

Results

The fish sampled were divided into the five weight categories usually used by rainbow trout farmers being 0.7-1.4 kg, 1.4-1.9 kg, 1.9-2.5 kg, 2.5-3.0 kg, and 3.0-4.0 kg (personal communication, Musholm A/S, Reersø, Denmark). The mean of gutted weight, content of fat, vitD3, and 25OHD3 in salmon and in rainbow trout are given in Table 1.

Farmed salmon (S. salar)					Farmed rainbow trout (O. mykiss)				
Weight kg	Fat g/100 g	Vitamin D3 µg/100 g	25OHD3 µg/100 g	Sample number	Weight kg	Fat g/100 g	Vitamin D3 µg/100 g	25OHD3 µg/100 g	Sample number
0.9 ± 0.3	2.9 ± 0.9	1.4 ± 0.5	0.10 ± 0.03	4	0.9 ± 0.0	7.6 ± 2.7	4.2 ± 1.3	0.09 ± 0.04	3
1.7 ± 0.2	5.5 ± 2.0	1.1 ± 0.4	0.14 ± 0.01	2	1.5 ± 0.2	7.6 ± 1.7	6.2 ± 3.2	0.09 ± 0.04	6
2.2 ± 0.2	4.9 ± 0.7	1.4 ± 0.6	0.14 ± 0.06	2	2.3 ± 0.1	7.5 ± 0.4	3.3 ± 2.1	0.09 ± 0.05	3
2.8 ± 0.2	8.5 ± 1.0	1.6 ± 0.1	0.26 ± 0.02	2	2.9 ± 0.0	10.9 ± 0.3	5.5 ± 0.5	0.11 ± 0.01	3
3.6 ± 0.3	7.9 ± 0.7	2.3 ± 0.3	0.24 ± 0.04	3	3.5 ± 0.5	12.0 ± 2.5	4.3 ± 1.1	0.17 ± 0.01	3
2.1 ± 1.1	5.6 ± 2.5	1.6 ± 0.5	0.17 ± 0.07	13	2.1 ± 0.9	8.9 ± 2.5	5.0 ± 2.3	0.10 ± 0.04	18

Table 1: Farmed salmon and farmed rainbow trout divided into five weight classes – content of vitamin D3, 25OHD3, and fat (mean ± SD).

In farmed salmon there was a linear correlation with fat being dependent on gutted weight (P<0.001). Additionally, content of vitD3 depends on gutted weight (P<0.05) and on content of fat (P<0.05). As for vitD3 content of 25OHD3 was found to depend on gutted weight (P<0.01) and content of fat (P<0.0001). The distribution of 25OHD3 amounted 11.0% ± 4.6% of the content of vitD3.

In farmed trout there was a linear correlation with fat being dependent on gutted weight (P<0.001). No association was found between vitD3 and gutted weight or vitD3 and fat, while for 25OHD3 there was an association with gutted weight (P<0.01), but not with fat content. The relative amount of 25OHD3 compared to vitD3 was 2.4% ± 1.1%.

Discussion

The farmed salmon and farmed rainbow trout were sampled at similar weights and ages. The analyses for farmed salmon were based on homogenization of the whole fillets, while for farmed trout the NQC was used. NQC is used in the assessment of vitamin D in fish for food composition data [17], and is a more practical approach in the homogenisation procedure. However, the difference in subsampling for homogenisation is a limitation in the interpretation of the results.

Salmon, farmed (S. salar)			Salmon, farmed, organic (S. salar)			Trout, farmed (O. mykiss)			Trout, farmed, organic (O. mykiss)			Year	Ref
Vitamin D3		25OHD3	Vitamin D3		25OHD3	Vitamin D3		25OHD3	Vitamin D3		25OHD3		
Mean	(R) ± SD1	Mean2	Mean	(R)/ ± SD1	Mean2	Mean1	(R)/ ± SD2	Mean1	Mean1	(R)/ ± SD2	Mean1		
						13	(8-16)	na				1994	23
						7.5	(7.3-7.8)	0.14				1995	15
						19	(10-23)	na				1999	24
7.6	(4.2-9.1)	na	5.4	(4.2-6.6)	na	8.1	(3.8-10.7)	na	8.2	(7.9-8.5)	na	2006	25
10		0.49				6.9		0,22				2012	17
						7.0		0.18				2013	26
5.9	± 3.6	na				8.0	± 3.36					2015	27
5.8	(3.6-8.2)	nd				1.6	(1.4-1.8)	nd				2016	28
6.7		0.38										2017	29
			1.6	± 0.5	0.17	5.0	± 2.3	0.1				2017	*
[1](R)/±SD: Range or ± standard deviation as provided in reference;													
[2]na: not analyzed; nd: not detected;													
*Data from this study.													

Table 2: Farmed salmon (*S. salar*) and farmed rainbow trout (*O. mykiss*) – published data on content of vitamin D3 and 25OHD3 (μg/100 g).

The mean content of vitamin D3 in investigated salmon and rainbow trout was 1.6 ± 0.5 and 5.0 ± 2.3 μg/100 g, respectively. The level of vitD in salmon is surprisingly low, which might be due to the content of vitD in feed. We did not focus on feeding, as the sampled salmon and trout were taken from healthy stocks sold for consumption.

Data in the literature for content of vitamin D in farmed salmonids which include the species i.e. taxonomic name are shown in Table 2. Only vitD3 has been included in the majority of studies, as quantification of 25OHD3 has been an analytical challenge. Data for vitD content in raw samples of farmed salmon and farmed trout are scarce, and primarily analysed for food databanks [15,17,23-29]. The sampling system is therefore systematic to cover the market in the specific countries i.e. Denmark, Norway, United Kingdom, and the sampling and homogenization strategy make use of whole fillets as well as NQC. Our aim was different, but might be compared to the data established for food databanks. The content of vitamin D found in our study is currently sufficient to label these farmed fish with health claim "High content of vitamin D" in Europe [30]. However, it should be noticed that recommended dietary intake for vitamin D has increased from 5μg/day to 15μg/day over the last years [31,32]. The data in Table 2 indicate that content of vitamin D in farmed salmon and farmed trout seems to have decreased within the last 25 years. The change of feed based on fishmeal to vegetable feed [33] could cause such decrease. Fish products are essential for the dietary intake of vit D e.g.

in French diet the estimate is that 33% of the dietary intake of vitamin D derives from salmon [34]. An estimate which is based on the content of 15μg vitamin D/100 g salmon, which is almost 10 times higher than we find. The studies which did include quantification of vit D3 and 25OHD3 show that the relative amount of 25OHD3 compared to vitD3 is 5-11% for salmon and 2-3% for trout. We hypothesize that the distribution between vitD3 and 25OHD3 is species specific.

In the whole fillet from farmed salmon we found that content of fat increases in a linear matter with increase in weight, and vitD3 increased with increased content of fat. For the NQC in farmed trout similar correlation for content of fat and gutted weight occurred, but the content of vitD3 showed no association with fat. Whether this difference is species dependent or caused by the difference in subsampling has to be confirmed. However, if this is due to the subsampling of NQC the use of NQC could introduce a BIAS if used for analyses of vitD in fish for establishing data for food databanks. Previously other differences have been reported for *S. salar* and *O. mykiss* in studies aiming to increasing content of vitamin D in the two species by increasing vitD3 in the fish feed. Trout were fed 89-539 μg/kg for four months up to a weight of 618 g-1282 g and 7-14% fat [35], while salmon were fed 40-28680 μg/kg for 11 weeks to a weight of 450 g [36,37]. No effect was observed in trout, while content of vitamin D in salmon increased linearly with increased vitamin D in feed.

The limitation in our study is, apart from the differences in sampling strategy, that only one production cycle of each farmed species was included. Furthermore, we did not focus on the content of vitamin D in feed, as the aim was to investigate the current production system, but the sampled salmon and trout were taken from healthy stocks sold for consumption.

Conclusion

The farmed produced salmon and the farmed rainbow trout were sampled at similar weights and ages. For analyses the whole filet was used for salmon while Norwegian Quality Cut was used for trout. The content of vitamin D3 was 1.1-2.3 µg/100 g and 3.3-6.2 µg/100 g, respectively. In farmed salmon a relationship was found for vitamin D3 to gutted weight as well as to fat content, while no similar relationship was found for farmed rainbow trout. Despite this, both species exhibit a linear correlation between fat and gutted weight. Compared to vitamin D3, the content of 25-hydroxyvitamin D3 was 11% and 3%, respectively. The low content of vitamin D found in the salmonids is challenging farmed salmon and farmed trout as essential vitamin D sources.

Acknowledgement

The authors thank Musholm A/S, Reersø, Denmark for the farmed rainbow trout, and Kirsten Pinndal and Sofie Rye Jønsson for performing excellent laboratory work. Danish Veterinary and Food Administration funded part of the project.

Conflict of Interest

None of the authors have conflicts of interest.

References

1. Lips P (2010) Worldwide status of vitamin D nutrition. Journal of Steroid Biochemistry and Molecular Biology 121: 297-300.

2. Jones G (2012) Metabolism and biomarkers of Vitamin D. Scandinavian Journal of Clinical and Laboratory Investigation 72: 7-13.

3. Kiely M, Black LJ (2012) Dietary strategies to maintain adequacy of circulating 25-Hydroxyvitamin D concentrations. Scandinavian Journal of Clinical and Laboratory Investigation 72: 14-23.

4. Søndergaard H, Leerbeck E (1982) The content of vitamin D in Danish foods. Danish Food Administration, Mørkhøj Bygade 19, Søborg Denmark.

5. Chen TC, Chimeh F, Lu Z, Mathieu J, Person KS et al. (2007) Factors that influence the cutaneous synthesis and dietary sources of vitamin D. Archives of Biochemistry and Biophysics 460: 213-217.

6. Lock EJ, Waagbø R, Bonga, SW, Flik G (2010) The significance of vitamin D for fish: a review. Aquaculture Nutrition 16: 100-116.

7. FAO (2016) In Brief The state of world fisheries and aquaculture.

8. Seafish (2012) Responsible sourcing guide: farmed Atlantic salmon. Grimsby, UK.

9. Woynarovich A, Hoitsy G, Moth-Poulson T (2011) Small-scale rainbow trout farming. Food and Agriculture Organization of The United Nations, Rome.

10. FAO (2015) Fishery Statistical Collections. Global Aquaculture Production.

11. NCR (2011) Nutrient requirements of fish and shrimp. Committee on the Nutrient Requirements of Fish and Shrimp. National Research Council. The National Academies Press, Washington, DC.

12. Woodward B (1994) Dietary vitamin requirements of cultured young fish, with emphasis on quantitative estimates for salmonids. Aquaculture 124: 133-168.

13. Official Journal of the European Union (2004) List of the authorised additives in feedingstuffs published in application of Article 9t (b) of Council Directive 70/524/EEC concerning additives in feedingstuffs. C50 1-144.

14. Hilton JW, Ferguson HW (1982) Effect of excess vitamin-D3 on calcium-metabolism in rainbow-trout salmo-gairdneri Richardson. Journal of Fish Biology 21: 373-379.

15. Mattila P, Piironen V, Uusi-Rauva E, Koivistoinen P (1995) Cholecalciferol and 25-hydroxycholecalciferol contents in fish and fish products. Journal of Food Composition and Analysis 8: 232-243.

16. Jakobsen J, Japelt RB (2012) Vitamin D. Handbook of Analysis of Active Compounds in Functional Foods. CRC Press, Taylor & Francis Group.

17. Myhre JB, Borgejordet A, Nordbotten A, Loken EB, Fagerli RA (2012) Nutritional composition of selected wild and farmed raw fish. Norwegian Food Safety Authority and Directorate of Health and Social Affairs. Nutrient analysis. Raw fish.

18. Jakobsen J, Clausen I, Leth T, Ovesen L (2004) A new method for the determination of vitamin D3 and 25-OH vitamin D3 in meat. Journal of Food Composition and Analyses 17: 777-787.

19. Jakobsen J, Maribo H, Bysted A, Sommer HM, Hels O (2007) 25-hydroxyvitamin D3 affect vitamin D status similar to vitamin D3 in pigs-but the meat produced has a lower content of vitamin D. British Journal of Nutrition 98: 908-913.

20. Burild A, Frandsen HL, Poulsen M, Jakobsen J (2015) Tissue content of vitamin D3 and 25-hydroxy vitamin D3 in minipigs after cutaneous synthesis, supplementation and deprivation of vitamin D3. Steroids 98: 72-79.

21. ISO (2005) General Requirements for the Competence of Testing and Calibration Laboratories. Geneva, Switzerland: International Organization for Standardization Central Secretariat.

22. NMKL (1989) Nordic Committee on Food Analysis. Fat Determination according to SBR in meat and meat products. Norwegian Veterinary Institute, Oslo, N.

23. Jacobsen J, Leth T (1994) Nutritional monitoring system. Levnedsmiddelstyrelsen, Mørkhøj Bygade Søborg, Denmark.

24. Jacobsen J, Knuthsen P (1999) Food monitoring system for nutrients. FødevareRapport. Food Directorate, Mørkhøj Bygade 19, 2860 Søborg, Denmark.

25. Ostermeyer U, Schmidt T (2006) Vitamin D and provitamin D in fish - Determination by HPLC with electrochemical detection, European Food Research and Technology 222: 403-413.

26. Roe M, Church S, Pinchen H, Finglas P (2013) Nutrient analysis of fish and fish products. Analytical Report. Department of Health.

27. Malesa-Ciecwierz M-C, Usydus Z (2015) Vitamin D: Can fish food-based solutions be used for reduction of vitamin D deficiency in Poland? Nutrition 31: 187-192.

28. Padula D, Greenfield H, Cunningham J, Kiermeier A, McLeod C (2016) Australian seafood compositional profiles: A pilot study. Vitamin D and mercury content. Food Chemistry 193: 106-111.

29. Knuthsen P, Hansen KS, Saxholt E, Jakobsen J (2017). Fish and fish products-Nutrient content. Technical University of Denmark, Lyngby, Denmark.

30. EP (2006) Regulation (EC) No 1924/2006 of the European Parliament and of the Council of 20 December 2006 on nutrition and health claims made on foods.

31. NRC (1989) Recommended dietary allowances, 10th Ed. National Research Council. Food and Nutrition Board, Commission of Life. National Academy Press, Washington, DC.

32. IOM (2010) Dietary reference intakes for calcium and vitamin D. Institute of Medicine.

33. Bell JG, Waagbø R (2008) Safe and nutritious aquaculture produce: Benefits and risks of alternative sustainable aquafeeds. Aquaculture in the Ecosystem.

34. Bourre JM, Paquotte P (2008) Contributions of marine and fresh water products to the French dietary intakes of vitamins D and B12, selenium,

iodine and docosahexaenoic acid: impact on public health. International Journal of Food Sciences and Nutrition 59: 491-501.

35. Mattila P, Piironen V, Hakkarainen T, Hirvi T, Uusi-Rauva et al. (1999) Possibilities to raise vitamin D content of rainbow trout (Oncorhynchus mykiss) by elevated feed cholecalciferol contents. Journal of the Science of Food and Agriculture 79: 195-198.

36. Horvli O, Lie O, Aksnes L (1998) The distribution of vitamin D3 in Atlantic salmon Salmo salar: effect of dietary level. Aquaculture Nutrition 4: 127-131.

9

Growth Performance, Feed Utilization and Body Composition of *Clarias gariepinus* (Burchell 1822) Fed Marine Fish Viscera-based-diet in Earthen Ponds

Vincent Oké[1], Youssouf Abou[1*], Alphonse Adité[1] and Jean-André T Kabré[2]

[1]*Laboratory of Ecology and Management of Aquatic Ecosystems (LEMEA), Department of Zoology, University Of Abomey-Calavi, PO Box 526, Republic of Benin*

[2]*Laboratory of Research and Training in Fishing and Wildlife, Institute of Rural Development, Polytechnic University of Bobo-Dioulasso, BP. 1091 Bobo 01, Burkina Faso*

*Corresponding author: Youssouf Abou, Laboratory of Ecology and Management of Aquatic Ecosystems (LEMEA), Department of Zoology, University Of Abomey-Calavi, PO Box 526, Republic of Benin, E-mail: y_abou@yahoo.com

Abstract

A 90-days experiment was conducted to study the effect of replacement of fishmeal (FM) with marine fish viscera (MFV) meal on growth performance, body composition and production of *Clarias gariepinus* fingerlings (mean weight 11.3 ± 0.1 g). Diets were three isonitrogenous (43% crude protein) and isoenergetic (20 KJ/g) diets containing 0% (D0), 30% (D30) and 50% (D50) of MFV, as FM substitute. Diet D0, without MFV, acted as a control. All these diets were compared to the commercial diet coppens developed for *C. gariepinus*. No significantly differences were found in final weight (range: 220.94-234.1 g), weight gain (range: 1937.2-1971.7%), specific growth rate (range: 3.30-3.37%/day), protein efficiency ratio (range: 1.93-2.09) and annual production (range: 378.3-415.0 kg/are/year) of fish fed coppens diet, D0 and D30 ($p > 0.05$). Fish fed D50 showed significantly lower growth and feed utilization performances ($p < 0.05$). Moisture and crude protein were similar among dietary treatments ($p > 0.05$). Lipid deposition in fish significantly increased with MFV level in diets, whereas ash content decreased ($p < 0.05$).

The study indicates that MFV meal can be used up to 30% in formulation fish feed for promotion of *Clarias gariepinus* rearing in rural areas.

Keywords: *Clarias gariepinus*; Fishmeal replacement; Marine fish viscera; Growth; Earthen ponds

Introduction

Fishmeal is the main ingredient for most fish diets because of its high protein content, balanced amino acid profile, high essential fatty acids content, minerals and vitamins [1-6]. As a consequence of rapid growth of aquaculture, fish meal prices have increased significantly in the past few years and are likely to increase further with continued growth in demand [7-10]. Considering the global increasing of human population, feeding FM to farmed fish on any significant scale is neither profitable nor sustainable, especially in developing countries where the use of FM in fish feed is often economically prohibitive [11,12]. Thus, studies on the use of other efficient and cheaper sources of protein as substitutes for fish meal are necessary for aquaculture development and durability [6,13].

African catfish *C. gariepinus* is a globally popular aquaculture species largely distributed throughout Africa and Asia [2,14-18]. It is widely cultured in freshwater ponds because of their easiness in reproduction, high growth rate, tolerance to high densities culture conditions, resistance to diseases, excellent flesh quality and ability to accept a wide variety of feed [15-17,19]. The technics culture for the full life cycle of African catfish has been well-established and the global production of this species has been increased from 11.8 tons in 2000 to 517.4 tons in 2010 [20]. However, its intensive culture is quite limited because of the high operational cost due to the high protein commercial diets which increased feed cost [19,21]. The economically feasible catfish farming can be achieved when it is based on cost-effective feed compound of locally available agricultural by-products [17,22,23]. Many alternatives resources such as feather meal [19], meat and bone meal, hydrolyzed feather meal, fleshings-meal and blood meal [24,25], dried fermented fish by-product silage [6], poultry silage [26], shrimp head waste meal [27,28], poultry by-product meal [2,29], skate meal and sablefish viscera meal [30] have been tried to replace fish meal either partially or fully, but even these meals of various animal sources are not sufficient to meet the growing demands of fish raising industry.

Appropriate use of local protein by-products could reduce feed costs and enhance environment and economic sustainability [30]. Marine fish viscera are non-edible parts produced as by-product in large quantities in Benin by fish processing industries. These wastes are being dumped in close vicinity to market and at sea. It is challenged to recycle these wastes into acceptable source of animal protein in diets for fish [31,32]. Marine fish viscera have likely similar nutritional qualities as the fish meals currently used in aqua feeds [33,34]. It includes significant quantities of lipids with long chain, highly digestible, well-balanced proteins and highly unsaturated (n-3) fatty acids [30,35]. Several works reported isolation and identification of polyunsaturated fatty acids especially eicosapentaenoic acid (EPA) and docosahexaenoic acid (DHA), enzymes and other bioactive compounds from marine fish viscera [36-42]. Indeed, EPA and DHA are important omega-3 polyunsaturated fatty acids of which human body needed but cannot produce [42]. They were confirmed to benefit the functions of various systems in human body, including cardiovascular health, brain health, eyesight health etc. [43-51]. The purposes of the present study were to feed *C. gariepinus* with marine fish viscera and to evaluate its efficacy in terms of growth performance, feed efficiency and change in whole-body carcass composition.

Materials and Methods

Fish viscera meal

Marine fish viscera was rendered from commercial fish processing industry in market place and stored frozen (-20°C). After, it was slightly heated and dried in oven at 55°C for 48 h. The dried product was grounded and meal was stored in a refrigerator in plastic bag until used.

Experimental ingredients and diets

FM used in this study was Sardinella aurita meal. Slaughter house blood was collected from Calavi town immediately after the slaughtering of oxen. The blood was allowed to clot and only the clotted portion was collected and immediately brought to the laboratory. At the laboratory, blood was heated and sun-dried for three days. Maize bran (Zea mays), palm oil and soybean oilcake (Glycine max) were purchased at the local market, the amount of the latter being kept at 10-15%, so as to minimize the effects of its anti-nutritional factors. Dried viscera, sun-dried blood, maize bran and soybean oilcake were grounded and separately stored in refrigerator at +4°C until used. Three isoproteic (43% crude protein) and isoenergetic (20 KJ/g) experimental diets (Tables 1 and 2) were formulated to meet the protein and energy requirements of the juvenile catfish. Diet D0 contained FM as the main animal protein and was considered as the control diet. In diets D30 and D50, MFV meal was incorporated to replace partially and completely the FM. All diets were compared with coppens diet in order to validate our experimental facilities and diets. The ingredients and diets were analyzed for the proximate composition using standard methods given in Millamena [52] and the results are presented in Tables 1 and 2, respectively.

All ingredients were grounded in grinding mill to desired particle size, weighed and mixed thoroughly in a food mixer for 30 min. The hot water (about 30% of dry weight diet) was progressively added to one kilogram of diet formulated and blended. The resulting dough was cut into paste and sun-dried for about three days at 32-35°C. After drying, the diets were broken into small particles (mm) and preserved in refrigerator (+4°C) until used. The formulation of the experimental diets is given in Table 2.

Ingredients	Dry matter	Crude protein	Crude lipid	Ash
Fish meal	92	66	7.88	15.77
Bood meal	90.9	71.9	1.7	6.4
Maize bran	91.4	6.2	3.1	1.4
Soybean oilcake	94.8	30	13.2	3.7
Marine Fish viscera	27	38.8	39	7

Table 1: Proximate composition (expressed as percent dry matter) of feeds ingredients.

Fish rearing, experimental design and feeding

The experiment was carried out during 90 days in wetland area at Louho, Porto-Novo, Benin. One-thousand and nine hundred (1900) C. gariepinus fingerlings (average weight 11.3 ± 0.1 g) were obtained from the Tonon fish farming foundation located at Calavi and were transported to the experimental station.

Ingredients (%)	Diets			
	Coppens*	D0	D30	D50
Fish meal		30	15	0
Blood meal		23	23	23
Maize meal		30	20	10
Soybean oilcake		15	10	15
Fish viscera meal		0	30	50
Palm oil		2	2	2
Proximate composition (%.MS)				
Dry matter	89.4	90	88.4	90.3
Crude protein	43	43	43.3	43.2
Lipid	13	10.8	12.3	12.9
Ash	9.9	13.1	12.6	12.7
Carbohydrate	34.1	31	31.8	31.2
Gross energy (KJ g-1)	21.2	20.2	20.6	20.7

Table 2: Formulation and proximate biochemical composition of experimental diets. D0, diet containing fish meal; D30, diet containing 30% fish viscera meal; D50, diet with 50% fish viscera meal.

Fish were randomly stocked into twelve earthen ponds (10 m×3 m×1 m) at a density of 5 fish m^{-2} (150 fish per pond). They were acclimated to experimental conditions for three days in ponds during which time all fish were fed a mixture of two experimental diets twice daily. At the start of the experiment, the acclimated fish were deprived of feed for 24 h. Ponds were grouped into four triplicate and each was randomly assigned an experimental diet. All ponds were filled naturally from water table. During the feeding trial, fish were hand-fed to apparent satiation at 09:00 and 17:00 hours daily. Care was taken to stop feeding as soon as the fish stopped eating. At each fortnight, 40% of fish in each pond were sampled out with a seine net (12.7 mm mesh size) and weighed [53,54].

Fortnightly, temperature, dissolved oxygen, pH, conductivity and total dissolved solid (TDS) were measured at a deep of 10 cm using multiparameter HANNA HI-9828. Water transparency was measured with Secchi disk. Nutrients such as nitrite and ammonium were determined by cadmium reduction and phenate methods respectively. Zooplankton abundance was also carried out.

Biochemical Analysis

One-hundred randomly chosen fish were sampled from the initial population to determine initial carcass composition. For final carcass composition analysis, twenty fish were randomly selected from each pond. Samples were analyzed according to standard method [52] for dry matter and total ash. Dry matter was evaluated from weight loss after drying in an oven at 105°C for 24 h. Crude protein was determined by the Kjeldhal technic (protein=N×6.25). Total lipid in fish carcass was extracted by chloroform-methanol method [55]. Ash value was evaluated from weight loss after incineration of samples in a muffle furnace for 24 h at 550°C. Total carbohydrates was estimated by subtracting crude protein, lipid and ash values from 100. Gross energy was then calculated on the basis of 23.7 $KJ/g_{protein}$, 39.5 KJ/g_{lipids}, 17.2 $KJ/g_{carbohydrate}$ [24].

Growth parameters

Growth performance, survival and feed utilization were evaluated as below : Survival (S,%)=100×(final count) / (initial count), weight gain (WG,%)=100×[(wf - wi)/wi], specific growth rate (SGR,% day^{-1})=100×[ln (wf) - ln (wi)]t^{-1}, Feed conversion ratio (FCR)=TFI (FB - IB)$^{-1}$, Protein efficiency ratio (PER)=(FB - IB) / DPI, Yield (Y, kg/are)=(FB - IB)/S, Production (P, kg/are/year)=([FB - IB) S^{-1}]×365) t^{-1} ;

where wi and wf=initial and final mean body mass (g); t is the duration of experiment (days); FB is the final biomass per pond (g); IB, the initial biomass per pond (g); TFI, the total food intake (g); DPI the dietary protein intake; S, pond superficies.

Statistical analysis

All data were subjected to a one-way analysis of variance (ANOVA) to test the effect of replacement of fishmeal. Differences between means were determined by Student-Newman-Keuls post hoc tests and were considered to be significant when P-values were <0.05. Before analysis, homogeneity of variance was checked using the Hartley statistical test [56,57] after log-transforming. All analyses were done using the statistical package SPSS version 22.0 for windows (SPSS, Chicago, Illinois, USA).

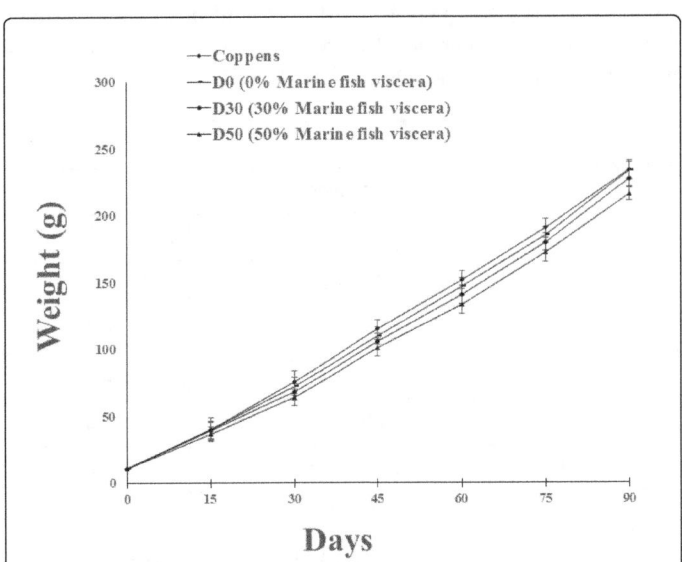

Figure 1: Growth of *Clarias gariepinus* fed marine fish viscera-diets in earthen ponds.

Parameters	Diets			
	Coppens	D0	D30	D50
1.Water quality				
Transparency (cm)	15.6 ± 2.8	16.4 ± 2.4	17.2 ± 1.5	16.8 ± 1.3
Temperature (°C)	29.7 ± 1.8	29.3 ± 1.5	29.6 ± 1.6	28.9 ± 1.3
pH	6.0 ± 0.5	5.7 ± 0.2	5.9 ± 0.4	6.2 ± 0.4
Dissolved oxygen (mg l^{-1})	4.3 ± 2.5	4.2 ± 1.7	4.2 ± 2.7	3.8 ± 2.3
Conductivity (μS cm^{-1})	115.6 ± 12.3	110.1 ± 15.3	109.4 ± 15.2	112.8 ± 14.0
Nitrite (mg l^{-1})	0.01 ± 0.00	0.01 ± 0.00	0.01 ± 0.00	0.01 ± 0.00
Ammonium (mg l^{-1})	0.05 ± 0.0	0.06 ± 0.01	0.05 ± 0.02	0.04 ± 0.01
2. Zooplankton (number l^{-1})	1531.3 ± 26.3[a]	600 ± 27.5[d]	1038.8 ± 608.6[b]	837.5 ± 705[c]
(%) Copepods	49	56.3	49.6	44.5
(%) Rotifers	35.7	39	36.6	39.6
(%) Cladocerans	15.4	4.8	13.8	16

Means values in the same row having different superscript are significantly different (p<0.05). D0, diet containing fish meal; D30, diet containing 30% fish viscera meal; D50, diet with 50% fish viscera meal.

Table 3: Values (Means ± SD) of water quality parameters and zooplankton density (number/l) in different treatments during 90-days trial. D0, diet containing fish meal; D30, diet containing 30% fish viscera meal; D50, diet with 50% fish viscera meal.

Results

Water quality characteristics in all ponds during the 90-days trial are summarized in Table 3. The water transparency ranged from 15.2 ± 2.9 to 16.4 ± 2.7 cm, temperature from 28.91 ± 1.31 to 29.73 ± 1.76°C, pH from 5.76 ± 0.23 to 6.17 ± 0.36, dissolved oxygen from 3.82 ± 2.32 to 4.26 ± 2.48 mg l^{-1}, conductivity from 109.4 ± 15.2 to 115.6 ± 12.3 (μS cm^{-1}), total dissolved solid from 50.28 ± 8.5 to 52.00 ± 6.6 ppm, ammonium 0.04 ± 0.01 to 0.06 ± 0.01 mg l^{-1}, nitrite 0.01 mg l^{-1}. There were no significant differences between all these parameters measured during the experimental period (p<0.05).

Parameters	Diets			
	Coppens	D0	D30	D50
Initial weight (g)	11.3 ± 0.1	11.3 ± 0.1	11.3 ± 0.1	11.3 ± 0.1
Survival (%)	90.0 ± 2.9	91.0 ± 2.5	91.0 ± 2.5	93.0 ± 0.6
Final weight (g)	234.1 ± 5.7[a]	233.3 ± 4.5[a]	230.2 ± 4.1[a]	220.9 ± 3.1[b]
Condition factor	0.96 ± 0.11[a]	0.95 ± 0.09[a]	0.97 ± 0.12[a]	0.91 ± 0.08[b]
Feed intake (g fish^{-1})	248.3 ± 3.2[b]	247.2 ± 3.6[b]	256.9 ± 3.4[a]	247.9 ± 3.4[b]
SGR (% days^{-1})	3.37 ± 0.03[a]	3.36 ± 0.02[a]	3.35 ± 0.02[a]	3.30 ± 0.02[b]
WG (%)	1971.7 ± 50.4[a]	1964.6 ± 39.8[a]	1937.2 ± 36.3[a]	1852.8 ± 27.4[b]
FCR	1.12 ± 0.03[b]	1.11 ± 0.02[b]	1.13 ± 0.02[b]	1.20 ± 0.03[a]
PER	2.08 ± 0.04[a]	2.09 ± 0.03[a]	2.05 ± 0.02[a]	1.93 ± 0.03[b]
Yield (kg are^{-1})	100.3 ± 2.6[a]	101.0 ± 2.0[a]	102.3 ± 2.8[a]	94.3 ± 2.4[b]
Production (kg are-1 year^{-1})	406.6 ± 10.4[a]	409.7 ± 8.3[a]	415.0 ± 7.8[a]	378.3 ± 5.6[b]

Values are means ± SD of three replications. Values in the same row having different superscript are significantly different (p<0.05). D0, diet containing fish meal; D30, diet containing 30% fish viscera meal; D50, diet with 50% fish viscera meal.

Table 4: Means (values ± S.D.) of growth parameters and annual production of *C. gariepinus* fed marine fish viscera-diets in earthen ponds for 90 days. D0, diet containing fish meal; D30, diet containing 30% fish viscera meal; D50, diet with 50% fish viscera meal.

Diets	Moisture	Crude Protein	Crude Lipid	Ash
Initial	10.79 ± 0.23	60.48 ± 0.88	15.13 ± 0.1	13.68 ± 0.23
Coppens	10.77 ± 0.09	61.39 ± 1.53	13.19 ± 0.07[c]	14.53 ± 0.50[a]
D0	10.55 ± 0.28	61.42 ± 0.98	14.32 ± 0.78[bc]	14.53 ± 0.10[a]
D30	10.32 ± 0.15	60.53 ± 0.43	17.48 ± 0.23[a]	13.63 ± 0.32[b]
D50	10.43 ± 0.40	59.56 ± 0.24	18.65 ± 0.57[a]	13.42 ± 0.47[b]

Values with different superscript are significantly different (p<0.05). D0, diet containing fish meal; D30, diet containing 30% fish viscera meal; D50, diet with 50% fish viscera meal.

Table 5: Proximate composition (%) of whole body of *Clarias gariepinus* fed the experimental diets. D0, diet containing fish meal; D30, diet containing 30% fish viscera meal; D50, diet with 50% fish viscera meal.

There were significant differences between absolute density and relative abundance of zooplankton of experimental ponds (p<0.05). The high absolute density value was obtained in ponds receiving coppens diets (1531 ± 26 individual l^{-1}) and the lowest value was observed in ponds receiving D0 (600 ± 27 individual l^{-1}). Furthermore, relative abundances of rotifers and copepods were of highest values in all ponds (Table 3).

Figure 1 presenting changes in fortnight mean weight of experimental fish showed growth overtime. Growth performances, survival rate and feed utilization parameters value were shown in Table 4. Survival rate was similar among dietary treatments. There were no significant differences (p>0.05) in final weight (220.9-234.1 g), weight gain (1937.2-1971.7%), specific growth rate (3.30%/day-3.37%/day),

feed conversion ratio (1.11-1.20), protein efficiency ratio (1.93-2.09) and annual production (378.3-415.0 kg/are/year) of fish fed with coppens, D0 and D30, the best performances being obtained with fish fed coppens diet. Significant differences (p<0.05) were found in feed intake. Fish consumed D30 (256.93 ± 3.41 g/fish) much more than those diets coppens (248.30 ± 3.19 g/fish), D0 (247.17 ± 3.63 g/fish) and D50 (247.89 ± 3.37 g/fish).

The whole-body composition of the experimental fish is presented in Table 5. Dietary replacement of FM by MFV meal did not affect (p>0.05) the moisture and body protein content of *C. gariepinus*. However, lipid deposition was significantly higher in fish fed with MFV-based-diets, whereas ash content significantly decreased (p<0.05).

Discussion

Water quality parameters were not significantly different between treatments and were within the acceptable ranges for *C. gariepinus* rearing [58]. The low zooplankton density observed in certain ponds could be attributed to fish predation [53]. Higher rotifers and copepods abundances reflect the optimal environmental conditions in ponds [51,53].

The present studies evaluate the potential of MFV to replace FM in *C. gariepinus* diet. To our known, there is no reliable study on the use of MFV as a protein source in diets for this fish. However, fisheries wastes recycling for fish farming is an economical and viable option for reducing environmental problems and simultaneously increasing animal protein production [35,59-61]. The results of this study indicated that it is possible to totally replace FM with MFV in African catfish diet without affecting growth performance, thus confirming previous studies findings that animal by-product meals are acceptable protein sources for replacement of fishmeal in catfish diet [2,6,11,19,29,34]. Previous studies have reported beneficial effects [34,62-64] but also adversely effects [65-69] of using MFV as protein sources in diets for several species. According to several studies, the poorest performance of fish fed alternative protein sources are due to the low feeding intake and low digestibility and imbalance of essential amino acids of diet [61,70]. The positive effect obtained in growth performance may be due to the increase protein digestibility and higher long chain polyunsaturated fatty acid content of MFV meal, as mentioned by Giri, et al. [11,32,34,71]. Indeed, according to Nwanna et al. [3,36,42], MFV meal is a good source of polyunsaturated fatty acid such as eicosapentaenoic acid and docosahexaenoic acid, which plays important roles in metabolism. Moreover, they are essential dietary nutrients as demonstrated in red sea bream Pagrus major and yellowtail Seriola quinqueradiata [72,73]. In this study, feed intake was significantly higher with fish fed D30 compared to that of fish fed other diets. This increasing in feed intake with fish fed diet D30 that contained some 15% FM could probably due to the presence of an adequate level of free amino acids in that diet containing lower level of MFV meal, as reported by Kotzamanis et al. [62]. Chotikachinda et al. [70] have reported significant inferior final weight, weight gain and specific growth rate in fish fed coppens diet compared to those of fish fed with the experimental diets, which is contrary with our findings. The weight gain and specific growth rate obtained here are higher than those reported by Sorensen et al. [30,33,74,75] with freshwater catfish Heteropneustes fossilis, Pacific threadfin Polydactylus sexfilis, *C. gariepinus* and Epinephelus fuscoguttatus fed respectively with fermented fish-offal, MFV meal, Agama agama meal and milkfish offal hydrolysate-based-diet. These results showed that MFV meal is better assimilated by fish species, including catfish than other those alternatives sources.

In the present study, whole body composition showed the inverse trend between lipid and ash content. Fish fed with D30 and D50 showed a significantly greater amount of body lipid and lower ash content, in comparison with those of fish fed coppens and D0. This trend was similar to that related in the earlier studies of Kristsanapuntu et al. [76] in red drum, Sciaenops ocellatus, and [32] in catfish *Clarias batrachus*. According to Luchtman et al. [32], the increased body lipid content may be due to increased energy content of diets containing MFV meal, which have a greater fat content (Table 1). The decreasing trend in ash content could be due to the reduction of FM and the inclusion of MFV meal in diet [77,78].

Conclusion

This study showed that up to 30% of marine fish viscera meal could be included in African catfish diet without adverse effects on growth performance and body protein composition. The use of marine fish viscera meal in *Clarias gariepinus* diet could reduce the cost of feed and increase the fish farmer incomes. This might enhance the expansion of the African catfish culture in Africa. We recommended that the further studies were carried out to determinate the optimal stocking density of *C. gariepinus* in order to improve the annual production.

Acknowledgements

The author wishes to thank the Higher Education and Scientific Research Board of Benin and HAAGRIM project for financial support. Thanks are also due to Mrs Appolinaire Effio and Bienvenu Chabi of Faculty of Agriculture Science of Benin for their technical assistance in drying and biochemical analysis. Sincere gratitude was extended to the fishermen from Louho village for their assistance throughout feeding trial.

References

1. Dasuki A, Auta J, Oniye SJ (2013) Effect of stocking density on production of Clarias gariepinus (Tuegels) in floating bamboo cages at Kubanni reservior, Zaria, Nigeria. Bajopas 6: 112-117.

2. Hernandez C, Sarmiento-Pardo J, Gonzalez-Rodriguez B, Parra AI (2004) Replacement of fish meal with co-extruded wet tuna viscera and corn meal in diets for white shrimp (Litopenaeus vannamei). Aquac Res 35: 1153-1157.

3. Nwanna LC (2003) Nutritional value and digestibility of fermented shrimp head waste meal by African catfish Clarias gariepinus. PJN 2: 339-345.

4. Cahu CI, Zambonino-Infante JL (1995) Maturation of the pancreatic and intestinal digestive functions in sea bass (Dicentrarchus labrax): effect of weaning with different protein sources. Fish Physiol Biochem 14: 431-437.

5. Nicholson T, Khademi H, Moghadasian MH (2013) The role of marine n-3 fatty acids in improving cardiovascular health: a review. Food Funct 4: 357-365.

6. Kang KY, Ahn DH, Jung SM, Kim DH, Chun BS (2005) Separation of protein and fatty acids from tuna viscera using supercritical carbon dioxide. Biotechnol Bioprocess Eng 10: 315-321.

7. Saidi SA, Azaza MS, Abdelmouleh A, Pelt JV, Kraiem MM, et al. (2010) The use of tuna industry waste in the practical diets of juvenile Nile tilapia (Oreochromis niloticusl, L.): effect on growth performance, nutrient digestibility and oxidative status. Aquaculture Res 41: 1875-1886.

8. Soltan MA, Hanafy MA, Wafa MIA (2008) An Evaluation of Fermented Silage Made from Fish By-Products as a Feed Ingredient for African Catfish (Clarias grariepinus). Global Veterinaria 2: 80-86.

9. Prasertsan P, Jitbunjerdkul S, Trairatananukoon, Prachumratana T (2001) In: Roussos S, Soccol CR, Pandey A, Augur C (ed.) Production of enzyme and protein hydrolysate from fish processing waste In: New horizons in biotechnology IRD editions, Kluwer Academic Publisher, India.

10. Kaushik SJ (1998) Whole body amino acid composition of European seabass (Dicentrarchus labrax), gilthead seabream (Sparus aurata) and turbot (Psetta maxima) with an estimation of their IAA requirement profiles. Aquat Living Resour 11: 355-358.

11. Giri SS, Sahoo GSK, Mohanty SN (2010) Replacement of by-catch fishmeal with dried chicken viscera meal in extruded feeds: effect on growth, nutrient utilization and carcass composition of catfish Clarias batrachus (Linn.) fingerlings. Aquacult Int 18: 539-544.

12. Tacon AGJ, Metian M (2008) Global overview on the use of fish meal and fish oil in industrially compounded aquafeeds: trends and future prospects. Aquaculture 285: 146-158.

13. Abdelhamid AM (2009) Recent trends in fish culture New Universal Office, Alexandria.

14. Nwanna LC, Balogoun MA, Ajenifuja YF, Enujiugha VN (2004) Replacement of fish meal with chemically preserved shrimp head in the diets of African catfish, Clarias gariepinus. JFAE 2: 79-83.

15. Goda AM, El-Haroun ER, Chowdhury MAK (2007) Effect of totally or partially replacing fish meal by alternative protein sources on growth of African catfish Clarias gariepinus (Burchell, 1822) reared in concrete tanks. Aquaculture Res 38: 279-287.

16. Khan MA, Abidi SF (2011) Dietary arginine requirement of Heteropneustes fossilis fry (Bloch) based on growth, nutrient retention and hematological parameters. Aquaculture Nutr 17: 418-428.

17. Giri SS, Shoo SK, Sahu AK, Mukhopadhyay PK (2000) Growth, feed utilization and carcass composition of catfish Clarias batrachus (Linn.) fingerlings fed on dried fish and chicken viscera incorporated diets. Aquaculture Res 31: 767-771.

18. Nyina-wamwiza L, Wathelet B, Kestemont P (2007) Potential of local agricultural by-products for the rearing of African catfish Clarias gariepinus in Rwanda: effects on growth, feed utilization and body composition. Aquaculture Res 38: 206-214.

19. Huisman EA, Richter CJJ (1987) Reproduction, growth, health control and aquaculture potential of African catfish, Clarias gariepinus (Burchell 1822). Aquaculture 63:1-14.

20. Hutchins-Wiese HL, Picho K, Watkins BA (2014) High-dose eicosapentaenoic acid and docosahexaenoic acid supplementation reduces bone resorption in postmenopausal breast cancer survivors on aromatase inhibitors: A pilot study. Nutr Cancer 66: 68-76.

21. Cahu CI, Zambonino-Infante JL (1995) Effect of the molecular from of dietary nitrogen supply in sea bass larvae: response of pancreatic enzymes and intestinal peptidases. Fish Physiol Biochem 14: 209-214.

22. Abdel-Warith AA, Russel PM, Davies SJ (2001) Inclusion of a commercial poultry by-product meal as a protein replacement of fish meal in practical diets for African catfish Clarias gariepinus (Burchell 1822). Aquaculture Res 32: 296-305.

23. Koshio S (2002) Red sea bream, Pagrus major. In: Webster CD, Lim CE (ed.) Nutrient requirements and feeding of finfish for aquaculture. CABI Publing, New York, NY pp: 51-62.

24. Nyina-Wamwiza L, Wathelet B, Richir J, Rollin X, Kestemont P (2009) Partial or total replacement of fish meal by local agricultural by-products in diets of juvenile African catfish (Clarias gariepinus): growth performance, feed efficiency and digestibility. Aquaculture Nutr 16: 123-129.

25. Cahu CI, Zambonino-Infante JL, Quazuguel P, le Gall MM (1999) Protein hydrolysate vs. Fish meal in compound diets for 10-day old sea bass Dicentrarchus labrax larvae. Aquaculture 171: 109-119.

26. Li P, Wang X, Hardy RW, Gatlin DM III (2004) Nutritional value of fisheries by-catch and by-product meals in the diet of red drum (Sciaenops ocellatus). Aquaculture 236: 485-496.

27. Oliva-Teles A, Luis Cerqueira A, Goncalves P (1999) The utilization of diets containing high levels of fish protein hydrolysate by turbot (Scophthalmus maximus) juveniles. Aquaculture 179: 195-201.

28. Chor WK, Lim LS, Shapawi R (2013) Evaluation of feather meal as a dietary protein source for African Catfish fry, Clarias gariepinus. J Fish Aquat Sci 8: 697-705.

29. Abou Y, Houssou E, Fiogbé ED (2010) Effets d'une couverture d'Azolla sur les performances de croissance et de production de Clarias gariepinus (Burchell) élevé en étangs. IJBCS 4: 201-208.

30. Sorensen LS, Thorlacius-Ussing O, Schmidt EB (2014) Randomized clinical trial of perioperative omega-3 fatty acid supplements in elective colorectal cancer surgery. Br J Surg 101: 33-42.

31. Degani G, Ben-Zvi Y, Levanon D (1989) The effect different protein levels and temperature on feed utilization, growth and body composition of Clarias gariepinus (Burchell, 1822). Aquaculture 76: 293-301.

32. Luchtman DW, Song C (2013) Cognitive enhancement by omega-3 fatty acids from child-hood to old age: findings from animal and clinical studies. Neuropharmacology 64: 550-565.

33. Mamauag REP, Ragaza JA (2016) Growth and feed performance, digestibility and acute stress response of juvenile grouper (Epinephelus fuscoguttatus) fed diets with hydrolysate from milkfish offal. Aquaculture Res 1: 1-10.

34. Hervoy EM, Espe M, Waagbo R, Sandnes K, Ruud M, et al. (2005) Nutrient utilization in Atlantic salmon Salmo salar fed increased levels of fish protein hydrolysate during a period of fast growth. Aquaculture Nutr 11: 301-313.

35. Kotzamanis PY, Alexis MN, Andriopoulou A, Castritsi-Cathariou I, Fotis G (2001) Utilization of waste material resulting from trout processing in gilthead bream (Sparus auratus L.) diets. Aquaculture Res 32: 288-295.

36. Adewolu MA, Ikenweiwe NB, Mulero SM (2010) Evaluation of an Animal Protein Mixture as a Replacement for Fishmeal in Practical Diets for Fingerlings of Clarias gariepinus (Burchell, 1822). ISR J Aquacult-Bamid 62: 237-244.

37. El-Beltagy AE, El-Adawy TA, Rahma EH, El-Bedawey AA (2004) Purification and characterization of an acidic protease from the viscera of bolti fish (Tilapia nilotica). Food Chem 86: 33-39.

38. Taufek NM, Raji AA, Aspani F, Razak SA, Muin H, et al. (2016) The effect of dietary cricket meal (Gryllus bimaculatus) on growth performance, antioxidant enzyme activities, and haematological response of African catfish (Clarias gariepinus). Fish Physiol Biochem 1: 1-13.

39. Ovissipour M, Kenari AMA, Motamedzadegan A, Rasco B, Nazari RM (2011) Optimization of protein recovery during hydrolysis of yellowfin tuna (Thunnus albacares) visceral proteins. J Aquat Food Prod T 20: 148-159.

40. Ramkumar HL, Tuo J, Shen DF (2013) Nutrient supplement with n-3 polyunsaturated fatty acids, lutein and zeaxanthin decrease A2E accumulation and VEGF expression in the retina of Cc12/C×3cr1-deficient mice on Crb 1rd8 background. J Nutr 143: 1129-1135.

41. AOAC (Association of Official Analytical Chemists) (2012) Official Methods of Analysis of AOAC International. (19th edition) Association of Official Analytical Chemists International, Arlington, VA.

42. Sulistiyarto B, Christiana I, Yulintine (2014) Developing production technique of bloodworm (Chironomidae larvae) in floodplain waters for fish feed. International Journal of Fisheries and Aquaculture 6: 39-45.

43. Ovissipour M, Kenari AA, Nazari R, Motamedzadegan A, Rasco B (2012) Tuna viscera protein hydrolysate: nutritive and disease resistance properties for Persian sturgeon (Acipenser persicus L.) larvae. Aquaculture Res 1: 1-11.

44. Masumoto T (2002) Yellowtail, Seriola quinqueradiata. In: Webster CD, Lim CE (ed.) Nutrient requirements and feeding of finfish for aquaculture. CABI Publing, New York, NY pp: 131-146.

45. Viveen WJR, Richter CJJ, Van PGWJ, Janssen JAL, Huisman EA (1985) Manuel pratique de pisciculture du poisson-chat africain (Clarias gariepinus) pp: 128.

46. Tiamiyu LO, Ataguba GA, Jimoh JO (2013) Growth performance of Clarias gariepinus fed different levels of Agama agama meal diets. Pakistan J Nutr 12: 510-515.

47. Harris WS, Dayspring TD, Moran TJ (2013) Omega-3 fatty acids and cardiovascular disease: new developments and applications. Postgrad Med 125: 100-113.

48. Guillaume J, Kaushik S, Bergot P, Métailler R (1999) Nutrition et Alimentation des poissons et crustacés, INRA-IPREMER Editions, Paris.

49. Middleton TF, Ferket PR, Boyd LC, Daniels HV, Gallagher ML (2001) An evaluation of co-extruded poultry silage and culled jewel sweet potatoes as a feed ingredient for hybrid tilapia (Oreochromis niloticus × O. mossambicus). Aquaculture 198: 269-280.

50. Janssen CI, Kiliaan AJ (2013) Long-chain polyunsaturated fatty acids (LCPUFA) from genesis to senescence: The influence of LCPUFA on neural development, aging, and neurodegeneration. Prog Lipid Res 53: 1-17.

51. Wu TH, Bechtel PJ (2008) Salmon by-product storage and oil extraction. Food Chem 111: 868-871.

52. Millamena OM (2002) Replacement of fish meal by animal meals in a practical diet for grow-out culture of grouper Epinephelus coioides. Aquaculture 204: 75-84.

53. Hartley HO (1959) Smallest composite designs for quadratic response surface. Biometric 15: 611-624.

54. Hernandez C, Hardy RW, Contreras-Rojas D, Lopez-Molina B, Gonzalez-Rodriguez B, et al. (2014) Evaluation of Tuna by-product meal as a protein source in feed for juvenile spotted rose snapper Lutjanus guttatus (2014). Aquaculture Nutr 20: 574-582.

55. FAO (2010) Cultured Aquatic Species Information Programme Clarias gariepinus. In: Cultured Aquatic Species Information Programme, Puomogne, V. (Ed.). FAO Fisheries and Aquaculture Department, Rome, Italy.

56. Guerard F, Guimas L, Binet A (2002) Production of tuna waste hydrolysates by a commercial neutral protease preparation. J Mol Catal B-Enzym 11: 1051-1059.

57. Tabinda AB, Butt A (2012) Replacement of fish meal with PBM meal (Chicken intestine) as a protein source in carp (grass carp) fry diet. Pak J Zool 44: 1373-1381.

58. Folch J, Lee M, Sloane-Stanley GH (1957) A simple method for the isolation and purification of total lipids from animal tissues. J Biol Chem 226: 497-509.

59. Russell FD, Bürgin-Maunder CS (2012) Distinguishing health benefits of eicosapentaenoic and docosahexaenoic acids. Mar Drugs 10: 2535-2559.

60. Ju ZY, Forster IP, Deng DF, Dominy WG, Smiley S, Bechtel PJ (2013) Evaluation of skate meal and sablefish viscera meal as fish meal replacement in diets for Pacific threadfin (Polydactylus sexfilis). Aquaculture Res 44: 1438-1446.

61. Hardy RW, Tacon AGJ (2002) Fish meal: historical uses, production trends and future outlook for sustainable supplies. In: Responsible marine aquaculture (eds. R. R. Stickney and J. P. McVey) CABI Publishing. Walling ford. UK.

62. Kotzamanis YP, Gisbert E, Gatesoupe FJ, Zambonino-Infante JL, Cahu CL (2007) Effects of different dietary levels of fish protein hydrolysates on growth. Digestive enzymes, gut microbiota, and resistance to Vibrio anguillarum in European sea bass (Dicentrarchus labrax) larvae. Comp Biochem Physiol A Mol Integr Physiol 147: 205-2014.

63. Minihane AM (2013) Fish oil omega-3 fatty acids and cardio-metabolic health, alone or with statins. Eur J Clin Nutr 67: 536-540.

64. Mondal K, Kaviraj A, Mukhopadhyay PK (2008) Evaluation of fermented fish-offal in the formulated diet of the freshwater catfish Heteropneustes fossilis. Aquaculture Res 39: 1443-1449.

65. Paul BN, Nandi S, Sarkar S, Mukhopahdyay PK (1997) Effects of feeding unconventional animal protein sources on the nitrogen metabolism in rohu Labeo rohita (Hamilton). Isr J Aquacult-Bamid 49: 183-192.

66. Arvanitoyannis SI, Kassaveti A (2008) Fish industry waste: treatments, environmental impacts, current and potential uses. Int J Food Sci Tech 43: 726-745.

67. Moon HYL, Gatlin III M (1994) Effects of dietary animal proteins on growth and body composition of the red drum (Sciaenops ocellatus). Aquaculture 120: 327-340.

68. Zhang DY, Xu XL, Shen XY, Mei Y, Xu HY (2016) Analysis of EPA and DHA in the viscera of marine fish using gas chromatography. Pak J Pharm Sci 29: 497-502.

69. Tacon AGJ (1996) Global trends in aquaculture and aquafeed production. In: International Milling Directory 1996, FAO, Turrest-RAI, Uxbridge pp: 90-108.

70. Chotikachinda R, Tantikitti C, Benjakul S, Rustad T, Kumarnsit E (2013) Production of protein hydrolysates from skipjack tuna (Katsuwonus pelamis) viscera as feeding attractants for Asian seabass (Lates calcarifer). Aquaculture Nutr 19: 773-784.

71. Cervera MAR, Venegas E, Bueno RPR, Medina MDS, Guerrero JLG (2015) Docosahexaenoic acid purification from fish processing industry by-products. Eur J Lipid Sci Tech 117: 724-729.

72. Glencross BD, Booth M, Allan GL (2007) A feed is only as good as its ingredients a review of ingredient evaluation strategies for aquaculture feeds. Aquaculture Nutr 13: 17-34.

73. Valle BCS, Dantas JR EM, Silva JFX, Bezerra RS, Correia ES, et al. (2015) Replacement of fishmeal by fish protein hydrolysate and biofloc in the diets of Litopenaeus vannamei postlarvae. Aquaculture Nutr 21: 105-112.

74. Cruz-Suarez LA, Tapia-Salazar M, Villarreal-Cavazos D, Beltran-Rocha J, Nieto-Lopez MG, et al. (2009) Apparent dry matter, energy, protein and amino acid digestibility of four soybean ingredients in white shrimp Litopenaeus vannamei juveniles. Aquaculture 292: 87-94.

75. Osman AGM, Wuertz S, Mekkawy IAA, Exner H, Kirschbaum F (2007) Embryo-toxic effects of lead nitrate of the African catfish Clarias gariepinus (Burchell, 1822). 23: 48-58.

76. Kristsanapuntu S, Chaitanawisuti N (2015) Replacement of fishmeal by poultry by-product meal in formulated diets for growing hatchery-reared juvenile Spotted Babylon (Babylonia areolata). J Aquac Res Development 6: 4-15.

77. Yang Y, Xie S, Cui Y, Lei W, Zhu X, et al. (2004) Effect of replacement of dietary fishmeal bymeat and bone meal and poultry by-product meal on growth and feed utilization of gibel carp, Carassius auratus gibelio. Aquaculture Nutr 10: 289-294.

78. Leal ALG, de Gastro FP, Lima JPV, Correia EDS, Bezerra RDS (2009) Use of shrimp protein hydrolysate in Nile tilapia (Oreochromis niloticus, L.) feeds. Aquacult Int 18: 635-646.

Effect of Natural and Hydroponic Barley Plant and Sprout on the Common Carp (*Cyprinus Carpio*) Growth Performances

Hazem S Abedalhammed[1*], Nasreen M Abdulrahman[2] and Haitham L Sadik[1]

[1]*Department of Animal Production, College of Agriculture, University of Al- Anbar, Iraq*

[2]*Department of Animal Production, Faculty of Agricultural sciences, University of Sulaimani, Iraq*

Corresponding author: Hazem S Abedalhammed, College of Agriculture, University of Al-Anbar, Ramadi, Al Anbar Iraq, E-mail: nasreenmar12@gmail.com

Abstract

This study was designed to investigate the effect of natural, sprout powder and hydroponic planting of Barley on some growth parameters of common carp *Cyprinus carpio* L. The trail was conducted for 56 days and for this purpose 175 fingerlings common carp, mean initial weight of 34.71 g were acclimated to laboratory conditions and fed with control pellets (30%crude protein) prior to the feeding trials for 21 days. Seven experimental diets were used and the control as 0% (T1), Natural planting 2.5 (T2) and 5 gm/kg diet (T3), Hydroponic Planting 2.5 (T4) and 5 gm/kg diet (T5), Barley sprout powder 2.5 (T6) and 5 gm/kg diet (T7). According to the results no significant differences observed in mean initial weight this was done in way to avoid differences attributed to fish initial weight, T4 (Barley sprout powder 2.5 g/kg diet) was significantly higher in each daily and relative growth rate, but the specific growth rate both T4 and T7 were significantly higher than other treatments. No significant differences observed in Food Conversion Ratio and Protein Efficiency Ratio but T4 (Barley sprout powder 2.5 g/kg diet), T6 (Natural planting 2.5 gm/kg diet) and T7 (Natural planting 5 gm/kg diet) differ significantly in Food Efficiency Ratio. As general conclusion the adding of germinated barley enhance common carp performance in any way of germination.

Keywords: Natural planting; Barley sprout powder; Hydroponic germination; Barley; Common carp; Growth performance

Introduction

World aquaculture has grown tremendously during the last years becoming an economically important industry. Today it is the fastest growing food-producing sector in the world with the greatest potential to meet the growing demand for aquatic food [1]. Globally, aquaculture is expanding into new directions, intensifying and diversifying. A persistent goal of global aquaculture is to maximize the efficiency of production to optimize profitability.

Barley is used for a wide range of traditional and novel end-uses. In most countries, the major portion of barley is fed to animals, particularly cattle and pigs. Human food uses of barley are more limited, although recent trends in the use of barley varieties, high in dietary fiber, have been identified. A significant high-value use is to produce malt as a raw material for the brewing industries, including beer and whiskey. An arabinoxylan-rich germinated barley product has been reported by Kanauchi [2] to induce the proliferation of bifidobacteria in the human intestine. However, as for all known and emerging prebiotics, convincing evidence of a consistent clinical benefit in the treatment of IBD remains to be demonstrated in large, randomised, double-blind, placebo-controlled studies [3,4].

Maltose is produced by hydrolysis of starch using the enzyme β-amylase. It occurs only rarely in nature and only in plants as a result of partial hydrolysis of starch. Maltose is produced during malting of grains, especially barley, and commercially by the specific enzyme catalyzed hydrolysis of starch using β-amylase from Bacillus species, although the β-amylases from barley seed, soybeans, and sweet potatoes may be used. Maltose is used sparingly as a mild sweetener for foods. Proteins of major cereals and legumes are often deficient in at least one of the essential amino acids. While proteins of cereals, such as rice, wheat, barley, and maize are very low in lysine and rich in methionine, those of legumes and oilseeds are deficient in methionine and rich or adequate in lysine. Oats, barley, and rye are examples of cereals that contain a relatively high percent (5%-25% of total carbohydrates) of non-starch polysaccharides in the flour. The pentosan fraction of cereals is a complex mixture of branched polysaccharides with an arabinoxylan backbone containing small amounts of glucose and ferulic acid [5].

So the objective of the study compare natural planting, Barley sprout powder and hydroponic germination of Barley of common carp growth performance in weight gain, Daily growth rate, Specific growth rate, Relative growth rate, Feed conversion ratio (FCR), Food efficiency ratio (FER) and Protein efficiency ratio (PER).

Materials and Methods

Experimental diet

Seven practical diets were formulated based on the proximate composition of the feed ingredients. Diet 1 (Control diet free of any barley), diets 2, 3, 4, 5, 6 and 7 contained 2.5 and 5 gm/kg diet of each of natural, Barley sprout powder and hydroponic planting respectively on an equivalent protein basis. Composition and proximate analysis of different experimental diet diets were shown in Table 1 and the chemical composition of the different diet by NRC et al. [6,7] explained in Table 2.

Chemical Composition	
Crude protein%	27.351
Crude fat%	2.584
Crude fiber%	6.155
Energy kgal/kg	2235.2
Ash%	87.61

Table 1: Chemical composition of fish diets used in the experiment.

Ingredients	Crude	Crude	Dry	Crude	Energy	% in diet
	Protein%	Fat%	Materials%	Fiber%	KG/kg	
Animal protein concentrate	40	5	92.9	2.2	2107	10
Yellow corn	8.5	3.6	89	2.2	3350	15
Soybean meal	44	1.1	89	7	2230	40
Barely	11	1.9	89	5.5	2640	15
Wheat bran	15.7	4	89	11	1300	18
Premix	---	---	---	---	---	2

Table 2: Chemical composition of the different diet by NRC et al. [5,6].

Animal concentrate commercial Brocon-5 Special W 40% imported by Wafi. B.V. Holland.

Premix: vitamins: Vit A: 6000 UI; Vit D3: 1000 UI; Vit E: 60 UI; Vit K: 12 UI; Vit B1: 24 mg/kg: Vit B2: 24 mg/kg; Pantothenic acid: 60 mg/kg; Niacin: 120 mg/kg; Vit B6: 24 mg/kg; Biotin: 0.24 mg/kg; Folic acid: 6 mg/kg; Choline chloride: 540 mg/kg; Vit B12: 0.024 mg/kg. Minerals include (mg/kg): Fe: 50; Cu: 3; Mn: 20; Zn: 50; I: 0.1; Co: 0.01; Se: 0.1.

Fish and feeding regime

Common carp *Cyprinus carpio* fingerlings with an average weight 34.71 g were brought from local fish farm located in Daqoq/HaftaGar Middle of Iraq were randomly allocated on the aquaria (7/aquarium). Each treatment was represented in three aquariums (3 replicates). A feeding regime of 3% body weight per day was employed throughout the experiment. The amount of food was calculated and readjusted weekly according to change in the body weight and distributed in three equal portions for 56 days.

Experimental diets

The different feeding combinations (seven formulas of isoenergy diets, (Table 1) were prepared as follows:

The control as T1 with 0% barley, Natural planting as T2 with 2.5 gm/kg diet, T3 with 5 gm/kg diet, Hydroponic Planting as T4 with 2.5 gm/kg diet, T5 with 5 gm/kg diet 5), Barley sprout powder as T6 with 2.5 gm/kg diet, and T7 with and 5 gm/kg diet.

Experimental system

The experimental facility consisted of 21 Aquaria (60 litters each). Each aquarium was supplied with aerated and dechlorinated tap water, which was stored in tanks for 24 hours and aerated by air pump (Model-Rina 301) during the experimental period. The water level was maintained to a fixed level by the addition of new well-aerated fresh water.

Growth parameters

The individual body weight (g) and total body length (cm) for all fish per treatment were measured weekly. The feed consumption of each treatment was recorded and readjusted according to the obtained biomass at every treatment weekly. The average body weight gain (WG) as (g/fish) was estimated according to the following equation:

Body weight gain (g/fish)=Mean of weight (g) at the end of the experimental period-weight (g) at the beginning of the experimental period

Daily weight gain (DWG)=Gain/experimental period

Relative weight gain (RWG%)=Gain/initial weight × 100

Specific growth rate (SGR)=(In W1-In W0)/T) × 100

W1: final weight W0: initial weight T: time between W1 and W0

Feed conversion ratio (FCR)=Total feed fed (g/fish)/total wet weight gain (g/fish)

Protein efficiency ratio (PER)=Total wet weight gain (g/fish)/ amount of protein fed (g/fish).

Statistical Analysis of Data

Statistical analysis was performed using the Analysis of variance (ANOVA) two-way classification and Duncan's multiple Range Test, to determine differences between treatments means at significance rate of $P < 0.05$. The standard errors of treatment means were also estimated. All statistics were carried out using Statistical Analysis System (SAS) program [8].

Results and Discussion

Common carp is fresh water fish that is distinct to the Northern Hemisphere. This species requires an optimal temperature range between (20°C to 28°C) according to [9]. The activity of the carps is affected by low water temperatures which minimize their moving and feeding activities [10]. The temperature of water demonstrated in the present study was approximately 20°C throughout the entire experimental period. Over the entire period of the experiment, no mortalities among the fish have been observed. Wang et al. [11,12] obtained a similar survival rate with *C. carpio* over 56 days of a feeding trial. This result reflects healthiness of the experiment fish.

According to the results in Table (3) no significant differences observed in mean initial weight this was done in way to avoid differences attributed to fish initial weight, T4 (Barley sprout powder 2.5 g/kg diet) was significantly higher in each daily and relative growth rate, but the specific growth rate both T4 and T7 were significantly higher than other treatments.

Treatment	Mean initial weight (gm)	weight gain (gm)	Daily growth rate (gm/fish/day)	Specific growth rate (/day)	Relative growth rate (gm/day%)
T1 (control)	34.775 ± 0.025 a	10.870 ± 0.100 c	0.194 ± 0.002 c	0.002 ± 0.000 b	31.255 ± 0.265 c
T2 Hydroponic Planting 2.5 g/kg diet	34.675 ± 0.050 a	10.920 ± 0.760 c	0.195 ± 0.012 c	0.002 ± 0.000 ab	31.395 ± 2.455 c
T3 Hydroponic Planting 5 g/kg diet	34.640 ± 0.020 a	11.265 ± 0.065 c	0.203 ± 0.001 bc	0.002 ± 0.000 ab	32.745 ± 0.055 bc
T4 Barley sprout powder 2.5 g/kg diet	34.670 ± 0.005 a	13.175 ± 0.010 a	0.231 ± 0.001 a	0.002 ± 0.000 a	37.355 ± 0.020 a
T5 Barley sprout powder 5 g/kg diet	34.730 ± 0.030 a	11.185 ± 0.345 c	0.201 ± 0.010 bc	0.002 ± 0.000 b	32.160 ± 1.675 bc
T6 Natural planting 2.5 gm/kg diet	34.700 ± 0.060 a	11.760 ± 0.165 bc	0.212 ± 0.001 abc	0.002 ± 0.000 ab	33.675 ± 0.490 abc
T7 Natural planting 5 gm/kg diet	34.765 ± 0.075 a	12.460 ± 0.050 ab	0.223 ± 0.006 ab	0.002 ± 0.000 a	35.840 ± 0.495 ab

Table 3: Effect of Natural planting, Barley sprout powder and hydroponic germination of Barley in common carp growth parameters of common carp reared in indoor aquaria. Mean values with different superscripts within a column differ significantly (P ≤ 0.05).

No significant differences observed in Food Conversion Ratio and Protein Efficiency Ratio as shown in table (4), T4 (Barley sprout powder 2.5 g/kg diet), T6 (Natural planting 2.5 gm/kg diet) and T7 (Natural planting 5 gm/kg diet) differ significantly in Food Efficiency Ratio.

Treatment	Food Conversion Ratio	Food Efficiency Ratio	Protein Efficiency Ratio
T1 (control)	3.085 ± 0.035 a	32.055 ± 0.005 b	1.180 ± 0.000 d
T2 (Hydroponic Planting 2.5 g/kg diet)	3.790 ± 0.510 a	32.700 ± 0.230 b	1.225 ± 0.000 c
T3 (Hydroponic Planting 5 g/kg diet)	3.530 ± 0.530 a	33.700 ± 0.020 b	1.230 ± 0.000 c
T4 (Barley sprout powder 2.5 g/kg diet)	2.595 ± 0.005 a	38.965 ± 0.030 a	1.410 ± 0.005 a
T5 (Barley sprout powder 5 g/kg diet)	3.025 ± 0.115 a	33.355 ± 1.535 b	1.240 ± 0.010 c
T6 (Natural planting 2.5 gm/kg diet)	2.590 ± 0.025 a	37.030 ± 0.345 a	1.340 ± 0.010 b
T7 (Natural planting 5 gm/kg diet)	2.695 ± 0.620 a	37.080 ± 0.100 a	1.355 ± 0.025 b

Table 4: Effect of Natural planting, Barley sprout powder and hydroponic germination of Barley in common carp feed utilization of common carp reared in indoor aquaria. Mean values with different superscripts within a column differ significantly (P ≤ 0.05).

Little information is known about the impact of barley in different germination way on feed utilization parameters in common carp. The findings of the current study may help to explain the improved feed utilization performance in this species. The most notable outcome of prebiotic supplementation, in general, is changes brought about to the intestine, both morphologically and microbiologically. Changes to the morphology of the intestine may be attributed to the production of short-chain fatty acids through the microbial fermentation of prebiotic substances.

The characteristics of proteinase enzyme are consistent with its being the predominant proteinase synthesized during barley germination. Such endoproteinases are important because they are responsible for transforming the grain endosperm storage proteins into soluble proteins, amino acids and peptides that can be metabolized and utilized by the growing plantlet. Commercially, this transformation is important because, during the malting of barley for brewing, the insoluble storage proteins must be reduced to low molecular mass nitrogenous compounds that can be utilized by brewing yeasts and this may be the reason of significant differences of the Barley sprout powder [13].

The entry of sugar from the endosperm is evidently very slow, at any rate during the greater part of this period, for it exercises very little influence on the carbohydrate metabolism. This is shown by a comparison between the changes which take place in embryos germinated on their endosperms and those which are excised and grown on sand moistened with a culture solution containing no carbohydrate The production of fresh cell wall material indicated by the increase in the insoluble fractions is as great in the excised embryos as in the germinating grains, both show a slight accumulation of maltose and in neither is there any marked accumulation of hexose. In both cases there is a rapid loss of sucrose and raffinose, which also contains a sucrose unit, and only here is there any clear indication of the entry of sugar derived from the reserves in the endosperm. Sucrose almost entirely disappears from the excised embryos, whereas those germinating on their endosperms still contain approximately 33% of the amount present after 2 hr. It will be seen that during the first 24 hr. the greater part of the sugar in the embryo has either been lost in respiration or used in the production of insoluble material [14].

Fish feed on 2.5 g (FOS) had better growth than those feed on 5 g (FOS) in the study of [15] Improved growth performance is likely to be brought about by elevated digestive enzyme activities, possible improvements of intestine morphology or via prebiotic fermentation by endogenous gut microbes to produce SCFAs as stated by Dimitroglou et al. [16]. The results of clearly demonstrate the association of improved growth and performance, gut health, immune status and resistance to disease in fish fed Bio-Mos.

As general conclusion the adding of germinated barley enhance common carp performance in any way of germination.

References

1. FAO (2014) The State of World Fisheries and Aquaculture. Opportunities and challenges. Food And Agriculture Organization Of The United Nations, Rome pp: 243.

2. Kanauchi O, Fujiyama Y, Mitsuyama K, Bamba T (1990) Increased growth of Bifidobacterium and Eubacterium by germinated barley foodstuff, accompanied by enhanced butyrate production in healthy volunteers. Int J Mol Med 3: 175-179.

3. Sweetman J, Davies S (2006) In: Lyons TP, Jackues K (Eds.) Improving growth performance and health status of aquaculture stocks in Europe through the use of Bio-Mos®. Nutritional Biotechnology in the Feed and Food Industries. Nottingham University Press, Nottingham, UK, pp: 445-452.

4. Gibson GR, Rastall RA (2006) Prebiotics: Development & Application. John Wiley & Sons Ltd, the Atrium, Southern Gate, Chichester, West Sussex PO19 8SQ, England.

5. Damosaran S, Parkin KL, Fennema OR (2008) Fennema's Food Chemistry, CRC Press, Boca Raton, FL. 4th Ed pp: 179-260.

6. NRC (1993) Nutrient Requirements of Fish. National Acad. Press, Washington, DC, pp: 114.

7. NRC (1994) National Academy of Science, Nutrient requirement of poultry (9th ed) Washington USA, pp: 157.

8. SAS (2004) SAS/STAT® 9.1 User's Guide, SAS Institute Inc, Cary, NC, USA.

9. Horvath L, Tamas G, Seagrave C (1992) Carp and Pond Fish Culture. Fishing News Books, Oxford pp: 158.

10. Bauer C and Schlott G (2006) Reaction of common carp (Cyprinus carpio L) to oxygen deficiency in winter as an example for the suitability of radio telemetry for monitoring the reaction of fish to stress factors in pond aquaculture. Aquaculture Res 37: 248-254.

11. Wang YB, Xu ZR (2006) Effect of probiotics for common carp (Cyprinus carpio) based on growth performance and digestive enzyme activities. Anim Feed Sci Tech 127: 283-292.

12. Al-Jammoor KMS (2012) Evaluating the effects of dietary immunostimulants on growth performance, survival, immune response and digestive enzymes activity of Koi (Cyprinus carpio Linnaeus 1758), University Perth.

13. Jones, Berne L, Poulle M (1990) A Proteinase from Germinated Barley II Hydrolytic Specificity of a 30 Kilodalton Cysteine Proteinase From Green Malt. Plant Physiol 94: 1062-1070.

14. Mahious AS, Ollevier F (2005) Probiotics and prebiotics in aquaculture: review", In: Urmia 1: 17-26.

15. Ahmed VM (2014) Comparative effects of probiotic (Saccharomyces cerevisiae), prebiotic (fructooligosaccharide fos) and their combination on growth performance and some blood indices in young common carp (cyprinus carpio). Faculty of Agricultural Sciences, University of Sulaimani.

16. Dimitroglou A, Merrifield DL, Spring P, Sweetman JMR, Davies SJ (2010) "Effects of mannan oligosaccharide (MOS) supplementation on growth performance, feed utilization, intestinal histology and gut microbiota of gilthead sea bream (Sparus aurata)", Aquaculture 300: 182e8.

Benthic Fish Fauna and Physicochemical Parameters of Otamiri River, Imo State, Nigeria

Ikenna Kelvin Obiyor, Christopher Didigwu Nwani, *Gregory Ejikeme Odo, Josephine Chinenye Madu, Doris Ulumma Ndudim and Ifeanyi Oscar Ndimkaoha Aguzie*

Department of Zoology and Environmental Biology, University of Nigeria, Nsukka, Enugu State, Nigeria

*Corresponding author:** Christopher Didigwu Nwani, Department of Zoology and Environmental Biology, University of Nigeria, Nsukka, Enugu State, Nigeria,
E-mail: chris.nwani@unn.edu.ng

Abstract

The study on the fish fauna and physicochemical characteristics of Otamiri River was carried out for six months (June to August and October to December 2015). Samples were collected monthly from three sampling stations along the river. Eckman grab, scoop net, line and hook, cast net, traps and dugout canoe were used to collect the fish samples. Samples were collected from three stations, station 1 (dumpsite in the river), station 2 (dredging section) and station 3 (vegetable farming section). A total of 129 fishes belonging to 5 species, Synodontis budgetti, S. soloni, Chrysicthys nigrodigitatus, Clarias gariepinus and Papyrocranus afer were collected. Station 3 had the highest species composition (n=4) and fish abundance, 108 (83.7%), while station 1 had the least species composition (n=2) and the least fish abundance, 10 (7.8%). Temperature variation from June to December was wide at station 1 (25-28°C), unlike station 2 (27-28°C) and station 3 (28-29°C). A similar trend was replicated by Dissolved Oxygen (DO). Significant correlation of S. budgetti abundance and temperature (r=0.696, p<0.05), depth of river (r=-0.615, p<0.01) and turbidity (r=0.595, p<0.01) was observed. Similarly, a significant correlation of C. nigrodigitatus abundance and temperature (r=0.473, p<0.05), C. gariepinus abundance and depth (r=-0.481, p<0.05), P. afer abundance and temperature (r=0.530, p<0.05) was observed. Fish species abundance and composition in Otamiri River was significantly affected by anthropogenic activities.

Keywords: Otamiri river; Fish; Benthic; Abundance; Composition; Physicochemical

Introduction

The benthic macro fauna are those organisms that live at the bottom of a water body and are used to detect changes in the natural environment [1,2]. Fishes are thus referred to as benthic if they occur within or around the bottom of water bodies. The use of macro-invertebrates and vertebrates diversity for bio-assessment provides a simpler approach compared to other environmental quality assessment procedures. This is because, macro-invertebrates and vertebrates can be sampled quantitatively and the relative sensitivity or tolerance of some of them to contamination is known [3]. Species vary in their degree of tolerance with the result that under polluted conditions, a reduction in species diversity is the most obvious effect [4-6].

The lotic and lentic inland waters, as well as brackish and marine waters in the tropics are habitats for a variety of vertebrate fauna. Work on the vertebrate fauna in the tropics has shown that the quantitative collection of key species from natural aquatic habitat or that modified by man can provide a means of estimating various ecological parameters, such as richness or evenness in diversity [7]. The occurrence and distribution of macro-invertebrates and vertebrates are governed mostly by the physical and chemical quality of water and immediate substrate of occupation. Water quality parameters such as temperature, pH, dissolved oxygen and nutrients have considerable effects on the life of aquatic organisms. They affect species composition and distribution, diversity, stability, production and physiological conditions of the organisms [6].

Humans are considered the principal driver of change on the earth's surface. Such impacts shape the earth in disparate ways which may be small and subtle or big and catastrophic [8]. These effects may result in multiple consequences felt by plants, animals and even humans alike. One major natural component of the earth is the aquatic environment which is home for a vast array of organisms from those with a planktonic existence through pelagic organisms to benthic species. Human activities also interfere with this environment. Freshwater bodies contain diverse habitats which support myriads of species of both plants and animals and are important sources of water for human activities.

In some instances freshwaters have been dammed to provide portable water for urban settlements and the Otamiri River is one of such freshwater bodies which are used for domestic purposes by the generality of Nekede and Ihiagwa communities. A lot of activities that may be detrimental to aquatic and human life take place at Otamiri River yet there is little to no existing study to assess the consequences. The water is intensively and perennially dredged for sand at some sections; at some others people dump domestic and industrial wastes into the river without concern for it consequences on human and aquatic lives. Also a bridge was recently constructed on a section of the river. In view of this, this study therefore investigated the physicochemical parameters and benthic fish fauna diversity of Otamiri River as markers of pollution status of the water.

Materials and Methods

Study area

The Otamiri River is one of the main rivers in Imo State, Nigeria and located on latitude 7° 06 E and longitude 5° 30′ N and at an elevation of 152 meters above sea level. The river takes its name from Ota Miri, a deity which owns all the waters that are called by its name, and who is often the dominating god of Mbari houses [9]. The river runs south from Egbu past Owerri, Nekede, Ihiagwa and through Ozuzu Etche, in Rivers State, from where it flows to the Atlantic Ocean [10]. The length of the river from its source to its confluence at Emeabiam with the Uramiriukwa River is 30 kilometres (19 mi) [10]. The vegetation of the sampled area is rain forest with the watershed mostly covered by depleted rain forest vegetation. Conversion of the tropical rainforest to grassland with slashes and bush burning practices constantly degrades the soil quality. Three sampling stations were used. Station 1 was located at Owerri urban along Nekede road, where the state's refuse dump site is located and no sand mining activities occurs, while Station 2 was at No 8 Bus stop Umugwueze, Nekede, where dredging activities take place and Station 3 was at Umuezerokam village which is the home of vegetable farming (Figure 1).

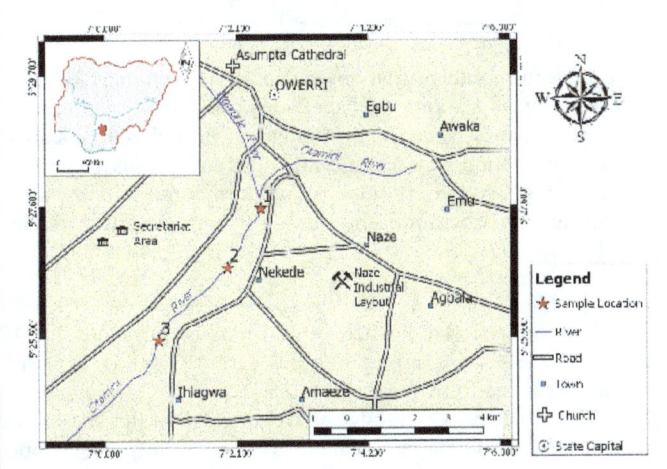

Figure 1: Map of Otamiri River showing the sampled stations. Station 1: Owerri Municipal, Station 2: No 8 Bus Stop Nekede and Station 3: Umuezerokam Village, Nekede.

Sample collection

Water samples were collected from the three stations (Stations 1, 2 & 3) within the river for six months (June-August and October-December, 2015). The samples were collected in triplicates from each sampling stations. Dugout canoes with paddles were used during sampling within the river. At the sampling stations, 250 ml Dissolve Oxygen containers to store water samples were rinsed several times with the river water. Water samples for physicochemical analysis were collected at 30 cm depth in bottles of 1000 ml capacity. Water sample was fixed separately with 2 ml each of Winkler's solution A and B (manganese sulphate, alkaliodide and conc. sulphric acid [11].

Benthic vertebrate samples were randomly collected from each sampling station with cast net of 0.5 mesh size, local trap (made from raffia palm) and hook and line, monthly for six months. Dugout canoes

with paddles were used during the sampling within the river. The samples were kept in a container and labeled properly. Samples were stored in 30-40% Ethanol [12,13] and were immediately transported to the laboratory of the Department of Zoology and Environmental Biology, University of Nigeria, Nsukka for identification. Benthic vertebrates were identified by means of identification keys provided by Bouchard [14] and Dailey [15].

Determination of physicochemical parameters

The pH of water samples were measured in the laboratory using the Hanna pH meter (Hi-1922 model; Hanna Instruments Inc., USA). Turbidity of water determination was conducted in-situ at the sampling stations using a 20 cm diameter Secchi disc attached to a calibrated rope. Depth of disappearance and re-appearance was recorded and average calculated. Dissolved oxygen was estimated using the Azide modified techniques of Winkler's method [16]. Water temperature of each station was measured in degree Celsius (°C) using laboratory thermometer calibrated from 0-100°C. On each sampling day, the thermometer was dipped in water for about 5 minutes at a depth of 5 cm. Temperature was recorded as soon as the reading stabilized. The depth of the water at each sampling station was measured using a graduated rope with sinker. The rope was immersed in the water until the sinker touched the substratum. The reading was taken and recorded in meters.

Statistical analysis

Data obtained were analyzed using Statistical Package for Social Sciences (SPSS) version 20 (IBM Corp., Armonk, New York, USA). Values were expressed as mean ± SE. Two-way Analysis of Variance (ANOVA) was used to determine significant differences among means of water quality parameters across stations and months. Species diversity was calculated using Shannon wiener's index (H) using the following equation;

$$H = \sum_{1}^{N} p_i \ln p_i$$

Where: H=Species diversity index

Pi=the proportional abundance of species

n=the number of species

In=Natural log

Σ=Sum of the calculation

Species dominance was also calculated using Simpson's dominance index (D) using the formula;

$$D = \frac{1}{\sum_{1}^{N} p_i^2}$$

Where: D=Species dominance index

Pi=the proportional number of species

N=the number of species

Σ=Sum of the calculation

Pearson correlation (r) was used to ascertain linear relationships between physicochemical characteristics of the river and benthic fish species diversity.

Results

Physicochemical parameters of Otamiri River

Temperature: The results of the temperature values of Otamiri River are presented in Figure 2. At station 1, temperature increased progressively from June and July till November and dropped slightly in December. The months of June, July and August had similar and least temperature values. At this station, temperature for June, July and August were significantly less than temperature for October to December ($p<0.05$). At station 2, a gradual increase in temperature from June to December was observed, though slight declines occurred in August and November. At this station, the peak temperature (29.03°C) for the duration of the study was recorded in December; it was also significantly higher than temperatures recorded in the other months ($p<0.05$). At station 3, unlike the other stations temperature was peak in June, dropped sharply in July and August, and rose gradually from October to similar levels as June by November and December. At this station, the least and highest temperature values of 27.67 ± 0.33°C and 29.30 ± 0.67°C were recorded in the months of August and June respectively. Temperature values of June, November and December were significantly higher than July and August values ($p<0.05$). Between stations, temperature of the river was significantly higher at station 3 than station 1 for all the month except October.

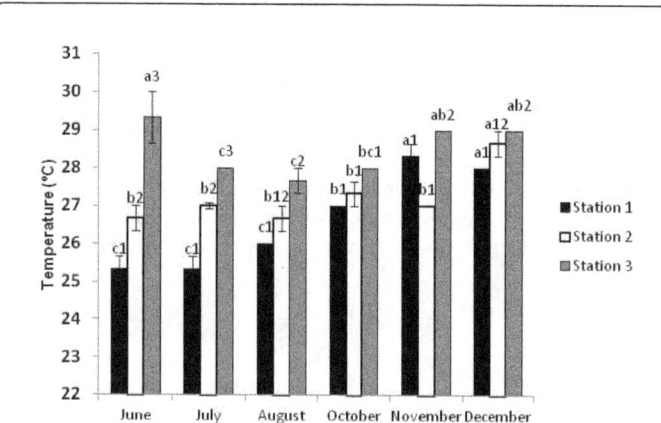

Figure 2: Mean-monthly variations in temperature (°C) of Otamiri River in the three sampled stations. Alphabet label compares variation in a given station between the months; months with different alphabet label for each station were significantly different ($p<0.05$). Number label compare between stations for each month; stations with different number label for each month were significantly different ($p<0.05$).

Depth: There were monthly variations in depth across the stations throughout the study period (Figure 3). At station 1, the highest depths were recorded in the months of June (4.21 ± 0.27 m) and July (4.19 ± 0.05 m) after which the water depth declined sharply from August attaining the least value for the duration of the study in December (0.46 ± 0.09 m). Still at station 1, depth of the river was significantly higher in June and July than the other months; August value was also significantly higher than October to December values ($p<0.05$). There was no significant difference in depth between October, November and December ($P>0.05$). At station 2, the river depth followed a similar trend observed at station 1. But the highest depth was recorded in June only (4.88 ± 0.64 m); thereafter, the depth declined sharply and

significantly in July (1.27 ± 0.18 m) and remained this low till December ($p<0.05$). At station 3, only August and October depths were significantly lower than the other months ($p<0.05$). Between the three stations, in July depth at station 2 was significantly less than those of station 1 and 3; in October station 3 had a significantly higher depth than the other two stations; in November depth between the stations were significantly different in the order station 1<station 2<station 3; and in December station 1 depth was significantly less than the other stations' ($p<0.05$).

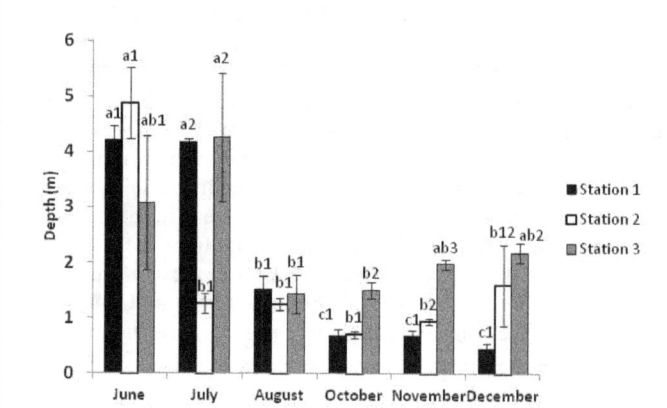

Figure 3: Mean-monthly variations in depth (m) of Otamiri River in the three sampled stations. Alphabet label compares variation in a given station between the months; months with different alphabet label for each station were significantly different ($p<0.05$). Number label compare between stations for each month; stations with different number label for each month were significantly different ($p<0.05$).

Dissolved Oxygen (DO): There were monthly variations in DO values across the stations during the study period (Table 1). In station 1, the values ranged from the least (4.57 ± 0.33 mgL^{-1}) in June to the highest (6.50 ± 0.06 mgL^{-1}) in July and October. DO at this station was significantly different between the months except for July and October. At station 2, the least DO (5.77 ± 0.33 mgL^{-1}) was recorded in the month of June and was significantly different from other months.

	Station 1	Station 2	Station 3
June	$4.57 + 0.33^{e1}$	$5.77 + 0.33^{d2}$	$5.67 + 0.33^{c2}$
July	$6.50 + 0.06^{a2}$	$6.47 + 0.03^{b2}$	$5.87 + 0.03^{c1}$
August	$5.70 + 0.00^{d1}$	$6.00 + 0.00^{c2}$	$5.80 + 0.10^{c1}$
October	$6.50 + 0.06^{a2}$	$6.20 + 0.06^{a1}$	$6.70 + 0.06^{a3}$
November	$6.00 + 0.00^{c1}$	$6.47 + 0.03^{b3}$	$6.37 + 0.03^{b2}$
December	$6.20 + 0.06^{b1}$	$6.00 + 0.00^{c1}$	$6.20 + 0.12^{b1}$

Table 1: Mean-monthly variations in Dissolved Oxygen (mgL^{-1}) of Otamiri River in the three sampled stations. Values with different alphabets superscript in a column are significantly different ($p<0.05$). Values with different numbers superscript in rows are significantly different ($p<0.05$).

In station 3, the lowest DO value of 5.07 ± 0.03 mgL^{-1} was recorded in the month of June which was not significantly different from July and August DO values. Still at station 3, the DO for October to December were significantly higher than DO for June to August (p<0.05). The DO in the month of October was significantly higher than other months in this station (P<0.05). Across the stations, the variations in DO were dependent on months but there was no consistent trend. In the month of June, the DO values (5.87 ± 0.03 mgL^{-1}) of stations 2 and 3 were similar and significantly higher than the value recorded at station 1 (P<0.05). In August, DO for stations 1 and 3 were similar but significantly less than the value for station 2. In October, the DO of station 3 was significantly higher than the values for stations 1 and 2.

pH: The results of the pH values of Otamiri River are presented in Table 2. pH of the river for the duration of the study varied between slight acidity to neutrality (5.70-7.50). At station 1, the highest and lowest pH was recorded in July and August respectively. pH values for July and August were significantly different from other months at this station (p<0.05). At station 2, the least pH value (5.70) was recorded in June and was significantly less than values for other months (p<0.05). Still at this station, pH values for the months of November (6.90) and October (6.87+0.03) were significantly higher than values for other months (p<0.05). At station 3, the months of June and August had the least pH value (5.90), which differed significantly from other months (P<0.05). Across stations, the variations of pH values were monthly dependent. Pattern of variation was not consistent.

	Station 1	Station 2	Station 3
June	6.73 + 0.03^{b3}	5.70 + 0.00^{e1}	5.90 + 0.00^{e2}
July	7.37 + 0.03^{a3}	6.60 + 0.00^{c1}	6.80 + 0.00^{d2}
August	6.00 + 0.00^{c2}	6.33 + 0.03^{d3}	5.90 + 0.00^{e1}
October	6.77 + 0.08^{b1}	6.87 + 0.03^{a1}	6.97 + 0.03^{c1}
November	6.80 + 0.00^{b1}	6.90 + 0.00^{a2}	7.47 + 0.03^{a3}
December	6.70 + 0.06^{b1}	6.80 + 0.00^{b1}	7.20 + 0.00^{b2}

Table 2: Mean-monthly variations in pH of Otamiri River in the three sampled stations. Values with different alphabets superscript in a column are significantly different (p<0.05). Values with different numbers superscript in rows are significantly different (p<0.05).

Turbidity: There were variations in turbidity across the months throughout the study period (Table 3). Turbidity was highest in the months of November and December for all the stations. Except for station 2 where October turbidity value was not significantly different from November value, the turbidity for November and December were significantly higher than other months for all stations (p<0.05).

	Station 1	Station 2	Station 3
June	2.97 + 0.03^{d2}	2.00 + 0.00^{d1}	3.00 + 0.00^{d2}
July	4.00 + 0.06^{c1}	5.00 + 0.00^{c2}	5.00 + 0.12^{c2}
August	4.17 + 0.17^{c1}	5.12 + 0.12^{c2}	5.00 + 0.00^{c2}
October	3.00 + 0.06^{d1}	6.02 + 0.16^{b3}	5.03 + 0.03^{c2}
November	7.03 + 0.04^{a2}	6.00 + 0.00^{b1}	6.00 + 0.01^{b1}
December	5.18 + 0.03^{b1}	8.00 + 0.06^{a2}	8.00 + 0.00^{a2}

Table 3: Mean-monthly variations in Turbidity (NTU) of Otamiri River in the three sampled stations. Values with different alphabets superscript in a column are significantly different (p<0.05). Values with different numbers superscript in rows are significantly different (p<0.05).

Seasonal variations in physicochemical parameters of Otamiri River

Seasonal variation in physicochemical parameters of the river is shown in Table 4. Turbidity was significantly different between wet and dry season at all the stations (p<0.05). The river had higher turbidity in the dry season. pH was significantly different between wet and dry season only in stations 2 and 3 (p<0.05); dry season pH were higher at these stations. DO was significantly different between the two seasons at stations 1 and 3 (p<0.05); the dry season also had higher values at the stations. Temperature readings for wet and dry seasons were significantly different at stations 1 and 2 (p<0.05); also dry season values were higher. Only at station 1 was water depth significantly different between the wet and dry season (p<0.05); where wet season values was higher.

	Stations	Wet	Dry	t-value	P
Temperature (°C)	1	25.60 ± 0.18	27.80 ± 0.22	-7.85 (15.19)	0.00*
	2	26.80 ± 0.22	27.70 ± 0.29	-2.44 (15.02)	0.03*
	3	28.30 ± 0.33	28.70 ± 0.17	-8.94 (11.77)	0.38ns
Depth(m)	1	3.31 ± 0.46	0.61 ± 0.07	5.82 (16)	0.00*
	2	2.46 ± 0.63	1.08 ± 0.25	2.03 (16)	0.06ns
	3	2.93 ± 0.64	1.89 ± 0.12	1.58 (16)	0.13ns
DO(MgL^{-1})	1	5.59 ± 0.28	6.23 ± 0.07	-2.21 (16)	0.04*
	2	6.08 ± 0.10	6.22 ± 0.07	-1.15 (14.06)	0.27ns
	3	5.78 ± 0.04	6.42 ± 0.08	-7.65 (11.49)	0.00*

pH	1	6.70 ± 0.20	6.76 ± 0.03	-0.28 (16)	0.79ns
	2	6.21 ± 0.13	6.86 ± 0.02	-4.78 (16)	
	3	6.20 ± 0.15	7.21 ± 0.07	-6.05 (16)	0.00*
					0.00*
Turbidity(NTU)	1	3.66 ± 0.17	5.33 ± 0.60	-2.68 (16)	0.02*
	2	4.00 ± 0.50	6.67 ± 0.33	-4.44 (16)	0.00*
	3	4.33 ± 0.33	6.33 ± 0.44	-3.62 (14.89)	0.00*

Table 4: Seasonal variations of physicochemical parameters of Otamiri River. DO=Dissolved Oxygen, pH=Hydrogen-ion concentration. *Significant difference, ns=No significant difference.

Species composition, abundance and diversity of benthic organisms in Otamiri River

A total of 129 benthic vertebrates belonging to 5 species and 4 families were collected (Table 5). Station 3 recorded more species, 4 and had highest abundance 108 (83.7%) of benthic organisms than the other sampled stations. Station 1 had only two benthic species, and the least abundance of benthic fish fauna, 10 (7.8%). Chrysichthys nigrodigitatus was the most abundant species (32.65%) recorded in the present study as against the least that was Synodontis soloni (2.18%) which occurred only in stations 1 and 2.

Family	Species	S 1 (%)	S 2 (%)	S3 (%)	Otamiri River (%)
Mochokidae	Synodontis budgetti	8 (80.0)	5 (45.5)	7 (6.5)	20 (15.5)
Mochokidae	Synodontis soloni	2 (0.0)	3 (27.3)	0 (0.0)	5 (3.9)
Calroteidae	Chrysichthys nigrodigitatus	0 (0.0)	0 (0.0)	83 (77.9)	83 (64.3)
Clariidae	Clarias gariepinus	0 (0.0)	2 (18.2)	11 (10.2)	13 (10.1)
Notopteridae	Papyrocranus afer	0 (0.0)	1 (9.1)	7 (6.5)	8 (6.2)
		10 (7.8)	11 (8.5)	108 (83.7)	129 (100)

Table 5: Species composition and percentage abundance of fish fauna in Otamiri River. S1=Station 1, S2=Station 2, S3=Station 3.

The abundance of benthic organisms in Otamiri River were dependent on season as more benthic organisms were recovered in dry season than in rainy season. More species were recorded and with high diversity indices in the dry season than the wet season in all the sampled stations (Table 6). The diversity indices yielded high diversity in station 3 than the other two stations.

		Station 1		Station 2		Station 3	
		H'	D	H'	D	H'	D
Overall		1.52	3.83	2.24	8.5	1.83	8.35
	Wet	0.84	1.97	1.38	4	1.7	3.5
Seasons	Dry	1.95	6.33	2.25	8.68	1.83	3.77
	June	0.5	1.47	0	1	1.01	2.57
	July	0	1	1.09	3	1.29	2.81
	August	0.64	1.8	0	0	0.9	2.1
	October	1.47	3.77	1.88	6.23	2.11	5.27
	November	1.33	3.57	2.02	7.2	1.52	2.82
Months	December	1.05	2.77	1.56	4.53	1.61	3.38

Table 6: Species diversity of fish fauna in Otamiri River. H'=Shannon Wienner index, D=Simpson's dominance index.

Pearson correlation (r) of species abundance of benthic organisms with physico-chemical parameters

From the Pearson correlation analysis of species abundance of benthic organisms with some physicochemical parameters, regression plots of only significantly linear relationships were plotted (Figure 4). There was significant positive linear relationship between abundance of *S. budgetti, C. nigrodigitatus and P. affer* and water temperature (r=0.696, 0.473 and 0.530 respectively). There was significant negative linear relationship between abundance of *S. budgetti and C. gariepinus* and water depth (r=-0.615 and -0.481 respectively). Only *S. budgetti* abundance had a significantly positive linear relationship with turbidity (r=0.595, p<0.01).

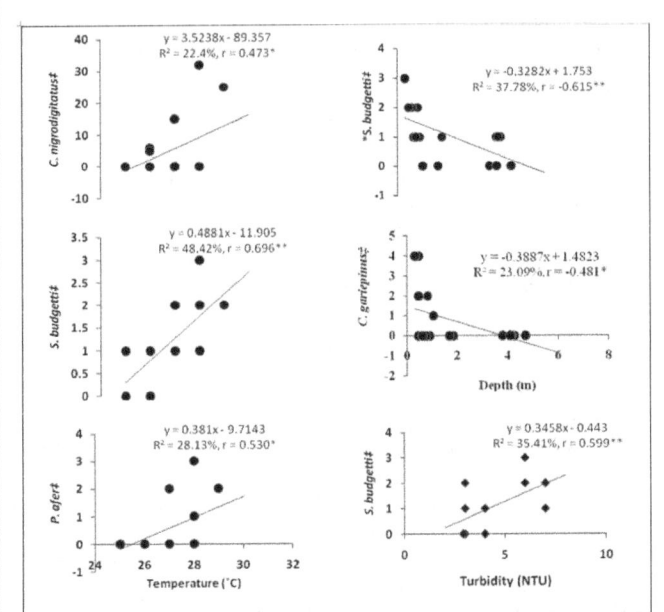

Figure 4: Regression plots of benthic fish species abundance against physicochemical parameters. ‡Species abundance of fish species, *Correlation significant at p<0.05, **correlation significant at p<0.01.

Discussion

The water chemistry of an aquatic ecosystem is dependent on the physical and geological features of its drainage basin [17,18]. The temporal variations in some of the physical and chemical parameters of the water samples at the study stations during the period of study were negligible. This is in agreement with the observations of Ajao and Fagade [19] and Nkwoji et al. [20].

In the present study, water temperature ranged between 25°C-29°C. This finding conforms to earlier works by Okeke and Adinna [21] in Otamiri River that reported a mean temperature of 27.4°C. Sharma et al. [22] and Yogesh [23] also reported similar fluctuations in Narmada River and Dahikhura Reservoir, both in India. Lower temperature during rainy season is expected; this is as increase rainfall during this period also marked by increase in the volume or depth of the river during the same season as observed from this study, results in greater requirement of heat radiation to lift water temperature.

Dissolved oxygen in natural and waste water depends on the physical, chemical and biological activities in the water body. Dissolved oxygen content plays a vital role in supporting aquatic life. It is an important limnological parameter which indicates the level of water quality and organic pollution causes in the water body. DO is susceptible to slight environmental changes. DO in water is dependent on oxygen transfer across air-water interface, water temperature, salinity and flow, wind velocity, rainfall and photosynthesis of aquatic biota [24,25]. In the present study, the concentration of DO in Otamiri River varied from 4.57 mgL^{-1} to 6.70 mgL^{-1}. The mean DO value recorded in Otamiri River was 5.33 mgL^{-1}. This meets the universal standard recommended minimum of 5 mgL^{-1} for aquatic life use [26,27]. This value is higher than earlier reported on the river by Johnbosco and Nnaji [28] who reported a mean DO value of 2.00

mgL^{-1}. The difference may have resulted from variation in sampling techniques employed. Our sampling was deeper into the water away from the shores; though it was not stated by Johnbosco and Nnaji [28], the exact distance away from the shore of their sampling point, one would expect it to be closer to the shore nearer pollution points due to their study objectives. Thus, while the points of high level of pollution noted by Johnbosco and Nnaji [28] may not support aquatic life, deeper away from it organisms live and probably flourish. Also, our study did not focus entirely at polluted sections of the river. Unpolluted areas were similarly sampled. In addition, the duration of sampling was not stated by Johnbosco and Nnaji [28]. Also, DO reported from this study are similar to reports for Otamiri River by Okeke and Adinna [21] who reported Do ranges of 2.1-6.8 mgL^{-1} in rainy season and 2.7-5.1 mgL^{-1} for the dry season.

pH is the measure of the concentration of hydrogen ions. It provides the measure of the acidity or alkalinity of a solution. In the present study the observed pH values which ranged from 5.70 to 7.47 showed that the Otamiri River varied from slight acidic to neutrality. The pH range is not detrimental to aquatic life. Adebisi [29], King and Ekeh [30] and Mama and Ado [31] attributed variations in pH to evapotranspiration process, rainfall causing dilution of chemical substances, and biological processes in water. The mean pH value of 6.73 ± 0.10 recorded during this study falls within the recommended range of 6.5-8.5 set by the WHO and National Standard for Drinking Water Quality in Nigeria [32,33]. This value is in agreement with the report of Akaahan et al. [2] which recorded a mean pH value of 6.63 ± 0.07 in River Benue at Makurdi, Benue State Nigeria. Narasimha and Benarjee [34] reported that pH range of 6.0-9.0 is suitable for fish culture and growth.

Turbidity in the water may be due to organic and inorganic constituents. Turbidity is often determined and used as surrogate measure of the total suspended solids [35]. During this study, the mean turbidity value obtained was 2.66 ± 0.16 NTU. This value is below the recommended standard maximum of 5.00 NTU for turbidity [32,33]. Dredging and sand mining activities only minimally affected turbidity as only slight difference were observed between turbidity at station 2 (sections of sand mining and dredging activities) and station 3 (section of vegetable farming). Also refuse dumpsites at the banks of the river did not affect river turbidity much a little deeper off shore. Fast flow of water probably caused the movement of waste away from areas where it was dumped thereby result in little impact on turbidity.

A total of 4 families of benthic fauna were recorded in Otamiri River. This is similar to the 4 families reported by Ogidiaka et al. [36] in Ogunpa River in Ibadan, but fewer than the 10 families reported by Adjarho et al. [37] in Ona River, Ibadan. Otamiri River being a freshwater body was dominated by Calroteidae (36.25%). The distribution pattern of Calroteidae showed that they were more abundant in Station 3 which is the home of vegetable farming. The dominance of C. nigrodigitatus at station 3 compared to other stations indicated pollution, stress and high level of anthropogenic activities such as dredging in Stations 1 and 2.

Nwankwo and Akinsoji [38] had attributed low species abundance and diversity at some sites in a river in southwestern Nigeria to the pollution of such sites. The relative abundance of benthic fauna in each station of this study is a reflection of the level of pollution of each station. Burger and Gochfeld [39] related the abundance and diversity of the benthic fauna to the health of the water body. Stations 1 and 2 in Otamiri River recorded relatively lower number of species; this could be attributed to the resultant effect of the bridge construction and other human activities (mainly sand mining, dredging and waste disposal) ongoing at these stations. Anthropogenic activities such as dredging often result in substratum instability and increased siltation. Edokpayi and Nkwoji [40] had reported in a previous study that suspended silt has the ability to reduce light penetration and primary productivity and could clog the gills of aquatic fauna thereby smothering them. The occurrence of relatively higher number of species and individuals in station 3 may be an indication of lower degree of anthropogenic activities at the station compared to other stations. Overall diversity had been reported to be the product of all dynamic spatial and temporal changes affecting an urban stream community in Nigeria [41,42].

The abundance and diversity of benthic fauna are generally affected by the physical and chemical characteristics of water, availability of food and substrate occupation [43]. In this study, such parameters like temperature, depth, DO, turbidity were observed to have influenced the community composition of Otamiri River. This is in agreement with earlier reports of Ajao and Fagade [19], Brown and Oyenekan [43] and Edokpayi and Nkwoji [40]. The relationship between water quality parameters and composition of benthic fauna of Otamiri River showed both direct and inverse relationships with some parameters. pH had no relationship with the species composition and abundance. Surface water temperatures, depth, DO and turbidity had significant relationship with some benthic fauna such as S. budgetti, C. nigrodigitatus, P. afer, C. gariepinus and S. soloni.

The study revealed that anthropogenic activities at Otamiri river affected the fauna diversity of the area. The anthropogenic activities dredging, sand mining, bridge construction, and dumping of waste into river probably caused stress to aquatic life at the two stations (1 and 2) where such activities take place therefore resulting in reduction in benthic fish species diversity and abundance. Proper management of the river should be put in place to preserve its water quality and biodiversity for sustainable development. Refuse dumping into the river should be stopped.

References

1. Idowu EO, Ugwumba AAA (2005) Physical, chemical and benthic fauna characteristics of a southern Nigeria Reservoir. The Zoologist 3: 15-25.

2. Akaahan TJA, Araoye PA, Azua ET (2015) Physico-chemical characteristics and macroinvertebrates of River Benueat Markurdi, Benue State, Nigeria. Intern J Innov Edu Innov Res 1: 43-54.

3. Adakole JA, Annune PA (2003) Benthic macroinvertebrates as indicators of environmental quality of an urban stream, Zaria, Northern Nigeria. Journal Aqua Sci 18: 85-92.

4. Emere MC (2000) A survey of macroinvertebrate fauna along River Kaduna, Kaduna, Nigeria. J Basic Applied Sci 9: 17-27.

5. Olomukoro JO, Egborge ABM (2003) Hydrobiological studies on Warri, River, Nigeria part I: The composition, distribution and diversity of macrobenthic fauna. Biosci Res Comm 15: 279-296.

6. Sharma S, Dubey S, Chaurasia R, Dave V (2013) Macroinvertebrate community diversity in relation to water quality status of Kunda River (MP), India. Discovery Journal 3: 40-46.

7. Odo GE, Inyang NM, Ezenwaji HMG, Nwani CD (2007) Macroinvertebrate fauna of a Nigerian Freshwater Ecosystem. Anim Res Intern 4: 611-616.

8. Karr JR (2005) Measuring biological conditions protecting biological integrity: Principles of conservation biology essay. Natural Academy Press: Washington DC.

9. Basden GT (1966) Among the Ibos of Nigeria. Frank Cass Publishers: London.

10. Anyanwu C (2009) A comparative evaluation of early rains phytoplankton productivity of Nworie and Otamiri Rivers, Owerri. Department of Agricultural Science, Alvan Ikoku College of Education, Owerri, Imo State, Nigeria.

11. APHA, AWWA, WEF (2005) Standard methods for examination of water and waste water. (21st edn), American Public Health Association: Washington, DC, USA.

12. Wangboje OM, Oronsaye JAO (2001) Bioaccumulation of iron, lead, mecury, copper, zinc and chromium by fish species from Ogba River Benin City, Nigeria. Afr J Zool Env Manag 3: 45-49.

13. Esenowo IK, Ugwumba AAA (2010) Composition and abundance of macrobenthos in Majidun River Ikorodu, Lagos State, Nigeria. Res J Biologl Sci 5: 556-560.

14. Bouchard RW Jr (2004) Guide to aquatic macro-invertebrates of the upper Midwest. Water Resources Centre, University of Minnesota Press, St. Paul, MN.

15. Dailey PJ (2006) Photographic atlas of aquatic macro-invertebrates of Paisa watershed Creeks and Ponds. Lewis and Clark Community College, St. Louis: USA.

16. APHA (1992) Standard methods for examination of water and waste water. APHA, AWWA, Washington, DC: USA.

17. Bishop JE (1973) Limnology of a small Malayan River, Sungai Grambak. In: Junk W (editor) The Hague.

18. Victor R, Al-Mahrouqi AIS (1996) Physical, chemical and faunal characteristics of a perennial stream in Arid Northern Oman. J Arid Envir 34: 465-476.

19. Ajao EA, Fagade SO (1990) The Ecology of Capitella capitata in Lagos Lagoon. Arc Hydrobiol 120: 229-239.

20. Nkwoji JA, Yakub A, Ajani GF, Balogun KJ, Renuer KO, et al. (2010) Seasonal variations in water chemistry and benthic macroinvertebrates of a South Western Lagoon, Lagos, Nigeria. J Amer Sci 6: 85-92.

21. Okeke PN, Adinna EN (2013) Water quality study of Otamiri River in Owerri, Nigeria. Univers J Envir Res Tech 3: 641-649.

22. Sharma S, Tali I, Pir Z, Siddique A, Mudgal LK (2012) Evaluation of physico-chemical parameters of Narmada River, MP, India. Researcher 4: 13-19.

23. Yogesh S, Pendse DC (2001) Hydrobiological study of Dahikhura Reservoir. J Envir Biol 22(1): 67-70.

24. Gaghia M, Surana R, Ansari E (2012) Seasonal variations in physicochemical charactreristics of Tapi Estuary in Hazira Industrial Area. Our Nature 10: 249-257.

25. Gwaski PA, Hati SS, Ndahi NP, Ogugbuaja VO (2013) Modelling parameters of oxygen demand in aquatic environment of Lake Chad for depletion estimation. ARPN J Science Tech 3: 116-123.

26. WHO (2002) Guideline for drinking water quality. Genva: WHO.

27. Kansas Department of Health and Environment (2011) Water quality standard, white paper allowance for low dissolved oxygen level for aquatic life use. KDHE, Kansas: USA.

28. Johnbosco EU, Nnaji AO (2011) Influence of landuse patterns on Otamiri River, Owerri and Urban Quality of life. Pakistan J Nutri 10: 1053-1057.

29. Adebisi AA (1981) The physico-chemical hydrology of a tropical seasonal river-upper Ogun River. Hydrobiologia 79: 157-165.

30. King RP, Ekeh SIB (1990) Status and seasonality in the physico-chemical hydrology of a Nigerian headwater stream. Acta Hydrobiol 32: 313-378.

31. Mama D, Ado G (2003) Urban lake system eutrophication- a case study. J Applied Sci Manag 7: 15-20.

32. World Health Organization (2004) Drinking Water Guidelines.

33. National Standard for Drinking Water Quality (2007) Nigerian industrial standard NIS 554, Standard Organization of Nigeria, Lagos.

34. Narasimha RK, Benarjee G (2013) Physicofactors influenced plankton biodiversity and fish abundance. A case study of Nagaram tank of Waragah Andhra Predesh. Intern J Life Sci, Biotech Pharma Res 2: 1-15.

35. Bilotta GS, Brazier RE (2008) Understanding the influence of suspended solids on water quality and aquatic biota. J Water Resources 42: 2849-2861.

36. Ogidiaka E, Esenowo IK, Ugwumba AAA (2012) Physico-chemical parameters and benthic macroinvertebrate of Ogunpa River at Bodija, Ibadan, Oyo State. Europe J Scientific Res 85: 89-97.

37. Adjarho UB, Esenowo IK, Ugwumba AAA (2013) Physico-chemical parameters and macro-invertebrate fauna of Ona River at Oluyole Estate, Ibadan, Nigeria. Res J Envir Earth Science 5: 671-676.

38. Nwankwo DI, Akinsoji A (1992) Epiphyte community on water hyacinth Eichhornia crassipes (Mart). Solms in Coastal Waters of Southwestern Nigeria. Arc Hydrobiol 124: 501-511.

39. Burger J, Gochfeld M (2009) On developing bioindicators for human and ecological health. Envir Monitor Asses 66: 23-46.

40. Edokpayi CA, Nkwoji JA (2007) Annual changes in the physicochemical and macrobenthic invertebrates characteristics of the Lagos lagoon sewage dump site at Iddo Southern Nigeria. Ecolog Envir Conserv 13: 13-18.

41. Victor R, Ogbeibu AE (1991) Macro-invertebrate communities in the erosional biotope of an urban stream in Nigeria. Trop Zool 4: 1-12.

42. Odo GE, Nwani CD, Ngwu GI, Eyo JE (2007b) Harnessing aquatic physico-chemical parameters influencing macroinverterbrate fauna in Anambra River Basin for sustainable fish productivity. Anim Res Intern 4: 705-712.

43. Brown CA, Oyenekan JA (1998) Temporal variability in the structure of benthic macrofauna communities of the Lagos lagoon and harbour, Nigeria. Arc Hydrobiol 45: 45-54.

12

Habitat Suitability Index Relationships for the Northern Clearwater Crayfish, *Orconectes Propinquus* (Decapoda: Cambaridae)

Thomas P Simon[*] and Nicholas J Cooper

School of Public and Environmental Affairs, 1315 E. Tenth Street, Indiana University, Bloomington, Indiana 47405, USA

[*]Corresponding author: Thomas P Simon, School of Public and Environmental Affairs, 1315 E. Tenth Street, PV 357, Indiana University, Bloomington, IN 47405, USA, E-mail: tsimon@indiana.edu

Abstract

We evaluated habitat models that determined relative abundance relationships among microhabitat, reach and watershed-scale factors important for predicting habitat selection. Thirty stream reaches in central Indiana were sampled to determine relationships between habitat associations and relative abundance, size, and age associations of the Northern Clearwater Crayfish, *Orconectes propinquus*. Females are significantly more abundant than males and the frequency of crayfish in gravel substrate was significantly higher than that of cobble substrate. The sizes of crayfish in cobble substrate were significantly larger than individuals found in gravel substrates, while females were significantly larger than males in gravel substrates. Watershed variables were not significantly related to crayfish abundance. The only reach scale variable that proved to be significant was boulder substrate. Microhabitat variables showed a significant increase between CPUE and cobble and gravel substrates. Habitat models provide valuable information on the conservation status and habitat parameters responsible for determining species preferences, life history strategies, and relative abundance.

Keywords: Habitat suitability; Crayfish; Microhabitat scale; Watershed-scale; Reach-scale

Introduction

Species responses to various habitat cues provide strong selection preferences that determine aquatic species relative abundance and ecological life history strategies. Crayfish are keystone species in many stream ecosystems since they create and establish habitat use of other sympatric aquatic species. Crayfish are among the largest macroinvertebrates and limit access for other taxa to aquatic food web energy transfer [1-4]. Crayfish are ecosystem engineers that provide a vital role in the structure and function of stream ecosystems [5,6] by determining species distribution and placement within the stream benthic habitat [7,8]. Crayfish are decomposers of organic material and contribute to energy transformation between trophic levels [6,9,10]. They are sensitive indicators of habitat degradation and respond to anthropogenic effects in streams. Species composition and relative abundance reflects anthropogenic response to water quality, habitat, and land use change and stressors [6,11].

The Northern Clearwater Crayfish, *Orconectes propinquus* (Girard 1852) [12] is ubiquitous in North America ranging from southern Ontario and Quebec, as far south as southern Illinois, Indiana, Ohio, Pennsylvania, and New York, and as far west as Iowa and Minnesota [13,14]. It is a tertiary burrowing crayfish [13] that dig simple depressions in the sediment beneath rocks during drought conditions and spend its entire 2-3 year life history [14-16] in surface water. The species is common in midwestern headwater streams [17] and is found in both stream and lake ecosystems [15,18]. The species' habitat is typically rocky riffle habitat in streams [14-16,19], and they prefer coarse habitat in lakes that can provide more cover from predators.

Watershed-, reach-, and microhabitat-scale land use can influence chemical and physical factors associated with a stream [20,21].

Watershed-scale land uses are known to have negative impacts on stream ecosystems [22]. For example, agricultural practices increase sediment inputs and nutrients into streams and may negatively affect water quality, habitat, and biological assemblages [20,23]. Excess sediments in streams can negatively affect macroinvertebrates, such as crayfish, by reducing food sources and filling in interstitial pore spaces in preferred habitats [23]. Urban land use has also been found to reduce stream habitat quality by the addition of chemical contaminants [21]. Alternatively, many types of land use can improve stream quality. Forest land use has been found to correlate with high quality habitat and also for bank stability and instream cover [21]. On a reach-scale, the channel morphology can be largely influenced by bank material, riparian vegetation, and the slope at which water and other inputs enter the stream [20,21]. The resulting channel morphology and substrate can determine the types of species that will likely inhabit that particular reach. On a microhabitat-scale cover and substrate particle size can influence individual strategies for feeding, reproduction, and establishment of territories. Factors known to influence crayfish distribution in stream reach scale include presence of predators, amount and stability of within stream cover, age and body size, food sources, and competition among other crayfish [10,24]. Larger crayfish are best able to defend themselves [25] and are then capable of obtaining preferred cover through competition [10].

The primary objective of this study is to determine habitat relationships between catch-per-unit of effort (CPUE) patterns with watershed-, reach-, and microhabitat-scale associations for *O. propinquus*. Relationships based on gender are correlated with habitat factors, including sediment particle, cover, and larger scale variables. Size, age, and gender patterns in *O. propinquus* habitat associations are examined in headwater streams to determine selection preferences.

Methods

Study area

All of the study sites occurred within the Interior Plateau Level III ecoregion of Indiana [26]. This ecoregion is characterized by rolling and heavily dissected, rugged terrain [26]. The underlying soil is composed of sandstone, siltstone, shale, and limestone [26]. The ecoregion consists of high hills and knobs and low and narrow valleys. The streams of this region are medium to high gradient [26]. Land use is mainly agricultural cropping and livestock pasturing, but includes several forest types. Forested areas were the most common land use surrounding study streams, composing 57% of the total land use.

A total of 30 sites were sampled in the counties of Brown, Monroe, Morgan and Lawrence in southcentral Indiana (Figure 1). Sites are located in the East Fork White River watershed, which is a primary tributary of the Wabash River drainage, Ohio River basin. The East Fork White River watershed is dominated by karst topography and limestone quarries [27].

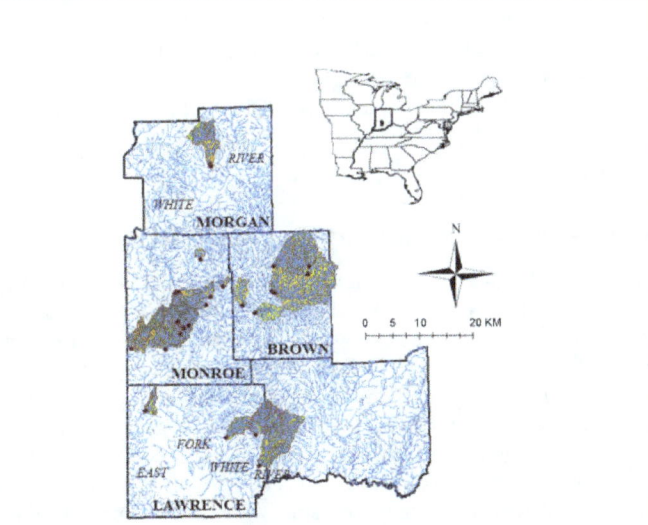

Figure 1: Study sites (red dots) sampled within the catchment land use during an investigation of *Orconectes propinquus* habitat suitability preferences in southcentral Indiana, USA, headwater streams. Land use key, Gray = forest; yellow = agriculture; orange = pasture; blue = water features.

Study design

Sites were selected using a random probability study design. Sites were classified by Strahler stream order [28] and selected without replacement from the universe of wadeable first through third order streams in the four county area of the Eastern Corn Belt Plain [29]. A variety of stream conditions were included in the study to determine response to both the highest quality streams and to those of lower qualities due to poor land management practices in southcentral Indiana [30]. Habitat relationships were tested to evaluate the association between relative abundance and CPUE of Northern Clearwater Crayfish based on gender, age, and watershed-, reach-, and microhabitat-scale factors. Our study investigates the relative abundance relationships with gender bias towards microhabitat, size,

and scale. We evaluated the percentage of cobble and gravel substrates at each site using a qualitative habitat procedure [31]. Within similar relative abundance categories, we evaluated if greater CPUE of *O. propinquus* individuals occurred in cobble compared to gravel substrates. We tested two size-related hypotheses, based on whether *O. propinquus* are larger in size in large coarse substrates, compared to smaller coarse substrates, and whether males are larger than females due to sexual dimorphism or intraspecific competition.

Scale and habitat association hypotheses are placed into three categories, including watershed-, reach-, and microhabitat-scales. For the watershed-scale associations, we determined does the CPUE of crayfish change with land use, while for reach scales, we evaluated whether increasing reach scale habitat heterogeneity led to greater CPUE. For microhabitat scale associations, we tested if CPUE increases with increasing size of substrate. Finally, age was evaluated to determine if ontogenetic differences existed in age class use of large coarse- compared to moderate-coarse substrates.

Field sampling

The stream reach length sampled was 15 times the wetted width [32]. Our study stream reach ranged from a minimum distance of 50 m and a maximum distance of 250 m during this study. Sampling proceeded in an upstream manner beginning at the downstream end of the stream reach, thereby reducing disturbance to upstream crayfish. The sampling events occurred between June 17, 2010 and July 18, 2010, and generally followed the method used by Simon [32]. A one-man common sense minnow seine (1 m x 1 m) with 3.1 mm standard mesh netting was used to sample crayfish by kick-seining a 1 m² area of substrate directly upstream of the seine [33]. Quantitative collection of crayfish were sampled from 20 one square meter (m²) of habitat randomly distributed in the stream reach, which represent the coarse substrate habitat portions of each stream reach. This was defined as our catch-per-unit of effort. Captured individuals of *O. propinquus* and all other crayfish species were counted, sexed, and placed into live wells until the completion of sampling. All individuals were released after all site data was obtained.

A total of twenty 1 m² seine samples were completed at each site [34], 10 samples were randomly located in both gravel-dominated and cobble-dominated substrates (total=20). Substrate size was classified following EPA physical habitat procedures [35] and seine samples were classified as either cobble or gravel when at least 50% of the substrate comprised the dominant substrate type. At each site 10 random m² samples in each of the two substrate types where sampled using a kick seine method [34] to collect individual crayfish. Stream width measurements at each site included wetted, active, and bank full widths [35]. The wetted width is the perpendicular measurement from shoreline to shoreline. The active width is the perpendicular stream measurement where the normal stream flows fluctuate by season and are delineated by the point where streambed vegetation ends, whereas bank full width measures the lateral extent of water during flooding.

Laboratory methods

Small individual *O. propinquus*, too small to sex or measure in the field, were taken to the laboratory where carapace length (CL, from tip of rostrum to the posterior border of the thoracic region, to nearest 0.1 mm), postorbital carapace length (POCL; [36]) and sex were recorded. Each individual crayfish was classified as either a male, female, or as a juvenile. A juvenile is defined as the size of the individual that prevented accurate determination of the sex based on the sexual organs not being fully developed. Age was determined based on length-frequency numerical count distributions plots of CL. Crayfish specimens were deposited in the Astacology collection at the Aquatic Research Center of the Indiana Biological Survey, Bloomington, Indiana.

Watershed-scale variables

ArcMap 10.0 was used to overlay the watershed boundary with stream hydrology and 2006 land cover to obtain site data. The stream and land cover data were obtained from IndianaMap.org [37]. The stream layer included the 2008 National Hydrology Data (NHD) streams, rivers, canals, ditches, artificial paths, coastlines, connectors, and pipelines layer. This layer was derived at 1:100,000 scale. The land cover layer included the 2006 USGS 30-meter resolution National Land Cover Data (NLCD). The percentage of each land use type was determined from the land use layer for each individual watershed.

Watershed-scale variables included 15 land cover types, i.e., open water, developed open space, developed low intensity residential, developed medium intensity residential, developed high intensity residential, deciduous forest, evergreen forest, mixed forest, shrub/scrub, grasslands/herbaceous, pasture/hay, cultivated crop, barren land, woody wetland, and emergent herbaceous wetlands [37]; and three additional variables (i.e., latitude, longitude, drainage area). The watershed boundaries were delineated using the Watershed Delineation Model [38], which uses the digital elevation associated with specific latitude and longitude. The drainage areas for each of the 30 sites sampled were obtained from US Geological Survey sources [39].

Reach-scale variables

Reach-scale variables were derived from qualitative habitat measures, i.e., Qualitative Habitat Evaluation Index (QHEI) [31]. The habitat measures include a variety of habitat qualities within the wetted stream width and the riparian area in the floodplain. The qualitative habitat measures include the following categories, i.e., substrate types, instream cover, channel morphology, riparian quality/bank erosion, pool/glide and riffle/run quality, and local stream gradient. The qualitative habitat information was collected from each stream reach. Each qualitative habitat category is ranked by a series of categories representing varying states of stream habitat condition. The total reach habitat score is the sum of each of the category scores, which provides a cumulative score for the entire stream reach. Each qualitative categorical score and the total reach habitat score was regressed against crayfish relative abundance to determine any significant relationships between the habitat category and crayfish relative abundance. Individual substrate particle size categories for each stream reach were compared to crayfish CPUE to determine if any significant relationships existed. The substrate types observed included boulder, cobble, gravel, sand, bedrock, detritus/muck, and artificial. The percentage of each specific substrate size class was determined for each site and used for the comparison. Several other physical reach-scale factors were evaluated including the total percentage of pool, run, and riffle habitat, and the wetted width, active width, and bank full width measurements for each reach. Each parameter was compared to the CPUE of crayfish at each site.

Microhabitat-scale variables

Two microhabitat-scale variables were examined and included the two primary coarse substrate types (i.e., cobble-dominated substrate and gravel-dominated substrate). A CPUE was calculated based on the 10 kick seine samples in each substrate size class and compared to evaluate associations between gender, size, and CPUE with each of the microhabitat substrate types.

Statistical methods

Basic statistics using Statistica were used for all analyses [40,41]. Each statistical analysis conducted used a significance value of $\alpha=0.10$ for field evaluation and a Tukey HSD post-hoc test. We selected a higher alpha to reduce beta error and reduce Type II errors in data interpretation since our study was based on a single watershed. Differences between relative abundance and CPUE of male and female *O. propinquus* among cobble and gravel habitats were determined using a Z-test. Differences in crayfish length were determined using an Analysis of Variance (ANOVA) single factor analysis to analyze variance between populations. A length-frequency distribution was developed to evaluate differences in age structure. To analyze each category of the habitat scale factor questions, a simple univariate linear regression was used. The regressions compared a specific watershed, reach, or microhabitat variable with the CPUE of crayfish at each site.

Results

Relative abundance and Catch-Per-Unit Effort

A total of 2,648 *O. propinquus* was collected from 29 of the 30 (96.7%) sites that were sampled during this study. No other secondary or tertiary burrowing species was sympatric; however, three other primary burrowers were collected during this study including *Fallicambarus fodiens* (Cottle, 1863), *Cambarus polychromatus* Thoma, Jezerinac, and Simon, 2005, and *C. diogenes* Girard, 1852. *Orconectes propinquus* individual CPUE effort ranged from 0 to 19.1 individuals/ m^2 at each stream reach. The number of males compared to females was consistent by site with males comprising 990 individuals and females 1,048 individuals.

A: Relative Abundance	Z	P -two-tail
Cobble vs. Gravel	-4.34	1.43E-05
Male vs. Female: All sites	-1.733	0.083
Male vs. Female Cobble	>.001	0.999
Male vs. Female Gravel	>.001	0.999
B: Carapace Length	**F**	**P-value**
Cobble vs. Gravel	70.643	>.001
Male vs Female: All Sites	0.547	0.46
Male vs. Female Cobble	0.601	0.438

Table 1: Comparisons between *Orconectes propinquus* relative abundance and size (CL) statistical significance for gender and substrate size ($\alpha=0.10$). A. Z-test statistical values for CPUE (individuals/m^2), and B. F-test P-values ($\alpha=0.10$) for size (CL). CL = Carapace length.

The sex ratio was 1:1.05 males to females. A total of 610 juveniles (range: 4.4 mm to 9.8 mm CL) were captured. The number of crayfish captured in cobble-dominated substrates was 989, while 1,049 were collected from gravel-dominated substrates. Females were significantly more abundant than males (Z-statistic=-1.733, P=0.083) within the stream reaches. CPUE of crayfish was significantly (P=<0.001) different in gravel-dominated substrates (Table 1A).

Length frequency distribution and age range

The mean CL for all of the 2,648 sampled *Orconectes propinquus* was 12.7 mm. Sampled *O. propinquus* ranged in CL from 4.4 mm to 39.8 mm. Mean CL was significantly larger for crayfish found in cobble-dominated substrates (P=<0.001) compared to gravel-dominated substrates. Females collected in gravel-dominated substrates had significantly larger CL than males (P=0.013; Table 1B). We observed three age classes based on our study (Table 2A). Both male and female *O. propinquus* individuals attained similar size at each age. Age 0 individuals were 3-18 mm CL; age I individuals were 18-33 mm CL; and age 2 indivudals were 33-42 mm CL. The length-frequency distribution showed the greatest number of individuals occurred at age 0 (cobble=973, gravel=1285; and males=1433, females=1434). The number of individuals decreased with increasing age group. Only 8 individuals were found in the 2 year age group, and no individuals reached age 3 (Table 2A and 2B). Large individuals (>18 mm CL) had a greater association with cobble substrates compared to smaller individuals (3-18 mm CL), which were associated with gravel substrates (Table 2B).

Habitat scale factors

None of the 18 watershed-scale variables showed a significant relationship with *O. propinquus* CPUE (Table 3). Only a single reach-scale variable, i.e., boulder substrate, was significantly correlated with crayfish CPUE (Table 4). An increasing amount of boulder habitat was associated with a decrease in crayfish relative abundance. Both microhabitat-scale factors, cobble- (R^2=0.103, F=3.232, P=0.083) and gravel dominated (R^2=0.094, F=2.904, P=0.099) substrates, were positively associated with *O. propinquus* CPUE.

Discussion

The relative abundance of crayfish is dependent on available stream substrate types [10,22,24]. The five lowest crayfish CPUE, i.e., 0, 2, 5 10, and 15 individuals/m^2, occurred at sites with reduced reach scale habitat. Bean Blossom Creek (n=2 individuals/m^2) was a stagnant stream with muck substrate, while Griffey Creek (n=5 individuals/m^2) was heavily impounded with an embedded substrate. These substrate factors are considered responsible for declining crayfish relative abundance due to reduced amounts of preferred substrate and instream cover. Linear regression models showed little significance between scale variables tested at watershed and reach-scales. This was a similar result observed by Burskey and Simon [22] and Stewart et al. [10]. All study area watersheds comprised relatively small drainage area size (range: 9.1 to 49,166 acres). We selected headwater streams to isolate potential impacts and increase the percentage of catchment forested land use. Forested areas provided a large amount of coarse particulate organic matter (CPOM), which are positive factors for stream ecosystems [42]. Forests provide large amounts of organic material and detritus, which are important for crayfish survival [43]. The high percentage of forested areas (mean: 57% for all sites) in these watersheds represent a least impacted condition for crayfish

populations. A large amount of forested area is considered to be the most important factor for explaining low significance in watershed scale analyses.

A. Sex/ Size class CL (mm)	Age	N
Male		
3-18	0	1433
18-33	1	164
33-42	2	3
Female		
3-18	0	1434
18-33	1	219
33-42	2	5
B. Substrate/Size class CL (mm)	Age	N
Cobble		
3-18	0	973
18-33	1	254
33-42	2	7
Gravel		
3-18	0	1285
18-33	1	129
33-42	2	1

Table 2: Age class frequency distribution. A. male and female by age class, and B. course substrate type. CL= carapace length.

Watershed land cover effects were not found to effect crayfish populations, while other studies linked various land use types to low crayfish abundance [10,11,22,44]. Row-crop agriculture, urban, and developed areas have been shown to negatively impact aquatic habitats and fish and macroinvertebrate communities [11]; however, agricultural land use was not a predominant component in the study streams. Reach scale stream variables scores showed increasing habitat condition levels in the study area (cumulative habitat score range: 37.5 to 91.0, mean=72.9). These relatively high reach scale habitat values show that streams represented relatively high overall ecological integrity. The only variable correlated with crayfish abundance included reach scale habitat substrate boulder proportion. Boulder presence showed a negative correlation with individual crayfish CPUE. This result seems contradictory; however, boulder substrates provide large interstitial spaces affording cover and habitat for predators. The univariate microhabitat-scale regression models showed a significant relationship between CPUE and both cobble and gravel substrates. This suggests that increasing amounts of coarse substrates correlates with increases in *O. propinquus* CPUE and may be differentially important for various life stages. Crayfish substrate preference is typically associated with the most overall cover and protection from predators [24,45]. Larger crayfish prefer larger substrate sizes since the larger substrates will provide more overall interstitial spaces. These interstitial spaces provide more areal coverage for protection from predators [45]. Crayfish size (CL) was significantly correlated with large substrate sizes compared to small substrates; however, since the study area was not glaciated during the latest Wisconsin glaciation event the dominant particle size is cobble in 321 the study area. *Orconectes propinquus* individuals associated with cobble substrates exhibited the highest CPUE (Table 2B).

Watershed Variables	R^2	F	P-value
Open water	0.0002	0.005	0.945
Developed open spaces	0.003	0.08	0.779
Developed low intensity residential	0.001	0.038	0.846
Developed medium intensity residential	3.28 E-05	0.0009	0.976
Developed high intensity residential	0.0006	0.017	0.896
Deciduous forest	0.002	0.044	0.835
Evergreen forest	0.009	0.247	0.623
Mixed forest	0.042	1.213	0.28
Shrub/Scrub	0.021	0.607	0.443
Grasslands/Herbaceous	0.088	2.716	0.111
Pasture/Hay	0.002	0.047	0.83
Cultivated crop	0.072	2.113	0.157
Barren land	0.034	0.99	0.328
Woody wetland	0.005	0.153	0.699
Emergent herbaceous wetland	0.003	0.094	0.761
Latitude	0.05	1.461	0.236
Longitude	1.90 E-05	0.0005	0.982
Drainage area	0.014	0.405	0.53

Table 3: Simple linear regression (R^2, F-test, Significant F, and P-value, $\alpha=0.10$) relationships between watershed-scale land use variables and *Orconectes propinquus* CPUE from headwater streams in south central Indiana.

Large individuals were associated with large substrate particle size and when mature adults are present, smaller individual crayfish typically were associated with small, gravel substrates [10,24]. *Orconectes propinquus* individuals were significantly more abundant in gravel substrates than in cobble substrates; however, this was based on the association between CPUE and high number of age 0 individuals. Overall, Age-0 crayfish comprised the largest proportion of individual crayfish at all sites (n=2258; 85.3%). A niche shift from small substrates to large substrate occurs at lengths greater than 18 mm CL. This niche shift demonstrates that individual crayfish select increasing substrate particle size proportional to increasing body size. Likewise, small age-1 individuals showed similar response as age-0 individuals with significantly increasing CPUE in the hypothesized less preferred gravel substrates.

Studies have shown that the dominance of many freshwater crayfish is based on size [10,24,46]. The study area male to female sex ratio is 1:1.05 (χ^2 (1, N=29), p >0.10), which is not statistically significant. We predicted that male CPUE would be significantly greater than female;

however, this assumption was based upon the premise that males would be significantly larger than females. Females were larger than males, but not statistically significant (P=0.460); however, females were significantly larger than males in gravel substrates. This suggests that females could have a slight numerical dominance over males during the early stages of their lives or be forced into smaller substrate particle sizes due to dominance and territoriality. This would explain females being significantly more abundant than males in the study streams sampled; however, females would be exposed to increased predation pressure affecting females CPUE with increasing age class.

Reach-Scale Variable	R^2	F	P-value
Stream width			
Wetted Width (m)	0.0004	0.01	0.921
Active Width (m)	0.0018	0.05	0.825
Bank Full (m)	0.03	0.866	0.36
Habitat			
Substrate	0.015	0.426	0.519
Instream Cover	0.0043	0.122	0.729
Channel Morphology	0.0072	0.204	0.655
Bank Erosion and Riparian Zone	0.048	1.412	0.245
Pool/ Current	0.0043	0.122	0.729
Riffle/Run	0.0213	0.608	0.442
Gradient	0.0016	0.046	0.831
QHEI Total Score	0.0006	0.017	0.897
Substrate			
Boulder	0.1026	3.202	0.084
Cobble	0.0544	1.611	0.214
Gravel	0.0222	0.637	0.431
Sand	0.0226	0.649	0.427
Bedrock	0.0026	0.074	0.787
Detritus/Muck	0.0698	2.102	0.158
Artificial	0.0287	0.828	0.371
Morphology			
% Pool	0.0374	1.09	0.305
% Run	0.0394	1.148	0.293
% Riffle	0.0003	0.008	0.93

Table 4: Simple linear regression (R^2, F-test, Significant F, and P-value, $\alpha=0.10$) relationships between reach-scale variables and *Orconectes propinquus* relative abundance from headwater streams in south central Indiana.

Primary habitat drivers effecting *O. propinquus* relative abundance, niche shift patterns, and age structure included microhabitat- and reach scale factors, but not watershed scale variables. Boulder presence negatively correlated with relative abundance, while cobble was selected instead of gravel substrates. Other cover types including various substrate particle sizes, woody debris, vegetation, and riparian channel factors did not correlate with relative abundance, gender, or size. No watershed scale variables effected *O. propinquus* relative abundance; however, our study was conducted in headwaters streams and only in a single watershed. Further study in multiple watersheds, larger order streams, or lakes may reveal differing life history strategies.

This study provides important understanding of life history strategies utilized by *O. propinquus,* which may be useful for management of other related imperiled crayfish species in need of conservation management. Conservation of the habitat heterogeneity and natural stream corridors in forested landscapes provide a unique opportunity to evaluate restoration goals that would promote stable relative abundance. Additional studies of other more restricted species would be necessary to confirm our results that the species is not responding to larger scale stressors.

Acknowledgements

We thank the Indiana Biological Survey Aquatic Research Center for professional courtesies, use of vehicles, and equipment. Special thanks to Wade Kimmon, Richard Barendt, and Alex Jackson for field assistance.

References

1. Bovbjerg RV (1952) Comparative ecology and physiology of the crayfish *Orconectes propinquus* and *Cambarus fodiens.* Physiol Zool 25:34-56.

2. Rabeni CF (1992) Trophic linkage 371 between stream centrarchids and their crayfish prey. Can J Fisheries Aquatic Sci 49: 1714-1721.

3. Parkyn SM, Rabeni CF, Collier, KJ (1997) Effects of crayfish (Paranephrops planifrons: Parastacidae) on in-stream processes and benthic faunas: a density manipulation experiment. New Zealand J Marine Freshw Res. 31: 685-692.

4. Flinders CA, Magoulick DD (2005) Distribution, habitat use and life history of stream-dwelling crayfish in the Spring River drainage of Arkansas and Missouri with a focus on the imperiled Mammoth Spring crayfish (*Orconectes marchandi*). Am Midl Nat 154: 358-374.

5. Momot WT (1995) Redefining the role of crayfish in aquatic ecosystems. Rev Fisheries Sci 3: 33-63.

6. Butler RS, DiStefano RJ, Schuster GA (2003) Crayfish: An overlooked fauna. Endang Species Bull 27: 10-11.

7. Taylor AT, Warren ML, Fitzpatrick JF, Hobbs HH, III, Jezerinac RF, Pflieger WL, et al. (1996) Conservation status of crayfishes of the United States and Canada. Fisheries 21: 25-38.

8. Creed RP, Reed JM (2004) Ecosystem engineering by crayfish in a headwater stream community. J North Am Benthol Soc. 23: 224-236.

9. Montemarano JJ, Kershner MW, Leff LG (2007) Crayfish effects on fine particulate organic matter quality and quantity. Fund. Appl. Limn/ Archiv für Hydrobiologie 168: 223-229.

10. Stewart PM, Miller MM, Heath WH, Simon TP (2010) Macrohabitat partitioning of crayfish assemblages in wadeable streams in the Coastal Plains of southeastern Alabama. Southeast Nat 9: 245-256.

11. Simon TP, Morris CC (2009) Biological response signature of oil brine threats, sediment contaminants, and crayfish assemblages in an Indiana watershed, USA. Arch Environ Contam Toxicol. 56:96-110.

12. Girard C (1852) A revision of the North American Astaci, with observations on their habits and geographical distribution. Proc Acad Nat Sci Phila 6.

13. Hobbs HH, Jr. (1989) Illustrated checklist of the American crayfishes (Decapoda: Astacidae, Cambaridae, and Parastacidae). Smithsonian Contrib Zool 480.

14. Crocker DW, Barr DW (1968) Handbook of the Crayfishes of Ontario. Royal Ontario Museum, University of Toronto Press, Toronto, CA, USA.

15. Page LM (1985) The crayfishes and shrimps (Decapoda) of Illinois. Ill Nat Hist Surv Bull 33: 406-412.

16. Taylor CA, Schuster GA (2004) Crayfishes of Kentucky. Ill Nat Hist Surv Spec Publ 28 Champaign, IL, USA.

17. Simon TP (2001) Checklist of the crayfish and freshwater shrimp (Decapoda) of Indiana. Proc Ind Acad Sci. 110: 104-110.

18. Hobbs HH, III, Jaas JP (1988) The crayfishes and shrimp of Wisconsin. Milwaukee Publ Mus,, Milwaukee, WI.

19. Momot WT (1966) Upstream movement of crayfish in an intermittent Oklahoma stream. Am Midl Nat. 75: 150-159.

20. Allan J (2004) Landscapes and riverscapes: The influence of land use on stream ecosystems. Annu Rev Ecol Evol Syst 35: 257-284.

21. Wang L, Lyons J, Kanehl P, Gatti R (1997) Influences of watershed land use on habitat quality and biotic integrity in Wisconsin streams. Fisheries 22: 6-12.

22. Burskey JL, Simon TP (2010) Reach- and watershed-418 scale associations of crayfish within an area of varying agriculture impact in west-central Indiana. Southeast Nat 9: 199-216.

23. Nerbonne BA, Vondracek B (2001) Effects of local land use on physical habitat, benthic macroinvertebrates, and fish in the Whitewater River, Minnesota, USA. 423 Environ Manage 28: 87-99.

24. Rabeni CF (1985) Resource partitioning by stream-dwelling crayfish: The Influence of body size. Am Midl Nat. 113: 20-29.

25. Stein RA (1977) Selective predation, optimal foraging, and the predator-prey interaction between fish and crayfish. Ecol 58:1237-1253.

26. Woods AJ, Omernik JM, Brockman CS, Gerber TD, Hosteter WD, et al. (2011) Ecoregions of Indiana and Ohio [map]. U.S. Environmental Protection Agency (US), Corvallis, OR, USA.

27. Rapid Watershed Assessment Lower East Fork White Watershed. (2011). US Depart Agriculture.

28. Strahler AN (1957) Quantitative analysis of watershed geomorphology. Trans Am Geophysical Union 38: 913-920.

29. Stevens D, Olsen AR (2004) Spatially-balanced sampling of natural resources. J Am Stat Assoc 99: 262-278.

30. Simon TP, Dufour R (1998) Development of index 440 of biotic integrity expectations for the ecoregions of Indiana: V. Eastern Corn Belt Plain. Environmental Protection Agency (US), Region 5, EPA 905-R-96-004, Chicago, IL, USA.

31. Rankin ET (1989) The Qualitative Habitat Evaluation Index (QHEI): Rationale, Methods, and Application. Ohio Environmental Protection Agency, Columbus, OH, USA.

32. Simon TP (2004) Standard Operating Procedures for the collection and study of burrowing crayfish in Indiana. I. Methods for the collection of burrowing crayfish in streams and terrestrial habitats. Misc Papers Ind Biol Surv Aquatic Res Center 2: 1-14.

33. Mather ME, Stein RA (1993) Direct and indirect effects of fish predation on the replacement of a native crayfish by an invading congener. Can J Fisheries Aquatic Sci 50: 1279-1288.

34. Barbour MT, Gerritsen J, Snyder BD, Stribling JB (1999) Rapid Bioassessment Protocols for Use in Streams and Wadeable Rivers: Periphyton, Benthic Macroinvertebrates and Fish, Second Edition. US Environmental Protection Agency; Office of Water; Washington, D.C, USA.

35. Kaufmann PR, Levine P, Robison EG, Seeliger C, Peck DV (1999) Quantifying physical habitat in wadeable streams. Environmental Protection Agency, Corvallis, OR, USA.

36. Hobbs HH Jr (1981) Crayfishes of Georgia. Smithsonian Contrib Zool. Number 318.

37. Indiana Map (2011) Indiana Geological Survey. Retrieved January 21, 2011.

38. Watershed Delineation Model (2013) 464 Watershed delineation software. retrieved June 27, 2013.

39. Hoggatt RE (1975) Drainage areas of Indiana. U.S. Depart Int, Geol Surv, Wat Res Div., Indianapolis, IN, USA.

40. Sokal RR, Rolf FJ (1995) The Principles and Practices of Statistics in Biological Research. 3rd 469 Edition. Freeman and Company, New York, NY, USA.

41. StatSoft Inc (2012) Electronic Statistics Textbook. Tulsa, OK: StatSoft.

42. England LE, Rosemond AD (2004) Small reductions in forest cover weaken terrestrial-aquatic linkages in headwater streams. Freshw Biol. 49: 721-734.

43. Saffran KA, Barton DR (1993) Trophic Ecology of Orconectes propinquus (Girard) in Georgian Bay (Ontario, Canada). Freshwat Crayfish. 9: 350-358.

44. Hrodey PJ, Sutton TM, Frimpong EA, Simon TP (2009) Land-use impacts on watershed health and integrity in Indiana warmwater streams. Am Midl Nat. 161: 76-95.

45. Stein RA, Magnuson JJ (1976) Behavioral response of crayfish to a fish predator. Ecol. 57: 751-761.

46. Pavey CR, Fielder DR (1996) The influence of size differential on agonistic behaviour in the freshwater crayfish, Cherax cuspidatus (Decapoda: Parastacidae). J Zool 238: 445-457.

13

Effects of Dietary Protein Levels on the Growth, Feed Utilization and Haemato-Biochemical Parameters of Freshwater Fish, *Cyprinus Carpio Var. Specularis*

Imtiaz Ahmed[*] **and Amir Maqbool**

Fish Nutrition Research Laboratory, Department of Zoology, University of Kashmir, Hazratbal, Srinagar

[*]**Corresponding author:** Imtiaz Ahmed, Fish Nutrition Research Laboratory, Department of Zoology, University of Kashmir, Hazratbal, Srinagar, India, E-mail: imtiazamu1@yahoo.com

Abstract

An 8-week feeding trial was conducted to study the effects of dietary protein levels on the growth, feed utilization and haemato-biochemical parameters of mirror carp, *Cyprinus carpio specularis* (1.50 ± 0.02 g; 4.5 ± 0.05 cm). Six casein-gelatin based isocaloric (367 kcal 100 g^{-1}, gross energy) diets containing graded levels of dietary protein (25%-50% CP) were formulated. 20 fish were randomly stocked in triplicate groups in 75L circular trough fitted with continuous flow-through system and fed experimental diets at 4% BW/day at 0800 and 1700h. Maximum live weight gain (258%), best feed conversion ratio (FCR) (1.63) and protein efficiency ratio (PER) (1.53) were obtained in fish fed diet containing 40% dietary protein. However, quadratic regression analysis live weight gain, FCR, PER and body protein deposition (BPD) data indicated requirements for dietary protein at 43.5%, 41.6%, 34.7% and 37.3% of dry diet, respectively. Significantly higher whole body protein, low moisture and intermediate body fat contents were recorded at 40% protein containing diet (P<0.05). While minimum ash content was recorded at 25% protein level. The highest HIS value (3.39%) was observed at the lowest protein level. Significant differences were also observed in Hb, HCT and RBC values of different groups fed with varying levels of dietary protein (P<0.05). Whereas, no significant differences were observed in their WBC count except at 25% protein level, where higher WBC count was recorded (P>0.05). Based on the above results, it is recommended that 41.5% protein level would be useful for optimum growth and efficient feed utilization of this fish species.

Keywords: *Cyprinus carpio specularis*; Dietary protein requirement; Growth; Blood; Biochemical parameters

Introduction

The global contribution of fish as a source of protein is indeed high, ranging from 10% to 15% of the human food basket across the world. It is estimated that around 60% of people in many developing countries depend on fish, for over 30% of their animal protein supplies [1]. The protein content of most fishes averages 15 to 20% on wet weight basis [2]. Fish also contains significant amounts of all essential amino acids, which are not available in plant protein sources and the digestibility of fish is approximately 5-15% higher than the plant-source foods [3].

Precise information on nutritional requirements of cultured species to provide appropriate amount of nutrients for optimal growth is essential to reduce feed cost, which accounts for a significant portion of the costs of an aquaculture enterprise. The development of cost-effective feeds that provide balanced diet to maximize growth, while minimizing environmental effects, depends on knowing the species' nutritional requirements and meeting those requirements with balanced diet formulations and appropriate feeding practices [4].

Among all the nutrients required by fish for growth and maintenance, protein is one of the most important and initial constituent, which comprises about 65-70% of the dry weight of fish muscle [5], and is also metabolized as an energy source by fish. Protein plays an important role in supporting fish growth [6-8]. Fish consume protein to obtain the essential and non-essential amino acids, which are necessary for muscle formation and enzymatic function and in part provide energy for maintenance [9]. Inadequate protein in the diet results in a reduction or cessation of growth and a loss of weight due to withdrawal of protein from less vital tissues to maintain the functions of more vital organs and tissues. Whereas, diet with excessive protein contents usually leads to extra energy costs, increased nitrogenous excretions and occasionally retarded fish growth [10,11]. Since protein constitutes in fish culture the single most expensive item in artificial feeds, it is logical to incorporate only that much, which is necessary for normal maintenance demand and growth. Any excess is considered wasteful, biologically as well as economically and therefore, it is important to minimize the amount of protein used for energy [12-14]. Thus an optimum dietary protein level in the diet is important for fish growth and maintenance of good farming environments [15].

Cyprinus carpio, as a freshwater fish species, has been one of the most widely cultured species all over the world due to its fast growth rate and easy cultivation [16]. Two varieties of common carp (*Cyprinus carpio*) viz: scale carp (*C.carpio var. communis*) and mirror carp (*C. carpio var. specularis*) are commercially cultured in Jammu and Kashmir. The mirror carp is detritus feeder feeding on decaying organic matter. This fish is herbivorous eating almost 80-85% plant food. It is column feeder [17]. The plant food consists of micro and macro phytes besides planktonic organisms. 15-20% of animal food consists of rotifers, annelids, crustaceans and insect larvae [18]. It is prolific breeder and has attained phenomenal population in all the lakes and rivers except for fast running cold hill stream. Almost 50% of fish population in valley lakes is mirror carp [19]. Mirror carp differs from other common carps in the development of back muscle (dorsal muscle) which is higher than the normal carp [20]. Due to this, mirror carp is also called as high back carp. Although some aspects of nutritional requirements of mirror carp have been worked out in the past by different workers [21-25], but no information related to the dietary protein requirement for the fingerling stage is available for this fish species. Keeping this in view, the present investigation was designed to study the effects of dietary protein levels on growth, feed utilization, whole body composition and haematological parameters of mirror carp, in order to determine the optimum dietary protein requirement of this fish, with a view to develop a nutritionally balanced diet for optimum production of this fish species through aquaculture.

Materials and Methods

Source of fish stock and acclimatization

Induced bred fingerlings of mirror carp, *Cyprinus carpio* var. *specularis* with the same batch and in apparent good health were procured from the 'State Government Fishery Department seed farm Manasbal'. The fingerlings were transported in polythene bags filled with water and oxygen and brought to the fish feeding trial laboratory (wet-lab) at the Department of Zoology, of . These fingerlings were first given a prophylactic dip in $KMnO_4$ (5 mgL^{-1}) to rule out any possible microbial infection and stocked in indoor circular aqua blue colored plastic fish tank (water volume = 600 L) for a fortnight. During this period, the fish were fed to satiation a mixture of soybean, mustard oil cake, rice bran, and wheat bran in the form of moist cake twice a day at 08:00 and 17:00 hours. These fingerlings were then acclimated for 2 weeks on H-440 diet [26] near to satiation twice a day at 08:00 and

17:00 h in the form of moist cake. A preliminary feed trial was conducted before the start of feeding trial to determine the appropriate feeding level and feeding schedule of the fish.

Preparation of experimental diets

Six casein-gelatin based isocaloric (367 kcal 100 g^{-1}, gross energy) diets containing graded levels of dietary protein (25%, 30%, 35%, 40%, 45%, and 50% crude protein) were formulated (Table 1). Diets were prepared taking into account the amount of protein contributed by casein and gelatin and made isocaloric by adjusting the amount of dextrin in the diet. Calculated quantities of dry ingredients were thoroughly mixed and stirred in a known volume of hot water (80°C) in a steel bowl attached to a Hobart electric mixer. Gelatin powder was dissolved separately in a known volume of water with constant heating and stirring and then transferred to the above mixture. Other dry ingredients and oil premix, except carboxymethyl cellulose (CMC), were added to the lukewarm bowl one by one with constant mixing at 40°C temperature. Carboxymethyl cellulose was added in last and the speed of the blender was gradually increased as the diet started to harden. The final diet, with the consistency of bread dough was poured into plastic Petri dishes and placed in a refrigerator to gel. The prepared diets were in the form of semi-moist cake, from which cubes were cut and packed in sealed polythene bags and then stored at -4°C until used. The composition of vitamin and mineral premixes were prepared as per Halver [26].

	Diet (%)					
Ingredients (g 100g^{-1}, dry diet)	(I)	(II)	(III)	(IV)	(V)	(VI)
Casein[1]	24.4	29.2	34	38.8	43.6	48.6
Gelatin[2]	6.1	7.3	8.5	9.7	10.9	12.15
Dextrin[3]	50.62	44.24	37.87	31.5	25.13	18.49
Corn oil	6	6	6	6	6	6
Cod liver oil	3	3	3	3	3	3
Mineral mix[4]	4	4	4	4	4	4
Vitamin mix[4,5]	3	3	3	3	3	3
Carboxymethyl cellulose	2	2	2	2	2	2
Alpha cellulose	0.88	1.26	1.63	2	2.37	2.76
Total	100	100	100	100	100	100
Calculated crude protein (g 100g^{-1})	25	30	35	40	45	50
Analysed crude protein (g 100g^{-1})	24.87	29.65	34.58	40.16	44.79	50.18
Gross energy[6] (kcal g100g^{-1}, dry diet)	367	367	367	367	367	367

Table 1: Formulation and proximate composition of experimental diets used for estimating the dietary protein requirement of mirror carp, *Cyprinus carpio var. specularis* fingerlings. [1]Crude protein (80%), [2]Crude protein (93%) Loba Chemie, India; [3]Loba Chemie, India. [4]Halver 2002 mineral ($AlCl_3$. $6H_2O$, 150; $ZnSO_4$. $7H_2O$, 3000; CuCl,100; MnSO4.$_4$-$6H_2O$, 800; KI,150; $CoCl_2$.$6H_2O$,1000 mg kg^{-1}; plus USP # 2 Ca (H2PO$_4$)$_2$. H_2O, 135.8; $C_6H10CaO_6$ 327.0; $C_6H_5O_7Fe$.5H_2O, 29.8; $MgSO_4$.7H_2O, 132.0; KH_2PO_4 (dibasic), 239.8; NaH_2PO_4.2H_2O, 87.2; NaCl, 43.5 (g kg^{-1}); [5]vitamin mix (choline chloride 5000: thiamin HCL 50; riboflavin 200; pyridoxine HCL 50; nicotinic acid 750; calcium pentothenate 500; inositol 2000; biotin 5.0; folic acid 15; ascorbic acid 1000; menadione 40; alpha-tocopheryl acetate 400; cyanocobalamine 0.1 (g kg^{-1}). [6]Calculated on the basis of physiological fuel values 4.5, 3.5 and 8.5 kcal g^{-1} for protein, carbohydrate and fat, respectively (Jauncey, 1982).

Experimental design and feeding trial

The fishes were sorted out from the acclimatized fish lots maintained in the wet laboratory and the desired number of *C. carpio* var. *specularis* fingerlings with almost similar body weight and size (1.50 ± 0.02 g; 4.5 ± 0.05 cm) were randomly selected in triplicate groups in 75 L high-density polyvinyl circular troughs (water volume 65 L) fitted with a continuous water flow-through system at the rate of 20 fish per trough for each dietary treatment levels. The water exchange rate in each trough was maintained at 1.0-1.5 L min^{-1}. The feeding schedule and feeding levels were chosen after carefully observing the feeding behaviour of the fish and their intake. For this purpose an 8-week preliminary feeding trial was also conducted under the same experimental setup in order to determine the appropriate ration size of the fish by feeding fish at the rate of 1%, 2%, 3%, 4%, 5% and 6% BW/day, results showed that the optimum ration size of the fish is approximately 4-4.5%. As per the result obtained in the preliminary feeding trial, the experimental fish were fed test diet in the form of moist cake at the rate of 4% of the body weight six days a week twice a day at 08:00 and 17:00 h, dividing into two equal feeding. The feeding trials lasted for eight weeks. Initial and weekly weights were recorded on a top loading balance (Sartorius CPA-224S 0.1 mg sensitivity, Goettingen, Germany). Fecal matter was siphoned before feeding and the daily feed offered was recorded. The uneaten feed (if any) was collected after active feeding approximately for 40 min with the help of siphoning pipe and collection tubes. The collected feed was then oven-dried at 100°C to calculate the final feed conversion ratio (FCR). No feed was offered to the fish on the day of weekly measurement. At the end of the experimental trial, desired numbers of fish were randomly sacrificed for the assessment of whole body composition.

Water quality analysis

The physico-chemical parameters of water (temperature, dissolved oxygen, free carbon dioxide, total alkalinity and pH) were recorded daily, following the standard methods [27]. The water sample for analysis was collected early in the morning before the feeding was done. Water temperature (23.6-24.5°C) was recorded using a mercury thermometer, dissolved oxygen (6.1-6.8 mg L^{-1}) was estimated by Winkler's iodimetric test, free carbon dioxide (3.9-5.7 mg L^{-1}), total alkalinity (91-112 mg L^{-1}) by titrimetric methods, respectively. While, pH (7.2-7.6) was measured by using a digital pH meter (pH ep-HI 98107, USA).

Chemical analysis

The Proximate composition of casein, gelatin, and experimental diet, initial and final carcass was estimated using standard [28] methods for dry matter (oven drying at 105 ± 1°C for 22 h), crude protein (N-Kjeldhal X 6.25), crude lipid (solvent extraction with petroleum ether B.P 40-60°C) by using Soxtec extraction technique (FOSS Avanti automatic 2050, Sweden), and ash (oven incineration at 650°C for 2-4 h) were determined. At the end of the experiment, eight fish were randomly pooled from each replicate of dietary treatment and three sub-samples of each replicate from the pooled sample (n=3×3) were analysed for final body composition. Similarly, three fish were randomly selected from each replicate of dietary treatment for organ index estimation and blood collection.

Hematological parameters

At the termination of feeding trial, blood samples for analysis were collected in heparinized (Na-heparinised) capillary tubes from the haemal arch after severing the caudal peduncle. Blood was pooled from each test group and stored in heparin coated vaccutainer plastic tubes for future tests. All the hematological analysis was carried out within 2hours after each extraction.

Haemoglobin (Hb)

Haemoglobin content of blood was estimated by the method of Drabkin [29]. 20μl of blood was mixed with 5 ml of Drabkin solution (Loba chemie, India) and left to stand for at least 15 minutes. Haemoglobin concentration was determined by measuring the absorbance at 540 nm and compared to that of haemoglobin standard (Ranbaxy, India). Prior to reading the absorbance, hemoglobin test samples were centrifuged to remove dispersed nuclear material.

Haematocrit (HCT)

Haematocrit (HCT%) was determined on the basis of sedimentation of blood. Heparinised blood (50μl) was taken in a micro-haematocrit capillary (Na-heparinised) and spun in a micro-haematocrit centrifuge (REMI RM-12C, India) at 12,000 rpm for 5 min to obtain haematocrit value. The haematocrit value was measured using a haematocrit reader and reported as percentage [30].

Red blood cell (RBC) and white blood cell count (WBC)

For RBC and WBC count, a blood sample (20 μl) was taken with a micro pipette (Finpipette, Finland), and diluted with Natt-Herrick's [31] diluent (1:200). The diluted sample was placed in a Neubauer improved haemocytometer (Marienfeld-Superior, Lauda-Konigshofen, Germany) and then the blood cells were counted using a light microscope (Magnus-MLM, India). RBC indices viz: mean corpuscular haemoglobin (MCH), mean corpuscular hemoglobin concentration (MCHC) and mean corpuscular volumes (MCV) were calculated according to Dacie and Lewis [32].

Growth parameters

Growth performance of the fish fed diets with different protein levels was calculated as a function of the weight gain by using the following formulae:

Weight gain (%) = Final body weight-initial body weight/initial weight x 100

Specific growth rate (SGR %) = 100 x
(In final wet weight (g)-In initial wet weight g)/duration (days)

Protein efficiency ratio (PER)=
Wet weight gain (g) / Protein consumed (g, dry weight basis)

Body protein deposition (BPD %)=
100 x (BW_f x BCP_f) - (BW_i x BCP_i) / [TF X CP]

Where BWi and BWf = mean initial and final body weight (g), BCPi and BCFf = mean initial and final percentage of muscle protein,   TF

=Total amount of diet consumed, and CP=Percentage of crude protein of the diet.

$$\text{Hepatosomatic index (HSI \%)} = \frac{\text{Liver weight (g)}}{\text{Body weight (g)}} \times 100$$

Statistical analysis

Responses of mirror carp fingerlings to graded levels of dietary protein were measured by weight gain (%) feed conversion ratio (FCR), protein efficiency ratio (PER), specific growth rate (SGR %) and body composition. These response variables were subjected to one-way analysis of variance (ANOVA) [33,34]. To determine the significant differences among the treatments, Duncan's Multiple Range Test [35] was employed. To predict more accurate responses to the dietary protein intake, the optimum dietary protein level was estimated using second-degree polynomial regression analysis ($Y = ax^2 + bx + c$) as described by Zeitoun et al. [36]. Statistical analysis was done using SPSS 11.5 (SPSS Inc., Chicago, IL, USA).

Results

Growth performance of mirror carp, *Cyprinus carpio* var. *specularis* fed diets containing graded levels of protein over the 8-week feeding trial are presented in Table 2. No mortality was observed among all the dietary treatment levels during the entire length of feeding trial. Live weight gain (LWG%), specific growth rate (SGR%), feed conversion ratio (FCR) and protein efficiency ratio (PER) were found to be significantly affected with the increase of dietary protein level in the diets. A linear relationship between the percentage of protein content in the diet and the increase in weight gain up to an incorporation rate of 40% was noted. The maximum weight gain (258%) for mirror carp was obtained with the diet containing 40% dietary protein level, although it was not significantly different from that achieved by the fish fed a 45% protein diet. However, an intermediate value of growth rate was observed in fish fed diet containing lower level of dietary protein i.e. <40% and higher level of dietary protein (>45%) diets, while the poorest growth rate was recorded for fish receiving diet with 25% protein followed by those receiving diet containing 30% protein in the diet, respectively. Feed conversion ratio decreased progressively with linearly increasing dietary protein level and was found to differ significantly among each dietary protein level (P<0.05). The best-FCR (1.63) was recorded with fish receiving diet at 40% dietary protein level, which was not significantly different to group that fed at 45% protein containing diet (P>0.05).

	Dietary protein levels (g 100 g⁻¹, dry diet)					
	25	**30**	**35**	**40**	**45**	**50**
Average initial weight (g)	1.557 ± 0.04	1.592 ±0.02	1.608 ± 0.01	1.599 ± 0.02	1.605 ± 0.02	1.614 ± 0.03
Average final weight (g)	3.142 ± 0.15	3.968 ± 0.07	4.909 ± 0.07	5.731 ± 0.05	5.623 ± 0.11	5.189 ± 0.13
Live weight gain (%)	101.64 ± 5.10d	149.25 ± 7.58c	205.37 ± 6.80b	258.48 ± 8.36a	250.43 ± 8.50a	221.48 ± 5.23b
Specific growth rate (SGR)	1.25 ± 0.04d	1.63 ± 0.04c	1.99 ± 0.03b	2.28 ± 0.03a	2.24 ± 0.04a	2.08 ± 0.02b
Feed conversion ratio (FCR)	2.93 ± 0.06a	2.45 ± 0.05b	1.92 ± 0.05d	1.63 ± 0.04e	1.70 ± 0.05e	2.08 ± 0.07c
Protein efficiency ratio (PER)	1.36 ± 0.03b	1.37 ± 0.02b	1.49 ± 0.04a	1.53 ± 0.04a	1.31 ± 0.03b	0.96 ±0.03c
Body protein deposition (BPD)	19.46 ± 0.53d	21.41 ± 0.44c	24.60 ± 0.63b	28.50 ± 0.69a	23.86 ± 0.70b	16.60 ± 0.69e
HIS (%)	3.39 ± 0.05a	3.08 ± 0.05b	2.80 ± 0.07c	2.41 ± 0.02d	2.33 ± 0.04d	2.72 ± 0.03c
Survival (%)	100	100	100	100	100	100

Table 2: Growth, FCR, protein deposition and percentage survival of mirror carp, *Cyprinus carpio var. specularis* fingerlings fed diets containing varying levels of dietary protein for 8 weeks (mean values of 3 replicates + SEM; n=3)*. *Mean values of 3 replicates ± SEM; Mean values sharing the same superscript are not significantly different (P>0.05).

The protein efficiency ratio in fish fed varying dietary protein levels differed significantly and showed an increasing tendency with increasing dietary protein level (P<0.05), which increased from 1.35 to 1.53 for fish fed 20% and 40% dietary protein, respectively. Whereas, a significant decline was observed in PER for fish fed 45% and 50% protein diets, with the lowest (0.96) PER being noted at 50% dietary protein level. Overall significantly highest PER (1.53) was recorded when fish were fed a diet containing 40% protein (P<0.05). The hepatosomatic index (HSI) value of mirror carp in the present study also showed some significant differences between the treatments, with maximum values observed with fish fed at lowest protein containing diet.

In order to get statistically more precise information, all the growth parameters were subjected to second-degree polynomial regression analysis. When live weight gain data (Y) and dietary protein levels (X) were analyzed using second-degree polynomial analysis, a break-point was evident at 43.5% dietary protein level (Figure 1). The relationship was described by the equation:

$$Y = -0.4568x^2 + 39.7203x - 616.0843 \ (r = 0.969)$$

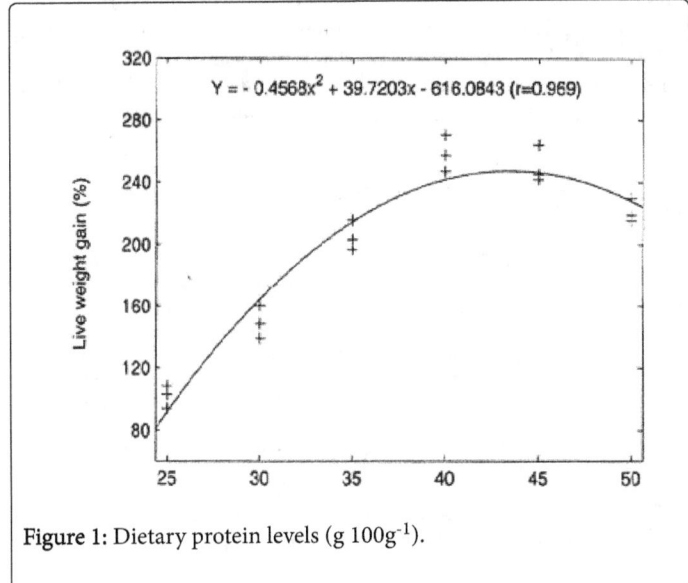

Figure 1: Dietary protein levels (g 100g⁻¹).

The specific growth rate of mirror carp fed varied levels of dietary protein also produced somewhat similar trends as obtained in the growth rate. The SGR (Y) to dietary protein level (X) was also analyzed by using second-degree polynomial regression analysis (Figure 2) and the break point was evident at 43.22% protein level. The mathematical equation was:

$$Y = -0.0031x^2 + 0.2662x - 3.5271 \ (r = 0.975)$$

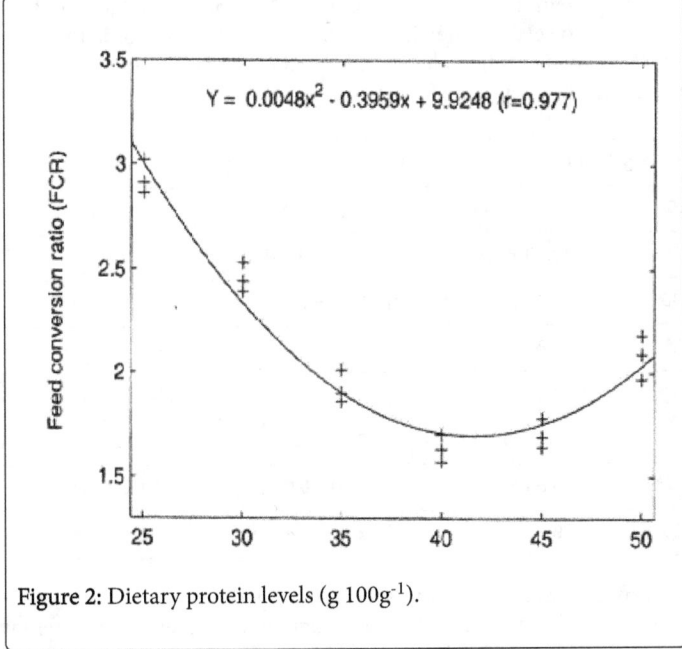

Figure 2: Dietary protein levels (g 100g⁻¹).

The FCR of mirror carp fed 40% and 45% dietary protein was significantly lower than those fed other dietary protein levels. The FCR (Y) to dietary protein levels (X) relationship was also best described using a second-degree polynomial regression analysis (Figure 3). The relationship being:

$$Y = 0.0048x^2 - 0.3959x + 9.9248 \ (r = 0.977)$$

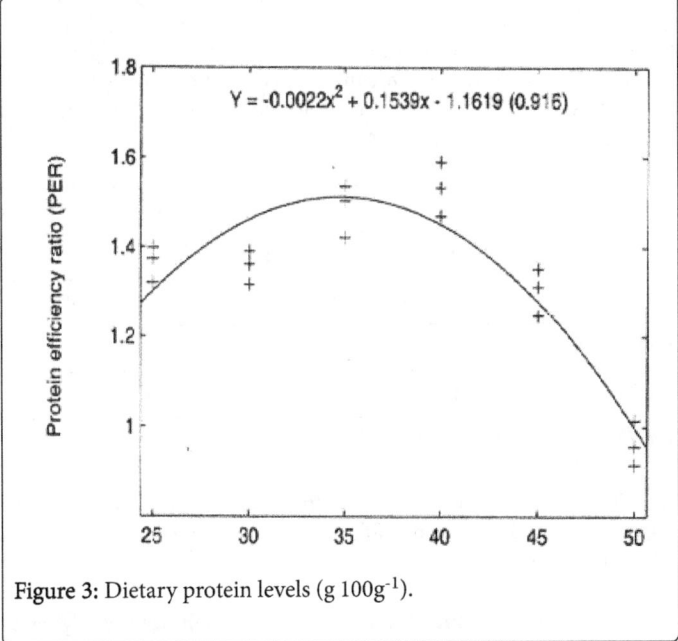

Figure 3: Dietary protein levels (g 100g⁻¹).

Significantly ($P<0.05$) highest PER was recorded with fish fed at 40% protein containing diet. The PER (Y) to dietary protein level (X) was also best described using second-degree polynomial regression analysis (Figure 4). The equation being as:

$$Y = -0.0022x^2 + 0.1538x - 1.1618 \ (r = 0.916)$$

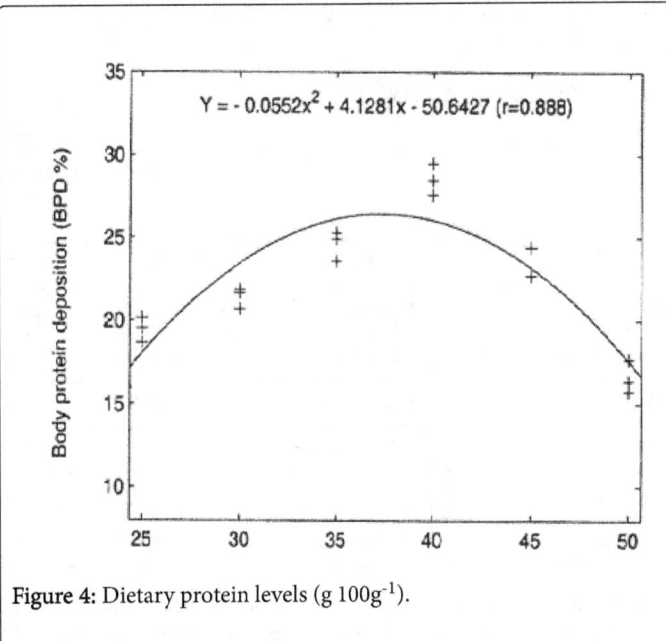

Figure 4: Dietary protein levels (g 100g⁻¹).

Based on the above polynomial equations the FCR and PER indicated that the optimum dietary protein requirement of mirror carp was estimated to be at 41.6% and 34.7%, respectively.

At the end of feeding trial, significant differences in whole body composition were observed among all the dietary groups ($P<0.05$) (Table 3). Generally, body composition was affected by increasing dietary protein levels. Whole body moisture content gradually decreased with the increase in the dietary protein content of the diet up to 45%. However, fish fed 45% protein containing diet produced significantly lowest whole body moisture content (P<0.05), which was insignificantly different to the group that fed 40% protein diet (P>0.05). Whole body protein content was significantly higher in fish fed diet containing 40% protein followed by those receiving diet containing 45% and 50%, respectively (P<0.05). Whole body fat content gradually increased with the increase of dietary protein level and significantly highest body fat content was recorded with fish group that were fed 45% protein diet (P<0.05), followed by those fed at 40% protein diet, while intermediate whole body fat values were recorded in those groups that fed 35% and 50% protein diets, respectively.

Dietary protein levels (g 100 g^{-1}, dry diet)							
	Initial	25	30	35	40	45	50
Moisture (%)	80.86 ± 0.44	77.49 ± 0.26^a	75.66 ± 0.040^b	74.34 ± 0.22^c	72.57 ± 0.17^d	72.25 ± 0.20^d	73.75 ± 0.14^c
Protein (%)	12.01 ± 0.15	13.15 ± 0.06^f	14.27 ± 0.05^e	15.06 ± 0.08^d	16.77 ± 0.07^a	16.49 ± 0.05^b	15.62 ± 0.06^c
Fat (%)	3.37 ± 0.13	4.81 ± 0.09^f	5.41 ± 0.11^e	6.13 ± 0.07^d	6.72 ± 0.08^b	7.61 ± 0.09^a	6.43 ± 0.07^c
Ash (%)	2.66 ± 0.05	3.16 ± 0.04^a	2.90 ± 0.02^b	2.61 ± 0.02^e	2.72 ± 0.04^d	2.65 ± 0.03d,^{e}	2.82 ± 0.02^c

Table 3: Whole body composition of fingerling, *Cyprinus carpio var. specularis* fed diets containing graded levels of dietary protein for 8 weeks (mean values of 3 replicates + SEM; n=3) *. *Mean values of 3 replicates ± SEM; Mean values sharing the same superscript are not significantly different (P>0.05).

Whole body ash content was found to be significantly higher at lower dietary protein containing diets i.e. 25% and 30%, whereas significantly lower ash content values were observed in fish fed the remaining dietary protein levels (P<0.05). Also fish fed diet containing 40% protein resulted in highest whole body protein deposition (BPD %), which was significantly highest among all the dietary groups. Second-degree polynomial regression analysis was also employed between the body protein deposition (Y) to dietary protein level (X) and a break-point was obtained at 37.3% protein level. The mathematical equation was:

$$Y = - 0.0552x^2 + 4.1281x - 50.6427 \ (r = 0.888)$$

The haematological parameters of mirror carp fed diets containing varied dietary protein levels also produced some significant differences (Table 4). The fish fed diet containing 40% and 45% protein diets had significantly highest haemoglobin (Hb) content, followed by those receiving 50% protein diet (P<0.05). Whereas, intermediate values of Hb content were recorded at 35% protein diet, while poorest Hb content was estimated at lowest level of protein diet i.e. 25%. Haematocrit (HCT) values increased significantly with the increase in dietary protein levels from 25% to 45% protein containing diets (P<0.05). However, higher HCT value (38.26%) was recorded for fish fed 45% protein diet, while the lowest HCT value (22.19%) was noted at the lowest protein level (25%).

Dietary protein levels (g 100g-1, dry diet)						
	25	30	35	40	45	50
Hb (gdl^{-1})[1]	6.86 ± 0.07^e	7.77 ± 0.08^d	9.27 ± 0.07^c	10.78 ± 0.10^a	10.92 ± 0.15^a	10.24 ± 0.07^b
HCT (%)[2]	22.19 ± 0.69^e	26.403 ± 0.83^d	30.83 ± 0.90^c	36.77 ± 0.72^a	38.26 ± 0.56^a	33.58 ± 0.50^b
RBC (×106/mm^3)[3]	1.23 ± 0.02^e	1.34 ± 0.03^d	1.52 ± 0.03^c	1.78 ± 0.04^b	1.90 ± 0.05^a	1.62 ± 0.04^c
WBC (×104/mm^3)[4]	2.42 ± 0.04^a	2.36 ± 0.05^{ab}	2.33 ± 0.03^{ab}	2.28 ± 0.04^b	2.30 ± 0.06^{ab}	2.24 ± 0.05^b
MCV (fl)[5]	180.33 ± 5.61^b	196.74 ± 8.74^a	202.62 ± 6.55^a	208.61 ± 4.73^a	201.78 ± 2.83^a	207.93 ± 6.38^a
MCH (pg)[6]	55.71 ± 0.53^d	57.86 ± 0.81^c	60.91 ± 1.02^{ab}	60.58 ± 0.83^b	57.60 ± 0.91^c	63.21 ± 0.13^a
MCHC (gdl^{-1})[7]	30.93 ± 0.98^a	29.46 ± 0.90^{ab}	30.10 ± 1.03^{ab}	29.34 ± 0.55^{ab}	28.94 ± 0.44^b	30.42 ± 0.64^{ab}

Table 4: Effect of experimental diets on hematological parameters of mirror carp, *Cyprinus carpio var. specularis* fingerlings for 8 weeks (mean values of 3 replicates + SEM; n=3)*. *Mean values of 3 replicates ± SEM; Mean values sharing the same superscript are not significantly different (P>0.05). [1]Haemoglobin concentration; [2]Haematocrit; [3]Red blood cell count; [4]White blood cell count; [5]Mean corpuscular volume; [6]Mean corpuscular haemoglobin; [7]Mean corpuscular haemoglobin concentration.

Red blood cell counts (RBC) in fish fed various dietary protein levels also produced significant differences. Significantly highest RBC value (1.9×106 mm^{-3}) was noted at 45% protein diet, followed by those receiving diet at 40% protein diet (P<0.05). Intermediate RBC values were recorded in fish fed other dietary protein levels, except those fed 25% and 30% protein diets, where significantly lowest RBC count values were obtained (P<0.05). Whereas, the fish fed varied levels of dietary protein could not produce any significant difference in their leukocyte (WBC) counts, except at lowest levels where slightly higher WBC counts were recorded.

No significant differences in mean corpuscular volume (MCV) values were observed in the present study, when fish were fed varied levels of dietary protein diets (P>0.05), except at lowest protein containing diet i.e. 25% where, significantly lowest MCV (180.33 fl) value was noted (P<0.05). Similar trends were also observed in mean corpuscular haemoglobin concentration (MCHC) with fish fed varied levels of dietary protein containing diets, while mean corpuscular haemoglobin (MCH) data in the present study showed significant differences among different groups. The highest MCH values were noted at 50% and 35% protein containing diets, which were not significantly different among each other. Whereas, the intermediate values of MCH were recorded in other dietary groups.

Discussion

Understanding the dietary protein requirement of fingerling stage of mirror carp becomes a pre-requisite for the development of nutritionally balanced, efficient and cost effective feed for culturing practice. In the present study, graded levels of dietary protein content had a significant effect on the growth rate, feed conversion ratio, protein efficiency ratio and specific growth rate. The growth and conversion efficiencies gradually increased with the increase of dietary protein levels from 25% to 45% protein containing diet. Although the maximum growth parameters were obtained when fish were fed at 45% protein containing diet, however, this growth rate was not significantly different to those groups that were fed at 40% protein diet. Whereas, the best-FCR, PER, SGR and BPD was recorded with fish fed 40% protein diet. Therefore, inclusion of 40% protein in the diet for fingerling mirror carp is more appropriate and economical. Also the growth rate significantly fell beyond the requirement level, especially at 50% protein diet, indicating that 40% protein diet (Diet IV) satisfied the protein requirement of the fish and is considered optimum for achieving maximum growth and efficient nutrient conversion efficiency. The decrease in growth rate at protein levels above the optimum requirement may be attributed to the fact that the fish body cannot utilize the dietary protein once after reaching the optimum protein level [37]. The excessive protein content in the diet could reduce the growth performance of fish due to higher energy requirement for catabolism rather than for protein deposition. The decrease in weight gain, when the fish were fed excess level of dietary protein may also be because of a reduction in available energy for growth and due to inadequate non-protein energy necessary to deaminate the high protein feed [38,39]. The reduced growth rate and decreased protein utilization beyond requirement of dietary protein level is well documented in the past by several workers [39-46].

In general, both, feed conversion ratio and protein efficiency ratio were poor in lower protein containing diets. However, improvement in FCR and PER was noticed with increasing incorporation of dietary protein levels. The best-FCR and highest PER values were recorded with fish fed at 40% protein containing diet. The BPD and PER increased with the increase in dietary protein content up to 40% and thereafter, a significant decrease was recorded with further elevation of dietary protein level i.e. 45% and 50% protein containing diets (Diet V and Diet VI). Similar trends in PER and BPD were also reported by other workers [47].

The whole body composition data showed that whole body moisture content gradually increased with the increase of dietary protein levels, with minimum moisture content was recorded at 40% protein diet. Whole body protein content linearly increased with the increase of dietary protein level up to 40% and thereafter, a decline in body protein content was noted. The highest protein content obtained in the present study, when fish fed at 40% protein diet could be due to the fact that at this particular level fish utilized the available protein content for growth more efficiently than those fed other dietary protein levels. Similar results on body protein content have also been reported by Kim et al. [39]. Kim and Lee [48] further reported that body protein content responded to dietary protein levels in a dose dependent manner and exhibited maximum protein content on that dietary protein level where maximum growth rate was also achieved.

Whole body fat content gradually increased with the increase of dietary protein levels and maximum body fat content was recorded at higher dietary protein containing diet (Diet VI). The higher whole body fat content beyond the optimum protein requirement level in the diet may be due to the fact that the excess dietary protein content in these diets gets deaminated and stored as body fat. The whole body ash did not show any significant difference among the treatment levels, except at lower protein containing diets where high body ash content was recorded. The fish fed varied levels of dietary protein produced some significant differences in HSI values. The highest HSI value (3.39%) was observed at the lowest dietary protein level, which was significantly higher compared to all the dietary groups. Higher values of HSI in lower protein diet could be due to the poor growth and health of the fish [49-51] and also due to more fat accumulations in the liver [11,15,47,52].

Besides biochemical analysis, haematological analysis was also carried out in the present study in order to find out the effects of dietary protein levels on these parameters, which are recognized as valuable tools for monitoring fish health, physiological responses, assessment of feed composition and nutritional status in relation to environmental stress [53-56].

In the present study, significant differences were observed in Hb, HCT and RBC values of different groups fed with varying levels of dietary protein, showing a general trend of linear increase with the increase of dietary protein levels. However, the haematological values obtained in the present study were within acceptable limits as reported by Svobodova et al. [56], for common carp. Fishes alter their metabolic profile to cope up with the different dietary conditions [57,58]. Hb and HCT values significantly increased with the increase of dietary protein levels from 25% - 45% protein containing diets. However, highest Hb (10.92 gdl^{-1}) and HCT values (38.26%) were recorded for fish fed 45% protein diet. An increase in RBC count was evident with the increase in dietary protein levels, which may have occurred due to its early release from the storage pool in the spleen [59,60], thus, causing a change in MCH values as well.

On the basis of second-degree polynomial regression analysis of growth parameters and body composition data, the optimum dietary protein level for optimum growth of mirror carp, *C. carpio var. specularis* fingerling is recommended to be at 41.50%. The protein requirement of fish varies from species to species and is reported with in the renge from 30 to 56% [12,13]. The protein requirement of mirror carp estimated during the present study in terms of percentage is comparable with the requirements reported for other fish species (Table 5). The differences in protein requirement among the fish species may be due to different dietary formulations, fish size and different methodologies adopted [61,62]. The variations may also be attributed to different lab conditions, experimental design e.g. feeding level and frequency, water quality, water flow rate, stocking density and protein sources in the diet [63]. Moreover, the protein requirement of fish may also vary with the feeding rate adopted. It has been reported that a decrease in the dietary protein requirement of juvenile carp and rainbow trout from 60-65% to as low as 30-32%, when feeding rate was increased from 2-4% body weight^{-1} [12].

Fish Species	Protein requirement (%)	References
Mirror carp, *C. carpio var. specularis*	41.5	Present study
Indian major carp, *Cattla catla*	40-47	Khan and Jafri [63], Singh and Bhanot [64]
Indian major carp, *Cirrhinus mrigala,*	36	Singh et al. [65]
Jian carp, *Cyprinus carpio*	34.1	Liu et al. [66]
Indian major carp, rohu, *Labeo rohita*	25-35	Satpathy et al. [67], Khan et al. [68] Debnath et al. [69]
Big head carp, *Aristichthys nobilis*	30	Santiago and Reyes [70]
African catfish, *Clarias. gariepinus,*	40-43	Degani et al. [71], Ali and Jauncey [72], Farhat and Khan [73]
Magur, *Clarias batrachus*	40	Khan and Jafri [74]
Malaysian catfish, *M. nemurus*	42	Khan et al. [75]
Juvenile sunshine bass, *M. chrysops x M. saxatilis*	41	Webster et al. [76]
Mangrove red snapper, *Lutjanus argentimaculatus,*	40	Catacutan et al. [77]
Juvenile masu salmon, *Oncorhynchus masuo*	40	Lee and Kim [78]
Mahseer, *Tor putitora,*	40	Hossain et al. [79]
African Cichlid, *Pseudotropheus socolofi*	40	Royes and Murie [80]
Milkfish, *Chanos chanos*	40	Jana et al. [81]
Juvenile blackspot sea bream, *Pagellus bogaraveo,*	40	Silva et al. [82]
Cuneate drum, *Nibea miichthioides*	40	Wang et al. [83]
Persian sturgeon, *Acipenser persicus*	40	Mohseni et al. [84]
Mexican silverside, *Menidia estor*	40.9	Martinez-Palacios et al. [85]
Tiger puffer, *Takifugu rubripes*	41	Kim and Lee [47]
Singhi, *Heteropnestus fossilis*	35-40	Akand et al. [86], Qamar and Khan [87]
Black sea bream, *Sparus macrocephalus*	41.4	Zhang et al. [88]
Pacific threadfin, *Polydactylus sexfilis*	41	Deng et al. [14]
Grey mullet, *Mugil capito*	24	Papaparaskera-Papoutsoglou and Alexis [89]
Nile tilapia, *Oreochromis niloticus*	25-45	Abdel-tawwab et al. [11], Siddiqui et al. [90], El-Saidy and Gaber [91]
Tilapia, *O. mossambicus*	28	De Silva et al. [92]
Juvenile silver perch, *Bidyanus bidyanus,*	42.15	Yang et al. [9]
Juvenile, *Spinibarbus hollandi,*	32.7	Yang et al. [51]
Black catfish, *Rhamdia quelen*	37	Salhi et al. [93]
Channel catfish, *Ictalurus punctatus*	28	Li et al. [94]
Golden shiner, *Notemigonus crysoleucas*	29	Lochmann and Phillips [95]
Blue streak hap, *Labidochromis caeruleus*	35	Ergun et al. [96]

Amazonian tambaqui, *Colossoma macropomum*	30	Oishi et al. [97]
Brown trout, *Salmo trutta*	57	Arzel et al. [98]
Grouper, *Epinephelus malabaricus*	44	Shiau and Lan [99]
Juvenile Florida pompano, *Trachinotus carolinus,*	45	Lazo et al. [100]
Discus, *Symphysodon spp.*	44.9-50.1	Chong et al. [101]
American eel, *Anguilla rostrata,*	47	Tibbetts et al. [40]
Spotted sand bass, *Paralabrax maculatofascinatus*	45	Alvarez-Gonzalez et al. [102]
Juvenile haddock, *Melanogrammus aeglefinus,*	49.9-54.6	Tibbetts et al. [60], Kim et al. [103]
Bagrid catfish, *Mystus nemurus*	44	Ng et al. [104]
Juvenile olive flounder, *Paralichthys olivaceus,*	46.4-51.2	Kim et al. [38]
Mahseer, *Tor putitora,*	45-50	Islam and Tanaka [105]
Juvenile turbot, *Scophthalmus maximus*	55	Cho et al. [106]
Pike perch, *Sander lucioperca*	43	Nyina-wamwiza et al. [107]
Black sea bass, *Centropristis striata*	45-52	Alam et al. [108]
Malaysian mahseer, *Tor tambroides,*	48	Ng et al. [109]
Silver pomfret, *Pampus argenteus,*	49	Hossain et al. [110]
Asian red-tailed catfish, *Hemibagrus wyckioides*	44.12	Deng et al. [111]
Sharpsnout sea bream, *Diplodus puntazzo*	43	Coutinho et al. [112]
Tongue sole, *Cynoglossus semilaevis*	55	Liu et al. [113]

Table 5: Dietary protein requirements of various cultivated fish species compared with *C. carpio var. specularis.*

The present study indicates that the dietary protein level influences fish growth, feed conversion ratio and haemato-biochemical composition of fish and therefore, it is recommended that the inclusion of 41.50% dietary protein in the diet is optimum for the growth, efficient feed utilization of mirror carp, *C. carpio* var. *specularis* fingerling. Data generated in the present study would be useful in developing nutritionally balanced diets for the intensive and semi-intensive culture of this fish species [64-113].

Acknowledgements

The author are grateful to the Head, Department of Zoology, University of Kashmir, Hazratbal, Srinagar, India for providing necessary laboratory facilities and also gratefully acknowledge the State Government Fishery Department Seed Farm Manasbal for provided fish seed for this experiment. We also gratefully acknowledge the Department of Science and Technology (DST), Govt of India, New Delhi for provided the financial support for the establishment of Fish Nutrition Research and Feed Technology Laboratory (Wet-Lab.) in the Department of Zoology.

References

1. Food Agricultural Organization (FAO) (2005) Nutritional elements of fish. Topics Fact Sheets. Text by Lahsen Ababouch. In: FAO Fisheries and Aquaculture Department, Italy, Rome.

2. Murray J, Burt JR (FAO) (1983) The Composition of Fish Torry Advisory Note: 38.

3. World Health Organization (WHO) (1985) Energy and protein requirements, Geneva: World Health Organization.

4. NRC (2011) Nutrient Requirements of Fish and Shrimp. National Academy Press, Washington, DC.

5. Wilson RP, Halver JE (1986) Protein and amino acid requirement of fishes. Ann Rev Nutr 6: 225-244.

6. Jones PL, De Silva SS, Mitchell DB (1996) The effect of dietary protein source on growth and carcass composition in juvenile Australian freshwater crayfish. Aquacult Int 4: 361-367.

7. Lee SM, Cho SH, KD (2000) Effects of dietary protein and energy levels on growth and body composition of juvenile flounder Paralichtys olivaceus. J World Aquac Soc 30: 306-315.

8. Luo Z, Liu YJ, Mai KS, Tian LX, Liu DH, et al. (2004) Optimal dietary protein requirement of grouper, Epinephelus coioides juveniles fed isoenergetic diets in floating net cages. Aquacult Nutr 10: 247-252.

9. Yang SD, Liou C, Liu F (2002) Effects of dietary protein level on growth performance , carcass composition and ammonia excretion in juvenile silver perch, Bidyanus bidyanus. Aquaculture 213: 363-372.

10. Monentcham SE, Pouomigne V, Kestemont P (2009) Influence of dietary protein levels on growth performance and body composition of African bonytongue fingerlings Heteriostis niloticus (Cuvier, 1829). Aquacult Nutr 16: 144-152.

11. Abdel-Tawwab M, Ahmad MH, Khattab YAE, Shalaby AME (2010) Effect of dietary protein level, initial body weight, and their interaction on the

growth, feed utilization, and physiological alterations of Nile tilapia, Oreochromis niloticus. Aquaculture 298: 267-274.

12. NRC (1993) Nutrient Requirements of Warmwater Fishes and Shellfishes, National Academy Press, Washington, DC pp: 102.

13. De Silva SS, Anderson TA (1995) Fish Nutrition in Aquaculture pp: 319.

14. Deng D, Yong Z, Dominy W, Murashige R, Wilson RP (2011) Optimal dietary protein levels for juvenile Pacific threadfin (Polydactylus sexfilis) fed diets with two levels of lipid. Aquaculture 316: 25-30.

15. Guo Z, Zhu X, Liu J, Han D, Yang Y, et al. (2012) Effects of dietary protein level on growth performance, nitrogen and energy budget of juvenile hybrid sturgeon, Acipenser baerii ♀× A. gueldenstaedtii ♂. Aquaculture 338: 89-95.

16. Guler GO, Kiztanir B, Aktumsek A, Citil OB, Ozparlak H (2008) Food chemistry determination of the seasonal changes on total fatty acid composition and n3/n6 ratios of carp (Cyprinus carpio L.) muscle lipids in Beysehir Lake (Turkey). Food Chem 108: 689-694.

17. Subla BA (1967) Studies on the functional anatomy of the alimentary canal. part III:on the functional anatomy of feeding apparatus and the food of some Kashmir fishes. Kashmir Sci 4: 148-166.

18. Das SM, Subla BA (1970) The Pamir-Kashmir theory of the origin and evolution of ichthyofauna of Kashmir. Ichthyologica 10: 8-11.

19. Fotedar DN, Qadri MY (1974). Fish and fisheries of Kashmir and the impact of Carp (Cyprinus carpio) on the endemic fishes. J Sci 2: 79-89.

20. Ivantcheva E, Todorov M (1989) Carcass evaluation of a different morphologic type of the mirror carp. Zhivotnovud Nauki 26: 58-64.

21. Ufodike EBC, Matty AJ (1983) Growth responses and nutrient digestibility in mirror carp (Cyprinus carpio) fed different levels of cassava and rice. Aquaculture 31: 41-50.

22. Kim JD, Kim KS (1994) Comparisons of commercial feeds on the growth and nutrient discharge into water by growing mirror carp (Cyprinus carpio). Korean J Anim Sci 36: 710-717.

23. Kim JD, Kim KS, Song JS, Jeong KS, Won CH, et al. (1994) Effects of microbial phytase supplementation to soybean meal-based diet on growth and excretion of phosphorus in mirror carp (Cyprinus carpio). Korean J Anim Nutr Feed 20: 109-116.

24. Kim JD, Kim KS, Song JS, Jeong KS, Woo YB, et al. (1996) Comparison of feces collection methods for determining apparent phosphorus digestibility of feed ingredients in growing mirror carp (Cyprinus carpio). Korean J Anim Nutr Feed 20: 201-206.

25. Halver JE (2002) The vitamins. In: Halver JE, Hardy RW (eds). Academic Press, San Diego, CA. Fish Nutrition, 3rd edn pp: 61-141.

26. APHA (1992) Standard methods for the examination of water and wastewater, 18th edn. APHA, Washington DC pp: 1268.

27. AOAC (1995) In: Cunniff P (ed.) Official methods of analysis of the association of official analytical chemists, 16th edn. Arlington, Virginia.

28. Drabkin DL (1946) Spectrometric studies XIV. The crystallographic and optimal properties of the hemoglobin of man comparison with those of other species. J Biol Chem 164: 703-723.

29. Del Rio-Zaragoza OB, Hernandez-Rodriguez M, Buckle-Ramirez LF (2008) Thermal stress effect on tilapia Oreochromis mossambicus (Pisces: Cichlidae) blood parameters. Mar Freshwater Behav Physiol 41: 135-145.

30. Natt MP, Herrick CA (1952) A new blood diluent for counting erythrocytes and leucocytes of the chicken. Poult Sci 31: 735-738.

31. Dacies S, Lewis S (1991) Practical hematology, 7th ed. Churchill Livingstone, London pp: 633.

32. Snedecor GW, Cochran WG (1967) Statistical methods (6th edn.) Iowa state university press, Iowa pp: 593.

33. Sokal RR, Rohlf FJ (1981) Biometry. Freeman New York, WH pp: 859.

34. Duncan DB (1955) Multiple range and multiple 'F' tests. Biometrics 11: 1-42.

35. Zeitoun IH, Ullrey DE, Magee WT, Gill JL, Bergen WG (1976) Quantifying nutrient requirements of fish. J Fish Res Bd Can 33: 167-172.

36. Phillips AM (1972) Calories and energy requirement. In: Halver JE (ed). Academic press, New York, Fish Nutrition pp: 28.

37. Jauncey K (1982) The effects of varying dietary protein level on the growth, food conversion, protein utilization and body composition of juvenile tilapias (Sarotherodon mossambicus). Aquaculture 27: 43-54.

38. Kim KW, Wang XJ, Bai SC (2002) Optimum dietary protein level for maximum growth of juvenile olive flounder, Paralichthys olivaceus (Temminck and Schlegel). Aquacult Res 33: 673-679.

39. Jobling M, Wandshik A (1983) Quantitative protein requirement of Artic Charr, Salvelinus alpines. J Fish Biol 22: 705-712.

40. Tibbetts SM, Lall SP, Anderson DM (2000) Dietary protein requirement of juvenile American eel (Anguilla rostrata) fed practical diets. Aquaculture 186:145-155.

41. Sales J, Truter P J, Britz PJ (2003) Optimum dietary crude protein level for growth in South African abalone (Haliotis midae L). Aquacult Nutr 9: 85-89.

42. Kalla A, Bhatnagar A, Garg SK (2004) Further studies on protein requirements of growing Indian major carps under field conditions. Asian Fish Sci 17: 191-200.

43. Cho SH, Lee SM, Lee JH (2005) Effect of dietary protein and lipid levels on growth and body composition of juvenile turbot (Scophthalmus maximus L.) reared under optimum salinity and temperature conditions. Aquacult Nutr 11: 235-240.

44. Kim LO, Lee SM (2005) Effects of dietary protein and lipid levels on growth and body composition of bagrid catfish, Pseudobagrus fulvidraco. Aquaculture 243: 323-329.

45. Sa R, Ferreira PP, Teles AO (2006) Effect of dietary protein and lipid levels on growth and feed utilization of White Sea bream (Diplodus sarus) juveniles. Aquacult Nutr 12: 310-321.

46. Lee SM, Kim DJ, Cho SH (2002) Effects of dietary protein and lipid level on growth and body composition of juvenile ayu (Plecoglossus altivelis) reared in seawater. Aquacult Nutr 8: 53-58.

47. Kim S, Lee K (2009) Dietary protein requirement of juvenile tiger puffer (Takifugu rubripes). Aquaculture 287: 219-222.

48. Brauge C, Medale F, Corraze G (1994) Effect of dietary carbohydrate levels on growth, body composition and glycaemia in rainbow trout, Oncorhynchus mykiss, reared in seawater. Aquaculture 123: 109-120.

49. Hamre K, Ofsti A, Naess T, Nortvedt R, Holm JC (2003) Macronutrient composition of formulated diets for Atlantic halibut (Hippoglossus hippoglossus L) Juveniles. Aquaculture 227: 233-244.

50. Moreira IS, Peres H, Couto A, Enes P, Teles AO (2008) Temperature and dietary carbohydrate level effects on performance and metabolic utilisation of diets in European sea bass (Dicentrarchus labrax) juveniles. Aquaculture 274: 153-160.

51. Yang SD, Lin TS, Liou CH, Peng HK (2003) Influence of dietary protein level on growth performance, carcass composition and liver lipid classes of juvenile Spinibarbus hollandi (Oshima). Aquavult Res 3: 661-666.

52. Bhaskar BR, Rao KS (1984) Influence of environmental variables on haematological ranges of milkfish, Chanos chanos (Forskal), in brackish water culture. Aquaculture 83: 123-136.

53. Schuett DA, Lehmann J, Goerlich R, Hamers R (1997) Haematology of swordtail, Xiphiphorus helleri I: blood parameters and light microscopy of blood cells. J Appl Icthyol 13: 83-89.

54. Jawad LA, Al-Mukhtar MA, Ahmed HK (2004) The relationship between haematocrit and some biological parameters of the Indian shad, Tenualosa ilisha (Family Clupeidae). Anim Biodiver Cons 27: 47-52.

55. Svobodova Z, Machova J, Drastichova J, Groch L, Luskova V, et al. (2005) Haematological and biochemical profiles of carp blood following nitrite exposure at different concentrations of chloride. Aquacult Res 36: 1177-1184.

56. Lundstedt LM, Melo JFB, Moraes G (2004) Digestive enzymes and metabolic profile of Pseudoplatystoma corruscans (Teleostei: Siluriformes) in response to diet composition. Comp Biochem Physiol 137: 331-339.

57. Melo JFB, Lundstedt LM, Meton I, Baanante IV, Moraes G (2006) Effects of dietary levels of protein on nitrogenous metabolism of Rhamdia quelen (Teleostei: Pimelodidae). Comp Biochem Physiol 145: 181-187.

58. Vijayan MM, Leatherland JF (1989) Cortisol-induced changes in plasma glucose, protein, and thyroid hormone levels, and liver glycogen content of coho salmon (Oncorhynchus kisutch Walbaum). Canadian J Zool 67: 2746-2750.

59. Pulsford AL, Lemaire-Gony S, Tomlinson M, Collingwood N, Glynn PJ (1994) Effects of acute stress on the immune system of the dab, Limanda limanda. Comp Biochem Physiol 109: 129-139.

60. Tibbetts SM, Lall SP, Milley JE (2005) Effects of dietary protein and lipid levels and DPDE-1 ratio on growth, feed utilization and heptosomatic index of juvenile haddock, Melanogrammus aeglefinus L Aquacult Nutr 11: 67-75.

61. Kim JD, Kim KS, Song JS, Lee JY, Jeong KS (1998) Optimum level of dietary monocalcium phosphate based on growth and phosphorus excretion of mirror carp, (Cyprinus carpio). Aquaculture 161: 337-344.

62. Kim KI, Kayes TB, Amundson CH (1992) Requirements for lysine and arginine by rainbow trout (Oncorhynchus mykiss). Aquaculture 106: 333-344.

63. Khan MA, Jafri AK (1991) Dietary protein requirement of two size classes of the Indian major carp, Catla catla Hamilton. J Aqua Trop 6: 79-88.

64. Singh BN, Bhanot KK (1988) Protein requirement of the fry of Catla catla (Ham.). In: Mohan M, Joseph M (eds.) Proceedings of the First Indian Fisheries Forum, AFS, India.

65. Singh RK, Chavan SL, Desai AS, Khandagale PA (2008) Influence of dietary protein levels and water temperature on growth, body composition and nutrient utilization of Cirrhinus mrigala (Hamilton, 1822) fry. J Thermal Biol 33: 20-26.

66. Liu Y, Feng L, Jiang J, Liu Y, Zhou X (2009) Effects of dietary protein levels on the growth performance, digestive capacity and amino acid metabolism of juvenile Jian carp (Cyprinus carpio. Var. jian). Aquaculture 40: 1073-1082.

67. Satpathy BB, Mukherjee D, Ray AK (2003) Effects of dietary protein and lipid level on growth, feed conversion and body composition in rohu, Labeo rohita (Hamilton), fingerlings. Aquacult Nutr 9: 17-24.

68. Khan MA, Jafri AK, Chadha NK (2005) Effects of varying dietary protein levels on growth, reproductive performance, body and egg composition of rohu, Labeo rohita (Hamilton). Aquacult Nutr 11: 11-17.

69. Debnath D, Pal AK, Sahu NP, Yengkokpam S, Baruah K, et al. (2007) Digestive enzymes and metabolic profile of Labeo rohita fingerlings fed diets with different crude protein levels. Comp Biochem Physiol Part B 146: 107-114.

70. Santiago CB, Reyes OS (1991) Optimum dietary protein level of growth of bighead carp (Aristichthys nobilis) fry in a static system. Aquaculture 93: 155-165.

71. Degani G, Yigal BZ, Levanon D (1989) The effect of different protein levels and temperatures on feed utilization, growth and body composition of Clarias gariepinus (Burchell 1822). Aquaculture 76: 293-301.

72. Ali MZ, Jauncey K (2005) Approaches to optimizing dietary protein to energy ratio for African catfish, Clarias gariepinus (Burchell, 1822). Aquacult Nutr 11: 95-101.

73. Farhat, Khan MA (2011) Growth, feed conversion, and nutrient retention efficiency of African catfish, Clarias gariepinus, (Burchell) fingerling fed diets with varying levels of protein. J Appl Aquacult 23: 304-316.

74. Khan MA, Jafri AK (1990) On the dietary protein requirement of Clarias batrachus Linnaeus. J Aqua Trop 5: 191-198.

75. Khan MS, Ang KJ, Ambak MA, Saad CR (1993) Optimum dietary protein requirement of a Malaysian catfish, Mystus nemurus. Aquaculture 112: 227-235.

76. Webster CD, Tiu LG, Tidwell JH, Wyk PV, Howerton RD (1995) Effects of dietary protein and lipid levels on growth and body composition of sunshine bass (Morone chrysops x Morone saxatilis) reared in cages. Aquaculture 131: 291-301.

77. Catacutan MR, Pagador GE, Teshima S (2001) Effect of dietary protein and lipid level and protein to energy ratio on growth, survival and body composition of the mangrove red snapper, Lutjanus argentimaculatus (Forsskal 1775). Aquacult Res 32: 811-818.

78. Lee SM, Kim KD (2001) Effects of dietary protein and energy levels on the growth, protein utilization and body composition of juvenile masu salmon (Oncorhynchus masou Brevoort). Aquacult Res 32: 39-45.

79. Hossain MA, Hasan N, Hussain MG (2002) Optimum dietary protein requirement of mahseer, Tor putitora (Hamilton) fingerlings. Asian Fisher Sci 15: 203-214.

80. Royes JB, Murie DJ, Francis-Floyd R (2005) Optimum dietary protein level for growth and protein efficiency without hepatocyte changes in juvenile African cichlids Pseudotropheus socolofi. North American J Aquacult 67: 102-110.

81. Jana SN, Garg SK, Patra BC (2006) Effect of inland water salinity on growth performance and nutritional physiology in growing milkfish, Chanos chanos (Forsskal): field and laboratory studies. J Appl Ichthyol 22: 25-34.

82. Silva P, Andrade CAP, Timoteo VMFA, Rocha E, Valente LMP (2006) Dietary protein, growth, nutrient utilization and body composition of juvenile blackspot seabream, Pagellus bogaraveo (Brunnich). Aquacult Res 37: 1007-1014.

83. Wang Y, Guo J, Bureau DP, Cui Z (2006) Effects of dietary protein and energy levels on growth, feed utilization and body composition of cuneate drum, Nibea miichthioides. Aquaculture 252: 421-428.

84. Mohseni M, Sajjadi M, Pourkazemi M (2007) Growth performance and body composition of sub-yearling Persian sturgeon, (Acipenser persicus, Borodin, 1987), fed different dietary protein and lipid levels. J Appl Ichthyol 23: 204-208.

85. Martinez-Palacios CA, Rios-Duran MG, Ambriz-Cervantes L, Jauncey KJ, Ross LG (2007) Dietary protein requirement of juvenile Mexican silverside (Menidia estor Jordan 1879) a stomachless zooplanktophagous fish. Aquacult Nutr 13: 304-310.

86. Akand AM, Miah MI, Haque MM (1989) Effect of dietary protein level on growth, feed conversion and body composition of chingi (Heteropneustes fossilis Bloch). Aquaculture 77: 175-180.

87. Siddiqui TQ, Khan MA (2009) Effects of dietary protein levels on growth, feed utilization, protein retention efficiency and body composition of young Heteropneustes fossilis (Bloch). Fish Physiol Biochem 35: 479-488.

88. Zhang J, Zhou F, Wang L, Shao Q, Xu Z, (2010) Dietary Protein Requirement of Juvenile Black Sea Bream, Sparus macrocephalus. J World Aqua Soc 41: 151-154.

89. Papaparaskera-Papoutsoglou E, Alexis MN (1986) Protein requirement of young grey mullet, Mugil capito. Aquaculture 52: 105-115.

90. Siddiqui AQ, Howlander MS, Adam AA (1988) Effects of dietary protein levels on growth, feed conversion and protein utilization in fry and young Nile Tilapia, Oreochromis niloticus. Aquaculture 70: 63-73.

91. El-Saidy DMSD, Gaber MMA (2005) Effect of dietary protein levels and feeding rates on growth performance, production traits and body composition of Nile tilapia, Oreochromis niloticus (L.) cultured in concrete tanks. Aquacult Res 36: 163-171.

92. De Silva SS, Gunasekera RM, Atapattu D (1989) The dietary protein requirements of young tilapia and an evaluation of the least cost dietary protein levels. Aquaculture 80: 271-284.

93. Salhi M, Bessonart M, Chediak G, Bellagamba M, Carnevia D (2004) Growth, feed utilization and body composition of black catfish, Rhamdia quelen, fry fed diets containing different protein and energy levels. Aquaculture 231: 435-444.

94. Li MH, Robinson EH, Oberle DF (2006) Effects of dietary protein concentration and feeding regimen on channel catfish, Ictalurus punctatus, production. J World Aquac Soc 37: 370-377.

95. Lochmann RT, Phillips H (1994) Dietary protein requirement of juvenile golden shiners (Notemigonus crysoleucas) and goldfish (Carassius auratus) in aquaria. Aquaculture 128: 277-285.

96. Ergun S, Guroy D, Tekesoglu H, Guroy B, Celik I (2010) Optimum dietary protein level for blue streak hap, Labidochromis caeruleus. Turkish J Fisher Aquat Sci 31: 27-31.

97. Oishi CA, Nwanna LC, Filho MP (2010) Optimum dietary protein requirement for Amazonian Tambaqui, Colossoma macropomum Cuvier, 1818, fed fish meal free diets. Acta Amazonica 40: 757-762.

98. Arzel J, Metailler R, Kerleguer C, Delliou HL, Guillaume J (1995) The protein requirement of brown trout (Salmo trutta) fry. Aquaculture 130: 67-78.

99. Shiau SY, Lan CW (1996) Optimum dietary protein level and protein to energy ratio for growth of grouper (Epinephelus malabaricus). Aquaculture 145: 259-266.

100. Lazo JP, Davis DA, Arnold CR (1998) The effects of dietary protein level on growth, feed efficiency and survival of juvenile Florida pompano (Trachinotus carolinus). Aquaculture 169: 225-232.

101. Chong ASC, Hashim R, Ali AB (2000) Dietary protein requirements for discus (Symphysodon spp). Aquacult Nutr 6: 275-278.

102. Alvarez-Gonzalez CA, Civera-Cerecedo R, Ortiz-Galindo JL, Dumas S, Moreno-Legorreta M, et al. (2001) Effect of dietary protein level on growth and body composition of juvenile spotted sand bass, Paralabrax maculatofasciatus, fed practical diets. Aquaculture 194: 151-159.

103. Kim JD, Lall SP, Milley JE (2001) Dietary protein requirement of juvenile haddock (Melanogrammus aeglefinus L). Aquacult Res 32: 1-7.

104. Ng WK, Soon SC, Hashim R (2001) The dietary protein requirement of a bagrid catfish, Mystus nemurus (Cuvier & Valenciennes), determined using semipurified diets of varying protein level. Aquacult Nutr 7: 45-51.

105. Islam MS, Tanaka M (2004) Optimization of dietary protein requirement for pond-reared mahseer, Tor putitora Hamilton (Cypriniformes: Cyprinidae). Aquacult Res 35: 1270-1276.

106. Cho SH, Lee SM, Lee JH (2005) Effect of dietary protein and lipid level on growth and body composition of juvenile turbot (Scophthalmus maximus L.) reared under optimum salinity and temperature conditions. Aquacult Nutr 11: 235-250.

107. Nyina-wamwiza L, Xu LX, Blanchard G, Kestemont P (2005) Effect of dietary protein, lipid and carbohydrate ratio on growth, feed efficiency and body composition of pike- perch Sander lucioperca fingerlings. Aquacult Res 36: 486-492.

108. Alam MS, Watanabe WO, Carroll PM (2008) Dietary protein requirements of juvenile black sea bass, Centropristis striata. J World Aquac Soc 39: 656-663.

109. Ng W, Abdullah N, De Silva SS (2008) The dietary protein requirement of the Malaysian mahseer, Tor tambroides (Bleeker), and the lack of protein sparing action by dietary lipid. Aquaculture 284: 201-206.

110. Hossain MA, Almatar SM, James CM (2010) Optimum dietary protein level for juvenile silver pomfret, Pampus argenteus (Euphrasen). J World Aquac Soc 41: 710-720.

111. Deng J, Zhang X, Bi B, Kong L, Kang B (2011) Dietary protein requirement of juvenile Asian red-tailed catfish Hemibagrus wyckioides. Anim Feed Sci Tech 170: 231-238.

112. Coutinho F, Peres H, Guerreiro I, Pousao-Ferreira P, Oliva-Teles (2012) Dietary protein requirement of sharpsnout sea bream (Diplodus puntazzo, Cetti 1777) juveniles. Aquaculture 356-357: 391-397.

113. Liu X, Mai K, Ai Q, Wang X, Liufu ZG, et al. (2013) Effects of Protein and Lipid Levels in Practical Diets on Growth and Body Composition of Tongue Sole, Cynoglossus semilaevis Gunther. J World Aquac Soc 44: 97-104.

De Novo Assembly and Analysis of the Testes Transcriptome from the Menhaden, *Bervoortia tyrannus*

Frank J Zadlock IV[1]*, **Satshil B Rana[1]**, **Zain A Alvi[1]**, **Ziping Zhang[2]**, **Wyatt Murphy[1]** and **Carolyn S Bentivegna[3]**

[1]*Department of Biological Science, Seton Hall University, South Orange, New Jersey, USA*

[2]*College of Animal Science, Fujian Agriculture and Forestry University, Fuzhou, China*

[3]*Department of Chemistry and Biochemistry, Seton Hall University, South Orange, New Jersey, USA*

*Corresponding author: Frank J Zadlock IV, Department of Biological Science, Seton Hall University, South Orange, New Jersey, USA, E-mail: Frank.Zadlock@student.shu.edu

Abstract

Background: The menhaden, *Bervoortia tyrannus*, is one of the most important fish within the oceanic ecosystem and a crucial species supporting major fisheries along the Atlantic and Gulf coasts. However, little is known about menhaden from a genetic aspect. The objective of this project is to apply high throughput sequencing to the testes of menhaden to provide the genetic tools required to further study their population dynamics

Result: We applied Illumina Next Seq 500 technology to two different testes and used Velvet/Oases to perform the de novo assembly that resulted in the construction of 254,462 contigs. Applying BLASTX to annotate the contigs against the non-redundant protein database resulted in 46.89% matches. To validate the accuracy of the assembly, the reads were mapped back to transcripts (RMBT) with a percentage of 87.83%. To experimentally verify the assembly results, primers were designed based on the assembled transcriptome, and PCR products were verified by Sanger sequencing. To enhance the functional categorization of the annotated contigs, they were further classified using Gene Ontology (GO), Kyoto Encyclopedia of Genes and Genomes (KEGG), and Clusters of Orthologous Groups (COG) databases.

Conclusion: This research is the first report of an annotated overview of the testes transcriptomes in *B. tyrannus*, resulting in the most comprehensive genetic resource available for menhaden to date. This work can provide a repository for future gene expression analysis, functional studies, and reproductive investigations in *B. tyrannus*. This will enhance the capabilities of population monitoring and can be used as a benchmark in comparative studies in other fish models. Overall, this research will open new opportunities and bring new insights for researchers studying *B. tyrannus*.

Keywords: Menhaden fish; De novo assembly; Assembly validation; Transcriptome analysis

Introduction

Menhaden (Family Clupeidae, Genus Brevoortia) are high fecundity, filter-feeding marine teleost fish that are considered to be one of the most economically and ecologically important species in North America [1]. In oceans and estuaries, they contribute to ecosystem health by clearing the water of excess algal biomass and detritus [2,3]. They are also the main food source for a wide variety of predatory invertebrates (jellyfish, squids, etc.) fish, (striped bass, bluefish, etc.,) birds, (osprey and brown pelican), and marine mammals [4].

From an economic standpoint, there are two established commercial fisheries for menhaden. The first is the reduction fishery that turns the menhaden into fish oil omega-3 supplements for example, and into fishmeal for livestock and aquaculture consumption [4,5]. The second is the bait fishery that uses the menhaden as bait for bluefish, crab, and lobster [6-15]. They have been rarely studied at the genetic level most likely due to the lack of genomic and transcriptome data.

In this study, we applied Illumina Next Seq 500 technology to two different menhaden testes and performed de novo assemblies on the generated raw reads using Velvet/Oases. BLASTX was utilized to annotate the assembled contigs against the NCBI non-redundant protein database. To validate the accuracy of the assemblies in silico, the contigs were mapped back to transcripts (RMBT) [16]. To experimentally verify the assembly results, primers were designed based on the assembled transcriptome and PCR products were verified by Sanger sequencing. To enhance the functional categorization of the annotated contigs, they were further classified using Gene Ontology (GO), Kyoto Encyclopedia of Genes and Genomes (KEGG), and Clusters of Orthologous Groups (COG) databases. This research also identified microsatellites, and various repetitive DNA elements between within the transcriptome.

To date, this research is the first report of an annotated overview for the testes transcriptome in *B. tyrannus*, resulting in the most comprehensive genetic resource available for the species. This work can provide a repository for future gene expression analysis, functional studies, and reproductive investigations in *B. tyrannus*. This will enhance the capabilities of population monitoring and can be used as a benchmark in comparative studies in other fish models. Overall, this research will open new opportunities and bring new insights for researchers studying *B. tyrannus*.

Materials and Methods

Fish collection and nucleic acid isolation

Male fish identified as *Brevoortia tyrannus* were collected off the coast of New Jersey in November 2013 by using a trolling net. The authorities who issued the permission for the capturing of the menhaden were NOAA, National Marine Fisheries Service, Northeast Regional Office (Permit #410087) and the State of New Jersey, Department of Environmental Protection (Permit #1333). In all instances the fish were alive when captured, and capture methods followed approved animal handling protocols reviewed by the authorities who issued the field permits. The vertebrate work was approved by Virginia Institutes of Marine Science's Institutional Animal Care and Use Committee (IACUC-2011-02-04-7125-jxgart) and NEAMAP Inshore Trawl Survey. The fish were sacrificed using spinal cord dislocation and the gonads were dissected from two separate male menhaden. Tissue samples were stored in RNAlater (Qiagen) at -20°C prior to RNA extraction.

Total RNA was extracted from each testis using TRIzol® Kit (Invitrogen™) following the manufacturer's instructions. RNA samples were then digested by DNase I to remove potential genomic DNA. A BioAnalyzer 2100 (Agilent Technologies) was used to validate that the RNA integrity number (RIN) for each sample had a value >7.0.

Illumina short-read library construction and sequencing

Waksman Genomics Facility at Rutgers University conducted all steps in transcriptome library preparation. Ambion MicroPoly A purist Kit was used to remove ribosomal RNA (rRNA) and recover high quality mRNA from the menhaden samples. Agilent Genomics BioAnalyzer mRNA Nano Kit was used to quantify the mRNA and confirm a successful rRNA depletion (<2% rRNA attained). The dUTP strand specific cDNA library preparation strategy was employed using the NEBNext Ultra Directional RNA Library Prep Kit for Illumina. Illumina Tru-Seq adapters were used to barcode the libraries and were amplified for 12-15 cycles of PCR. Illumina NextSeq 500 High Output with 300 cycle kit was used to sequence the 155×155 bp pair end libraries. Agilent Technologies and ThermoFisher Scientific's Qubit DNA HS Assay Kit was used to quantify the completed RNA-seq libraries.

Sequence data processing

The quality assessment and adapter trimming was performed using FastQC v0.10.1 and Trimmomatic v0.32 on the raw Illumina sequence reads [17,18]. All low quality reads with a Phred score value below 20 were removed. After trimming, FastQC analysis was performed again to verify the quality of the remaining raw sequence data. The data sets were further cleaned from contaminating sequences using Deconseq with the parameters set to 90% of the contig length with an identity of 94% [19].

De novo transcriptome assembly and annotation

The high quality filtered reads were assembled using Velvet (v1.2.07), which assembles short reads using the de Bruijn algorithm [20] along with Oases (v0.2.08), which operates on the output of Velvet [21]. Firstly, to create the input files for Velvet, the paired-end fastq files were interleaved. Then a multiple k-mer assembly strategy was applied to assemble k-mer sizes 35, 41, 51, 61, 71, 81, and 91 [22]. The seven k-mer assemblies were merged with Oases followed by CD-HIT-EST (v 4.6.1) to further remove the redundancy and cluster the contigs for annotation [23]. Lastly, the two separate transcriptomes were merged together with CD-HIT-EST.

Annotation of the contigs was accomplished using local BLASTX (v2.2.29+). Homologous sequences were searched for against the NCBI non-redundant (Nr) protein database using an E-value <1e-5 [24]. Gene annotations were assigned based on the top BLASTX hit. Functional annotation was performed by merging the results from Blast2GO (v 2.7.2) with the results from InterProScan to expand the number of annotated sequences [25-27]. These annotations were used to assign putative functionalities, level-two GO terms, and KEGG (Kyoto Encyclopedia of Genes and Genomes) based metabolic pathways via BLAST2GO [28]. To further elucidate possible functions, the contig sequences were aligned to the Clusters of Orthologous Groups (COG) database using BLASTX with an E-value <1e-5 [29].

Assembly and annotation assessment

In order to demonstrate that the de novo assembly was performed properly, an in silico method was used to match the assembled contigs to protein sequences of related species. To accomplish this, we utilized PhyloT (http://phylot.biobyte.de/contact.html) to generate a phylogenetic tree between menhaden (B. tyrannus) and the eleven publically available fish genomes (Poecilia formosa, Amazon molly; Astyanax mexicanus, Mexican tetra; Gadus morhua, Atlantic cod; Takifugu rubripes, Japanese pufferfish; Oryzias latipes, medaka; Xiphophorus maculatus, southern platyfish; Lepisosteus oculatus, spotted gar; Gasterosteus aculeatus, stickleback; Tetraodon nigroviridis, green spotted pufferfish; Oreochromis niloticus, Nile tilapia; and Danio rerio, zebrafish) on Ensembl. This program creates trees based on the NCBI taxonomy database, and it was visualized by the web based tool, Interactive Tree of Life (v2) [30]. Based on the analysis, zebrafish (D. rerio) and Mexican tetra (A. mexicanus) were shown to be the closest related. All assembled contigs were then compared to the Ensembl proteins of zebrafish and Mexican tetra using BLASTX with an E-value cut-off of 1e-5. Secondly, to determine the accuracy of the assembly, the percentage of raw reads that could be mapped back to transcripts (RMBT) was determined [16]. To accomplish this, indexes were generated using bowtie2-build followed by Bowtie2 (v 2.2.5) to map the reads against the assembly [31].

To experimentally verify the assembly results, primers were designed based on the assembled transcriptome and PCR was performed on 8 genes (CYP17a1, 3-β-HSD, β-actin, GAPDH, HIF-1a, StaR, ARNT, and EGR). From the isolated RNA, a cDNA library was constructed using oligo-dT primers (Applied Biosystems, Foster City, CA). The cDNA was used as a template to amplify the genes of interest. The PCR amplification was performed at 94°C for 30 sec, 60°C for 30 sec, and 72°C for 30 sec using the designed primers pairs for each gene. These primers were designed using Primer3Plus (http://www.bioinformatics.nl/cgi-bin/primer3plus/primer3plus.cgi) and are shown in Table 1. The presence of a unique PCR product of appropriate size was verified by agarose gel electrophoresis and the gene of interest was confirmed using commercially available Sanger sequencing (Genewiz Inc. Plainfield, NJ, USA) (data not shown).

Repetitive element investigation and microsatellite identification

Assembled sequences were scanned with RepeatMasker (v 4.0.5) to identify all repetitive elements using zebrafish as a reference [32]. Microsatellite motifs were identified using the program Msatfinder (v 2.0.9) [33]. The repeat thresholds for di-, tri-, tetra-, penta-, hexa-nucleotide motifs were set as 8, 5, 5, 5 and 5. The mononucleotide repeats were removed by modifying the perl script. For future PCR validation, 80 microsatellite sequences that met the selection criteria of having flanking sequences longer than 50 bp on both sides have been designed as seen in Table 2.

Results

To globally profile the two testes transcriptomes of menhaden, we employed Illumina NextSeq 500 technology to sequence the libraries generating 10,765,249 pair-end short reads encoding 1,560,309,410 bases for Menhaden 1 and 17,910,601 pair-end short reads encoding 2,491,172,965 bases for Menhaden 2 (Table 1). All the raw sequencing reads were deposited into the Short Read Archive (SRA) of the National Center for Biotechnology Information (NCBI), and can be accessed under the accession numbers SRX892008 and SRX994942, respectively.

	Menhaden 1	Menhaden 2
Number of nucleotide bases	1,56,03,09,410	2,49,11,72,965
Number of raw reads	1,07,65,249	1,79,10,601
Number of clean reads for assembly	89,13,332	1,29,57,697
Percent of used reads	83%	72%

Table 1: Statistics of the raw reads after Illumina sequencing and processing.

To improve the accuracy of the assembly, the raw sequence reads were cleaned to remove Illumina adaptor sequences, low quality reads with a Phred score value less than 20, and contaminating sequences.

	Menhaden Testes
Contig Number	2,54,462
N50 Length	1,324 bp
Minimum Contig Length	200 bp
Largest Contig Length	30,617 bp
Average contig length	831.68 bp
GC (%)	46
RMBT %	87.83
Nr Database Match	46.89%
Zebrafish Genome Match	42.56%
Mexican tetra Genome Match	41.34%

Table 2: Statistics of the Velvet/Oases Assembly and Annotation.

The filtering of the raw sequence reads resulted in 8,913,332 (83%) clean reads for Menhaden 1 and 12,957,697 (72%) clean reads for Menhaden 2 (Table 1). Velvet and Oases, common assembly programs successfully used in previous Illumina based de novo transcriptome studies, were employed to perform the assembly [7,10,14,21,22,34,35]. The paired-end sequence reads were assembled into 254,446 contigs with an N50 length of 1,324 bp and average contig length of 831.68 bp (Table 2). The contig length distribution ranged from 200 bp to more than 3,000 bps as shown in Figure 1.

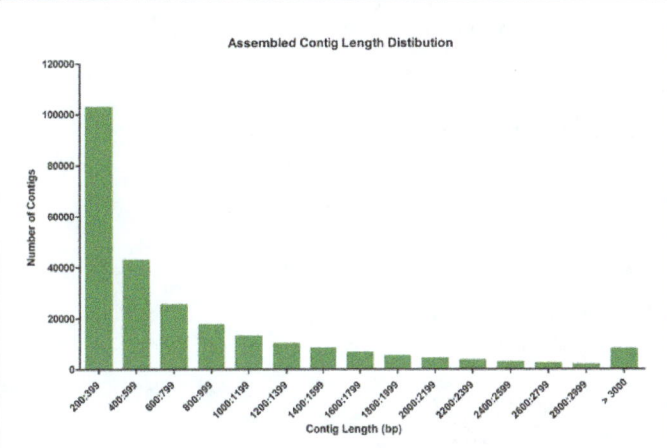

Figure 1: Contig length distribution results. The distribution of the contig lengths after the assembly.

Assessment of transcriptome assembly

It is crucial to assess whether or not a de novo assemblies is reliable.

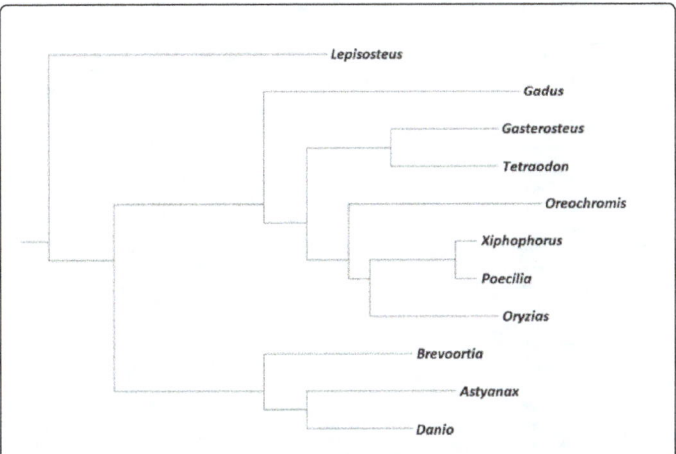

Figure 2: Phylogenetic tree analysis. A phylogenetic tree analysis of the 11 publically available fish genomes on Ensembl reveals that *D. rerio* (zebrafish) and *A. mexicanus* (Mexican tetra) are the closest relatives to *B. tyrannus*.

One of the options to gauge whether or not the de novo assembly was properly performed is to align the assembled contigs to a phylogenetically related fish genome when a reference genome, cDNA, or EST sequences are not available. Based on a phylogenetic tree comparison between menhaden and the eleven publically available fish genomes on Ensembl, it was determined that zebrafish along with Mexican tetra were the closest related genera to menhaden (Figure 2). Therefore, to validate the accuracy of the Velvet/Oases assemblies, the contigs from each transcriptome were aligned to the zebrafish and Mexican tetra genomes independently using BLASTX with an E-value of <1e-5. The menhaden had 42.56% matches to the Zebrafish and 41.34% to the Mexican tetra database (Table 2). The RMBT percentage which infers the accuracy of the assembly was 87.83% (Table 2). A total of eight gene-specific primers were verified by the presence of a unique PCR product followed by Sanger sequencing to further legitimize the accuracy and annotation of the assembly (Figure 3 and 4).

BLAST analyses

To annotate the menhaden testis transcriptomes.

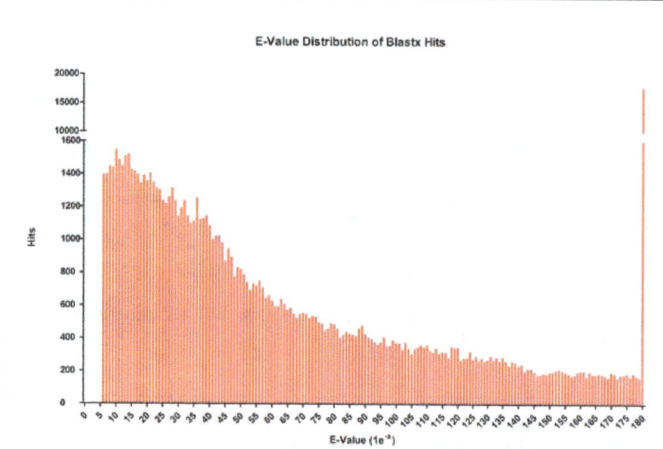

Figure 3: The E-value distribution of the BLASTX hits. The contig distribution based on BLASTX E-Values.

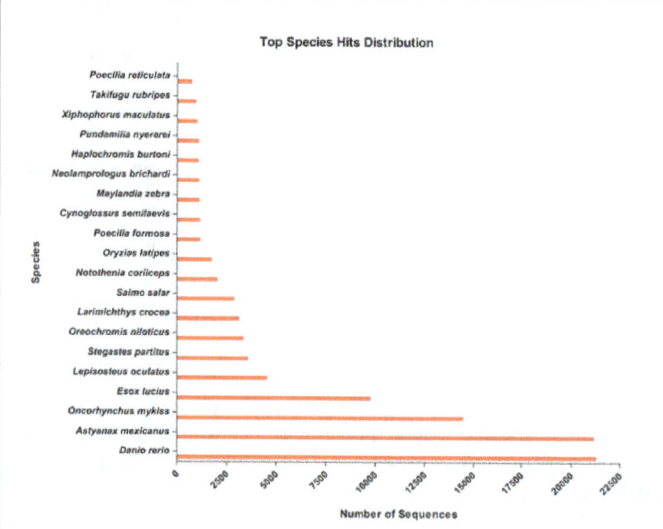

Figure 4: The top hit species distribution. The top species distribution based on the BLASTX results. The top 21 species shown all belong to various fish species.

The contigs were first searched against the NCBI non-redundant (Nr) protein database by using BLASTX with an E-value <1e-5 resulting in 46.89% matches as seen in Table 2. An E-value of <1e-5 was chosen because there is no available annotated menhaden genome nor a close relative that can provide a reference genome. Although we used an E-value threshold of <1e-5, the majority of the matches were below this threshold as graphically depicted in Figure 3. The species distribution of the top BLASTX hits against the Nr database for the transcriptome revealed that the top hits were contributed from other fish species as shown in Figure 3. The zebrafish and Mexican tetra provided the most matches with 21,598 (8.41%) and 21,541 (8.38%).

Go ontology assignment and functional classification

Functional annotation is imperative to assign biological information to the transcriptomic data of non-model organisms. Therefore, Gene Ontology (GO) terms were subsequently assigned to the menhaden contigs based on their sequence matches to known protein sequences in the Nr database. By merging the Blast2GO annotations with the InterProScan results, 70,919 contigs out of 254,446 (27.87%) were assigned at least one GO term (Table 3). The majority of these GO assignments belonged to molecular function (101,730, 40.05%) followed by biological processes (100,885, 39.71%) and cellular component (51,378, 20.22%) (Table 2). The top level 2 GO terms for the molecular function category were binding, catalytic activity, transporter activity, signal transducer activity, molecular transducer activity, and molecular function regulator. For the biological process category, the top level 2 GO terms were cellular process, metabolic process, single-organism process, biological regulation, regulation of biological process, and response to stimulus. As for the cellular component category, cell, cell part, organelle, membrane, and macromolecular complex were the top level 2 GO terms (Figure 5).

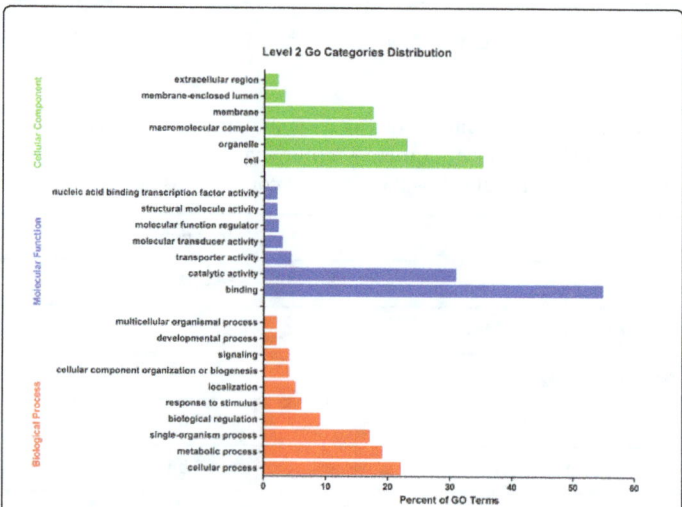

Figure 5: Functional classification based on the three categories of Gene Ontology. The GO-terms were categorized at level 2 into three major functional categories: biological process, cellular component, and molecular function. A detailed summary is listed in supplementary table.

In order to further resolve the functionality of the menhaden testis transcriptomes, the annotated contigs were categorized into different functional groups based on the Cluster of Orthologus Groups (COGs) database. This database contains proteins generated by comparing the protein sequences of complete genomes from bacteria, algae, and eukaryotes where each cluster has proteins of paralogs from at least three lineages [36]. Of the 119,326 contigs that had BLASTX matches, 30,829 (25.83%) could be classified into one of the 26 COG categories (Figure 6). Among these categories, the majority of clusters were "General function prediction only" (26,264, 85.19%), "Signal transduction mechanisms" (17,150, 55.62%), "Transcription" (4,663, 15.12%), "Posttranslational modification, protein turnover, chaperones" (4,537, 14.71%), and "Translation, ribosomal structure and biogenesis" (4,219, 13.68%) respectively.

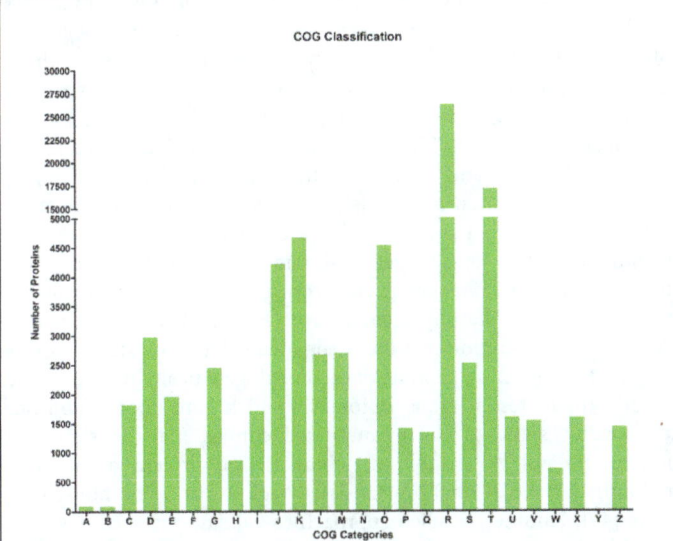

Figure 6: Histogram of the COG classification (A) RNA processing and modification; (B) Chromatin structure and dynamics; (C) Energy production and conversion; (D) Cell cycle control, cell division, chromosome partitioning; (E) Amino acid transport and metabolism; (F) Nucleotide transport and metabolism; (G) Carbohydrate transport and metabolism; (H) Coenzyme transport and metabolism; (I) Lipid transport and metabolism; (J) Translation, ribosomal structure and biogenesis; (K) Transcription; (L) Replication, recombination and repair; (M) Cell wall/membrane/envelope biogenesis; (N) Cell motility; (O) Posttranslational modification, protein turnover, chaperones; (P) Inorganic ion transport and metabolism; (Q) Secondary metabolites biosynthesis, transport and catabolism; (R) General function prediction only; (S) Function unknown; (T) Signal transduction mechanisms; (U) Intracellular trafficking, secretion, and vesicular transport; (V) Defense mechanisms; (Y) Nuclear structure; (Z) Cytoskeleton.

To further enhance the functional categorization of the testis in menhaden, we utilized the contigs with BLASTX matches to search through the Kyoto Encyclopedia of Genes and Genomes (KEGG) database to discover the active metabolic pathways at the time of collection. Based on the results, 967 enzymes were mapped to 125 different metabolic pathways (Table 3). The top five pathways with KEGG annotations were purine metabolism (56, 17.2 5.79%), Arginine and proline metabolism (30, 3.1 2.99%), Amino sugar and nucleotide sugar metabolism (29, 2.99%), Pyrimidine metabolism (27, 2.79%),

and Cysteine and methionine metabolism (25, 2.58%) as seen in Figure 7.

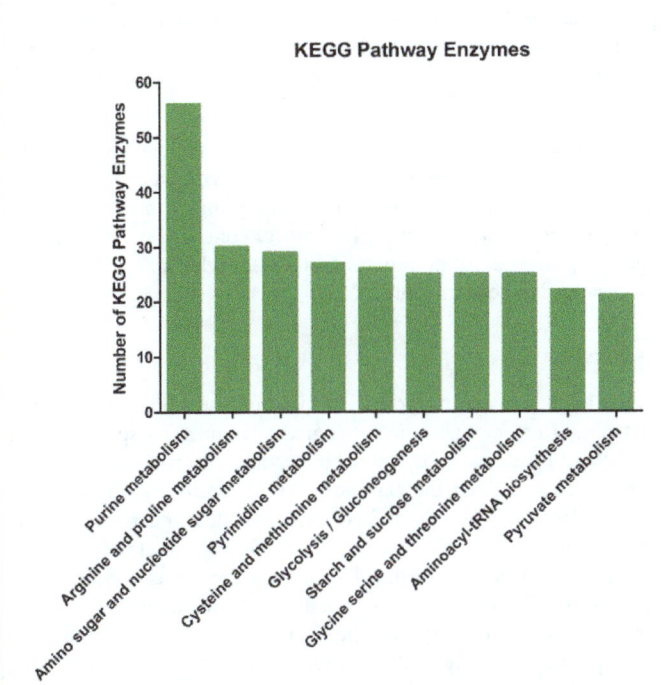

Figure 7: Summary of *B. tyrannus* KEGG pathways. Distribution of the top ten KEGG pathway enzymes related to *B. tyrannus*. A detailed summary is listed in Table 3.

Repetitive Element Analysis and Microsatellite Identification

The extent of the repetitive elements in the menhaden's testes transcriptome was assessed by using Repeatmasker with the Zebrafish Repeat Database as a reference.

Menhaden Testes	
Total number of sequences surveyed	2,54,462
Number of sequences containing repeats	34,486
Total number of bp searched	21,16,32,176
Total number of microsatellites found	46,864
Di-nucleotide repeats	28,566
Tri-nucleotide repeats	13,373
Tetra-nucleotide repeats	4,079
Penta-nucleotide repeats	708
Hexa-nucleotide repeats	139

Table 3: Statistics of the microsatellite distribution within the transcriptome.

Searching through the 211,632,176 bps within the 254,446 contigs resulted in the detection of 11,017,712 bps (5.21%) of repeated sequences. The distribution and classification of the identified repetitive elements are shown in Table 3. The most abundant types of repetitive elements in the sequences were simple repeats (2.18%), retroelements (1.19 %), DNA transposons (0.93%), and small RNA (0.30%).

To determine the presence of microsatellites, the assembled contigs were scanned with the selection criteria of having sufficient flanking sequences of at least 50 bps for primer designing (Table 2). A total of 46,865 microsatellites within 34,489 contigs consisted of either di-,tri-, penta- or hexa-nucleotide repeats (Table 3). Eighty primer pairs have been designed as seen in Table 2).

Discussion

Menhaden are an important species for the commercial fishing industries. They are also an ecologically important species within oceanic and estuarine ecosystems by providing a crucial link between phytoplankton and marine carnivores [1]. Despite the importance of this fish, detailed genetic information for it is currently lacking. Transcriptome analyses are a cost-effective method for the characterization of species that lack a fully sequenced genome [7,8]. The short reads from Illumina paired-end sequencing can provide a deep sequencing coverage for the accurate base calling in de-novo transcriptome assembly and has been utilized to characterize many non-model species [37-40]. However, no transcriptome sequencing data is available for menhaden.

In a recent study comparing 454 GS-FLX (Roche Diagnostics Corporation) to Illumina (Illumina, Inc.) for the utility of RNA-seq in a non-model bird, Illumina assemblies performed best for de novo transcriptome characterization in terms of contig length, transcriptome coverage, and complete assembly of gene transcripts [41]. Therefore, Illumina HiSeq 500 was utilized in this de novo study of this non-model fish. For the sequencing strategy, we elected to perform the paired-end sequencing approach because it facilitates the assembly process compared to the common one way sequencing approach. This is because the paired-end approach has the ability to produce longer contigs by filling gaps in the consensus sequence and longer contigs are better for mapping.

Due to the absence of a suitable reference genome for menhaden, the de novo assembly approach was performed. Velvet and Oases, common assembly programs in de novo transcriptomic studies using Illumina reads, were employed to assemble the processed reads [7,10,14,21,22,34,35,42,43]. It should be noted that the quality of the de novo transcriptome assembly is reliant on the user-defined k-mer value characterized as the length parameter defining the sequence overlap between two reads forming a contiguous sequence (referred as a contig) [44,45]. Low k-mer values will recover less abundant transcripts by producing a large number of contigs, but this will lead to the assembly of numerous and highly fragmented transcripts due to sequencing errors and lack of overlap [44,45]. Meanwhile, high k-mer values will theoretically result in a more contiguous assembly of high coverage transcripts and splice variants. However, this approach will produce fewer contigs and cause lower transcript representation due to capturing only highly represented reads [44,45]. Interestingly, the single k-mer approach is still a popular choice for the de novo assembly of transcriptomes even though this might result in the loss of relevant biological information due to the inadequate representation of

various k-mer lengths [45]. Therefore, we employed a multiple k-mer approach in this study because clustering multiple single k-mer assemblies takes advantage of the characteristics of both the low and high k-mer lengths, resulting in a better transcript diversity of the assembly.

It is imperative to assess whether or not de novo assemblies are reliable. This can be investigated with either in silico or in vitro approaches. One in silico approach is to match the contigs to known cDNA or EST sequences to confirm the accuracy of the de novo assembly [8,46]. However, due to the scarcity of cDNA or EST sequences for menhaden, another alignment technique was performed for the validation. In this method, the contigs for the species of interest was aligned to its closest phylogenetic relative with a reference genome utilizing BLASTX [9]. Based on a phylogenetic tree analysis with the eleven publically available fish genomes on Ensembl, it was determined that zebrafish along with Mexican tetra were the closest related genera to menhaden. The BLASTX alignments for the menhaden against the zebrafish (42.56%) and Mexican tetra (41.34%) yielded analogous results with other non-model fish de novo transcriptome assembly studies signifying that the assemblies are reliable [9,12,13,47,48]. The low number of hits for both zebrafish and Mexican tetra could be due the presence of menhaden-specific contigs, as menhaden belong to separate clades in the phylogenetic tree (Figure 2). It also should be taken into consideration that the menhaden testis transcriptomes were compared to the entire genome of each phylogenetically related species instead their testis transcriptomes, which are not published. Additionally, some of the unannotated contigs may be short and therefore 1) not consist of well characterized protein domains, 2) predominately be 3' or 5' untranslated regions, or 3) be non-coding RNAs [49]. Another in silico approach to validate the accuracy of an assembly is the percentage of BLASTX matches against the Nr database. In this study, the percentage of matches for the menhaden was 46.23%. This percentage is comparable to other non-model fish de novo assemblies, further establishing the credibility of the assemblies [9,12,13,47,50]. It should be noted that the number of matches is affected by the amount of available sequence data, and there is a limited number of publically available fish genomes to contribute to the overall success of the annotation [49]. To validate that the BLASTX matches were relevant to the non-model organism of interest, the species distribution of the top BLASTX hits against the Nr database was performed using Blast2GO. Results showed that the top twenty one contributing species for the annotation were from other fish species, further strengthening the validity of the assembly. The top two contributors, zebrafish and Mexican tetra, correlated with the phylogenetic tree results that classified the reference genomes independently based on NCBI classification. To further legitimize the accuracy of the assembly along with its annotation, an in vitro approach of validating primers designed based off the assembly was performed. A total of eight gene specific primers were validated with Sanger sequencing (Table 1). All together, these successful in silico and in vitro validation steps established the reliability of the Velvet/Oases assemblies in both transcriptomes.

The GO project is an international collaborative effort to standardize and use ontologies to support biologically meaningful annotation of genes and their products in any organism [51]. GO ontologies provide the standardized description of attributes of genes along with their products in three key biological domains that are shared by all organisms: molecular function, biological process and cellular component [51]. By merging the Blast2GO annotations with the InterProScan results, 70,919 contigs out of 254,446 (27.87%) were assigned at least one GO term (Table 3). These results are comparable to other annotation efforts in non-model fish studies that utilized the de novo assembly approach [9,13,47,50]. Again, it should be taken into consideration that the limited number of publically available fish genomes inhibits the overall successful rate of annotating non-model fish [49].

The GO analysis in this research was similar to the testes transcriptome analysis for the channel catfish by Sun et al., [52] and the yellow catfish by Lu et al., [53]. Four of the top five GO terms in the biological process category for the channel catfish (cellular process, metabolic process, biological regulation, and response to stimulus) and the top four GO terms for the yellow catfish (cellular process, single-organism process, metabolic process, and biological regulation) matched the top five GO terms in this study [52,53]. Within all three testis studies, the cellular process, metabolic process, and biological regulation GO terms were observed as being with in the top four contributors for the biological category. For the molecular function category, four of the top five GO terms for the channel catfish (binding, catalytic activity, molecular transducer activity, and transporter activity) and four of the top five GO terms for the yellow catfish (binding, catalytic activity, transporter activity, and molecular transducer activity) matched the top four GO terms in this study [52,53]. The binding, catalytic activity, transporter activity, and molecular transducer GO terms were represented within the top five GO terms of the molecular function category for all three testis transcriptome studies. For the cellular componentcategory, four of the top five GO terms for the channel catfish (cell, organelle, macromolecular complex, and extracellular region) and four of the top five GO terms for the yellow catfish (cell, organelle, membrane, and macromolecular complex) matched the top five GO terms in this study [52,53]. The cell, organelle, and macromolecular complex GO terms were observed within the top five GO terms in all three testis studies. Overall, the GO categorization results show similarities between three different fish species studies interested in profiling their testis transcriptomes.

The menhaden were caught during their reproductive season in November [4]. During this time of year, the sexually activated testes are in the process of growing and becoming enlarged [54]. Typically, environmental cues and metabolic signals control the reproductive state of fish by inducing the hypothalamus to release gonadotropin-releasing hormone (GnRH) [55]. This results in stimulating the pituitary gland to secrete gonadotrophic hormones (GTH) along with growth hormones (GH) [56-58]. The GH controls several complex processes including growth and metabolism of proteins, fats, along with carbohydrates [59]. These activities lead to an increase in cellular uptake of amino acids, lipolysis, and blood sugar that promote the exponential growth of the testes. The sexual activation also requires the biosynthesis of purines and pyrimidines for DNA synthesis that is essential for the growth of the testes along with spermatogenesis [60]. Therefore, it is not surprising that the top Go Ontology assignments, functional characterizations, and metabolic analysis all share a similar trend of growth promotion within the testes. This is represented by the general themes of metabolic processes, nucleotide biosynthesis, replication, translation, and energy results from the GO, COG and KEGG analysis.

Conclusion

This research is the first to assign functional annotations to the testes transcriptome in menhaden, resulting in the most comprehensive biological information available for the species to date. This transcriptomic data can provide the ground work for studying the menhaden's population dynamics, biomonitoring, and reproductive health. Additionally, this data can be utilized in comparative studies in other fish models to further enhance breeding programs and evolutionary studies. Overall, this research will open new opportunities to use menhaden as a model organism for the oceanic and estuary ecosystems instead of relying on traditional fish model organisms that are not indigenous to the same environment.

Acknowledgement

This project was sponsored by Funding provided by the Louisiana Department of Wildlife and Fisheries- PI, Ralph Portier (LSU), Co-PI, Carolyn S. Bentivegna.

References

1. Franklin HB (2007) The Most Important Fish in the Sea: Menhaden and America. Washington: Island Press/Shearwater Books.

2. Deegan LA, Peterson BJ, Portier R (1990) Stable isotopes and cellulase activity as evidence for detritus as a food source for juvenile gulf menhaden. Estuaries 13: 14-19.

3. Durbin AG, Durbin EG (1998) Effects of menhaden predation of plankton populations in Narragansett Bay, Rhode Island. Estuaries 21: 449-465.

4. Ahrenholz DW (1991) Population biology and life history of the North American menhadens, Brevoortia spp. Marine Fisheries Review 53: 3-19.

5. Kristofersson D, Anderson JL (2006) Is there a relationship between fisheries and farming? Interdependence of fisheries, animal production and aquaculture. Marine Policy 30: 721-725.

6. NOAA, Chesapeake Bay Office.

7. Garg R, Patel RK, Tyagi AK, Jain M (2011) De Novo Assembly of Chickpea Transcriptome Using Short Reads for Gene Discovery and Marker Identification. DNA Research 18: 53-63.

8. Fan H, Xiao Y, Yang Y, Xia W, Mason AS, et al. (2013) RNA-Seq Analysis of Cocos nucifera: Transcriptome Sequencing and Subsequent Functional Genomics Approaches. PLoS ONE 8: e59997.

9. Coppe A, Agostini C, Marino IAM, Zane L, Bargelloni L, et al. (2013) Genome Evolution in the Cold: Antarctic Icefish Muscle Transcriptome Reveals Selective Duplications Increasing Mitochondrial Function. Genome Biol Evol 5: 45-60.

10. Feldmeyer B, Wheat CW, Krezdorn N, Rotter B, Pfenninger M (2011) Short read Illumina data for the de novo assembly of a non-model snail species transcriptome (Radix balthica, Basommatophora, Pulmonata), and a comparison of assembler performance. BMC Genomics 12: 317.

11. Finseth FR, Harrison RG (2014) A Comparison of Next Generation Sequencing Technologies for Transcriptome Assembly and Utility for RNA-Seq in a Non-Model Bird. PLoS ONE 9: e108550.

12. Ji P, Liu G, Xu J, Wang X, Li J, et al. (2012) Characterization of Common Carp Transcriptome: Sequencing, De Novo Assembly, Annotation and Comparative Genomics. PLoS ONE 7: e35152.

13. Huth TJ, Place SP (2013) De novo assembly and characterization of tissue specific transcriptomes in the emerald notothen, Trematomus bernacchii. BMC Genomics 14: 805.

14. Nandety RS, Kamita SG, Hammock BD, Falk BW (2013) Sequencing and De Novo Assembly of the Transcriptome of the Glassy-Winged Sharpshooter(Homalodisca vitripennis). PLoS ONE 8: e81681.

15. Van Belleghem SM, Roelofs D, Van Houdt J, Hendrickx F (2012) De novo Transcriptome Assembly and SNP Discovery in the Wing Polymorphic Salt Marsh Beetle Pogonus chalceus (Coleoptera, Carabidae). PLoS ONE 7: e42605.

16. Rana SB, Zadlock FJIV, Zhang Z, Murphy WR, Bentivegna CS (2016) Comparison of De Novo Transcriptome Assemblers and k-mer Strategies Using the Killifish, Fundulus heteroclitus. PLoS ONE 11: e0153104.

17. Andrews S (2010) FastQC: a quality control tool for high throughput sequence data.

18. Bolger AM, Lohse M, Usadel B (2014) Trimmomatic: A flexible trimmer for Illumina Sequence Data. Bioinformatics 30: 2114-2120.

19. Schmieder R, Edwards R (2011) Fast Identification and Removal of Sequence Contamination from Genomic and Metagenomic Datasets. PLoS ONE 6: e17288.

20. Zerbino DR, Birney E (2008) Velvet: Algorithms for de novo short read assembly using de Bruijn graphs. Genome Res 18 : 821-829.

21. Schulz MH, Zerbino DR, Vingron M, Birney E (2012) Oases: robust de novo RNA-seq assembly across the dynamic range of expression levels. Bioinformatics 28: 10861092.

22. Haznedaroglu BZ, Reeves D, Rismani-Yazdi H, Peccia J (2012) Optimization of de novo transcriptome assembly from high-throughput short read sequencing data improves functional annotation for non-model organisms. BMC Bioinformatics 13: 170.

23. Li W, Godzik A (2006) Cd-hit: a fast program for clustering and comparing large sets of protein or nucleotide sequences. Bioinformatics 22: 1658-1659.

24. Altschul SF, Madden TL, Schaffer AA, Zhang J, Zhang Z et al. Gapped BLAST and PSI-BLAST: a new generation of protein database search programs. Nucleic Acids Res 25: 3389-3402.

25. Conesa A, Gotz S, Garcia-Gomez JM, Terol J, Talon M, et al (2005) Blast2GO: A universal tool for annotation, visualization and analysis in functional genomics research. Bioinformatics 21: 3674-3676.

26. Götz S, García-Gómez JM, Terol J, Williams TD, Nagaraj SH, et al. High-throughput functional annotation and data mining with the Blast2GO suite. Nucleic Acids Res 36: 3420-35.

27. Quevillon ESV, Pillai S, Harte N, Mulder N, Apweiler R, et al. (2005) InterProScan: protein domains identifier. Nucl. Acids Res 33: W116-W120.

28. Kanehisa M, Goto S (2000) KEGG: kyoto encyclopedia of genes and genomes. Nucleic Acids Res 28: 27-30.

29. Tatusov RL, Galperin M, Natale DA, Koonin EV (2000) The COG database: a tool for genome-scale analysis of protein functions and evolution. Nucleic Acids Res 28: 33-36.

30. Letunic I, Bork P (2011) Interactive tree of life v2: online annotation and display of phylogenetic trees made easy. Nucl Acids Res 39: 475-478.

31. Langmead B, Salzberg SL (2012) Fast gapped-read alignment with Bowtie 2. Nature methods 9: 357-359.

32. Smit AFA, Hubley R, Green P (2013-2015) RepeatMasker Open-4.0.

33. Thurston MI, Field D Msatfinder: Detection and characterization of microsatellites.

34. Barrero RA, Chapman B, Yang Y, Moolhuijzen P, Keeble-Gagnère G, et al. (2011) De novo assembly of Euphorbia fischeriana root transcriptome identifies prostratin pathway related genes. BMC Genomics 12: 600.

35. Fox SE, Geniza M, Hanumappa M, Naithani S, Sullivan C, et al. (2014) De Novo Transcriptome Assembly and Analyses of Gene Expression during Photomorphogenesis in Diploid Wheat Triticum monococcum. PLoS ONE 9: e96855.

36. Sharma N, Jung C-H, Bhalla PL, Singh MB (2014) RNA Sequencing Analysis of the Gametophyte Transcriptome from the Liverwort, Marchantia polymorpha. PLoS ONE 9: e97497.

37. Collins LJ, Biggs PJ, Voelckel C, Joly S (2008) An approach to transcriptome analysis of non-model organisms using short-read sequences. Genome Inform 21: 3-14.

38. Li R, Fan W, Tian G, Zhu H, He L, Cai J, et al. (2010) The sequence and de novo assembly of the giant panda genome. Nature 463: 311-317.

39. Wang XW, Luan JB, Li JM, Bao YY, Zhang CX, (2010) De novo characterization of a whitefly transcriptome and analysis of its gene expression during development. BMC Genomics 11: 400.

40. Wu T, Qin Z, Zhou X, Feng Z, Du Y (2010) Transcriptome profile analysis of floral sex determination in cucumber. J Plant Physiol 15: 905-13.

41. Finseth FR, Harrison RG (2014) A Comparison of Next Generation Sequencing Technologies for Transcriptome Assembly and Utility for RNA-Seq in a Non-Model Bird. PLoS ONE 9: e108550.

42. Gordo SM, Pinheiro DG, Moreira EC, Rodrigues SM, Poltronieri MC, et al. (2012) High-throughput sequencing of black pepper root transcriptome. BMC Plant Biology 12: 168.

43. Ashrafi H, Hill T, Stoffel K, Kozik A, Yao J, et al (2010) De novo assembly of the pepper transcriptome (Capsicum annuum): a benchmark for in silico discovery of SNPs, SSRs and candidate genes. BMC Genomics 13: 571.

44. Surget-Groba Y and Montoya-Burgos JI Optimization of de novo transcriptome assembly from next-generation sequencing data. Genome Res 20: 1432-1440.

45. Chopra R, Burow G, Farmer A, Mudge J, Simpson CE, et al. (2014) Comparisons of De Novo Transcriptome Assemblers in Diploid and Polyploid Species Using Peanut (Arachis spp.) RNA-Seq Data. PLoS ONE 9: e115055.

46. Xia Z, Xu H, Zhai J, Li D, Luo H, et al. (2011) RNA-Seq analysis and de novo transcriptome assembly of Hevea brasiliensis. Plant Mol Biol 77: 299-308.

47. Shin SC, Kim SJ, Lee JK, Ahn DH, Kim MG, et al. (2012) Transcriptomics and Comparative Analysis of Three Antarctic Notothenioid Fishes. PLoS ONE 7: e43762.

48. Salisbury JP, Sîrbulescu RF, Moran BM, Auclair JR, Zupanc GKH, et al. (2015) The central nervous system transcriptome of the weakly electric brown ghost knifefish (Apteronotus leptorhynchus): de novo assembly, annotation, and proteomics validation. BMC Genomics 16: 166.

49. Gao J, Wang X, Zou Z, Jia X, Wang Y, et al. (2014) Transcriptome analysis of the differences in gene expression between testis and ovary in green mud crab (Scylla paramamosain). BMC Genomics 15: 585.

50. Windisch HS, Lucassen M, Frickenhaus S (2012) Evolutionary force in confamiliar marine vertebrates of different temperature realms: adaptive trends in zoarcid fish transcriptomes. BMC Genomics 13: 549.

51. The Gene Ontology Consortium (2008) The Gene Ontology project in 2008. Nucl Acid Res 36: D440-D444.

52. Sun F, Liu S, Gao X, Jiang Y, et al. (2014) Male-Biased Genes in Catfish as Revealed by RNA-Seq Analysis of the Testis Transcriptome. PLoS ONE 8: e68452.

53. Lu J, Luan P, Zhang X, Xue S, Peng L, et al. (2014) Gonadal transcriptomic analysis of yellow catfish (Pelteobagrus fulvidraco): identification of sex-related genes and genetic markers. Physiol Genomics 21: 798-807.

54. Maugars G, Schmitz M (2008) Expression of gonadotropin and gonadotropin receptor genes during early sexual maturation in male Atlantic salmon parr. Mol Reprod Dev 75: 403-413.

55. Tena-Sempere M (2006) KiSS-1 and reproduction: focus on its role in the metabolic regulation of fertility. Neuroendocrinology 83: 275-281.

56. Popa SM, Clifton DK, Steiner RA (2008) The Role of Kisspeptins and GPR54 in the Neuroendocrine Regulation of Reproduction. Annual Review of Physiology 70: 213-238.

57. Klausena C, Chang JP, Habibi HR (2001) The effect of gonadotropin-releasing hormone on growth hormone and gonadotropin subunit gene expression in the pituitary of goldfish, Carassius auratus. Comparative Biochemistry and Physiology 129: 511-516.

58. Li WS LH, Wong A (2002) Effects of gonadotropin-releasing hormone on growth hormone secretion and gene expression in common carp pituitary. Comparative Biochemistry and Physiology pp: 335-341.

59. Møller N, Jørgensen JO (2013) Effects of Growth Hormone on Glucose, Lipid, and Protein Metabolism in Human Subjects. Endocr Rev 30: 152-77.

60. Singh K, Deepika J (2009) One Carbon Metabolism, Spermatogenesis, and Male Infertility. Reprod Sci 20: 622-30.

Fisheries of Jemma and Wonchit Rivers: As a Means of Livelihood Diversification and its Challenges in North Shewa Zone, Ethiopia

Erkie Asmare[1*], **Sewmehon Demissie**[2] **and Dereje Tewabe**[1]

[1]*Bahir-Dar Fisheries and Other Aquatic Life Research Center, P.O. Box: 794, Bahir-Dar, Ethiopia*

[2]*Amhara Regional Agricultural Research Institute, P.O. Box: 527, Bahir Dar, Ethiopia*

[*]**Corresponding author:** Erkie Asmare, Bahir-Dar Fisheries and Other Aquatic Life Research Center, P.O. Box: 794, Bahir-Dar, Ethiopia, E-mail: erkie.asmare@yahoo.com

Abstract

Fishing plays a critical role as a 'bank in the water' for local populations that largely rely on this activity to access cash quickly. This study aimed: (1) to assess the importance of fisheries in improving farmer's livelihood in the study area. (2) to assess households and individual's involvement in inland fisheries in terms of utilization and management, and (3) to recommend means of interventions for sustainable use of the resource and enhance benefits from the river fishery. This activity was conducted by using a combination of monitoring of fish catch, focus group discussions, and key informant interviews. Fishing is seasonal and intensively carried out during the dry seasons starting from February up to April. The most popular fishing gears used for fishing are the seed of *Millettia ferruginea* (in Amharic called Birbira) and barks of *Balanites aegyptiaca* (locally called Bedeno). In the area the main fish type consumed by the community are *Clarias gariepinus* [catfish] and *Labeobarbus intermedius* [Barbus] fish species in fresh and sun dried forms but *Oreochromis niloticus* is not known as it is edible. The farmers have a good fish consumption habit which is by far greater than the town's inhabitants. Hence, Farming and fishing are overwhelmingly the most important activities for household food supply and means of income generation. Fish catches from the rivers have declined significantly because of the destructive way of fishing, water pollution, and resource encroachment, thereby threatening the sustainability of Jemma and Wonchit river fisheries as well as the river's ecosystem.

Keywords: *Milletia ferruginia*; Destructive way of fishing; Livelihood diversification; River fisheries

Introduction

Ethiopia is uniquely rich in water resources. It has numerous water bodies including ponds, lakes, rivers, reservoirs and wetlands [1]. As a landlocked country following the secession of Eritrea in 1993, fisheries in Ethiopia come exclusively from inland sources [2]. The inland water body of Ethiopia is estimated to encompass about 7,400 km^2 of lake area and a total river length of about 7,000 km [3-6].

Fish is an important food item that has significant socioeconomic contribution as a source of income, employment and cheap protein for marginal people in developing countries including Ethiopia [7]. Inland fisheries are particularly important for the food security of poor people, as most inland fish production goes for subsistence or local consumption [8]. It was estimated that more than 56 million people were directly involved in inland fisheries in the developing world in 2009 [9-10]. Fisheries are one of livelihood strategies that have contributed much to people in developing countries. It is one of the vital strategies for the poor to achieve food, income and other social benefits. For instance, it serves as an important source of diet for over one billion people [11-13].

Migrant fishers may employ agricultural workers as crew, providing seasonal employment and contributing to village economies. Fisheries contribute to livelihoods in a range of ways: Directly as food, as a source of income and through other social benefits, such as source of supplementary income [14]. Fisheries play particularly an important role among disadvantaged groups as a main or supplementary source of income, employment, and livelihood [15].

Fisheries of Wonchit and Jemma rivers are highly impacted by irresponsible fishing practices, which result in reduced potential benefits and loss of aquatic biodiversity. Resource potential, uses and socioeconomic benefits from Wonchit and Jemma rivers have not been studied yet and there are no reports on their fisheries. This study aimed to assess the importance of fisheries in improving farmer's livelihood in the study area; assess household and individual involvement in fisheries in terms of utilization and management; to recommend means of interventions for sustainable use of the resource and enhance benefits from the river fishery.

Materials and Methods

Description of the study area

Wonchit and Jemma rivers, which are one of the most flowing rivers to the lower course of Blue Nile, are mainly found in North Shewa zone of Midaworemo and Merhabetie districts respectively. For these rivers, major beneficiaries of the fisheries resource are Midaworemo, Merhabetie, Muteranajiru, Ensaro and neighboring districts of south Wello of the same region and districts from Oromia region. But in most cases fishing is practiced on Midaworemo, Merhabetie and Muteranajiru districts.

Methods of data collection

Both primary and secondary data has been collected to make this paper successful. The qualitative data were collected between November 2012 and December 2014. The qualitative approach employing different data collection tools including transect walk, key informant interview, focus group discussion, stakeholder consultation, and document analysis were used to collect most of the qualitative data. The main data collected included information on fishing related activities, the market situation for fish, and major fishing gears used by fishers. Secondary data was collected from literature and district agricultural and rural development experts.

Sampling procedures

Based on the existence of fishing activities, two districts were purposely chosen. A purposive sampling technique was followed for the selection of districts and fishers. At the first stage Midaworemo and Merhabetie districts were selected purposively to represent Wonchit and Jemma rivers respectively. At the second stage fishers were purposively selected from non-fishers. Finally, simple random selection of fishers was done for key informant interview, focus group discussion and stakeholder consultation. The collected data was analyzed by qualitative approaches. In addition, SWOT (strength, weakness, opportunities and threats) analysis was used to assess the situation of Jemma and Wonchit fisheries.

Result and Discussion

Fishing and fishing gear used

Fisheries provide trade, employment, nutrition and recreation for people throughout the world, and particularly in the developing world. However, the sector is impacted by numerous other uses of water, as well as by irresponsible fishing practices. The open access nature of the rivers in which there is no restriction of entry into the fisheries has resulted in heavy fishing pressure on stocks and attempt at using chemicals [16]. These practices result in a loss of fishery production, reduced food security and loss of aquatic biodiversity. Many fishers flee from all sides of the river gathered somewhere and move to downstream where fishes are found in mass at low altitudinal sites by carrying plant poisoning materials, locally made netlike sack, and panga. The main function of locally made netlike sack and panga are to collect poisoned fish from the surface of the water and to kill the fish that are weakened due to use of poisoning plant materials respectively.

In the study area, modern fishing gears such as cast nets, gill nets and hooks are not known. The most popular fishing material that are extensively used for fishing are *Millettia ferruginea* seed, a tree that is endemic to Ethiopia (Figure 1) and the bark of *Balanites aegyptiaca* (locally called Bedeno). These plant materials are used by crashing and diluting with water, squeeze and then spread over the surface of the pooled water bodies starting from the post rainy seasons up to just pre-rainy seasons of a year. However, fishers' knowledge towards the negative impact of poisoning plants is very minimal, and they believed that fishes have come into existence along the incidence of clouds during the rainy season. Using of extracts of certain poisonous plants for fish exploitation could damage some useful macro- and microorganisms which are essential for the stabilization of the ecosystem. These plant extracts pollute the aquatic ecosystem and reduce the fish stocks through uncontrollable mortality.

Figure 1: *M. ferruginea* seed sold at the local market by youth and adult men.

The matured seeds in the powder form are used for catching fish due to its toxicity is a common practice in the country. The powder is spread over the water surface for stunning fish. In agreement with this study Karunamoorthi [17] and Choudhury and Shiferaw reported that seed extracts of *M. ferruginea* are extremely toxic to the fishes as well as the environment [17,18]. The seeds are pulverized and are used to take care of the external parasitic and poison the fish by the [19-22]. It possibly affects other beneficial organisms in the aquatic ecosystem and ultimately disrupts the food chain due to their toxic nature. The effect of the two poisoning plant materials used for fishing are quite different. The crushed *M. ferruginea* seed has strong toxicity against aquatic macroinvertebrates in general and damages the nervous system and eye of the fish in particular.

The fishes poisoned with the seed of *M. ferruginea* are unable to see the environment it lives and make them float to the water surface for allowing easy catches. A study by Ameha also revealed that solutions of *M. ferruginea* seed powder affect oxygen uptake by the fish and the fertilized eggs [23]. In addition, solutions resulted in abnormal activities such as restlessness, sudden quick movements, rolling movements, swimming on the back, and settling at the bottom. The experiment of Ameha confirmed that when using *M. ferruginea* concentrations of 0.02 to 0.4 g/l, all of the fish died in about 30 to 60 minutes [23]. When we see the fish killed with *Balanites aegyptiaca* poisoning plant, the fish exhibited stressful behaviours such as unusual swimming and loss of balance which is due to the bark damage the nervous system and general metabolism of the fish. As a result, the body cavity of the fish becomes bad smell and changed its normal color to black color.

Seasonality of fishing

Fisheries represent a supplementary livelihood in the study area, as local people generally consider themselves farmers, with fishing as a part-time and seasonal activity. Nevertheless, fishing was ranked as very important for income generation, the most important activities for household food supply is agriculture (cropping) especially sorghum production. Jemma and Wonchit river fisheries are exploited largely by local communities which are open access fisheries for all. River basin communities and their traditional livelihoods are intimately linked to the seasonal cycle and the mixture of fishing and agricultural cropping. Hence, one of the most important contributions of Jemma and Wonchit river fisheries as a source of cash for households, not only for families of full-time fishers but also for a large number of rural households that partly and seasonally engaging fishers.

There are three main periods in the fishery: dry season (February to April), early-wet and wet season (May to August) and late-wet/recession season (September to January). During the early-wet season, the farmers prepare their land for crop production and therefore; fishers concentrate less on fishing and fisheries related activities. During the early wet and wet season water level become at peak and flow rapidly. At these times, farmers become more concentrated on agricultural activities and fishing in the down courses of Jemma and Wonchit rivers is impossible. Fishing is most intense from December to April during the dry season since agricultural activities are reduced. These seasons are very conducive for fishing because the water volume becomes decreased and make important for plant poisoning materials to be concentrated and not washed out by running water.

Household fish consumption and preference for fish species

Fisheries provide a crucial source of animal protein and essential micronutrients for local communities. The contribution of fish to household food and nutrition security depends on availability, access and cultural and personal preferences. Access is largely determined by location, seasonality and price [24]. In Jemma and Wonchit river fisheries, the farmers have a good fish consumption habit which is by far greater than the town's inhabitants. The household uses their catches for both home consumption and generating income by selling at the nearby local market only.

In the area the main fish type consumed by the community are *Clarias gariepinus* [catfish], *Heterobranchus longifilis* [catfish] and *Labeobarbus intermedius* [barbus] fish species in fresh and sun dried forms. Surprisingly, *Oreochromis niloticus* is not known as it is edible by the surrounding community. Although *O. niloticus* is a healthy source of protein as well as Omega-3 fatty acids, it needs excessive care while consuming. This is because of the narrow and thin bones that line the meat may get stuck in and piercing the consumer's throat. Moreover, the bone is quite thin, it might not pass along the throat easily and making it more difficult to remove. This situation refrains fishers and local consumers from eating this delicious food.

Fish market in Wonchit and Jemma rivers

The catches of fishers used for both home consumption and generating income by selling at the nearby local market. Hence, the Meragna town from Midaworemo and Alemketema from Merhabetie are the towns where the fish market operates. Most of the catches are sold in fresh, gutted whole fish and sun-dried form. Catches are brought to the market in fresh, whole gutted fish and sun-dried form.

During fish market survey 1.2 m *Heterobranchus longifilis* gutted fresh whole fish was registered (Figure 2).

Figure 2: Gutted *Clarias gariepinus* fish at the marketing place.

Many other medium sized with the range of 70 and 80 cm whole fresh gutted and washed *Clarias gariepinus* and *Heterobranchus longifilis* were also recorded. Large *Labeobarbus intermedius* fish species appear at the market in sun dried forms filled with sacks. Big sized gutted fresh whole fish sold from 40 to 50 Ethiopian birr. During market transaction period most of the time customers are women from peasant associations. This simply shows the livelihood and income contribution of Jemma and Wonchit river fisheries for the smallholder farmers. A study by Okeowo also showed that the business of artisanal fishing is profitable in both locations [25].

Socio economic role of fisheries in the Jemma and Wonchit rivers

Farming and fishing are overwhelmingly the most important activities for household food supply and means of income generation in the study area. In particular, the poorest rely in a larger proportion on fishing activities while the better off mainly rely on farming. The study shows clearly that fishing is of considerable importance for people living in the study area including crop producer and part-time or seasonal fishers.

According to Moni & Khan, fisheries has an important implication for ensuring emergency cash flow in terms of urgent medical expenses, financing children's education and supporting household economy in times of maintaining social and family occasions [15]. It also alters households' protein consumption level and income, expenditure and savings pattern of the households. Andersson and Ngazi also reported that fisheries can provide an important contribution to household cash income [26]. This cash income gives access to other benefits such as education, health services, clothing, and other foodstuffs etc. It also allows investment in other assets or enterprises such as land, livestock and fishing gear.

Likewise, fisheries of Jemma and Wonchit have a profound role for food and income generation including for women who mainly participate in post-harvest processing. Rural farmers in the study area employ casual workers for agricultural activities when they go to fishing; this provides seasonal employment for the poor and landless dwellers. People often turn to fishing when other livelihood options are limited, thereby; fisheries reduce vulnerability to hunger by providing a complementary food source as part of diversified livelihood strategies. This shows fisheries of Jemma and Wonchit can act as a 'safety net' for the poor. For example, people who have not agricultural lands could participate in fishing to meet their basic needs (Figure 3).

Figure 3: Local fresh and dried fish market at Meragna town from Midaworemo district.

In agreement with our finding, a report by Welcomme RL confirmed that small-scale fisheries play a role as a 'safety-net' in that fishing can provide alternative or additional sources of income, employment and food for the poor and near-poor households whose livelihoods have been temporarily reduced or affected by unexpected shocks or in periods of individual or collective economic crisis [10]. Similarly, Bene, reported that fishing plays a critical role as a 'bank in the water' for local populations that largely rely on this activity to access cash quickly [27].

Different studies showed that the foods we eat can influence our health seriously. On the other hand, fish is the major source of omega-3 fatty acids in the diet and has long been known to lower cholesterol. Therefore, fish can protect against heart disease, newborn development, combatting depression, reduces the risk of Alzheimer's disease, reduced risk of prostate cancer, longevity, and decrease the risk of sudden cardiac death. Fisheries of Jemma and Wonchit in this regard have a substantial nutritional role for the local community and the fishers themselves. Due to absence of market linkage and transportation problems, fisher sale their catches around the vicinity alone. This increases fish consumption habit around the rural area and it might have high contribution for improve nutrition status of smallholder farmers.

In addition to financial and nutritional benefits, fisheries of Jemma and Wonchit have a meaningful social and cultural role. In the study area, many fishers flee from all sides of the river gathered somewhere and move to downstream where fishes are found in mass. In the dry season, the farmer goes far from their home to fishing by holding their food for many days until they get enough amounts. Because of speedy water flow to the down course of the river and group fishing by using *M. ferruginea* seed and bark of *B. aegyptiaca*, fishing is rarely carried out alone and is often a very social activity in nature. During their stay, fishers share fishing and other household experiences each other. This has a paramount role in strengthening bonds between people and community cohesion. Increased production from fisheries provides greater community income; this enables them to invest in community projects such as school, road, and support poorer community members.

During market transaction period catches brought to market by both men and women. Age structures are not clearly observed rather both youths, middle classes, and older age groups are involved in the marketing and fishing. In fisheries, men and women often have distinct roles. Similar to a study by Erkie Asmare around Lake Tana, in Jemma and Wonchit fisheries only men go to fishing, but women are often involved in marketing and post-harvest processing [11]. In general, women's participation in the fishery sector is restricted especially, fishing is unthinkable. Pre-fishing activities like logistical functions, picking up equipment, crashing seeds of *M. fergunia*, purchasing seeds of *M. fergunia* and barks of *B. aegyptiaca*, and post-harvest processing are executed by both men and women. However, caring child, preparing food, fetching water and fuel wood, cleaning house, shopping, washing close and utensils, grinding are the main tasks of women in the study area.

Opportunities from Wonchit and Jemma river fisheries

Attractive fish prices at local market for better profit; the presence of diversified fish species; and inhabitants' traditional knowledge for fisheries and good consumption habit. Because of fishers let small fishes out to the water body while they are collecting their catches, Gotera/kefo a locally made fishing gear which has a hive like structure is the best practice. This system enables fishers to be either selective or non-selective which depends on the size and preference of the fishers. Fishers in the study area have a good practice in the post-harvest processing, which is either fresh or gutted when there is demand for fish or sun-dried form during fasting season and surplus of production.

Challenges for Jemma and Wonchit river fisheries

- Inaccessibility for transportation and marketing which only delimited to local areas
- Lower awareness of the community about the wise way of utilization and sustainability of the resource
- Non-existence of aquaculture production to supplement the river fisheries and habitat degradation
- Fishers don't get much help from the government in terms of training and other forms of capacity building. That is why fishers use traditional and devastating method of fishing only and the don't care about the resources' sustainability.
- Due to climate change and activities, the water volume of the rivers declines each year dramatically.
- Poisoning plant material added to the upper part of the river flow with the water to the down course of the rivers by poisoning or damaging all the aquatic organisms non-selectively. This is by violating regional fisheries proclamation No. 315/2003 article 5(7) which states "fishing by using illegal fishing materials, such as poisons, and fish narcotizing plant is forbidden except for the purpose of research" [28]. Therefore, poor implementation regional fisheries regulation was clearly observed.

Destructive way of fishing by the inhabitant is the major threats to the sustainability of aquatic organisms including fingerlings.

In addition, some fish species, such as *O. niloticus* considered as inedible in the study area.

Conclusions and Recommendations

From the study it was found that fisheries are ranked as a very important activity for income generation; it is the most important activities for household food supply. Fishing is seasonal in the study area and the most popular fishing gear used is *M. ferruginia* seed and barks of *B. aegyptiaca* which poses a great threat to fish and other non-target organisms. Fishers' knowledge towards the negative impact of poisoning plants is very minimal, and they believed that fishes have come into existence along the incidence of clouds during the rainy season. In the area, the main fish type consumed by the community are *C. gariepinus* (catfish) and *Labeobarbus intermedius* fish species in fresh and sun dried forms. Surprisingly, O. niloticus is not known as it is edible. In Jemma and Wonchit rivers fishery activities are sex oriented. Transportation and market problem, nonexistence of aquaculture production, destructive way of fishing, and high algal population are the main threats to Jemma and Wonchit river fisheries.

Taking into account the above issues we recommend the followings: awareness creation on promoting aquaculture to supplement river fisheries; awareness creation for the inhabitants for sustainable use of fisheries resource and its management; provision of appropriate fishing gears; and train ways and means of using the fishing materials. Prohibiting use of poisoning plant materials like *M. ferrugunia* and *B. aegyptiaca* by implementing regional fisheries proclamation No. 315/2003 article 5(7). Support the fishermen in terms of fishing and processing equipment which are eco-friendly techniques; aware the residents on how O. niloticus is important for human food so that can be one of commercially important fish species in the area; train and demonstrate how *O. niloticus* and *Heterobranchus longifilis* (locally called Gilgel) fish species are important for fish farming; and integrate fish farming with the existing irrigation scheme of the area.

Methods for fish preservation and transportation should be designed to allow fishers to sell their catch in areas where the price of fish is attractive. Fish dried by direct sun often results in low quality as a result of slow drying, insect infestation and contamination from airborne dust etc. However, drying fish by solar tent fish dryer enables to produce hygienic, high quality, organoleptically good dried fish with low cost [7]. Therefore, Introducing and disseminating solar tent fish drier technology to keep the hygiene of catch will have a vital role. Further study is also recommended on the health implication of fish caught by using *B. aegyptiaca* and *M. ferrugunia*. Finally, the effect of poisoning plant materials on the biology of fish must be studied.

Acknowledgment

The author gratefully acknowledges the contribution of fishers who participate in research activity, researchers, research assistants, car drivers and extension workers who make tremendous efforts to make the research successful. The financial support of Ethiopian Institute of Agricultural Research (EIAR) and Amhara Regional Agricultural Research Institute (ARARI) to conduct this research is also very much appreciated.

References

1. Tessema A, Mohammed A, Birhanu T, Negu T (2014) Assessment of Physico-chemical Water Quality of Bira Dam, Bati Wereda, Amhara Region, Ethiopia p: 5.

2. Amare A, Alemayehu A, Aylate A (2014) Prevalence of Internal Parasitic Helminthes Infected Oreochromis niloticus (Nile Tilapia), Clarias gariepinus (African Catfish) and Cyprinus p: 5.

3. Mitike JA (2014) Fish Production, Consumption and Management in Ethiopia. International Journal of Economics and Management Sciences 3: 460-466.

4. Wood R, Talling J (1988) Chemical and algal relation-ships in a salinity series of Ethiopian waters. Hydrobiologia 158: 29-67.

5. Tewabe D (2015) Status of Lake Tana Commercial Fishery, Ethiopia. Int J Aquac and Fish Sci 1: 12-20.

6. Goshu G, Tewabe D, Tefera B (2010) Spatial and temporal distribution of commercially important fish species of Lake Tana, Ethiopia. Ecohydrology and Hydrobiology 10: 231-240.

7. Asmare E, Tewabe D, Mohamed B, Hailu B (2015) Pre-scaling Up of Solar Tent Fish Drier in Northern and North Western Part of Lake Tana, Ethiopia. International Journal of Aquaculture and Fishery Sciences 1: 48-53.

8. FAO (2004) The State of World Fisheries and Aquaculture (SOFIA). Part 1: World review of fisheries and aquaculture, Fishers and fish farmers. Rome.

9. BNP (2009) Big number program. Intermediate report. Rome/Penang, Italy/Malaysia: Food and Agriculture Organization and World Fish Center.

10. Welcomme RL, Cowx IG, Coates D, Béné C, Funge-Smith S, et al. (2010) Inland capture fisheries. Philos Trans R Soc Lond B Biol Sci 365: 2881-2896.

11. Erkie A, Sewmehon D, Dereje T, Mihret E (2016) Impact of climate change and anthropogenic activities on livelihood of fishing community around Lake Tana, Ethiopia. Journal of EC Cronicon Agriculture 3: 548-557.

12. Manasi S, Latha N, Raju KV (2009) Fisheries and livelihoods in Tungabhadra Basin, India: Current status and future possibilities. The Institute for Social and Economic Change, Bangalore p: 24.

13. Gebremedhin S, Budusa M, Mingist M, Vijverberg J (2013) Determining factors for fisher's income? The case of Lake tana, ethiopia. Intn J Cur Res 5: 1182-1186.

14. Hortle KG, Suntornratana U (2008) Socio-economics of the fisheries of the lower Songkhram River Basin northeast Thailand. Mekong River Commission, Vientiane p: 85.

15. Moni NN, Khan NN (2014) Fish Cultivation as a Livelihood Option for Small Scale Farmers-Study in Southwestern Region of Bangladesh. IOSR J Human and Soci Sci 19: 42-50.

16. Oruonye ED (2014) The Challenges of Fishery Resource Management Practices in Mayo Ranewo Community in Ardo Kola Local GovernmentArea (LGA), Taraba State Nigeria. Glob J Sci Front Rese: D Agriculture and Veterinary p: 14.

17. Karunamoorthi K, Bishaw D, Mulat T (2009) Toxic effects of traditional Ethiopian fish poisoning plant Milletia ferruginea (Hochst) seed extract on aquatic macroinvertebrates. Eur Rev Med Pharmacol Sci 13: 179-185.

18. Choudhury MK, Shiferaw Y (2015) Toxicity of Millettia ferruginea (Hochst) Baker against the Larvae and Adult Ticks of Boophilus decoloratus a One-Host Tick in Cattle. J Nat Rem.

19. Banerjee S, Demo K, Abebe A (2013) Some Serum Biochemical and Carcass Traits of Arsi Bale Rams Reared on Graded Levels of Millettia ferruginea Leaf Meal. Worl App Sci Jour 28: 532-539.

20. Bekele A (2007) Useful trees and shrubs for Ethiopia: identification, propagation and management for 17 agro climatic zones. RELMA in ICRAF Project, Nairobi p: 552.

21. Banouzi JT, Prost A, Rajemiarimiraho M, Ongoka P (2008) Traditional uses of the African Millettia species (Fabaceae). Int J Bot 4: 406-420.

22. Negash L (1995) Indigenous trees of Ethiopia: Biology, uses and Propagation Techniques. SLU Reprocentralen, Umea, Sweden p: 285.

23. Ameha A (2004) The Effect of Birbira, Milletia Ferruginea (Hochst). Baker on Some Barbus Spp. (Cyprinidae, Teleostei) in Gumara River (Lake Tana), Ethiopia. ADDIS ABABA.

24. Beveridge MC, Thilsted SH, Phillips MJ, Metian M, Troell M, et al. (2013) Meeting the food and nutrition needs of the poor: the role of fish and the opportunities and challenges emerging from the rise of aquaculture. J Fish Biol 83: 1067-1084.

25. Okeowo TA, Bolarinwa JB, Ibrahim D (2015) Socioeconomic Analysis of Artisanal Fishing and Dominant Fish Species in Lagoon Waters of EPE and Badagry Areas of Lagos State. Inter J Res Agri Fores 2: 38-45.

26. Andersson J, Ngazi Z (1998) Coastal community's production choices, risks diversification and subsistence behaviour responses in periods of transition. Ambio 27: 686-693.

27. Bene C, Steel E, Kambala Luadia B, Gordon A (2009) Fish as the bank in the water-evidence from chronic-poor communities in Congo. Food Policy 34: 104-118.

28. Federal Negarit Gazeta (2003) Fisheries Development and Utilization Proclamation, Ethiopia p: 315.

Dietary Encapsulated Butyric Acid (Butipearl™) and Microemulsified Carotenoids (Quantum GLO™ Y) on the Growth, Immune Parameters and their Synergistic Effect on Pigmentation of Hybrid Catfish (*Clarias macrocephalus* × *Clarias gariepinus*)

Edwin Pei Yong Chow*, Kah Heng Liong and Elke Schoeters

Kemin Industries (Asia) Pte Limited, 12 Senoko Drive, Singapore 758200, Singapore

*Corresponding author: Edwin Pei Yong Chow, Kemin Industries (Asia) Pte Limited, 12 Senoko Drive, Singapore 758200, Singapore, E-mail: edwin.chow@kemin.com

Abstract

A 12-weeks feeding trial was conducted to evaluate the effects of encapsulated butyric acid (ButiPEARL™) and microemulsified yellow carotenoid (Quantum GLO™ Y) on growth, immune parameters and their synergistic effect on pigmentation of hybrid catfish. In the experiment, the catfish was randomly divided into 12 groups of 15 fishes and then fed with four experimental diets containing 0.5 kg/t ButiPEARL™, 0.7 kg/t Quantum GLO™ Y, 0.5 kg/t ButiPEARL™ + 0.7 kg/t Quantum GLO™ Y, or none of these supplements (control diet). The results showed that the ButiPEARL™ + Quantum GLO™ Y fed group gave the highest yellowness (b*) score of 18.43 in the back muscle and almost double the total carotenoid measured in the fish muscle (151.35 mg/kg) compared to the Quantum™ Y fed group. This suggests that butyric acid in the diet had a synergistic effect on carotenoid absorption and pigmentation performance of catfish. The body weight of all treatment groups was significantly different from the control group and the catfish fed ButiPEARL™ alone had the highest body weight gained followed by the ButiPEARL™ + Quantum GLO™ Y fed group with an FCR improvement of 25 points and 13 points respectively over the control. There was no adverse effect on the immune system after feeding both butyric acid and carotenoid to the catfish and the immune parameters (number of leucocytes, erythrocyte, percent of hematocrit, haemoglobin and total protein) of ButiPEARL™ + Quantum GLO™ Y fed group were improved compared with the other groups. In conclusion, ButiPEARL™, Quantum GLO™ Y and the combination have positive effects on performance and pigmentation in catfish aquaculture.

Keywords: Encapsulated butyric acid; Microemulsified carotenoids; Hybrid catfish; Synergistic; Growth; Immune parameters; Pigmentation

Introduction

Hybrid catfish (Female *Clarias macrocephalus* × Male *Clarias gariepinus*) have become one of the most important protein sources for Thai people because of its low cost, rapid growth, high availability, and high nutritional value. To match the increasing demand, the production of hybrid catfish in Thailand has dramatically increased every year for more than ten years. However, hybrid catfish grown in intensive aquaculture are often exposed to stressful conditions which have a negative impact on their growth and immunity. Therefore, hybrid catfish in such environments usually have low growth rate and high tendency to develop diseases, especially bacterial infectious diseases. Currently, bacterial infections in aquaculture are mainly controlled by antibiotics. However, recently, the use of antibiotics in aquaculture has received considerable attention because their use can lead to the development of drug resistant bacteria, thereby reducing drug efficacy. Because the usage of antibiotics in aquaculture is discouraged, it is necessary to find an alternative solution to prevent bacterial infection. Organic acids are among the most promising substances as they have been reported to increase survival rate of catfish. Carotenoids can also improve survival rate, enhance resistance to several stress conditions as well as pigmentation. Therefore, both organic acids and carotenoids have the potential to be used in catfish farming as feed additives [1].

Butyric acid has been shown to have bactericidal activity on some enteric bacteria as well as to stimulate villi growth [2]. Researchers have shown that butyrate is quickly absorbed in the upper digestive tract, which makes it less ideal as a feed additive [3]. However, butyrate efficacy has been shown to increase when it is fed in a protected form such as encapsulation [4]. Researchers have shown that encapsulation can effectively deliver butyric acid throughout the intestinal system, providing energy for intestinal proliferation and differentiation and thereby ensuring that absorption of important nutrients is optimized [5]. In aquaculture, maintenance of the natural skin pigmentation is of great importance from a commercial point of view, as it has a direct impact on consumer acceptance or rejection [6] as well as product market price. A variety of natural and/ or synthetic carotenoids are available to enhance coloration in the flesh of salmonid fish and in the skin of others such as European red porgy (Pagrus pagrus). Both synthetically produced pigments, astaxanthin and canthaxanthin, either alone or in combination, have been efficiently used as dietary additives for muscle pigmentation in salmonids. While synthetic carotenoid pigments are commercially available as feed additives, they are expensive and up-take levels are poor, estimated between 5% and 10% [7]. Moreover, there is increasing consumer awareness about synthetic feed additives and safety because these substances may exert effects within the body. This has recently promoted increase interest in the use of natural carotenoid sources for some fish and shrimp species of economic interest. In our earlier works [8,9], we have showed the enhanced solubility and dissolution of carotenoids can be achieved in bicontinuous microemulsions (liquid system) containing polyethoxylated sorbitan ester (Tween 80), water, limonene, ethanol and glycerol. This system is able to prepare stable bicontinuous carotenoid microemulsions of droplet size ~0.25 μm upon mild agitation in liquid media. These fine droplets of microemulsions have the advantage of presenting the carotenoids in a dissolved form, with a large interfacial surface area for absorption, which will result in an enhanced, more uniform and reproducible absorption. Indeed, we have been able to demonstrate the possibility of using the microemulsified carotenoids in enhancing the bioavailability over corresponding regular preparations, leading to greater yolk pigmentation at lower inclusion rate in layers. With this success, an attempt has been made to look at using the microemulsified carotenoids for skin and flesh pigmentation in aquaculture. In this present work, we evaluated the effects of encapsulated butyric acid (ButiPEARL™) and microemulsified yellow carotenoid (Quantum GLO™ Y) on growth, immune parameters and their synergistic effect on pigmentation of hybrid catfish under stress-free condition.

Materials and Methods

Production of products

The encapsulated source of butyric acid (ButiPEARL™; Kemin Cavriago, Italy used in the experiment consisted of min 45% butyrate salt. The microemulsified source of yellow carotenoid (Quantum GLO™ Y; Kemin Industries (Asia) Pte Limited, Singapore) used in the experiment consisted of min 20 g/kg total xanthophyll. The carotenoids were found to be approximately 0.25 μm in size, as analyzed by electron microscopy and light-scattering diffraction study [9]. The dosages for each product were expressed in kg/t of feed.

Trial specifications

The experimental trial was conducted in the Laboratory of Nutrition and Aqua feed, Department of Aquaculture, Faculty of Fisheries, Kasetsart University, Bangkok, Thailand. Hybrid catfish was produced under captivity, with an initial body weight of 45-55 g, were randomly divided into twelve groups of 15 fishes each (four treatments, three replicates). Each group was stocked in a 500 L tank and the fishes were allowed to acclimatise one week prior to the start of the experiment. The fishes were hand-fed to apparent satiation thrice daily, corresponding to 3%-3.5% of the body weight, (08:30, 12:30 and 16:30) for 12 weeks. During the feeding trial, the water temperature ranged from 27°C to 30°C, pH and dissolved oxygen content of water was greater than 7.2 mg/L and 76 mg/L respectively for the duration of the study. These water quality parameters are crucial for normal fish mortality and will have an impact on the growth and pigmentation. All use of experimental animals was in compliance with guidelines from the Kasetsart University, animal care and use committee.

Four treatment diets containing 0.5 kg/t ButiPEARL™, 0.7 kg/t Quantum GLO™ Y, 0.5 kg/t ButiPEARL™ + 0.7 kg/t Quantum GLO™ Y, or none of these supplements (control diet) were prepared. Ingredients and proximate composition of the experimental diets are given in Table 1. The experimental diets were formulated and pelletized using a 3-mm pellet press. The amount of feed consumed per tank and per treatment was recorded and monitored throughout the feeding trial. No mortality was registered during the experiment. At the end of the feeding trial, fish of each tank were collectively weighed and the fish

weight gain, feed conversion ratio and specific growth rate were determined.

Ingredients (%)	Experimental diets			
	Control	T1	T2	T3
Soybean	25	25	25	25
Poultry meal	10	10	10	10
Wheat gluten	3	3	3	3
Canola/rapeseed	5	5	5	5
Deoil-rice bran	5	5	5	5
Tapioca	26.67	26.62	26.62	26.57
Dehull full fat SB	15	15	15	15
Soy protein concentrate	5	5	5	5
Crude fish oil	1	1	1	1
Choline	0.3	0.3	0.3	0.3
Vitamin C	0.2	0.2	0.2	0.2
Lysine	0.2	0.2	0.2	0.2
Methionine	0.23	0.23	0.23	0.23
Di-calcium phosphate	2.3	2.3	2.3	2.3
Limestone	0.1	0.1	0.1	0.1
Vitamin premix	1	1	1	1
ButiPEARL	0	0.05 (0.5 kg/t)	0	0.05 (0.5 kg/t)
Quantum GLO Y	0	0	0.07 (0.7 kg/t)	0.07 (0.7 kg/t)

Table 1: Details of experimental treatments and dosages of test additives.

Colorimetric and total carotenoids analysis

Colour analysis was performed every four weeks by reflective spectroscopy with a Minolta colour reader CR-10 colorimeter in accordance with the system CIE Lab (CIELAB) for lightness, redness, and yellowness, respectively [10]. The measurements were performed on skin and muscle areas of the fish's body. Total carotenoid content in the fish's muscle was determined after extraction with acetone. For carotenoid extraction, sample was weighted and 60 ml acetone and some sodium sulphate anhydrous were added. The mixture was ground and filtered through glass microfiber filters (GF/A, whatman paper) and rinsed with chloroform to increase the boiling point of the mixture. After mixing and phase separation between diethyl ether and water in separatory funnel, the upper layer was taken and placed in a round bottle flask to evaporate in a rotary evaporator at 35°C. The extract was concentrated and dissolved in benzene. Total carotenoids concentration in the muscle was determined spectrophotometerically in benzene using E (1%, 1 cm)=2500 at 460 nm for yellow carotenoid.

Haematological assay

Blood was analysed with routine methods used in fish haematology [11,12]. In short, blood was collected from the caudal vein with 1 mL non-heparinized disposable syringes fitted with 0.55 × 25 mm disposable needles. Blood samples (approximately 1 mL/fish) were centrifuged at 300 × g, 25°C for 10 min. A volume of 500 µL of the serum was removed and vortexed with 1 mL of ethanol for 30 s, then 2 mL of petroleum ether was added, and the mixture was vortexed for 1 min. The petroleum ether was separated by centrifuging at 300 × g, 25°C for 10 min. Red and white blood cell count was determined with chamber method using Neubauer's haemocytometer; haemoglobin concentration with cyanmethemoglobin method [13] and haematocrit in capillary tubes of 75 µL volume, which were centrifuged in a microhematocrit centrifuge and the haematocrit values were read with a reader. The total protein was determined using the following method [14].

Statistical analysis

Significant differences among treatment groups were tested by one-way analysis of variance (ANOVA) and the comparison of any values was made by Duncan's multiple range tests. A significance level of p<0.05 was used. The statistical analysis was performed by Stat graphics 5.1.

Results

Growth performance and feed utilization

The effects of both encapsulated butyric acid and microemulsified carotenoid diets on the growth parameters for the fishes throughout the experimental periods are given in Table 2. There were significant differences among the fish growth parameters measured (Figure 1). All fish grew normally and no specific signs of disease were observed. No mortality occurred throughout the experiment. Feed intake among treatments showed no significant differences (p>0.05) except for the ButiPEARL™ fed group. The results showed that ButiPEARL™ addition

to the control diet at 0.5 kg/t significantly improved the body weight gain by 95.81 g (p<0.05) with an FCR improvement of 25 points (p<0.05). This group also gave the best specific growth rate among other groups. However, growth performance of the ButiPEARL™ + Quantum GLO™ Y fed group was not significant different from the Quantum GLO™ Y fed group (p>0.05).

Figure 1: Digital image of hybrid catfish fed with experimental diets over 12 weeks: (C) Control, (T1) 0.5 kg/t ButiPEARL, (T2) 0.7 kg/t Quantum GLO Y and (T3) 0.5 kg/t ButiPEARL + 0.7 kg/t Quantum GLO Y.

Rearing parameter	Experimental diets			
	C	T1	T2	T3
Weight gain (g)	120.20 ± 4.72[a]	139.96 ± 1.22[c]	126.39 ± 6.18[ab]	133.65 ± 4.62[ab]
Average weight gain (g/fish/day)	75.79 ± 5.49[a]	95.81 ± 1.35[c]	84.17 ± 5.0[ab]	87.65 ± 4.03[ab]
Specific growth rate (%/day)	1.17 ± 0.05[a]	1.35 ± 0.01[bc]	1.23 ± 0.03[ab]	1.30 ± 0.04[b]
Feed conversion ratio (FCR)	1.58 ± 0.11[a]	1.33 ± 0.01[c]	1.42 ± 0.08[ab]	1.45 ± 0.06[ab]
Daily feed consumed (g/fish/day)	1.42 ± 0.02[a]	1.52 ± 0.01[b]	1.42 ± 0.02[a]	1.52 ± 0.01[b]
Survival rate (%)	100	100	100	100

Table 2: Growth performance parameters of hybrid catfish fed with experimental diets for 12 weeks.

Colorimetric and total carotenoids analysis

Colour intensity of hybrid catfish fed with experimental diets throughout the experimental periods is shown in Table 3 and Figure 2. Lightness (L*) was not affected by carotenoid supplementation (p>0.05), although the white skin of these groups changed slightly from a white hue to a yellow hue. There were, however, significant

(p<0.05) differences in yellow (b*) among the treatment groups. The group fed the control diet showed a weak redness and yellowness, which differed significantly from values found for groups, fed the other diets. Yellow tonality for abdominal skin was observed to be the best for fish fed the diet supplemented with Quantum GLO™ Y followed by ButiPEARL™ + Quantum GLO™ Y. However, for the back muscle colour score, the ButiPEARL™ + Quantum GLO™ Y combination gave the highest numerical value of 18.43 which is significantly different from the other treatments. This group also gave the highest total carotenoid measured in the back muscle (151.35 mg/kg) which is almost two times higher than what was measured for the Quantum GLO™ Y fed fish.

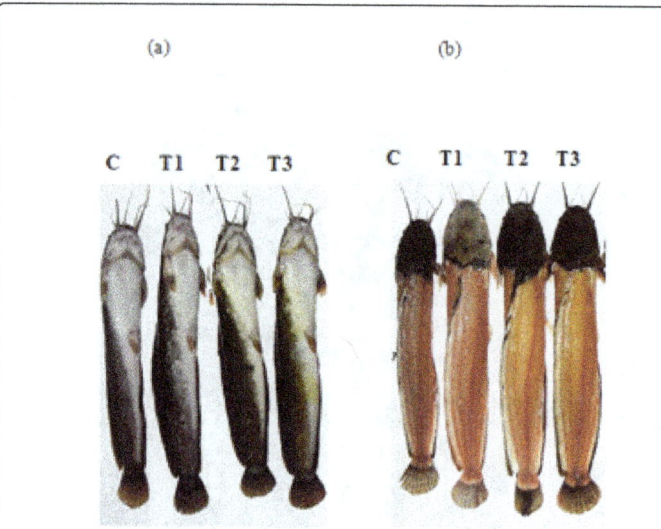

Figure 2: Digital image of (a) abdominal skin and (b) muscle for colour measurement of hybrid catfish fed with experimental diets over 12 weeks: (C) Control , (T1) 0.5 kg/t ButiPEARL, (T2) 0.7 kg/t Quantum GLO Y and (T3) 0.5 kg/t ButiPEARL + 0.7 kg/t Quantum GLO Y.

Experimental diets	L	a	b
	Abdominal skin		
C	67.23 ± 3.16[a]	0.07 ± 0.15[ab]	0.57 ± 0.42[a]
T1	66.67 ± 3.07[a]	0.37 ± 0.21[b]	1.60 ± 0.72[a]
T2	70.53 ± 1.26[a]	-0.73 ± 0.35[a]	8.30 ± 0.78[bc]
T3	70.08 ± 1.74[a]	-0.10 ± 0.36[ab]	7.20 ± 0.75[b]
	Back muscle		
C	45.13 ± 0.80[a]	2.43 ± 0.45[a]	10.57 ± 1.25[a]
T1	48.63 ± 1.40[a]	3.57 ± 0.42[ab]	13.30 ± 0.70[bc]
T2	49.30 ± 0.75[a]	4.53 ± 0.85[b]	16.33 ± 1.37[bc]
T3	49.47 ± 0.85[a]	4.47 ± 0.93[ab]	18.43 ± 1.11[c]
	Total carotenoid (mg/kg) in muscle		
C	22.80 ± 6.31[a]		
T1	35.00 ± 4.71[a]		
T2	88.27 ± 21.18[b]		
T3	151.35 ± 31.06[c]		

Table 3: Body color intensity and total carotenoid of hybrid catfish fed with experimental diets over 12 weeks (L=Lightness, a*=Red and b*=Yellow). Mean with different superscripts in the same column are significantly different (p<0.05).

Haematological analysis

Fishes fed on carotenoid, butyric acid or combination diets exhibited numerically increased RBC and WBC counts (p<0.05; Table 4) when compared to the control. Dietary carotenoids and butyric acid significantly affected the haemoglobin and hematocrit of fishes as compared to the control. Haemoglobin varied from 7.23 g/dl to 7.33 g/dl, hematocrit from 39.33% to 41.67%. The total protein was significantly increased for the ButiPEARL™ + Quantum GLO™ Y fed fish compared to the control while numerically increased for the other treatments.

Hematological parameter	Experimental diets			
	C	T1	T2	T3
Red blood cell (RBC) (× 10⁶ cell/ml)	1.59 ± 0.06[a]	1.67 ± 0.04[a]	1.64 ± 0.06[a]	1.67 ± 0.13[a]
White blood cell (WBC) (× 10⁵ cell/ml)	1.20 ± 0.08[a]	1.26 ± 0.09[a]	1.25 ± 0.08[a]	1.27 ± 0.12[a]
Hemoglobin (Hb) (g/dl)	6.64 ± 0.13[a]	7.28 ± 0.17[b]	7.23 ± 0.37[b]	7.33 ± 0.18[b]
Hematocrit (HCT) (%)	36.00 ± 1.00[a]	40.67 ± 0.58[b]	39.33 ± 1.50[b]	41.67 ± 0.58[b]
Total protein (mg/dl)	6.18 ± 0.13[a]	6.48 ± 0.13[ab]	6.44 ± 0.10[ab]	6.50 ± 0.07[b]

Table 4: Haematological factors of hybrid catfish fed with experimental diets for 12 weeks. Mean with different superscripts in the same rows are significantly different (p<0.05).

Discussions

Organic acids are mainly used as feed additives for improving growth performance of pigs and poultry [15-17]. There are also reports on the benefit of organic acids in aquatic animals, including red hybrid tilapia [18], yellowtail [19] and Pacific white shrimp [20]. The organic acids and salts used are formic acid/calcium formate, acetic acid/ sodium acetate, propionic acid/ calcium propionate or butyric acid/ sodium butyrate that are in the free form and they can enhance the growth performance and health status of fish. However, in this study, we would like to demonstrate the additional beneficial effect of feeding an encapsulated source of organic acid. ButiPEARL™ used in this study is an encapsulated butyric acid where the encapsulation can lead to a slow release of the butyric acid alongside the gastro-intestinal tract. The encapsulation of the butyric acid is beneficial because it is a method to prevent leaching. This is especially important in animals that do not swallow whole feed particles but masticate their feed. Furthermore, the encapsulation leads to a slow release effect of free butyric acid throughout the gut. The result obtained from this study showed that fish fed ButiPEARL™ has significant increase in body weight gain and FCR improvement by 25 points compared to fish fed with other treatments. This indicated that ButiPEARL™ had a positive effect on performance and the encapsulation prevented the butyrate from being absorbed too fast. As a result, the butyrate would have been more available in the small intestine to enhance the villi growth for better nutrient digestibility and growth performance.

In addition to growth, addition of dietary ButiPEARL™ also helped to significantly increase the carotenoid deposition in the catfish muscle and enhance yellowness in muscle colour when compared to using Quantum GLO™ Y alone. Quantum GLO™ Y is a microemuslified pigment that belongs to the xanthophyll class and from our earlier studies [8,9], the bioavailability of the carotenoids prepared using this system was significantly better than other carotenoid preparation due to the increased ratio of surface area to volume of the smaller carotenoid structures after the emulsification (size reduction from 20 μm to 0.25 μm that is 80 order of magnitude smaller). This enabled the carotenoid molecules to better penetrate the intestinal epithelium, increasing their residence time and enhancing the absorption. Also, ButiPEARL™ has shown to have diverse modes of action, such as increased villi height and crypt depth, leading to increased absorptive surface of the small intestine and resulting in better nutrient utilization [5]. Based on our results, ButiPEARL™ might have activated the intestinal function and allowed increased intestinal absorption of the microemulsified carotenoids, resulting in richer yellowness observed for the muscle and almost doubling the amount of carotenoid in the muscle from the fishes treated with the ButiPEARL™ + Quantum GLO™ Y combination. This further suggests that dietary butyric acid had a synergistic effect on the carotenoid absorption and pigmentation performance of catfish.

In fact, many immune parameters of Quantum GLO™ Y, ButiPEARL™ or combination-fed fish were improved as shown in Table 4, including the RBC, WBC, haemoglobin, haematocrit and total protein. These outcomes suggest that carotenoid had an immunostimulatory property preventing disease infection in catfish. Antioxidant activity of carotenoids may be involved in the immunomodulatory effect; by quenching singlet oxygen and free radicals, carotenoids can protect white blood cells from oxidative damage [21]. Given that carotenoids also possess an antioxidant property this suggests that such mechanism can take part in immunomodulation. Furthermore, the effects of carotenoids on enhancing cell-mediated and humoral immune responses of vertebrates are also documented [21,22]. Several studies have reported that dietary carotenoids can increase the immune parameters, enhance the survival rate, or act as a prophylactic to pathogens for many aquatic animals such as common carp [23,24] and rainbow trout [25]. Even if butyric acid and carotenoid have different modes of action to the catfish, both had positive effects on their immune responses. Also, our results showed that dietary ButiPEARL™ and Quantum GLO™ Y synergistically enhance the immune parameters of catfish.

Conclusions

Disease resistance, growth improvement and colour intensity are important quality criteria and market value determinants for hybrid fishes. The use of butyric acid source alone may contribute to stimulate villi growth and enhance growth but the butyrate efficacy can be increased when it is fed in encapsulated form (ButiPEARL™). The findings as reported in this paper clearly demonstrated and proved that ButiPEARL can synergistically enhance the microemulsified carotenoids (Quantum GLO™ Y) absorption and significantly improve the pigmentation in catfish when used in combination. In addition, ButiPEARL + Quantum GLO™ Y fed fish also showed improvement in many immune parameters compared to other treatments. In conclusion, ButiPEARL™, Quantum GLO™ Y and their combination have positive effects on performance and pigmentation in catfish aquaculture.

Acknowledgements

The authors would like to thank Dr Orapint, Department of Aquaculture, Faculty of Fisheries, Kasetsart University for conducting the trial and the valuable comments and suggestions.

References

1. Chuchird N, Rorkwiree P, Rairat T (2015) Effect of dietary formic acid and astaxanthin on the survival and growth of Pacific white shrimp (Litopenaeus vannamei) and their resistance to Vibrio parahaemolyticus. Springerplus 4: 440.

2. Guilloteau P, Martin L, Eeckhaut V, Ducatelle R, Zabielski R, et al. (2010) From the gut to the peripheral tissues: The multiple effects of butyrate. Nutr Res Rev 23: 366-384.

3. Van der Wielen P (2002) In: Blok MC, Vahl HA, De Lange L, Van De Braak AE, Hemke G, et al. (Eds). Dietary strategies to influence the gastrointestinal microflora of young animals and its potential to improve intestinal health. Nutrition and Health of the Gastrointestinal Tract. Wageningen, the Netherlands: Wageningen Academic Publishers, pp: 37-60.

4. Smith DJ, Barri A, Herges G, Hahn J, Yersin AG, et al. (2012) In vitro dissolution and in vivo absorption of calcium [1-14C] butyrate in free or protected forms. J Agric Food Chem 60: 3151-3157.

5. Levy AW, James WK, Lorraine F, Susan W, Greg FM, et al. (2015) Effect of feeding an encapsulated source of butyric acid (ButiPEARL) on the performance of male Cobb broilers reared to 42d of age. Poult Sci 94: 1864-1870.

6. Shahidi F, Metusalach A, Brown JA (1998) Carotenoid pigments in seafoods and aquaculture. Crit Rev Food Sci Nutr 38: 1-67.

7. Smith P (1990) Innovations in salmon and shrimp feed. Aquaculture International Congress Proceedings. Aquaculture International, pp: 121-126.

8. Chow PY, Gue SZ, Leow SK, Goh LB (2014) The bioefficacy of microemulsified natural pigments in egg yolk pigmentation. Br Poult Sci 55: 398-402.

9. Chow PY, Gue SZ, Leow SK, Goh LB (2015) Solid self-microemulsifying system (S-SMECS) for enhanced bioavailability and pigmentation of highly lipophilic bioactive carotenoid. Pow Tec 274: 199-204.

10. Skrede G, Storebakken T (1986) Characteristics of colour in raw, baked and smoked wild and pen-reared Atlantic salmon. J Food Sci Technol 51: 123-134.

11. Blaxhall PC, Daisley KW (1973) Routine haematological methods for use with fish blood. Journal of Fish Biology 5: 771-781.

12. Vallada K (1986) Manual de técnicas hematológicas. Rio de Janeiro: Livraria Atheneu.

13. Larsen HN, Snieszko SF (1961) Comparison of various methods of determination of hemoglobin in trout blood. The Progressive Fish-Culturist 23: 8-17.

14. Lowry OH, Rosenbrough NJ, Farr AL, Randall RJ (1951) Protein measurement with the folin phenol reagent. J Biol Chem 193: 265-275.

15. Dibner JJ, Buttin P (2002) Use of organic acids as a model to study the impact of gut microflora on nutrition and metabolism. J Appl Poult Res 11: 453-463.

16. Franco LD, Fondevila M, Lobera MB, Castrillo C (2005) Effect of combinations of organic acids in weaned pig diets on microbial species of digestive tract contents and their response on digestibility. J Anim Physiol Anim Nutr (Berl) 89: 88-93.

17. Papatsiros G, Billinis C (2012) In: Bobbarala V (Ed). The prophylactic use of acidifiers as antibacterial agents in swine. Antimicrobial agents. InTech, Rijeka, Croatia.

18. Ng WK, Koh CB, Sudesh K, Siti-Zahrah, A (2009) Effects of dietary organic acids on growth, nutrient digestibility and gut microflora of red hybrid tilapia, Oreochromis sp., and subsequent survival during a challenge test with Streptococcus agalactiae. Aquacult Res 40: 1490-1500.

19. Sarker MSA, Sato S, Kamata K, Haga Y, Yamamoto Y (2012) Supplementation effect(s) of organic acids and/or lipid to plant protein-based diets on juvenile yellowtail, Seriola quinqueradiata Temminck et Schlegel 1845, growth and nitrogen and phosphorus excretion. Aquacult Res 43: 538-545.

20. Su X, Li X, Leng X, Tan C, Liu B, et al. (2014) The improvement of growth, digestive enzyme activity and disease resistance of white shrimp by the dietary citric acid. Aquacult Int 22: 1823-1835.

21. Bendich A (1989) Carotenoids and the immune response. J Nutr 119: 112-115.

22. Chew BP, Park JS (2004) Carotenoid action on the immune response. J Nutr 134: 257S-261S.

23. Anbazahan SM, Mari LS, Yogeshwari G, Jagruthi C, Thirumurugan R, et al. (2014) Immune response and disease resistance of carotenoids supplementation diet in Cyprinus carpio against Aeromonas hydrophila. Fish Shellfish Immunol 40: 9-13.

24. Sowmya R, Sachindra NM (2015) Enhancement of non-specific immune responses in common carp, Cyprinus carpio, by dietary carotenoids obtained from shrimp exoskeleton. Aquacult Res 46: 1562-1572.

25. Amar EC, Kiron V, Satoh S, Watanabe T (2001) Influence of various dietary synthetic carotenoids on bio-defence mechanisms in rainbow trout, Oncorhynchus mykiss (Walbaum). Aquacult Res 32: 162-173.

17

Have Centuries of Inefficient Fishing Sustained a Wild Oyster Fishery

Stephen Long[1,2,3*]**, Richard Ffrench-Constant**[1]**, Kristian Metcalfe**[2] **and Matthew J Witt**[1]

[1]*Centre for Ecology and Conservation, University of Exeter, Penryn Campus, Cornwall, TR10 9FE, UK*

[2]*Environment and Sustainability Institute, University of Exeter, Penryn Campus, Cornwall, TR10 9FE, UK*

[3]*Department of Geography, University College London, Pearson Building, Gower Street, London, WC1E 6BT, UK*

*****Corresponding author:** Stephen Long, Department of Geography, University College London, Pearson Building, Gower Street, London, WC1E 6BT, UK, E-mail: stephen.long.16@ucl.ac.uk

Abstract

The native European flat oyster (*Ostrea edulis*) has declined throughout its range, due to over-exploitation, a situation mirrored in oyster stocks globally. There are three remaining oyster fisheries in England (Fal, Solent, and Thames Estuary). The Fal oyster fishery though employs traditional methods, using hand-hauled dredges from rowing punts or under sail and is home to the last commercial sailing fleet in Europe. Against a backdrop of temporary closures to protect dwindling stocks in the Solent and Thames Estuary, this study considers whether the longevity of the Fal oyster fishery is linked to the traditional methods that have been employed for centuries. Using GPS tracking in combination with on board observers, we demonstrate that dredging under sail is inefficient compared to more modern mechanically powered methods that are utilised elsewhere. A review of historical landings suggests that both overall landings and fishing effort have declined. The fishery appears to have gone through cycles of over-exploitation and one closure due to disease. However, the key to the long-term survival of the Fal oyster fishery may be linked to the traditional method of dredging. It is estimated that a switch from traditional methods to modern techniques would result in a greater than 9 fold increase in effort per season. The data presented highlight this unique fishery as a counterfactual to the increases in power seen in commercial fisheries over the last century and serve as a reference point for future studies.

Keywords: Oysters; Shellfish; Artisanal; Small scale fisheries; Traditional; Fishery

Introduction

Despite recognition that centuries of marine exploitation have drastically reduced stocks of formerly abundant marine species [1], unsustainable fishing continues on a global scale [2]. There is now significant scientific and economic interest in rebuilding collapsed fisheries and ensuring marine resources are used sustainably [3,4]. Achieving such sustainability will be particularly challenging given the wide range of fishing practices, their target stocks and host ecosystems. The sedentary adult phase of bivalves, such as scallops, oysters and mussels, make local populations highly vulnerable to over exploitation leading to the collapse of fisheries [5].

Oysters are environmentally and economically important; they act as ecosystem engineers producing biogenic reef habitat [6], as well as providing ecosystem services including water filtration, food provision, habitat and coastal defence [7]. Oysters are of significant economic value particularly to coastal communities, with global production from aggregated country level data of 5.3 million tonnes in 2014, of which the vast majority (5.1 million tonnes) was from aquaculture [8,9]. However, long term exploitation has edged oyster reefs to extinction, with an estimated 85% of oyster reefs lost globally [10].

The native or European flat oyster (*Ostrea edulis*, L.) is a sessile, filter feeding, bivalve mollusc, once highly abundant, ranging throughout the Mediterranean and Atlantic coasts of Europe [11]. The species and the oyster beds they form are on the OSPAR List of Threatened and/or Declining Species and Habitats [12], and are typically included as biodiversity action plan species where they occur (e.g. UK BAP). Historically, over-exploitation and other factors have led to declines since the early 18th century, with wild populations reportedly scarce around Europe by the 1940s [13,14]. Landings of native oysters have significantly declined in the United Kingdom and across Europe [11,13,15]. In England and Wales landings have fallen by 98%, from more than 2000 tonnes in the 1920s [16] to 39 tonnes in 2015 [17]. In England and Wales only three oyster fisheries remain, these are found in the Solent, Thames Estuary and River Fal [16].

The Fal oyster fishery has been in operation since the Roman times [18], and still employs traditional fishing methods. Hand-hauled dredges are operated from two forms of non-mechanically powered boat. These are sailing boats and haul tow punts, the latter being where boats are hand winched to a previously deployed anchor whilst towing dredges. These practices are dictated by the management authority, and so the fishery is considered home to the last commercial sailing fleet in Europe [18]. The fishery lies within the Fal and Helford Special Area of Conservation (SAC), as per the EC Habitats Directive [19], where all forms of bottom dredging are now banned with the exception of traditional oyster dredging methods which continue under a special dispensation. Historically the fishery was principally a brood stock fishery, providing stock for on-growing to areas including the Helford estuary in Cornwall and Essex oyster beds [20,21]. However, the oyster disease *Bonamia*-caused by the parasite *Bonamia ostreae* has had a significant impact on local stocks. Introduced to Europe via spat from the USA in the late 1970s it was first identified in the UK in the Fal and Helford estuaries in 1982 [20] and rendered fishing unprofitable during the 1980's [22,23]. At the time the spread of the disease was managed though the use of movement controls [24], which would have prevented operation as a brood stock fishery. Stock levels remained low throughout the 1980's, with increasing catches through the 1990's [24,25], presumably associated with increased fishing effort. Despite the unique nature of the fishery and the ecological complexity of the Fal estuary it has attracted relatively little scientific investigation. The Fal Oyster currently has Protected Designation of Origin status [23] and is sold to both the food industry and into aquaculture as brood stock. Annual landings of approximately 50-60 tonnes [18], makes an economic contribution to an economically deprived region of the UK.

The fishery offers a sharp contrast to the industrialisation and increased fishing power of almost all other fisheries throughout the North Atlantic Ocean over the last two centuries, which has been linked to stock declines and fishery collapse [26,27]. Set against this historic back-drop the current study aims to highlight the Fal as a fishery that has maintained traditional practices and not exhibited the increases in fishing power through technological advances. More specifically the comparative inefficiency of these traditional practices is considered. Historic and new data are combined to offer insights into the long-term trends, current operation of the fishery and to provide a reference point. Whilst establishing strict causality is not possible in a data-deficient context the comparatively inefficient dredging methods may have preserved viable stocks accounting for the longevity of this traditional fishery.

Methodology

Description of the Fal fishery

The total area of the fishery is estimated as 1,101 hectares [28] this being the total dredgable area below the mean low water mark between the southerly limit (a line between St. Mawes Castle and Trefusis Point) and the northerly limit (Malpas) of the fishery (Figure 1a). For the purposes of annual surveys the UK Centre for Environment Fisheries and Aquatic Sciences (Cefas) divided the fishery into three sections, the River, Harbour and Outer Harbour Sections (Figure 1a) which are also referred to in this study [25,29].

Regulation and operation of the current fishery

The Truro Port Fishery Order 1936 (amended 1975 and 1984) regulated the fishery until expiry of the order on 31st July 2014 [28,30-32]. Due to delays in the development of a new regulating order, an emergency byelaw was enacted from 13th March 2015 for a maximum of 12 months [32,33]. The emergency byelaw, was replaced by The Fal Fishery Order 2016 [34]. The emergency byelaw and subsequent 2016 order adopted the principal management measures from the original expired order, specifically: i) that fishing must not use mechanical power (neither to propel vessels or haul dredges), ii) a Minimum Landing Size (MLS) of 2 5/8' (67 mm), iii) a season (1st October to 31st March; 182 days), iv) fishery hours (dredging between 09:00 to 15:00 Monday to Friday, 09:00 to 13:00 on Saturdays) and v) fishermen must purchase a licence (£165.00 in the 2016/17 season) for each dredge they operate [33]. Sail boats operate in the Harbour, and to a lesser extent Outer Harbour (Figure 1a). In practice, only the northern most portion of the Outer Harbour Section is dredged, with no activity on Falmouth Bank and St. Mawes Bank, as the substrate here is unsuitable for oyster settlement [25]. Sail boats drift down wind and tide whilst towing dredge gear, before sailing back up to repeat the process. Sail boats are crewed by 1 or 2 men, operating 2 or 4 dredges

respectively. Dredges are hauled and contents sorted alternately through the course of each drift. Haul-tow boats are small punts and operate in the River section. The dredge is towed as the fishermen winches, by hand, back to an anchor deployed a short distance away. Upon hauling, the contents of the dredge, known as *cultch* (term also applies to spat settlement substrate), is sorted and oysters ≥ MLS are retained. Owing to constraints of time and access the focus of this study is the sail fishery and Harbour section.

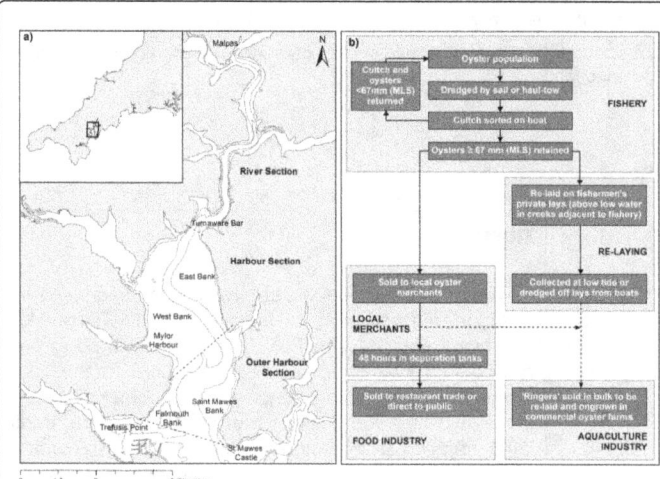

Figure 1: Overview of the Fal wild native oyster fishery. Map a) shows the extent of fishery, constrained by mean low water mark, from southerly limit (line from Trefusis Point to St Mawes Castle) and northerly limit (Malpas), 0 m and 5 m contour lines are drawn. Schematic b) shows movement of oysters from fishery to market. Note it is not clear what proportion of *ringers* (oyster ≥ MLS but too small for food industry) are sold directly to aquaculture industry by fishermen or via local merchants or both (indicated with dashed line).

Historical data on fishing fleet

A historic record of the number of sail and haul-tow boats working the Fal fishery was identified, covering the period 1910 to 1973 [35]. Available data on the estimated landings and number of dredge licences was obtained from the Port of Truro, this covered the period from 1995 to 2013. Estimated landings are the sum of total weights purchased in the season by individual merchants known to purchase Fal oysters from fishermen. Oysters purchased by merchants may include both direct landings from the fishery and oysters gathered from re-laid stocks.

Dredging survey

'Dredge gear' refers to the gear towed across seabed, whilst dredge refers to the activity of deploying, towing and hauling the dredge gear. The two dredge gears used on board the sailing boat from which data were gathered were typical of those used in the fishery. Constructed from stainless steel with the widths of the front edge of 75 cm and 70 cm, 60 mm belly rings (to resist abrasion and provide free passage to material <60 mm) and a nylon mesh bag. Data were collected by an observer on board one sailing boat on 6 days in February and March 2014. Upon hauling, fishers sort the dredge contents (*cultch*) and select oysters to retain, depositing the rest of the dredge contents overboard.

For each dredge the number of oysters retained was recorded. For each retained oyster the wet weight (± 5 g, Pesola spring balance) and maximum diameter of each oyster (± 0.5 mm, Vernier callipers) was determined. Prior to weighing, any material, organic or otherwise, attached to, or to which the oyster was attached, was removed. For a sub-sample of 15 dredges all the oyster in each dredge were examined; counting, weighing and measuring the oysters (using the above procedure) returned, in addition to those retained by the fisherman. The start time (dredge gear deployment) and finish time (hauling the dredge gear) were noted for each dredge. Catch per Unit Effort (CPUE) is the number of oysters caught per m^2 of seabed dredged. Area of seabed covered by each dredge was determined by multiplying the mean dredge gear width (0.725 m) by the distance covered by the dredge between the start and finish time of each tow.

Vessel monitoring

GPS data loggers (i-gotu gt-100 and i-gotu gt-600 GPS) combined with external battery packs were placed on two sailing boats based at Mylor Harbour from February 19th 2014 to the end of the season March 31st 2014. Locations of individual dredges and vessel tracks were all obtained using GPS logger data. Positions of vessels were logged every 30 seconds during the hours of fishery operation. Spatial analysis was conducted using ArcGIS 10.1 [36] in conjunction with Geospatial Modelling Environment [37]. Number of sailing vessels was surveyed from Weir Point, Fal estuary (N50° 11' 25", W005° 3' 23"), on 24 days between February 5th 2014 and February 31st 2014 inclusive. Counts took place between ~09:30 and 14:30 each day. Mean wind speed between 09:00 and 15:00 was calculated from weather station observations at Culdrose, Cornwall (UK Met Office station; N 50° 5' 2.4", W005° 15' 32.4") [38]. The total number of sailing boats was modelled using a Generalised Additive Model (GAM) using a smoothing spline fit, with mean wind speed as the explanatory variable and a Poisson error structure. This allowed an estimate of the number of fishable days per season based on suitable weather conditions.

Effort comparison

Data from the dredging survey and vessel monitoring activities were used to estimate effort expended per day (m^2 fisher^{-1} day^{-1}) and per season by an individual fisher. No comparable figure for a motorised vessel was available in the literature. Thus an equivalent estimate of effort was developed for a motorised vessel operating dredges in this fishery. For the purpose of this estimation, dredge width was the maximum dictated by existing byelaw in the Thames estuary, which is a motorised fishery [39]. The dredge speed used was the averaging towing speed used in dredge surveys in the Fal undertaken by Cefas [25], as this is the only known example of motorised dredging in the Fal. The percentage of time spent dredging during fishery hours using a motorised vessel (95%) was set based on discussion with local fishermen in relation to observations during dredging survey. Note journeying to and from oyster beds is excluded from the fishery hours. The percentage of fishable days was determined from the number of days in the 2012/13 and 2013/14 where the observed mean wind speed was <30 mph, note the fishery is more sheltered than the wind observation site (Culdrose, see above).

Results

Fleet statistics and landings

The total number of boats and the ratio of sail to haul-tow boats varied greatly between 1910 and 1972 (Figure 2). At its height, in 1922, the fishery supported 189 boats, 144 haul-tow and 45 sail. Unfortunately, a more recent continuous record is not available; however, in 2009 there were approximately 12 sailing boats in the fleet [18]. In the 2013/14 season there were 16 registered sailing boats, and 14 registered haul-tow boats. Of these only 6 sailing boats and 5 haul-tow boats fished most suitable days, with the rest being used occasionally or rarely (P. Ferris, pers. comm.). Landings and licences peaked in 1996 at 117 tonnes and 87 licences. The subsequent trend is a decline to 47 tonnes and 55 licences in 2012. Although the lowest landings and number of licences was recorded in 2005, with 24 tonnes and 41 licences (Figure 3).

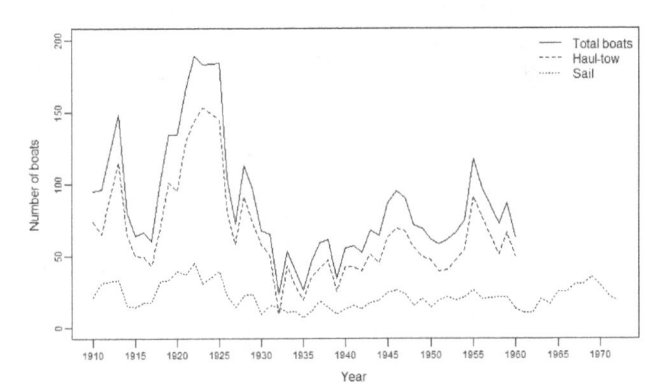

Figure 2: Historic numbers of boats operating in the Fal oyster fishery. With total (solid line), sail (short dashes) and haul-tow (long dashes) boat numbers between 1910 and 1972. Between 1910 and 1960 only years where both number of sail and haul-tow were available are plotted. From 1961 to 1972 only the number of sail boats was available. Note year refers the year in which the season starts, thus 1910 is the 1910/11 season. Data: Davies, 1989.

Dredging survey

Data were collected from 238 separate dredges. The median distance dredged was 98 m (IQR=56.7-149.8, n=238), representing a median area per dredge of 71 m^2 (IQR=41.1-108.6, n=238). The median number of oysters retained was 2 oysters dredge^{-1} (IQR=1-4, n=238), which equates to a median CPUE of 0.03 oysters per m^2 dredged (IQR=0.012-0.061, n=238; Figure 4a). The mean dredge duration was 5.7 minutes (sd=2.3, n=238). The median wet weight of retained oysters was 80 g (IQR=70-105, n=652; Figure 4b), with a median maximum diameter of 78 mm (IQR=74-86, n=652; Figure 4c). CPUE was shown to vary spatially.

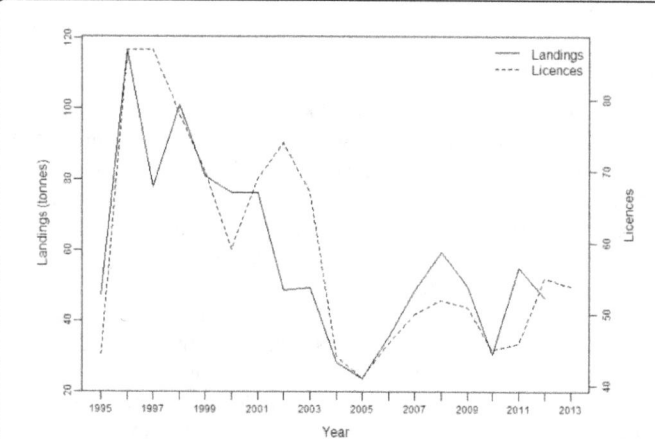

Figure 3: Recent trends in landings and licences, in the Fal oyster fishery. Between 1995 and 2013, note year refers the year in which the season starts, thus 1995 is the 1995/96 season. Landings are indicated with and solid line and licences with a dashed line. Data: P. Ferris, Port of Truro.

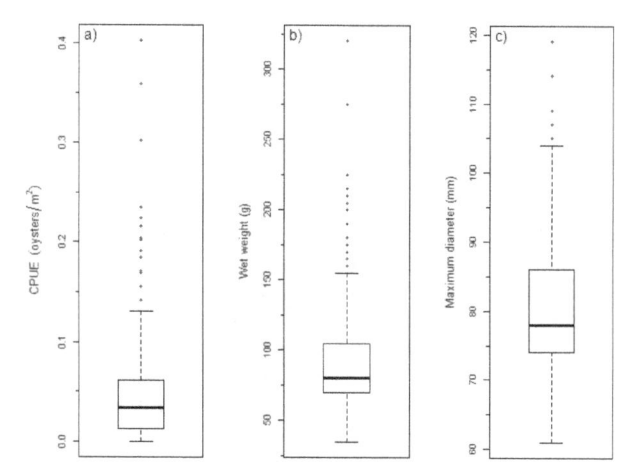

Figure 4: Box-whisker plots of metrics from dredges from observer on board a sailing vessel in the Fal oyster fishery, Cornwall. Where a) shows the Catch per Unit Effort (oyster m^{-2}) (n=238), b) shows the wet weight (g) (n=652) and c) shows the maximum diameter (mm) (n=652). Median is indicated by a thick back line. The box represents the interquartile range (IQR) bounded by the 25th (Q1) and 75th (Q3) percentiles. The lower whisker is drawn at the larger of: the minimum value x and Q1-1.5*IQR. The upper whisker is drawn at the smaller of: the maximum value of x and Q3+1.5*IQR . Outliers are indicated by open circles.

For a sub-sample of 15 dredges, all oysters in the haul were weighed and measured. The median number of oysters ≥ MLS was 3 oysters per dredge (IQR=1.5-3.5, n=15), with a median CPUE of 0.04 oysters m^{-2} dredged (IQR=0.019-0.097, n=15). The median wet weight of oysters ≥ MLS was 63 g (IQR=46.3-90.0, n=54), with a median maximum diameter of 74 mm (IQR=70.3-80.0, n=54).

Vessel monitoring

Towards the latter end of the season, including the period GPS loggers were installed, sail boat dredging effort was concentrated on the East bank (Figure 5a). Using data collected from sailing vessels the dredging effort (km^2 $fisher^{-1}$ day^{-1}) exerted by one fisher per day was estimated (Table 1). This was extrapolated to estimate effort expended in a season. This was achieved by estimating the percentage of fishable days, using a modelled relationship between mean wind speed and number of sailing vessel fishing from observed wind speeds in the last two seasons. Visual census of the fleet highlighted dependency on suitable wind conditions, with only 62.1% of days fishable by at least one boat in the previous two seasons (Table 1). For comparison the effort expended by a fisher in a motorised vessel is estimated (Table 1). Fishing is not continuous, dredging ceases whilst sailing back up tide and wind (Figure 5b), with dredges only deployed 87% of the time, excluding sailing to and from harbour (Table 1).

Metric	Vessel Type:		Unit
	Sailing	Motorised	
Width of dredge	0.7	2§	m
Number of dredges	2	2	
Speed	1.3*	3**	km h^{-1}
Hours in fishery day	6	6	hours
Percentage of time spent dredging	86.8†	95††	%
Effort per day	9,479	68,400	m^2 $fisher^{-1}$ day^{-1}
Number of days in season	182	182	days
Percentage of fishable days	62.1‡	84.6‡‡	%
Number of fishable days	113	154	days
Total effort per season	1.1	10.5	km^2 $fisher^{-1}$ $season^{-1}$

* Mean dredging speed recorded by observer on two full days of fishing (19/02/14 and 21/03/14)
† Percentage of time one or more dredges were deployed during one full day of fishing (19/02/14)
‡ Percentage of days, in the last 2 seasons, where GAM (Figure 1, Supporting Information) predicts at least one sailing vessel in operation, from observed wind speeds
§ As per maximum aggregate blade width (4 m) dictated by byelaw in Thames estuary oyster fishery, represented here as two 2 m dredges
** Average speed of motorised vessel (7.9 m, 60 kW) used by Cefas during oyster surveys for towing dredges in the Fal oyster fishery.
†† allowing 5% of time for hauling, emptying and re-deploying dredges
‡‡ Percentage of days in last 2 season where mean wind speed <30 mph

Table 1: Comparison of effort between sail and motorised dredging. Based on GPS and observer data from sailing vessels and stated assumptions for the operation of motorised vessels.

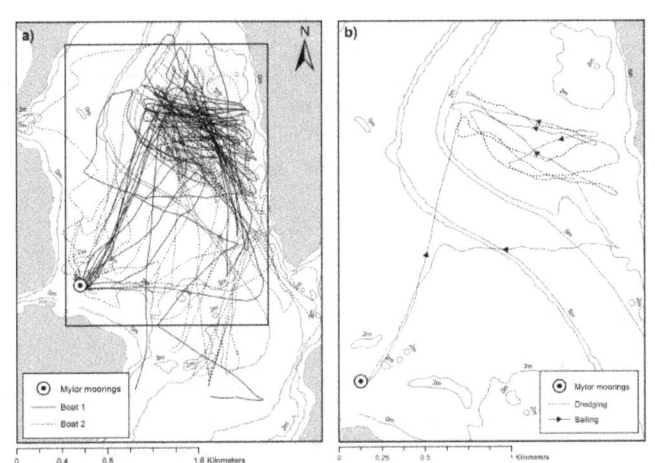

Figure 5: GPS tracks of a) Two sailing boats in fleet and b) a typical dredging trip. Where a) shows 13 oyster dredging trips between 19/20/2014 and 31/03/2014 inclusive, undertaken by boat 1 (7 trips) and boat 2 (6 trips) and b) a single trip on 19/02/2014 differentiating between periods of travel 'Sailing' and oyster fishing 'Dredging'. For b) periods of oyster fishing 'Dredging' determined by board observer recording times when dredges were in operation. GPS data loggers recorded position every 30 seconds. Tracks smoothed using Polynomial Approximation with Exponential Kernel (PAEK) smoothing with 100 m tolerance.

Effort comparison

The estimated effort expended using a sailing vessel was 9,479 m^2 fisher^{-1} day^{-1} and 1.1 km^2 fisher^{-1} season^{-1}, compared with 68,400 m^2 fisher^{-1} day^{-1} and 10.5 km^2 fisher^{-1} season^{-1} for a motorised vessel, based on the predicted number of fishable days for the respective vessels (Table 1).

Discussion

Long-term trends

Historically levels of exploitation were much higher, with the fishery supporting 189 boats at its peak in 1922. Davies (35) attributes the expansion and contraction of the fleet to changes in stock levels, with variation in the ratio of sail to haul-tow boats likely reflecting changes in oyster abundance in the Harbour and River sections of the fishery. The onset of *Bonamia* in 1982 reduced landings, from 120-250 tonnes year^{-1} in the years immediately preceding the first record, to an average of 20 tonnes year^{-1} between 1983 and 1992 [24]. Surveys in 1986 and 1988 found stocks were very low [25], though historical data suggests there was a recovery of stocks and consequently the fishery in the 1990s. Following a low-level mortality event in 2010, testing of *O. edulis* from the Fal confirmed the presence of *B. osteae* and *B exitiosa*, although subsequent testing for *B. exitiosa* has been negative suggesting it has not become established [40].

Unfortunately, reliable and comprehensive historical data on landings are unavailable. However, Davies [35] reports the largest quantity of oysters from a single season to be approximately 3.5 million oysters in 1953/54 season. The basis of this estimate is not clear, although it is known at this time oysters were sold by number rather than weight. Assuming that the wet weight of individual oysters landed was 74 g (mean weight of oyster ≥ MLS from 15 dredge sub-sample), this equates to an estimated landing of 258 tonnes in the 1953/54. This estimate would represent a 5.5 fold decrease in landings from 1953/54 to 2012/13. This, in conjunction with greater levels of effort, strongly suggests historic landings exceed current catches. Overall, annual licences granted, whilst indicative, cannot be used as a direct proxy for effort. Some dredge licences are bought and used infrequently (P. Ferris, *pers. comm.*). Current landings figures are also not necessarily an accurate record of the total weight of oysters landed from the fishery in a season. The landings figures presented in this study were obtained from the previous management authority (Port of Truro), who produced the data annually by contacting known merchants and enquiring as to the total weight purchased each season. The landing figures do not differentiate between oysters dredged directly from the fishery and those gathered having previously been re-laid. Furthermore, fishermen may use routes to markets that exclude the merchants surveyed each year. However, data is sufficient to conclude there has been a decline in both the landings and licences. Landings in the 2012/13 season of 46.6 tonnes are 40% lower than the recent peak of 116.7 tonnes in 1996/97 and 82% lower than the historic peak, estimated to be 258 tonnes in 1953/54. It is not clear whether the trend in recent years is a decline in landings due to reduced effort, or a decline in landings due to declining stocks.

Inefficiency of traditional dredging methods

A unique feature of the Fal oyster fishery is the traditional methods employed. The mean dredge duration of 5.7 minutes (sd=2.34, n=238) observed in this study, is approximately the same as the 6 minute duration reported almost eight decades ago [41]. This likely indicates fishing practices have remained unchanged, in sharp contrast to the advances in gear and technology deployed in other fisheries in the last century. There are a number of factors pertaining to the traditional methods that limit the comparative efficiency of the fishery. Dredge gear size is physically limited to what can be hauled by hand. The reliance on suitable weather conditions means it is estimated only 62.1% of days (113 of 182 days, per season) were fishable in the 2012/13 and 2013/14 two seasons. Sub-optimal wind conditions typical result in fewer boats fishing and sub-optimal dredge speeds, further reducing efficiency. These factors limit the effort in a fishery that exhibits low CPUE.

Consideration should be given to the fact that the fisherman observed in this study has not been dredging for as many years as some others operating boats in the fleet. Fishermen operating within the fishery for longer may benefit from greater CPUE. Experience would improve boat handling to optimise dredge speed, better target productive areas and minimise time spent re-positioning the boat. The efficiency of the dredge gear used in the fishery is not known and likely varies with factors identified by Gosling [42] including; the nature of the seabed, operating conditions and the skill of the fisherman. No applicable estimate of hand dredge efficiency could be found in the literature. A good estimate could be obtained by dredging an area such as Turnaware Bar having first determined actual density by a quadrat based survey. This would allow a better understanding of the dynamics of the fishery. An estimate of dredge efficiency would allow catch-per-dredge figures, from this study and annual surveys [25,29], to be converted to actual abundance in order to estimate the total population.

Could inefficiency account for the longevity of the fishery?

Given the highly variable and sporadic nature of recruitment, severe population reduction caused by *Bonamia* and the fluctuation in fleet size it is difficult to determine whether levels of exploitation are, or have been, sustainable. Some authors suggest there have been periods of over-fishing. Orton [41] concluded that 'the population of oysters... continues to decrease as a result of more oysters being taken from the beds than are being replaced by natural means'. Cove [43] reported that there were episodes of over-fishing in previous decades. However, despite potential episodes of over-fishing and the severe population reduction caused by *Bonamia* the fishery has avoided collapse. It is worth noting both remaining wild native oyster fisheries in England, Thames estuary and Solent, which deploy modern methods, effected temporary closures in the 2013/14 season to protect severely depleted stocks (Kent and Essex IFCA, 2013; Sussex IFCA, 2013) [39,44]. The Thames estuary oyster fishery closure was subsequently extended to cover the 2015/16 to 2017/18 seasons [45], whilst the Solent fishery has extended closures for the 2014/15, 2015/16 and 2016/17 seasons [46-48]. The longevity and survival of the Fal fishery is therefore notable.

Motorised vessels would dramatically increase the total units of effort per day and increase the number of fishable days in the season, resulting in a greater than 9 fold increase in the level of effort per season (Table 1). Reviewing declines in *Crassostrea virginica* fisheries of the USA, Mackenzie [49] highlights the increases in productivity brought by the motorization of fleets, allowing fishing on days when wind conditions prevented sailing. Further, mechanical power would allow heavier dredge gear, which would likely increase CPUE. This means that the actual difference in effort between sail and motor is likely greater than the conservative estimate made here. Heavy mechanical gears would also likely damage benthic habitat and structure with potential effects on recruitment. This has been reported in *C. virginca* fisheries in the USA, examples include studies in the Chesapeake Bay fishery [50] and Neuse River fishery [6]. The unique nature of the fishery means comparisons with other cannot be readily drawn, thus the ecological and physical impact of gears currently used is not known. Nevertheless, it is worth noting that Lenihan and Peterson [51] report that a 1m wide 25 kg dredge deployed from a powered vessel reduced the height of experimentally restored *C. virginica* reefs by 34%, in the Neuse River, North Carolina.

The authors are not aware of any other examples where the prohibition of powered vessels is currently used as a management measure in an oyster fishery. This management measure was formerly applied in the Chesapeake Bay oyster fishery, USA. Rothschild, et al. [50] report that the restriction of dredging to sail-powered vessels to constrain effort was of limited effectiveness; identifying the sheer number of vessels, >1000 by 1890 [52], being responsible for the unsustainable level of effort in spite of the restriction. The review identified intense fishing pressure leading to mechanical destruction of habitat-the degradation of oyster bar structure, shallowing of bar profiles, removal of settlement substrate and resulting siltation-as one of the primary causes of decline but state this pre-dated industrialisation [50]. A similar explanation may apply to the Fal oyster fishery where significantly higher levels of effort in the early part of the 20th century ultimately led to declines in catch and effort. In the present day Fal estuary fishery restriction to non-powered caps efficiency and renders profitability marginal.

There is currently no limit imposed by the management authority on landings or licenses. It is the marginal profitability of traditional fishing which limits the size of the fleet. If over-exploitation induced stock declines have occurred it has likely resulted in decreased effort before the point of fishery collapse could be reached. There is increasing recognition that the industrialisation of fishing and resulting increase in fishing power has dramatically reduced stocks, altered marine ecosystems and led to fishery collapses throughout the North Atlantic Ocean [26,27]. The key to the longevity of this fishery may therefore be the inherent inefficiency of the capture method, which ensures a low CPUE and renders over-exploitation to collapse economically unfeasible.

Conclusion

This study aimed to gain insights into sustainability of the Fal oyster fishery. The timing is pertinent as the long standing regulating order has expired. Management is now in the remit of the Cornwall Inshore Fisheries and Conservation Authority (CIFCA) and regulated by the 2016 order. With increasing interest in the restoration of oyster beds in the UK [15] and globally, the Fal oyster fishery may offer a unique 'model' approach to the management of wild oyster fisheries. Whilst causality is difficult to establish in a data deficient fishery and is beyond of the scope of the data presented here, the fishery's longevity is noteworthy and could be considered an indicator of sustainability. Although a more accurate description could be a state of flux between over-exploitation and reduced effort allowing recovery. Given the stock is demonstrably vulnerable to additional pressures as demonstrated by the impacts of *Bonamia* in the 1980s, management measures should aim to increase the resilience of the fishery as well as recruitment to increase productivity. The crucial feature of the existing regulation of this fishery, and hence its long term survival, seems to be the inherent inefficiency of the traditional methods of dredging employed, which it is conservatively estimated results in at least 9 times less effort per season than if the fleet were motorised.

Acknowledgements

The authors would like to thank the following without whom the above would not have been possible. A. Spargo, who initiated the project, provided access to his boat, offered candid insights into the fishery and introduced an amateur to the art of sailing. R. Chapman for agreeing to install a GPS logger on his boat. C. Ranger for sharing data and providing insight from a merchant's perspective. Port of Truro, in particular P. Ferris for sharing knowledge of the fishery and providing data. K. Vaasteen at Cefas for granting access to the annual survey of the oyster population. Officers of the Cornwall Inshore Fisheries and Conservation Authority who provided guidance. J. Beswick who assisted with data collection. The manuscript was significantly improved by the comments of anonymous reviewers.

References

1. Jackson JB, Kirby MX, Berger WH, Bjorndal KA, Botsford LW, et al. (2001) Historical overfishing and the recent collapse of coastal ecosystems. Science 293: 629-637. Srinivasan UT, Cheung WW, Watson R, Sumaila UR (2010) Food security implications of global marine catch losses due to overfishing. Journal of Bioeconomics 12: 183-200.

2. Pauly D, Christensen V, Guénette S, Pitcher TJ, Sumaila UR, et al. (2002) Towards sustainability in world fisheries. Nature 418: 689-695.

3. Worm B, Hilborn R, Baum JK, Branch TA, Collie JS, et al. (2009) Rebuilding global fisheries. Science 325: 578-585.

4. Nielsen EE, Kenchington E (2001) A new approach to prioritizing marine fish and shellfish populations for conservation. Fish Fish 2: 328-343.

5. Lenihan HS, Peterson CH (1998) How habitat degradation through fishery disturbance enhances impacts of hypoxia on oyster reefs. Ecol Appl 8: 128-140.

6. Coen LD, Brumbaugh RD, Bushek D, Grizzle R, Luckenbach MW, et al. (2007) Ecosystem services related to oyster restoration. Mar Ecol Prog Ser 341: 303-307.

7. FAO (2015) FIGIS (Fisheries Global Information System). Fisheries Commodities and Trade 1976-2012, Rome.

8. FAO (2015) FIGIS (Fisheries Global Information System). Global capture production 1950-2013, Rome.

9. Beck MW, Brumbaugh RD, Airoldi L, Carranza A, Coen LD, et al. (2011) Oyster reefs at risk and recommendations for conservation, restoration, and management. Bioscience 61: 107-116.

10. Lallias D, Boudry P, Lapegue S, King JW, Beaumont AR (2010) Strategies for the retention of high genetic variability in European flat oyster (Ostrea edulis) restoration programmes. Conserv Genet 11: 1899.

11. OSPAR Commission (2008) OSPAR List of Threatened and/or Declining Species and Habitats.

12. Kennedy R, Roberts D (1999) A survey of the current status of the flat oyster Ostrea edulis in Strangford Lough, Northern Ireland, with a view to the restoration of its oyster beds. Proceedings of the Royal Irish Academy B Biology & Environment Proceedings of the Royal Irish Academy 99: 79-88.

13. Smyth D, Roberts D, Browne L (2009) Impacts of unregulated harvesting on a recovering stock of native oysters (Ostrea edulis). Mar Pollut Bull 58: 916-922.

14. Laing I, Walker P, Areal F (2006) Return of the native is European oyster (Ostrea edulis) stock restoration in the UK feasible. Aquat Living Resour 19: 283-287.

15. Laing I, Walker P, Areal F (2005) A feasibility study of native oyster (Ostrea edulis) stock regeneration in the United Kingdom. Centre for Environment, Fisheries & Aquaculture Science.

16. MMO (2015) Monthly UK Sea Fisheries Statistics-Reported Landings. Marine Management Organisation.

17. Royal Haskoning (2009) Falmouth Cruise Project Environmental Statement. Marine Ecological Survey.

18. Council of the European Commission (1992) Council Directive 92/43/EEC of 21 May 1992 on the conservation of natural habitats and of wild flora and fauna. Official Journal of the European Community Series L 206: 7-49.

19. Hudson E, Hill B (1991) Impact and spread of bonamiasis in the UK. Aquaculture 93: 279-85.

20. Laing I, Spencer BE (2006) Bivalve cultivation: criteria for selecting a site. Centre for Environment Fisheries and Aquaculture Science pp: 1467-5609.

21. Héral M, Deslous-Paoli J (1991) Oyster culture in European countries. Estuarine and Marine bivalve mollusc culture. Boca Raton, Florida pp: 154-190.

22. DEFRA (2006) Council Regulation (EC) No 510/2006 on protected geographical indications and protected designations of origin "Fal Oyster". Department for Environment, Food and Rural Affairs.

23. Laing I, Dunn P, Peeler E, Feist S, Longshaw M (2014) Epidemiology of Bonamia in the UK 1982 to 2012. Dis Aquat Organ 110: 101-111.

24. Cefas (2012) Truro Oyster Fishery Annual Survey 2012 Fal Estaury, Cornwall. Centre for Environment, Fisheries and Aquaculture Science.

25. Thurstan RH, Roberts CM (2010) Ecological meltdown in the Firth of Clyde, Scotland: two centuries of change in a coastal marine ecosystem. Plos One 5: e11767.

26. Thurstan RH, Brockington S, Roberts CM (2010) The effects of 118 years of industrial fishing on UK bottom trawl fisheries. Nat Commun 1: 15.

27. HM Government (1975) Truro Port Fishery (Variation) Order.

28. Cefas (2013) Truro Oyster Fishery Annual Survey 2013 Fal Estuary, Conrwall. Centre for Environment, Fisheries and Aquaculture Science.

29. HM Government (1936) Truro Port Fishery Order.

30. Port of Truro (2013) Ports of Truro, Penryn, Prince of Wales Pier, Portscatho and Newquay Harbours Fees and Charges.

31. CIFCA (2015) Fal Shellfish Fishery Emergency Byelaw. Impact Assessment: Cornwall Inshore Fisheries and Conservation Authority.

32. CIFCA (2009) Fal Shellfish Fishery Emergency Byelaw. Cornwall Inshore Fisheries and Conservation Authority.

33. HM Government (2016) The Fal Fishery Order 2016.

34. Davies A (2002) The History of Falmouth Working Boats. Falmouth, UK.

35. ESRI (2013) Relands, CA. Environmental Systems Research.

36. Beyer HL (2010) Geospatial modelling environment.

37. Met Office (2014) Culdrose Observations.

38. KEIFCA (2013) Shellfish Beds Byelaw: Temporary prohibition on taking native oysters. Kent & Essex Inshore Fishery and Conservation Authority.

39. Longshaw M, Stone DM, Wood G, Green MJ, White P (2013) Detection of Bonamia exitiosa (Haplosporidia) in European flat oysters Ostrea edulis cultivated in mainland Britain. Diseases of Aquatic Organisms 106: 173-179.

40. Orton J (1927) Observations on the Fal estuary oyster beds during 1926, including a study in over-fishing. Journal of the Marine Biological Association of the United Kingdom 14: 923-934.

41. Gosling E (2008) Bivalve molluscs: biology, ecology and culture. John Wiley & Sons.

42. Cove J (1973) Hunters, trappers, and gatherers of the sea: a comparative study of fishing strategies. Journal of the Fisheries Board of Canada 30: 249-259.

43. SIFCA (2013) Sussex IFCA introduce adaptive management measures to protect oyster stock. Emergency Temporary Closure of Shellfish Fisheries Byelaw. Sussex Inshore Fisheries and Conservation Authority.

44. KEIFCA (2015) Byelaw Area A. Kent & Essex Inshore Fishery and Conservation Authority.

45. SIFCA (2014) Temporary closures of shellfish bed byelaw. Sussex Inshore Fisheries and Conservation Authority.

46. SIFCA (2015) Temporary closures of shellfish bed byelaw. Sussex Inshore Fisheries and Conservation Authority.

47. SIFCA (2016) SIFCA Solent oyster temporary closure notice: Sussex Inshore Fisheries and Conservation Authority.

48. Mackenzie CL (2007) Causes underlying the historical decline in eastern oyster (Crassostrea virginica Gmelin, 1791) landings. J Shellfish Res 26: 927-938.

49. Rothschild B, Ault J, Goulletquer P, Heral M (1994) Decline of the Chesapeake Bay oyster population: a century of habitat destruction and overfishing. Mar Ecol Prog Ser 111: 29-39.

50. Lenihan HS, Peterson CH (2004) Conserving oyster reef habitat by switching from dredging and tonging to diver-harvesting. Fish B Noaa 102: 298-305.

51. Stevenson CH (1894) The oyster industry of Maryland. Causes underlying the historical decline in eastern oyster landings Journal of Shellfish Research pp: 927-938.

52. U.S. Commission of Fisheries (1901) Decline of the Chesapeake Bay oyster population: a century of habitat destruction and overfishing. Marine Ecology Progress Series pp: 29-39.

18

Profitability of Selected Ventures in Catfish Aquaculture in Ondo State, Nigeria

Thompson OA and **Mafimisebi TE**

Department of Agricultural and Resource Economics, the Federal University of Technology, Akure, Nigeria

Corresponding author: Thompson OA, Department of Agricultural and Resource Economics, the Federal University of Technology, Akure, Nigeria, Tel: E-mail: tonyfav22@gmail.com

Abstract

The need to provide information to prospective investors on the decision to invest led to this study on profitability analysis of selected ventures in catfish aquaculture in Ondo State, Nigeria. A multi-stage sampling technique was used to select 144 fish farmers rearing fingerlings, juveniles and table size (full) fish in the study area. The results of the analysis of socio-economic characteristics showed that the mean age of the three groups of farmers was about 35.0 years, while about 88.0% was male and about 83.0% was married. All the respondents had western education while about 88.0% of farmers had tertiary education. Fish farming was a secondary occupation for about 61.0% of farmers while the mean farming experience was 6.3 years. The result of Benefit-Cost Ratio (BCR) was 1.46 for fingerlings, 1.29 for juveniles and 1.26 for full fish. Profitability and efficiency ratio of 0.85 and 1.85, 0.71 and 1.71, and 0.55 and 1.55 were recorded for fingerlings, juveniles and full fish, respectively. Comparing these values indicated that fingerlings production is the most profitable catfish enterprise in the study area. It is recommended that governments at all levels explore the possibility of using the various ventures in catfish farming as a solution to the worrisome unemployment problem in Nigeria.

Keywords: Full fish; Fingerlings; Juveniles; Efficiency ratios; Profitability; Catfish aquaculture; Nigeria

Introduction

Fish is an important protein food for Nigerian households and is supplied to large urban markets of Nigeria [1]. At the national level, fisheries provide a significant part of the national income. The processed fish (mainly dried clariid catfish) is a highly valuable trade item (Central Bank of Nigeria) [2] and thousands of people are employed in the marketing chain (gear manufacturers, processors, transporters, merchants, etc). The importance of fisheries to the Nigerian economy is indicated by its contribution to the Gross Domestic Product which stood at 4.4% in 2008 [3]. There have been empirical findings by Mafimisebi and Thompson that the fisheries sub-sector inherently contributes more to the Nigerian economy than is apparent in this paltry 4.4% [4].

The fisheries sub-sector of the agricultural sector in Nigeria is classified on the bases of type and structure (Federal Department of Fisheries (FDF) [5]. The industry is divided into three (3) sub-sectors; artisanal, industrial and aquaculture. In the last 3 decades, both production and consumption of fish have risen drastically and the national demand for fish also continues to increase [6].

Nigeria is a food deficit nation and it is obvious that protein intake is grossly inadequate in both qualitative and quantitative terms [7]. Although, fish is generally regarded as a cheap source of animal protein [8-10], the shortfall in domestic production due to the neglect of the sub-sector and environmental impact of crude oil exploration on fish production, has resulted in increased importation of fish in Nigeria [10,11]. However, because of its remarkable profitability, there is a growing aquaculture industry that has come to the rescue in an attempt to bridge the gap between supply and demand [12-15].

Therefore, this study is focussed on the profitability analysis of various ventures in catfish production in Ondo State, Nigeria. This is with the aim of providing informed guidance to prospective investors seeking to invest their funds in profitable fish farming enterprises. Comparing profitability across ventures will reveal the extent to which each of the various ventures is attractive. This will serve to encourage more investment in aquaculture business and ascertain the most profitable enterprises with regard to catfish aquaculture. Increased investment in aquaculture has become very important in boosting domestic fish production which will subsequently reduce the annual fish import bill in Nigeria. Also, this study hopes to provide prospective fish farmers with information on the various cost items and how best to invest their limited competitively utilizable resources in a bid to maximize profit.

Methodology

Study area

The study was carried out in Ondo State, Nigeria. The state is one of the six states in South-West of Nigeria. The state is bounded in the West by Osun and Ogun States and in the North by Ekiti and Kogi States. Ondo State also shares boundaries with Edo and Delta States in the East and in the South by the Atlantic Ocean [10]. The state is made up of 18 Local Government Areas (LGA) with a total population of about 3.4 million inhabitants (National Population Commission) [16]. Ondo State has three distinct ecological zones; the mangrove forest to the south, the rain forest in the middle and the guinea savannah to the north.

The state is well suited for the production of both permanent and arable crops and fishery products from both artisanal and aquaculture sub-sectors. Ondo State has about 180 km coastline which is the longest in the Nigeria. The coastline harbours Ilaje Local Government Area, which is inhabited by three ethnic nationalities which are Ilaje,

Apoi and Arogbo Ijaws. The major occupation of these riverine or coastline ethnic groups is fishing either at the artisanal or motorized levels with minor occupations which are also related to fishing such as related lumbering and production of local gins [11,12]. The fact that Ondo State is one of the highest producers of fish in Southwest Nigeria justified the reason for selecting it as the study area [17].

Data collection

A multi-stage sampling technique was used to select respondents for the study. In the first stage, Ondo State was chosen based on the fact that the state is the highest producer of fish in South west Nigeria. In the second stage, two Local Government Areas (LGAs); Akure South and Akure North which accounted for 18.11% of the total population of the state Ondo State Ministry of Information [18] and 48% of the fish farms in the state (Ondo State Agricultural Development Programme (OSADP) [19] were purposively selected. In the third stage, random sampling technique was used to select the fish farmers. Seventy-Two (72) respondents divided into 24 each of fingerlings, juveniles and table size fish farmers, respectively) were randomly selected in each LGA. A set of 24 questionnaires was administered to each category of farmers in each of the two LGAs giving a total of 144 respondents. In the farms surveyed, data were collected with the aid of structured questionnaire.

Data analysis

Descriptive statistics comprising of frequency distribution, mean and percentage was used to summarize the socio-economic characteristics of fish farmers. The Benefit-Cost Ratio (BCR) analysis and ordinary least squares regression were also used to analyze the data. Benefit-Cost model is calculated as the NPV of benefits divided by the NPV of costs. It is shown as follows

$$BCR = \frac{\sum\limits_{r=1}^{r} \frac{Bt}{(1+r)^t}}{\sum\limits_{r=1}^{r} \frac{Ct}{(1+r)^t}}$$

Where Bt is the benefit in time t and Ct is the cost in time t. Where t is the first five years of which fish farmers has been running the fish farming business based on their records and r is 9% interest rate which is the prevailing rate at which agricultural loan is given to farmers by the financial institutions as directed by the Central Bank of Nigeria.

Therefore, if the BCR exceeds one, then the fish farming venture is considered profitable. The BCR was constructed to determine (and compare) the benefits and costs of producing fingerlings, juveniles and full size fish in the study area. Also, the profitability and efficiency ratios of each fish venture (fingerlings, juveniles and full fish) was calculated and compared.

Thus:

Profitability Ratio=NP/TC

Where NP=Net Profit

TC=Total Cost

Efficiency Ratio=TR/TC

Where:

TR =Total Revenue

TC =Total Cost

Furthermore, in estimating the parameters of socio–economic and operational characteristics postulated as explanatory variables, the explicit production function relating income realized from the sales of fingerlings, juveniles and full fish was estimated using OLS regression. Various functional forms of multiple linear regression models were fitted to data collected to reveal the best fit.

The explicit regression equation for fingerlings production is presented as follows:

$y=b_0 +b_1X_1 +b_2X_2+b_3X_3+b_4X_4+b_5X_5 +b_6X_6+b_7X_7+U_i$

Y=Gross revenue realized from fingerlings production (Naira)

X_1=Age of respondent (years)

X_2=Educational status (years of formal schooling)

X_3=Major occupation (1=fish farming, 0=otherwise)

X_4=Initial stock (number)

X_5=Cost of feeds (Naira)

X_6=Veterinary Cost (Naira)

X_7=pond size

U_i=Error term

Where b_0=Intercept or constant

b_1=Parameter estimates

For juvenile production, the explicit regression equation is as follows:

$Y=b_0 +b_1X_1 +b_2X_2+b_3X_3+b_4X_4+b_5X_5 +b_6X_6+b_7X_7+b_8X_8+b_9X_9+U_i$

Y=Gross revenue realized from juvenile production (Naira)

X_1= Age of fingerlings stocked (weeks)

X_2=Cost of fingerlings stocked (Naira)

X_3=Cost of feeds (Naira)

X_4=Veterinary cost (Naira)

X_5=Cost of equipment used (Naira)

X_6=Educational status of respondents (years of formal education)

X_7=Number of family members involved in production

X_8=Years of fish production experience (years)

X_9=Pond Size

U=Error term.

Where b_0 and b_i are as earlier defined.

For full fish production, the explicit regression equation is as written hereunder:

$Y=b_0 +b_1X_1 +b_2X_2+b_3X_3+b_4X_4+b_5X_5 +b_6X_6+b_7X_7+b_8X_8+b_9X_9+U_i$

Y=Gross revenue realized from table size fish production (Naira)

X_1=Age of juveniles stocked (weeks)

X_2=Cost of juveniles stocked (Naira)

X_3=Cost of feeds (Naira)

X_4=Veterinary cost (Naira)

X_5=Cost of equipment used (Naira)

X_6=Educational status of respondents (years of formal education)

X_7=Number of family members involved in fish production

X_8=Years of fish production experience (years)

X_9=Pond Size

U=Error term.

Where b_0 and b_i are as previously defined.

The estimated functional form that yielded the best fit for each farm category was selected using statistical, economic and econometric criteria.

Results and Discussion

Table 1 revealed that over 85% of the respondents were less than 40 years for all the categories of fish farmers (i.e. fingerlings, juveniles and full fish) with a mean age of 33.5 years, 38.7 years and 31.4 years, respectively. This is in line with the findings of Aderinola and Adeyemo that most fish farmers are in their active productive ages in the study area [13]. The sex distribution of the respondents' showed that majority of the respondents was male. The value recorded was about 88.0%, 85.0% and 90.0% for fingerlings, juveniles and full fish producers, respectively). Thus, more males were involved in fish production than females. This is also in line with the findings of Ajayi and Fagbenro who described fish farming as "a totem of masculinity" [20].

Again from Table 1, the marital statusof respondents revealed that majority (81.3%, 83.3% and 85.4% for fingerlings, juveniles and full fish producers, respectively) were married. Those that were single were less than 20% for all the categories of fish respondents. The farmers could therefore be expected to strive to make rational production decisions that will enhance returns from the business since they may be relying on the farm to provide for their family members [21]. Education is important for the adoption of new innovations according to Olarinde and Kuponiyi [22]. All the respondents had western education. Majority (89.6%) of the fingerlings farmers had tertiary education while 85.4% and 87.5% of juveniles and full fish farmers had tertiary education.

Socio-economic Characteristics	Fingerlings Farmers		Juvenile Farmers		Table Size Farmers	
	Frequency	Percentage	Frequency	Percentage	Frequency	Percentage
Sex						
Male	42	87.5	41	85.4	43	89.6
Female	6	12.5	7	14.6	5	10.4
Total	**48**	**100**	**48**	**100**	**48**	**100**
Age in years						
20-29	15	31.2	8	29.3	23	47.9
30-39	27	56.3	21	58.3	19	39.6
40-49	2	4.2	10	6.2	4	8.3
50-59	4	8.3	9	6.2	2	4.2
Total	**48**	**100**	**48**	**100**	**48**	**100**
Fish Farming Experience in years						
9-Jan	44	91.7	43	89.6	38	79.2
19-Oct	4	8.3	5	10.4	10	20.8
Total	**48**	**100**	**48**	**100**	**48**	**100**
Marital Status						
Single	8	16.7	6	12.5	7	14.6
Married	39	81.3	40	83.3	41	85.4
Widowed	1	2	2	4.2	0	0
Total	**48**	**100**	**48**	**100**	**48**	**100**

Solely Fish Farming Business	28	58.3	29	60.4	31	64.6
Self employed	9	18.8	7	14.6	7	14.6
Civil Servant	11	22.9	12	25	10	20.8
Total	**48**	**100**	**48**	**100**	**48**	**100**
Access To Credit	6	12.5	4	8.3	5	11.6
No Access To Credit	42	87.5	44	91.7	43	88.4
Total	**48**	**100**	**48**	**100**	**48**	**100**
Education Attained						
Completed Secondary School	5	10.4	7	14.6	6	12.5
Tertiary	43	89.6	41	85.4	42	87.5
Type of Ponds						
Concrete Pond	45	93.8	43	89.6	42	87.5
Earthen Pond	3	6.2	5	10.4	6	12.5
Total	**48**	**100**	**9**	**100**	**100**	**100**
Size of Pond						
100 - 250 m2	41	85.4	42	87.5	39	81.3
251 -999 m2	5	10.4	3	6.25	5	10.4
Above 1000m2	2	4.2	3	6.25	4	8.3
Total	**48**	**100**	**9**	**100**	**100**	**100**
Mean Age of Fish Farmers	33.5 years		38.7		31.4 years	
Mean Farming Experience	5.8 years		6.0 years		7.0 years	

Table 1: Socio-Economic Characteristics of Fish Farmers (i.e. Fingerlings, Juvenile and Table Size) in Ondo State. Source: Field Survey, 2012.

Furthermore, Table 1 shows that most (58.3%) of the fingerlings producers engaged in farming as a major occupation while the balance were civil servants who took to these ventures on part-time basis. Few (18.8%) of the fingerlings farmers were self-employed. About 60.4% of juvenile farmers were engaged in farming as the main occupation while some (25.0%) engaged in civil service. Some (14.6%) of juveniles farmers were self-employed. Also, 64.6% of the full fish farmers had farming as their major occupation, 20.8% was in the civil service and 14.6% was self-employed. This implies that engagement in a secondary income source is a very popular practice among fish farmers in the study area. Again, according to Table 1, over 87.0% of the respondents confirmed that they do not have access to credit for all the categories of fish farmers while less than 15% reported having access to credit through informal sources.

From Table 1, majority (91.7%, 89.6% and 79.2%) of fingerlings, juveniles and full fish farmers, respectively, had less than ten years of fish production experience while 8.3%, 10.4% and 20.8%, respectively, had over ten years of fish production. The mean farming experience of 5.8 years, 6.0 years and 7.0 years, respectively,also attested to their years of experience in the fish farming in the study area. On management practices, a greater proportion (93.8%) of fingerlings

farmers made use of concrete ponds, while majority (89.6% and 87.5%) of both juveniles and full fish farmers also made use of concrete ponds. This might be due to the fact that concrete ponds are more secured and reliable than the earthen ponds as observed by Kudi et al. [23].

Furthermore, the average frequency distribution of ponds sizes in the sampled area as given in Table 1 shows that 84.7% of the ponds were small sized ponds of 100 to 250 m^2 and 9.0% was medium sized ponds of 251 to 999 m^2. Big ponds of size more than 1000 m^2 had the lowest percentage of 6.3%. Small-sized ponds of 100 to 250 m^2 may have be preponderant because of lack of skills and infrastructural facilities to accommodate large scale fish farming as well as limited data and information on research and development requirements for fish farming. Large ponds will require modern pond engineering techniques and advanced management methods about which little is known at present in the study area [10]. Hence, most of the farmers prefer to have relatively small ponds, which they can manage.

Benefit-cost ratio to fish farming production per hectare of fish farm

	Fingerlings	Juvenile	Table Size
BCR Value	1.46	1.29	1.26

Note: The Benefit- Cost Ratio analysis was based on the average of five years of operation of the three fish farming enterprises.

Table 2 revealed that the fingerlings enterprise is the most profitable fish farming enterprise in the study area since its value of BCR is greater than that of juvenile (1.29) and full fish (1.26). However, from the field survey, it was discovered that the cost of juvenile and full fish enterprises was higher because they will require more ponds for sorting and also, more brood-stock. This may likely be the reason why the fingerlings enterprise is more profitable than the other two fishing enterprises.

	Fingerlings Amount (N m)	Juvenile Amount (N m)	Table Size Amount (N m)
A. Variable Cost:	1.15	10.00	18.00
Cost of Stock	1.38	2.56	126.32
Cost of Feeds	1.20	1.45	42.15
Labour	0.20	0.25	0.35
Veterinary Cost	3.93	14.26	186.82
Total Variable Cost			
B. Fixed Cost	1.70	2.36	33.33
Depreciation Cost	1.70	2.36	33.33
Total Fixed Cost	5.63	16.62	220.15
Total Cost of Production			
C. Total Revenue	10.00	18.00	350.00
D. Gross Margin	6.07	3.74	163.18
E. Net Profit	4.37	1.38	129.85
F. Profitability Ratio	0.78	0.08	0.59

Table 3: Average Cost and Returns in Fish Production per Hectare of Fish Farm per Annum. Source: Field Survey, 2012.

Table 3 showed that total production cost per fingerlings in the study area was N 5.63 k while the revenue was N 10.00 k per fingerling. The analysis also revealed that fingerlings farmers earned an average of N 4.37 k as net profit per fingerling. For juvenile production, the production cost per juvenile was N 16.62 k while the revenue per juvenile sold was N 18.00 k. Thus, farmers producing juveniles earned an average of N 1.38 k as net profit per juvenile. Also, for full fish, the production cost N 220.15 k while the revenue per full fish was N 350.00 k. The results indicated that full fish farmers earned an average of N 129.85 k as net profit per full fish sold. The profitability ratio analysis which measures the ratio of revenue to expense and which gives room for comparison between two or more firms [24] revealed that fingerlings production was more profitable than production of both juveniles and full fish in the study area.

Production function for fish farms

The R2 for the estimated regression implied that 89.0% of the variations in the revenue from sales of fingerlings are explained by the explanatory variables. It was found from the regression result that age, cost of feeding, veterinary cost and pond size were the major determinants of the gross revenue from fingerlings production in the study area.

Variables	Coefficient	T-Values
Constant	-	-
Age	0.101	4.21*
Cost of feeding	-0.312	2.514*
Veterinary cost	-0.021	3.016*
Pond size	0.055	5.215*

Table 4: Estimated Production Function for Fingerlings Farms. Source: Field Survey, 2012. R^2=0.89, F =6.14*, * Significant at 5% level.

Age had a positive and significant relationship meaning that the older the fingerlings farmers are, the higher their productivity is. This may be owing to the fact that older fingerlings farmers are more patience and thorough as a hindsight of experience. According to Shimang, fingerlings farming requires the quality of resilience because, it is a fragile farming venture. In a single day, a fingerlings farmer may lose half of his/her fingerlings [25].

Also, from Table 4, cost of feeds had a negative coefficient which meant that the higher the gross revenue (and hence output), the lower the cost of feeds in raising fingerlings. This may have been due to the possibility of bulk purchase of feeds directly from dealers which leads to reduced costs. This will be the case especially when the feeding is done according to prescription. Veterinary cost had negative coefficient, which implied that the average veterinary cost for larger fingerlings farms will be lower than average cost on smaller fingerlings farm. Pond size had a positively significant coefficient with the implication that the larger the size of the pond for the production of fingerlings, the higher the gross revenue especially in a situation in which mortality rate is highly reduced.

Variables	Coefficient	T-Values
Constant	-	-
Cost of feed	0.605	5.841*
Cost of equipment	-0.091	2.625*
Production experience	0.056	2.312*
Pond size	0.145	6.185*

Table 5: Estimated Production Function for Juveniles Farms. Source: Field Survey, 2012. Notes: R^2=0.86, F =8.61*, *Significant at 5% level.

From Table 5, it was shown that cost of feeds, cost of equipment, production experience and pond size were the major determinants of income from juvenile production. With an F- value of 8.61 which is significant at 5% level, it is shown that most of the postulated variables influenced the income from juvenile production. The R^2 for the estimated regression was 0.86 implying that about 86.0% of the variations in gross revenue from sales of juveniles are explained by the explanatory variables. Cost of feeds had positive and significant regression coefficient. This meant that the higher the quantity of feeds used, the higher the revenue from juveniles production.

Cost of equipment had a negative but significant relationship with gross revenue from juveniles production. This connotes that the higher the output and by implication, gross revenue, the lower the cost of equipment per juvenile. Production experience had a positively significant regression coefficient which is interpreted to mean that the more experienced the juvenile producer is, the higher the output and hence, gross revenue, other things being equal. From Table 5, pond size is a very crucial determinant of juvenile production in the study area. The results revealed that the larger the size of the pond, the more the output of juveniles. At juvenile production level, the size of the pond is very crucial to the output level. In juvenile production, the pond with larger dimension gives rise to more output compared with a smaller pond.

Variables	Coefficient	T-Values
Constant	-	-
Cost of feed	0.716	3.418[*]
Cost of equipment	-0.011	3.425[*]
Educational level	0.191	2.018[*]
Production experience	0.077	4.952[*]

Table 6: Estimated Production Function for Table Size Fish Farms. Source: Field Survey, 2012. R^2=0.84, F =8.91[*], [*]Significant at 5%.

The R2 for the estimated regression implied that 84% of the variations in the revenue from sales of full fish is explained by the postulated explanatory variables. From the result of the regression model, it was observed that cost of feeds, cost of equipment, educational level, production experience and pond size were the major determinants of gross revenue from full fish production. Cost of feeds had positive and significant regression coefficient, which implies that the higher the cost of feeds, the higher the revenue from full fish production. The growth of the full fish is essentially determined by the quantity and quality feed [26].

Therefore, the more feed consumed, the more weight of full fish that will be produced. Also, the coefficient of the cost of equipment used in full fish production is significant and negative. It is understandable that fixed cost like cost of equipment will always be high if the quantity of fish produced is low. The study revealed that most of the full fish farmers in the study area were not into large scale production, and then the fixed cost like cost of equipment will be high as there exists an inverse relationship between the cost of equipment and gross revenue realized from full fish production.

Again, the educational level has a positive relationship with the gross revenues from full fish. Thus, the higher the level of education of full fish farmer, the higher the gross revenue. This corroborates the findings of Olarinde and Kuponiyi, that "education is an important factor that determines adoption of new innovations [22]. It provides readability consciousness and awareness, which enable decisions to be made. Therefore, the higher the level of farmer's education, the better is his/her decision making ability, especially in the adoption of new technologies and innovation". Such decision will enhance output.

Furthermore, the coefficient of production experience is positive and significant, because the farmer is able to make wise economic decisions in production by drawing on first hand farm experience which is better when compared with relying on theoretical knowledge [10]. The pond size coefficient is also positive because the size of the pond also determines the size of the full fish as it ensures enough space for growth without antagonism from other fishes in the same pond (reduced cannibalism).

Conclusion

The study revealed that the fish farmers were in the active working age bracket, they are well educated while the business of fish production is male dominated. Fish production was profitable in the study area but for the three farm ventures, fingerlings production was more profitable than production of juveniles and full fish. Since all ventures of catfish farming were discovered to be profitable in the study area, government at all levels can adopt them as an employment scheme to solve the pervasive and worrisome unemployment problem in the country by providing the enabling environment for school leavers to go into catfish production and exportation.Also, this step to encourage school leavers to go into aquaculture can also boost fish supply and subsequently bridge the demand-supply gap of fish in Nigeria.

References

1. Amiengheme P (2005) The Importance of Fish in Human Nutrition. A paper delivered at a Fish Culture Forum Federal Department of Fish Farmers Abuja Nigeria.

2. CBN (2007) Central Bank of Nigeria Annual publication 56: 137-142.

3. FAO (2008) Fish for Food and Employment. Food and Agricultural Organization annual report. 35:56-66.

4. Mafimisebi TE, Thompson OA(2012) Empirical Evidence of Fisheries Sub-sector's Contribution to the Nigerian Economy. International Journal of Agricultural Science Research and Tech 2: 31-35.

5. FDF (2007) Federal Department of Fisheries. Quarterly Bulletin 23: 13-18.

6. FDF (2010) Federal Department of Fisheries. Press Report of the Federal Ministry of Agriculture and Water Resources Abuja. Delivered by the Director of Fisheries at NiconNoga Hilton Hotel on 24 December 2010.

7. Olukoya O (2007) The Agricultural Sector and Nigeria's Development: Comparative Perspectives from the Brazilian Agro-industrial Economy 1960-1995. Federal Ministry of Agricultural and Water Resources Annual Bulletin 31: 26-29.

8. Fagbenro OA (1987) A Review of Biological and Economic Principles Underlying Commercial Fish Culture in Nigeria. Journal of West African Fisheries 3: 171.

9. Shang YC (1992) The Role of Aquaculture in the World Fisheries. Paper presented at the World Fish Congress Athens UNESCO 63.

10. Fapounda OO (2005) Analysis of Bio-Technical and Socio-Economic Factors Affecting Agricultural Production in Ondo State Nigeria. Unpublished PhD. Thesis, The Federal University of TechnologyAkure Nigeria 1–10.

11. Fagbenro OA, Akinbulumo MO, Ojo SO (2004) Tilapia: Fish for Aquaculture in Nigeria: Past Experience Present Situation and Future Outlook. World Aquaculture 35: 23-28.

12. Mafimisebi TE (1995) Profitability and Yield Performance of Selected Fish Ponds in IlajeEse-Odo Local Government Area of Ondo State Nigeria. Unpublished M.Sc Dissertation University of Ibadan 83.

13. Aderinola EA, Adeyemo AA(2001) A Socio-economic Study of Freshwater Fish Production in Osun State of Nigeria. Applied Tropical Agric 6: 57-62.

14. Williams CO, Onabanjo EO(2007) Artisanal Fisheries Development in Nigeria: Status Constraints and the Way Forward. Paper Presented at the National Stakeholders Workshop on Inland Captured Fisheries Development Kaduna Nigeria.

15. Kareem RO, Dipeolu AO, Aromolaran AB, Williams SB(2008) Economic Efficiency in Fish Farming: Hope for Agro-allied Industries in Nigeria. Chinese Journal of Oceanology and Limnology 26: 104–115.

16. NPC (2007) National Population Commission. Nigeria Annual Census 2006.

17. Mafimisebi TE, Okunmadewa FY (2006) Comparative Yield Performance of Upland and Mangrove Aquacultural Farms in Selected Maritime States of Southwest Nigeria. Conference CD-Rom of the 13th biennial conference of International Institute of Fisheries Economics and TradePortsmouth UK 12.

18. OSMI (2010) Ondo State Ministry of Information. Statistical Annual Report for the year 2009. Published by Ondo State Press.

19. OSADP (2010) Ondo State Agricultural Development Project. Ondo State Ministry of Agriculture Annual Bulletin 2: 13-14.

20. Ajayi AE, Fagbenro OA(2005) Studies on Protein Quality Methods for Seafood. Journal of Fisheries Society of Nigeria 13: 12-15.

21. Amos TT (2006) Analysis of Backyard Poultry Production in Ondo State Nigeria. International Journal of Poultry Science 5: 247-250.

22. Olarinde LO, Kuponiyi FA(2004)Resource Productivity Among Poultry Farmers in Oyo State Nigeria. Journal of Sustainable Development 1: 20–26.

23. Kudi TM, Bako FP, Atala TK(2008) Economics of Fish Production in Kaduna State Nigeria ARPN. Journal of Agricultural and Biological Sciences. Asian Research Publishing Network 3: 17-21.

24. Gelles GM, Mitchell DW(1996) Returns to Scale and Economies of Scale: Further Observations. Journal of Economic Education 27: 259-261.

25. Shimang GN (2005) Fisheries Development in Nigeria Problems and Prospects. A presentation by the Federal Director of Fisheries in the Federal Ministry of Agriculture and Rural Development on Homestead Fish Farming Training for Serving and Retiring Public Servants in the Federal Ministry of Agriculture and Rural Development FCT Abuja.

26. Samson ST (2006) Fishery and Nigeria economy: Journal of Fisheries Society of Nigeria 15: 31-35.

19

Isolation and Identification of *Edwardsiella tarda* from Lake Zeway and Langano, Southern Oromia, Ethiopia

Kebede B[1]* and Habtamu T[2]

[1]*Wacale District Livestock and Fisheries Development Office, Oromia, Ethiopia*

[2]*Veterinary Drug and Animal Feed Administration and Control Authority, Ethiopia*

*Corresponding author: Bedaso Kebede, Veterinary Drug and Animal Feed Administration and Control Authority, Ministry of Livestock and Fisheries Development, Addis Ababa, Oromia 251, Ethiopia, E-mail: kebede.bedaso@yahoo.com

Abstract

A study was carried out from October, 2009 to April, 2010 with the objective of isolating *Edwardsiella tarda* an important fish pathogen from fish harvested for human consumption from Lake Zeway and Langanoo. A total of 372 tissue samples (three from each fish) comprising liver, intestine and kidney were collected from 124 fish (*Clarias gariepinus* and *Oreochromis niloticus* originated from Lake Langanoo and Zeway. Distribution of *E. tarda* infection among the three organs examined indicated that *E. tarda* was isolated most frequently from liver (6.5%) followed by intestine (2.4%) and kidney (0.8%) with significant difference among organs. Statistical significant differences ($P < 0.05$) were found in *E. tarda* infection with respect to site although the bacterium was isolated from fish originating from both Lake Zeway and Langanoo with *E. tarda* being more prevalent in fish sampled from lake Zeway. *E. tarda* was isolated more frequently from male fish, the differences in the occurrence of *E. tarda* infection with respect to sex were not significant ($P > 005$) indicating that both sexes are equally susceptible. The isolation of *Edwarsiella* from wild fish population of Lakes Zeway and Langano destined for human consumption in the current study is indicates that *E. tarda* is a potential threat of both the fishery sector/aquaculture and public health. Finally, as is the case for any infectious fish pathogen, there is limited information on *E. tarda* of fish in Ethiopia and hence further study to have comprehensive information on the agent is forwarded.

Keywords: Catfish; *Edwardsiella tarda*; Intestine; Isolation; Kidney; Langanoo; Liver; Tilapia; Zeway

Abbrevations:

µm: Micrometer; BHI: Brain Heart Infusion Agar; CHO: Carbohydrate; EIM: Edwardsiella Isolation Media; *E. tarda:* *Edwardsiella tarda;* FAO: Food and Agricultural Organization; FISH: Fluorescence *in Situ* Hybridization; g/l: Gram per liter; H_2O_2: Hydrogen Peroxide; H_2S: Hydrogen Sulfide; HPCE: Higher Performance Capillary Electrophoresis; LAMP: Loop-Mediated Isothermal Amplification; LFDP: Lake Fisheries Development Working Paper; ml: Milliliter; PCR: Polymerase Chain Reaction; SIM: Sulfur Indole and Motility Test Media; TSA: Tryptic Soya Agar; TSIA: Triple Sugar Iron Agar; USA: United State of America; V/V: Volume by Volume; WHO: World Health Organization; XLD: Xylose Lysine Deoxycholate

Introduction

Aquaculture is growing rapidly worldwide with fish being the primary sources of animal protein in many countries. The fishery sector plays a significant role in food security through supplementation of food for developing countries. As a whole fish currently make up about 19% of the total protein consumption or just over the 5% of proteins from both plants and animals origin [1].

As in all animal production systems, however, fish are possibly susceptible to microbial diseases which are one of the major problems hampering production, development and expansion of the aquaculture industry. Fish diseases are global problem affecting fresh water, marine water, cultured, sport and even ornamental fish. The problem is extremely important when fish are subjected to intensive culture practices [2].

The control of fish diseases is particularly difficult because fish are often farmed in system where production is dependent on natural environmental conditions. Changes or deterioration in the aquatic environment cause the occurrence of most fish disease and also environmental effects play a great role in influencing the health status of fish. Therefore, the multidisciplinary approaches involving the characteristics of potential pathogenic microorganisms for fish, aspects of the biology of fish as well as a better understanding of the environmental factors affecting such cultures will allow the application of adequate measures to prevent and control the diseases limiting fish production [3,4].

Edwardsiellosis is among the most important bacterial diseases causing severe economic losses in fish farms of many countries. The disease is caused by *E.tarda* which is a gram negative, motile, facultative anaerobic, short rod shaped bacterium (1µm in diameter and 2-3 µm long) pathogenic to a wide range of fish hosts such as Channel cat fish (*Ictaluri punctatus*), Striped bass (*Morone saxatili*), eel (*Anguilla anguilla*), Tilapia (*Oreochromis niloticus*), carp (*Cyprinus cyrpio*) and Flounder (*Paralichthys olivaceus* [5]. The organisms is frequently found in organically polluted water, poor quality water and affect fish stressed by this situations [6,7]. *E.tarda* can be isolated on Edwardsiella Isolation Media (EIM), Brain Heart Infusion (BHI), Tryptic Soya Agar (TSA), Xylose Lysine Deoxycholate(XLD) and MacConkey. It is seen as small, circular, raised, whitish with black center on XLD and pale on MacConkey agar, grow best at a temperature between 25°C-37°C, PH 7-8 and 0.5%NaCl

[7,8] and characteristically, catalase positive, cytochrome oxidase negative, glucose fermentative, indole positive, citrate negative, lysine positive, mannitol, dulcitol, sorbitol, inositol, xylose, rhamnose negative, produce hydrogen sulfide, alkaline slant and acid but on Triple sugar iron Agar [9].

Edwardsiella tarda is a health threat not only to fish and other animals but also to humans [10] with the risk factors being exposure to aquatic environment, pre-existing liver diseases, iron over load and raw sea food ingestion [11].

In humans the bacteria usually cause diarrhea, gastroenteritis, wound infection and even death [5,12]. There are reports of extra intestinal infection with the clinical pictures including a typhoid like illness, peritonitis with sepsis and cellulites with occasionally liver abscecss [13] and meningitis [5]. The infection is more severe in immunocompromised individuals.

The practice of consuming partially cooked fish meals, manual handling of fish and unhygienic practice during filleting in Ethiopia indicate that the public is at higher risk of contracting the disease. Therefore, the disease deserves attention due to its impact on the fishery sector as well as its potential threat to future aquaculture industry and public health [14,15]. In Ethiopia, the bacterium has been isolated from apparently healthy fish of Lake Zeway [16] and Tana [17]. However, there is no further work done in covering the different fish species and aquatic environments. Therefore, this study was conducted with the aim of Isolating *E. tarda* from cat fish and Tilapia slauthered at zeway fishery resource center originating from Lake Zeway and Langanoo and elucidates the safety of fish products with respect to *Edwardsiella tarda* contamination.

Materials and Methods

Study site

The study site comprised fish species harvested from lakes Zeway and Langanoo.

Lake zeway: Lake zeway is located on the Eastern side of Zeway town, 163 km South east of Addis Ababa it lies in northern part of the rift valley between 7°51N to 8°7'N and 38°43' E 38°57' E with an open water area of 422 km^2 and shoreline length of 137 km. The lake is fed by two major rivers, i.e. Ketar and Meki River and has one out flow in the South, Bulbula river which flow into Abiyata [18]. Five bigger islands are situated in the lake Viz Tulu Gudo (4.8 km^2), Tsedecha (2.1 km^2), Funduro (0.4 km^2), Debresina (0.3 km^2) and Galila (0.2 km^2). While the latter two have few inhabitants, the three bigger ones are populated with several hundreds of people [19].

The catch from Lake Zeway consists of almost exclusively Tilapia (*Oeochromis niloticus*). Since recent years, however, Cat fish (*Carias gariepinus*) and Crucian carp (*Carcasius gracius*) have appeared in small amounts of the total catch [20]. There are a number of landing points around the lake from where fish is collected either by boat or trucks and brought to the major landing points adjoining Zeway town.

Lake langanoo: Lake langanoo is located 200 km South of Addis Ababa lying between 7°36' N;38°45' E. It is 18 km long and 16 km wide with an open water area of 230 km^2, 7.5 km shoreline and 1600 km^2 catchments area [18]. The main fish species in the lake include *Barbus* species, Clarias species and *Oreochronis niloticus* [14] with the total annual catch of 1000 tones.

Study animals

Study animals used in this study included African cat fish (*Clarias gariepinus*, N=30) and Nile tilapia (*Oreochromis niloticus*; N=94) which were harvested from Lake Zeway and Langanoo for human consumption. The fish were physically examined for any external lesions before necropsy and collecting tissue samples.

Necropsy and Tissue Sampling

In dissecting fish, ventral approach to kidney was employed. The fish sample was cut along the midline of the abdomen starting from the anus up to the mouth using sterile dissecting scissor followed by another dissection from the anus to the lateral line and further along the lateral line up to the gills cover to remove the lateral side of the abdominal wall and expose the internal organs. Internal organs were examined for any gross pathology and the findings recorded. Tissue samples were then taken from intestine (N=124), liver (N=124) and kidney (N=124) aseptically using sterile scalpel blade and forceps kept in sterile universal bottles of 100 ml capacities. All necropsy and tissue sampling procedures were carried out under asceptic condition. The bottles containing the samples were then kept in ice box containing ice packs all the way to School of Veterinary Medicine, Debre Zeit where they are processed for bacterial isolation and identification.

Laboratory Examination

Isolation of *Edwardsiella tarda*

Tissue samples from kidney, liver and intestine of cat fish and tilapia were homogenized in physiological saline. The homogenate was then taken by sterile loop and streaked on xylose lysine deoxycholate agar plate (Titan Biotech) and then incubated at 37°C for 24 hours. Colonies showing or resembling with morphological characteristics of E.tarda were further subcultured on MacConkey agar plates and incubated at 37°C for 24 hours. All lactose non-fermented colonies were further subcultured on tryptic soya agar containing 0.5% NaCl and incubated at 37°C for 24 hours. Presumptive identification of the resulting isolates (colonies) was done employing different tests which included primary bacteria identification techniques and biochemical identification tests.

Primary identification of isolates

Primary Identification of pure culture of the isolates was done based on gram reaction, motility tests, and catalase and oxidase tests. Overnight cultures of pure colony on tryptic soya agar (TSA) plates were used in all of these tests.

Gram reaction: Gram staining was done according to the procedure described by Rowland et al. [21] accordingly, colonies that were gram negative, short rods were considered for further tests.

Catalase test: Catalase test detects whether the bacterium has the enzyme catalase that converts hydrogen peroxide to water and gaseous oxygen. The test was carried out on pure fresh colony on tryptic soya agar plates [22]. Since Edwardsiella tarda is catalase positive, colony showing an elaborated bubble formation was considered positive and taken for further tests.

Oxidase test: This test detects the presence of cytochrome oxidase enzyme in a bacterial cell and characterized by purple colour formation within 10 seconds when the bacterial sample is made in contact with 1 percent aqueous solution of tetramethyl-p-phenylenediamine dihydrochloride [22]. In this study, oxidase test was conducted employing filter paper method for each bacterial isolates.

Motility test: Motility test was conducted using sulfur, indole and motility test media. The slant of the medium in test tubes was stab-inoculated with fresh colony (isolate) using a straight sterile wire followed by incubation at 37°C for 24 hours [21]. Turbidity of the medium or outgrowths from the line of inoculation was considered indicative of motility and the results recorded.

Biochemical tests

Secondary biochemical identification of bacterial isolates was conducted employing conventional biochemical tests according to the standard procedures described previously [23,24]. These tests are based on the ability of the bacterium to utilize a sugar, an amino acid or an alcohol or any carbon source in the medium where by the byproducts of such reaction if any is detected using appropriate indicators incorporated into the medium. The detailed procedures of the biochemical tests are presented. In all of the tests, care was taken to maintain aseptic procedures to avoid contamination.

Triple sugar iron (TSI) test: Triple sugar iron agar (TSI) test shows hydrogen sulfide production, gas production, fermentation of lactose, sucrose and glucose. E. tarda is expected to show red slant and yellow butt with hydrogen sulfide production. In this work, TSI test was carried out by inoculating (by stabbing the butt and streaking the slant) of the test tube of TSI agar slant using straight inoculating wire after which the inoculated tube was loosely capped and the findings recorded after 24 hours of incubation at 37°C [21].

Indole production, motility and H₂S production tests: SIM media (BBL) was used to demonstrate indole production, motility and hydrogen sulfide (H_2S) production. SIM media in test tubes were inoculated with pure overnight grown colonies on TSA plates followed by incubation at 37°C for 24 hours after which the findings were recorded [22]. To demonstrate if there is any indole production, kovac's reagent was added to SIM media after 24 hours of incubation and deep red colour was developed.

Simmon's citrate test: Simmon's citrate slants in test tubes were stab inoculated in same way as SIM media and incubated at 37°C for a week after which the findings were recorded [22,23]. The test detects the ability of the bacterium to utilize citrate as the only carbon source which imparts blue in case of positive cases.

Lysine decarboxylase test: Test tubes with lysine broth were inoculated with pure overnight isolate followed by incubation at 37°C for 4 days. The findings were recorded after 96 hours of incubation. Absence of color change, i.e. the maintenance of purple color indicates the ability of the isolate to utilize the amino acid lysine and produce alkaline PH in the medium [21,23].

Sugar or carbohydrate fermentation tests: Conventional biochemical tests comprising four alcohols (dulcitol, mannitol, inositol and sorbitol) and two sugars (rhaminose and xylose) were used for adequate presumptive identification of E. tarda. Phenol red basal broth in Durham tubes containing 1% dulcitol, mannitol, sorbitol, inositol, xylose and rhamnose was prepared and inoculated with the isolates. The inoculates were then incubated at 37°C for 24 hours after which the results were recorded [21]. Bacterial isolates with consistent characteristics of E. tarda based on primary and secondary identification criteria were considered presumptively as Edwardsiella

tarda fish isolates and preserved in 15% glycerol (V/V) at -20°C for further characterization studies.

Data Analysis

Descriptive statistics such as proportions and frequency were employed in summarizing the data. Chi-square test of independence was employed in comparing the prevalence/occurrence of E. tarda infection with respect to site, sex, fish species and organ of isolation. A confidence interval of 95% was used to interpret the statistical association and significance was considered when P-value is less than 0.05 [25].

Results

Of the total of 372 tissue samples comprising kidney, liver and intestine collected from 124 fish, E. tarda was isolated from 12 tissue samples (8 from liver, 3 from intestine and 1 from kidney). The isolates appeared as small punctuate grayish white colonies on xylose lysine deoxycholate agar after 24 hrs of incubation at 37°C. Except few of the isolates, most showed typical characteristics of E. tarda isolated elsewhere, which were gram negative short rods, motile, catalase positive and oxidase negative. In biochemical tests, these typical isolates were positive for indole, H₂S production, and lysine decarboxylase and unable to utilize Simmon's citrate and the different sugars used in this study (Table 1). However, some of the isolates showed variation from the typical characteristics. One isolate was negative for indole and able to utilize Simmon's citrate while the remaining was able to ferment mannitol, rhaminose, xylose and inositol (Table 1). Although these isolates divert in some of the biochemical tests from the typical characteristics expected of E. tarda they were considered as variants of E. tarda strain due to the fact that the cultural characteristics, growth requirements and the different tests suggested the isolates to be within E. tarda species.

Distribution of E. tarda infection among the three organs examined indicated that E. tarda was isolated most frequently from liver (6.5%) followed by intestine (2.4%) and kidney (0.8%) with statistical significant difference (P<0.05) among organs (Table 2).

Statistical significant differences (P<0.05) were found in E. tarda infection with respect to site although the bacterium was isolated from fish originating from both lake Zeway and Langanoo with E. tarda being more prevalent in fish sampled from lake Zeway (Table 3).

Although, E. tarda was isolated more frequently from male fish, the differences in the occurrence of E. tarda infection with respect to sex were not significant (P>0.05) indicating that both sexes are equally susceptible (Table 4).

There was no statistical significant difference (p>0.05) in isolation of Edwardsiella tarda from Catfish (Clarias gariepinus) and Tilapia (Oreochromis niloticus) indicating that both fish species are susceptible to the infection.

Discussion

The major species of Edwardsiella those infect fish are E. tarda and E.ictaluri. E. tarda infects fish and other animals including human being, while E. Ictaluri being a pathogen of fish only [26]. In this study, E. tarda were isolated from intestine, kidney and liver of fish indicating that the bacterium is a potential threat to aquaculture as well as to public health. The organism has been isolated from several sources previously such as from intestine of fish, humans faeces with sporadic cases of diarrhea [12] and from dressed fish samples [27,28].

Parameter	Results	Remark
Cultural characteristics on XLD agar	**Small, circular, grayish white Colonies**	
Morphological characteristics	Gram negative, motile short rods	Two isolates, non-motile
Biochemical characteristics		
Indole production	+	One isolate, indole -ve
H2S production	+	
Oxidase	−	
Catalase	−	
Citrate	−	One isolate, citrate +ve
Lysine	+	
Mannitol	−	Four isolates,mannitol +ve
Dulcitol	−	
Inositol	−	Four isoltes,inositol +ve
Sorbitol	−	
Xylose	−	Four isolates, xylose +ve
Rhaminose	−	Four isolates, rhamnose +ve

Table 1: Phenotypic and biochemical characteristics of *E. tarda* strains.

Organ	Results		Total
	Positive	Negative	
Intestine	3	121	124
Liver	8	116	124
Kidney	1	123	124
Total	12	360	372

Table 2: Distribution of *E. tarda* isolates among the organs. (X^2=6.5, df=2, P<0.05).

Parameters		Positive	Negative	Total	X^2 value	P- value
Site	Zeway	7	27	34	6.38	0.012
	Langanoo	5	85	90		
Total		12	112	124		
Sex	Female	3	27	30	0.005	0.945
	Male	9	85	94		
Total		12	112	124		

Table 3: Occurrence of *E. tarda* isolates with respect to site and sex of fish.

Species	Result		Total
	Positive	Negative	
Catfish	1	19	20
Tilapia	11	93	104
Total	12	112	124

Table 4: Occurrence and distribution of *E. tarda* with respect to fish species. (X^2=0.59, df=1, P>0.05).

Concerning the morphological and biochemical characteristics of *Edwardsiella tarda* isolates, that showed typical characteristics, the results were consistent with those reported previously [29-31]. The finding of one isolate negative for indole production and positive in Simmons citrate test, however, indicates atypical strain which was also reported in previous studies where variation in Simmons citrate utilization [7,32,33] and indole production [34] was reported among *E. tarda* strains.

The present study showed two isolates were found non motile and this fact is matched with that of Okuda et al. [35]. Although most of the phenotypic characteristics of the isolates were similar as claimed by [36], some isolates showed, however, variation in some of the biochemical tests particularly in the utilization of sugars which included mannitol, rhamnose, xylose, inositol. The finding of such variation contradicts with the study of Baya et al. [37] where no variation was observed with respect to these biochemical tests among fourty four E tarda isolates studied. Variation among *E. tarda* isolates, however, was reported with respected to utilization of rhaminose [7], mannitol [38].

The occurrence of variation in phenotypic characteristics among *E. tarda* isolates may due to the presence or absence of plasmid that control metabolic activities. Generally, the significance of the incidence of *Edwardsiella tarda* in catfish and tilapia couldn't be substantiated phenotypic characteristics of *E. tarda* [32].

Nowadays, more rapid, relatively accurate and simple molecular based identification techniques have been employed unlike the traditional phenotypic identification methods which rely on bacterial morphology and biochemical characteristics, Nucleic acid probes and the polymerase chain reaction (PCR) have been developed for the identification of pathogens of aquatic animals including *E. tarda*. Molecular methods such as PCR may sometimes yield false positive results as they can be highly subject to laboratory contamination. . Polymerase chain reaction based diagnosis of *Edwardsiella tarda* infection in blood samples of oyster toad fish and more sensitive real time PCR methods were reported as the major way in identifying the genes responsible for virulence of *Edwardsiellosis* [39].

Loop- mediated isothermal amplification (LAMP) is another rapid and sensitive method used for the diagnosis of *Edwardsiellosis* [40].

A fluorescence *in situ* hybridization (FISH) technique using twenty four mer oligonucleotide probe has also been used for detection of bacterial cells belonging to *Enterobacteriaceae* including *E. tarda* without giving false positive reaction [41]. The use of higher performance capillary electrophoresis (HPCE) has also been reported [42] as another technique which identifies, separate and quantifies intact bacteria. They identified *E. tarda* in fish and traced the bacteria in less than ten minutes after injection in to fish fluid using blue light emitting diode induced fluorescence and a cell permeable green nucleic acid strain.

The absence of significant differences in the occurrence of *E. tarda* between males and females indicates that both sexes are equally susceptible to the bacterium. This is an agreement several works where both sexes are equal chance of being infected with *E. tarda* [40,42]. The significant differences in the rate of isolation of *Edwardsiella tarda* between the study lakes may be attributed to differences in the nutritional status of the fish, the environmental condition (Salinity and bacterial load of the water), water quality, changes in temperature, PH and fluctuation in dissolved oxygen which are believed to affect the occurrence of *E. tarda* infection [17,43,44]. Although, *Edwardsiella tarda* affect intestine, liver and kidney of catfish and tilapia, the highest percentage of the pathogen was isolated from liver. This could be due to the metabolic activities of the organs [43].

In conclusion, *Edwardsiella tarda* is one of the most important bacterial diseases among *Oreochromis niloticus* in fish sampled from Lake Zeway and Langanoo. The severity of *Edwardsiella tarda* may have immune suppressive effect which proved by lymphoid depletion induced in spleen.

Conclusion and Recommendations

Edwardsiellosis is the most important bacterial disease causing severe economic loss and hindrance in aqua farming. Apart from veterinary health importance, *Edwardsiella tarda* has also public health significance in people engaged in fishery industry and those depend on fish products for their annual income. The isolation of *Edwarsiella tarda* from wild fish population of Lakes Zeway and Langanoo

destined for human consumption is, therefore, indicates that *E. tarda* is a potential threat of both the fishery sector/aquaculture and public health. The finding of certain isolates that divert in their biochemical characteristics warrants further investigation using more advanced methods of bacteria characterization. Generally, assessment of environmental condition, management and other stress factors enhancing the occurrence, distribution and severity of *Edwardsiella tarda* in fish is essential to design an effective disease control and prevention. Since the current state of knowledge of *E. tarda* infection in fish and humans in Ethiopia is almost nil, further study on the epidemiology of *E. tarda* in different hosts and environments as well as comprehensive information on the strains involved should be established for better fish productivity and public health.

Acknoweldgement

We authors would like to express thanks to all parasitology and pathology, microbiology and physiology laboratory staff of College Veterinary Medicine and Agriculture, Zeway Fishery Resource and Research Center and National Veterinary Institute staff for their assistance.

Competing Interests

The authors declare that they have no competing interests.

References

1. Dugenci SK, Candan A (2003) Isolation of Aeromonas Strains from the Intestinal Flora of Atlantic Salmon. Turk J Vet Anim Sci 27: 1071-1075.

2. Trust TJ (1986) Pathogenesis of Infectious Diseases of Fish. Annu Rev Microbiol 40: 479-502.

3. WHO (1999) Food Safety Issues Associated with Products from Aquaculture. World Health Organ Tech Rep Ser. Geneva pp: 4-8.

4. Toranzo AE, Magarinos B, Romalde JL (2005) A Review of the Main Bacterial Agriculture 246: 37-61.

5. Plumb JA (1999) Edwardsillea Septicemias. In: Woo PTK, Bruno DW (ed.) Fish Disease and Disorders. CAB International, New York, NY, USA 3: 479-525.

6. Novotony L, Dvoska L, Loremcova A, Beran V, Pavlik I (2004) Fish: A Potential Sources of Bacterial Pathogens for Human Beings. International Journal of Vet Med 49: 343-358.

7. Wei LS, Musa N (2008) Phenotyping, Genotyping and Whole Cell Protein Profiling of Edwardsiella tarda Isolated from Cultured and Natural Habitat Fresh Water Fish. American Eurasian, S. Agric and Environ Sci 3: 681-691.

8. Zheng D, Mai Ku, Liu S, Limin C, Liufu Z, et al. (2004) Effect of Temperature and Salinity on Virulence of Edwardsiella tard to Japanese Flounder Paralichthys Olivaceus (Temminik Et Schlegel). Aqua Res 35: 494-500.

9. Carter GR (1984) Diagnostic Procedures in Veterinary Bacteriology and Mycology. 4th (ed.) Charles C Thomas Publication. USA pp: 3-160.

10. Clarridge JE, Musher DM, Fainstein V, Wallace RJ Jr (1980) Extraintestinal human infection caused by Edwardsiella tarda. J Clin Microbiol 11: 511-514.

11. Wang SM, Liu CL, Cheng SP, Kao PT (2008) Mixed Tuberculosis and Edwardsiella tarda Infection of the Abdomen. A Case Report and Litreture Review. J Emerg Critcare Med 19: 23-27.

12. Vandamme LR, Vandepitte J (1980) Frequent Isolation of Edwardsiella tarda and Plesiomonas Shigelloides from the Healthy Zaire Fresh Water Fish. A Possible Sources of Sporadic Diarrhea in the Tropics. Appl Environ Microb 39: 475-479.

13. Zighelboim J, Williams TW, Bradshaw MW, Harris RL (1992) Successful Medical Management of A Patient with Hepatic Abscess due to Edwardsiella tarda. Clin Infect Dis 14: 117-121.

14. FAO (1995) Review of the Fisheries and Aquaculture Sector of Ethiopia. Rome pp: 1-3.

15. WHO (1999) Food Safety Issues Associated with Products from Aquaculture. Report of Joint FAO /WHO Group. Geneva pp: 4-8.

16. Yimer E (2000) Preliminary Survey of Parasites and Bacterial Pathogens of Fish Lake Zeway. Ethiop J Sci 3: 25-33.

17. Nuru A (2007) Study on Bacterial Pathogens of Fish in Southern Gulf of Lake Tana with Special References to Aeromonas hydrophila and Edwardsiella tarda. Addis Ababa University. FVM. Debere Zeit Ethiopia pp: 10-30

18. LFDP (1993) Fisheries Base Line Survey. Lake Ziway. Lake Fisheries Development Working Paper No. 7. Addis Ababa. Ministry of Agriculture pp: 134-165.

19. Anon (1999) Regional Government of Oromia. Oromia Economic Study Project Office. Agricultural Sector Study Draft Final Report. Addis Ababa pp: 123-135.

20. LFDP (1994) Preliminary Estimation of the Maximum Sustainable Yield of the Lakes pp: 196-204.

21. Rowland S, Walsh SR, Teel LD, Carnahan AM (1994) Pathogenic and Clinical Microbiology. A Laboratory Manual Little Brown Company. 1st (ed.) New York pp: 71-107.

22. Quinn PJ, Carter ME, Markey B, Carter GR (1999) Clinical Veterinary Microbiology. International Limited Company. New York pp: 48-617.

23. Baron JE, Peterson RL, Gold MSF (1994) Diagnostic Microbiology. 9th Ed. Clarinda. USA pp: 362-384.

24. Woodland J (2006) National Wilds Fish Health Survey. Laboratory Producers. 3rd Fish and Wild Life Service. USA pp: 73-117.

25. Agrawa BL (1996) Basic Statistics 3rd (ed.) New Age In + 1 (P): Limited Published Delhi, India pp: 215-232.

26. Woo PTK, Bruno DW (1999) Fish Disease and Disorders. CABI Publishing. UK 3: 267-577.

27. Wyatt LE, Nickelson R 2nd, Vanderzant C (1979) Edwardsiella tarda in freshwater catfish and their environment. Appl Environ Microbiol 38: 710-714.

28. Noga EJ (1990) Fish Diseases. Diagnosis and Treatment. Amazon Inc. New York pp: 79-172.

29. Roberts RJ (1989) Fish Pathology. 2nd (ed.) Bailliere Tindal, London, England pp: 263-274.

30. Stoskopf KM (1993) Fish Medicine. WB. Saunders Company. Harcoart Bruce. Javanrich Incist pp: 45-63.

31. Ling SH, Wang XH, Lim TM, Leung KY (2001) Green Fluorescent Protein Tagged Edwardsiella tarda. Reveals Portal of Entery in FEMS Microbial. Lett 15: 239-243.

32. Acharya M, Maiti NK, Mohanty S, Mishra P, Samanta M (2007) Genotyping of Edwrdsiella tarda Isolated from Fresh Water Fish Culture System. Comp. Immunol. Microbiol Dis 30: 33-40.

33. Kumar G, Rathore G, Sengupta U, Singh V, Kappor D, et al. (2007) Isolation and Characterization of Outer Membrane Proteins of Edwardsiella tarda and its Application in Immunoassays. Aquaculture 272:98-104.

34. Ewing WH, Mcwhorter AC, Escobar MR, Lubin AH (1965) Edwardsiella, A New Genus of Enterobacteriacea based on A New Species. Int Bull Bacteriol Nomenel Taxon 15: 33-38.

35. Okuda J, Murayama F, Yamanoi E, Iwamoto E, Matsuoka S, et al. (2007) Base changes in the fliC gene of Edwardsiella tarda: possible effects on flagellation and motility. Dis Aquat Organ 76: 113-121.

36. Holt JG (1997) Bergey's Manual of Determinative Bacteriology. 9th (Edn) William and Wilkins, Baltimore.

37. Baya AM, Romalde JL, Green DE, Navarro RB, Evans J, et al. (1997) Edwardsiellosis in wild striped bass from the Chesapeake Bay. J Wildl Dis 33: 517-525.

38. Stock I, Wiedemann B (2001) Natural antibiotic susceptibilities of Edwardsiella tarda, E. ictaluri, and E. hoshinae. Antimicrob Agents Chemother 45: 2245-2255.

39. Horenstein S, Smolowitz R, Uhlinger K, Roberts S (2004) Diagnosis of Edwardsiella tarda Infection in Oyster Toadfish (Opsanus tau) Held at the Marine Resources Center. Biol Bull 207: 171.

40. Savan R, Kono T, Itami T, Sakai M (2005) Loop-mediated isothermal amplification: an emerging technology for detection of fish and shellfish pathogens. J Fish Dis 28: 573-581.

41. Ootsubo M, Shimizu T, Tanaka R, Sawabe T, Tajima K, et al. (2002) Oligonucleotide probe for detecting Enterobacteriaceae by in situ hybridization. J Appl Microbiol 93: 60-68.

42. Yu L, Yuan L, Feng H, Li S (2004) Determination of the Bacterial Pathogenic Edwardsiella tarda in Fish Species by Capillary Electrophoresis with Blue Light Emitting Diode Induced Fluorescence. Electrophoresis 25: 3139-3144.

43. Cahill MM (1990) Bacterial flora of fishes: A review. Microb Ecol 19: 21-41.

44. Ringoa E, Olsenb RE, Mayhew TM, Myclebusted R (2003) Electro Microscopy of the Intestinal Micro flora of Fish. Aquaculture 36: 377-386.

20

Myeloperoxidase Inactivation Affects Neutrophil Recruitment in Zebrafish Injury-Induced Model

Yajuan Li[1,2*], Ren DL[1], Chen M[1] Ge SC[1] and Bing Hu[1*]

1Chinese Academy of Sciences Key Laboratory of Brain Function and Disease, China

2Laboratory of Structural Immunology and School of Life Sciences, University of Science and Technology of China, P R China

*Corresponding author: Yajuan Li, Bing Hu School of Life Sciences, University of Science and Technology of China, No.96 Jinzhai Road, Hefei, Anhui Province, 230026, P. R. China, E-mail: lyj106@mail.ustc.edu.cn; bhu@ustc.edu.cn

Abstract

Uncontrolled migration and excess recruitment of neutrophils can lead to persistent inflammation, tissue damage and disease. Myeloperoxidase is a remarkable target for further understanding the immune cell migration. This study tests the hypothesis that myeloperoxidase may regulate the immune cell activity and provides a new perspective on the treatment of immune diseases by exploring the mechanism of neutrophil migration. Studies of leukocyte migration *in vivo* in the zebrafish model, which set a collection of advantages both mammalian and cell lines, have gained widespread attention. In this study, we used tg (*coro1a*: eGFP; *lyz*: dsred2) and tg (*lyz*: eGFP) lines labelling both macrophages and neutrophils to study the effect of mpx on neutrophil chemo taxis to wounds. We found that myeloperoxidase was required for neutrophil migration to the wound site in injury-induced inflammation, but not required for neutrophil migration to the infection site in the infection-induced inflammation. Further, the regulation of myeloperoxidase was specific to neutrophil migration to wound inflammation, but was not necessary for macrophage migration. Thus, myeloperoxidase activity shows therapeutic potential for inflammatory disease related to neutrophil migration.

Keywords: Zebrafish; Leukocyte; Migration; Inflammation; *In vivo*

Introduction

Immune cell migration is a key aspect of inflammatory response. In recent years, it has been the focus to explore the pathogenesis of inflammation and find appropriate treatment methods among a great deal of research in the life sciences. In the early stages of autoimmune disease, the infiltration of innate immune cells is very important. Neutrophils exert an immediate effect via antimicrobial products and reactive oxygen, and through the secretion of cytokines with more long-lasting effect, influencing other immune cells [1]. Neutrophils can also regulate the activity of antigen-presenting B and T cells, thereby regulating adaptive immune responses [2-4]. All of these effects depend on neutrophils arriving at sites of inflammation through a specific mechanism. However, neutrophils cannot precisely distinguish between host and foreign antigens; this nonspecific reaction is the main mechanism leading to normal tissue damage. So targeting the migration function of neutrophils is one of important therapeutic strategies to treat immune disease.

In recent years, studies of leukocyte migration *in vivo* in the zebrafish model, which set a collection of advantages both mammalian and cell lines, have gained widespread attention [5-8]. The use of *in vivo* imaging of transgenic zebrafish whose neutrophils express green fluorescence protein (GFP), as well as increased real-time tracking of immune cell migration, provides a new perspective on how chemicals and genetic interference operate on neutrophils' chemotactic response. After injury, wound epithelial cells release H_2O_2, which forms a decreasing concentration gradient. This gradient is required for rapid recruitment of leukocytes to the wound [8,9]. The inhibition of hydrogen peroxide production only inhibits the first wave of neutrophil wound infiltration. The factors that mediate the later sustained phase of neutrophil recruitment are still unknown. When inflammation occurred, immune cells swiftly migrated to the inflammation site in response to signalling by cytokines, chemokine, and ROS. Many experiments in vitro have demonstrated that IL-8 induced neutrophil migration [10,11]. There is IL-8 homologue in zebrafish, cxcl8, play an important role in acute inflammation, and is the major regulator of neutrophil chemotactic response [12]. The zebrafish inflammation model will be instrumental in elucidating mechanisms that regulate leukocyte migration *in vivo* and factors that affect neutrophil migration, and will continue to provide new insight into the onset and resolution of inflammation.

Myeloperoxidase (mpo), and 146 kDa heme peroxidase, is mainly derived from neutrophils. Increased mpo activity is a potential sign of inflammation [13], and mpo regulates inflammation through positive or negative feedback [14,15]. In sum, due to its effect on cascade amplification of inflammatory signals, mpo plays an important role in inflammatory processes. It is dependent or independent of activity characteristics that have functional effects on regulating immune cells and are involved in inflammatory disease. However, some hypotheses have been based on in vitro experiments, and they need further validation based on exploration of the regulatory effect of mpo on innate immune cells.

Zebrafish has the conserved ortholog of mammalian mpo, myeloperoxidase (mpx), is expressed specially in neutrophils in the developmental stage [16]. Using a new neutrophil-replete but myeloperoxidase-deficient mutant (durif), Pase et al. [17] showed that myeloperoxidase-deficient zebrafish had abnormally sustained high concentrations of H_2O_2 in wounds despite similar numbers of arriving neutrophils. Another study reported that inhibiting myeloperoxidase activity resulted in decreased inflammatory cell recruitment in a mouse model of multiple sclerosis [18]. Thus in this study, we aim to model mpx inactivity in zebrafish using a pharmacologically irreversible mpx activity inhibitor, 4-amino-benzoyl (acid) hydrazine (abah), to explore innate immune cell migration and test a neutrophil-targeted drug *in vivo*, and to try to further understand how mpx regulates neutrophil recruitment behavior. This can provide insight into the mechanism of neutrophil migration.

Materials and Methods

Drugs

An mpx inhibitor, 4-amino-benzoyl (acid) hydrazine (abah, Sigma, Cat #A41909), was ordered from Sigma-Aldrich. The concentration for immersion was 10 mm, and the concentration for microinjection was 50 mm Recombinant Human IL-8 (77aa) cytokine was purchased from PEPROTECH (USA).

Experimental Animal

Four days post fertilization (dpf) AB/wild type zebrafish strains and tg(*lyz*:GFP), tg(*coro1a*:eGFP; *lyz*:DsRed2) transgenic lines were used in this study [19].

All the treatment of experimental animals was approved by the Committee on the Ethics of Animal Experiments of the USTC (License Number: USTCACUC1103013), and was in accordance with the guidance and provisions of the Committee on Laboratory Animal Resource Centre and the Animal Care and Use Guidelines of the University of Science and Technology of China.

Experimental bacterial strains

GFP-expressing strains of *Staphylococcus aureus* (*S. aureus*, NCTC8325) were used and diluted to a uniform concentration of 1,000 colony-forming units (cfu)nl-1.

Mechanical damage model

Tail fin and ventral fin injury were used to induce chemotactic response. To make a tailfin amputation, a sterile scalpel was used to cut an incision along the spinal cord ends as straight as possible [20,21]. To make the ventral fin acupuncture injury, a sterile pin was used at in the ventral fin area, 4-5 somites from the cloaca.

Muscle injection

Four dpf larvae were pre-incubated with abah, then performed an injection in the muscle at the 4th-6th somite from the cloaca with 20 nl PBS, 13 nl IL-8 recombinant protein at 5.6 µM, or 13 nl Staphylococcus aureus at 1,000 cfu nl-1, respectively. The control group was not pre-incubated with abah. Images were taken at 2 hours post amputation (hpa).

Zebrafish *in vivo* imaging

For *in vivo* observation of macrophages and neutrophils behavior, we used 4 dpf tg (*coro1a*:dGFP;lyz:DsRed2) hkz04t;nz50, tg (*lyz*:eGFP) transgenic lines and tg (*flk*:GFP;*lyz*:DsRed) hybrid lines. All larvae prior to imaging were treated in 0.01% MS222 (3-amino benzoic acid ethyl ester). In order to prevent damage to the living body samples, the exposure time was as short as possible when imaging. Imaris software

(Bitplane, Switzerland), ImageJ software (NIH, USA) and Adobe Photoshop CS2 (USA) were used for image processing.

Long-term continuous imaging: The leukocyte in the whole body or damaged area was observed using OLYMPUS FV-1000 (BX61WI, Japan) confocal microscopy with 10x OLYMPUS Plan Fluor objective (NA 0.30) and 60x OLYMPUS Plan water-dipping objective (NA 0.9). For observation of the dynamic migration of neutrophils to the wound area, we continuously photographed from 15 minutes post amputation (minpa) to 4 hpa, 1 min/z-stack, then made the photographs into a real-time video.

For analysis of neutrophil trajectory, we used data obtained from Imaris software (Bitplane, Switzerland) with an OLYMPUS-FV1000 (BX61WI) continuous imaging confocal microscope, 1 min every one z-stack, from 0.75 hpa-1.25 hpa or 1.5 hpa-2 hpa, to analyze the motion characteristics of neutrophils. Only the neutrophils that speed was greater than the 0.015 µm s-1, and migration along the X-axis direction of the wound were included in the statistics.

Fluorescence images obtained: An OLYMPUS (BX60WI) fluorescence microscope with 10x Plan Fluor objective (NA 0.40) was used to obtain fluorescence images for counting the number of neutrophils and macrophages recruited to the wound site.

Gene expression analysis

At the time points indicated, using a sterile scalpel, the body portion between the cloaca and the wounded tail tip was excised from 100 larvae at each point. Tail tissue samples were then pooled and frozen in RNAiso Plus (TAKARA, China). RNA was extracted according to standard procedures. RT-PCR was performed by one-step RT-PCR kit (TAKARA, China).

Real-time PCR was performed with a BioRadicuclermyIQ2 instrument using a two-step real-time quantitative PCRSYBR Green Supermix kit (Bio-Rad, USA). The primers used are shown in Table 1. In all cases, PCR was performed with triplicate samples and repeated at least twice.

Statistical analysis

Data were analysed in Graph Pad Prism 5.0 (Prism, USA) using one-way ANOVA and t-test. The results were shown as mean ± SEM (performed as three independent experiments). $P < 0.05$ was considered statistically significant. *, **, and *** represent $P < 0.05$, $P < 0.01$, and $P < 0.001$, respectively.

Results

Inhibition of mpx activity reduces the number of neutrophils recruitment to the wound site in zebrafish injury-induced inflammation model

Inflammation is divided into two types: sterile and infection inflammation. To study the role of mpx in both types of inflammation, we pharmacologically inhibited mpx activity with abah using 4 dpf tg(*lyz*:eGFP) transgenic lines. Tail amputation (Figure 1A) and bacterial injection were then performed in both the abah-treated and untreated groups. At 3 hpa of the tail fin, about 17 ± 2 neutrophils were found in the region of 200 µm away from the wound site of 4 dpf normal zebrafish. However, a significantly lower number of neutrophils were observed in the abah treatment group (Figure 1B and

1C, p<0.0001). The results showed that inhibition of mpx activity reduces the number of neutrophils recruitment to the wound site.

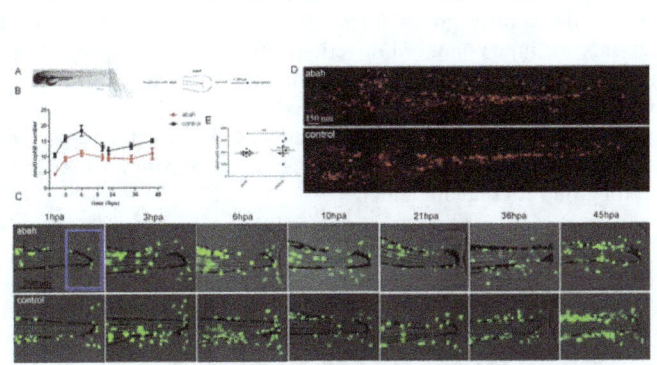

Figure 1: Inhibitory effect of abah on neutrophil migration toward the wound site. Neutrophil migration assay was carried out using tg (*lyz*GFP) larvae. (A) A schematic view of zebrafish tail fin amputation and operation process. (B) Quantification of neutrophils recruited to the region of 200 μm away from injury site in abah-treated and control group. N=30-50. (C) A schematic of zebrafish tail fin at the indicated time points; blue boxes indicate the statistical area of neutrophil numbers. (D) The distribution of neutrophils in the whole zebrafish is shown in the abah treatment and control larvae groups, respectively. (E) Quantification of (D). N=8.

To confirm that abah acts to block the recruitment rather than the ontogeny of neutrophils, whole body neutrophil counts were performed in the presence of the abah and the vehicle-only control. No differences in morphology, number or distribution of neutrophils were observed after abah treatment (Figure 1D and 1E), and the average total number of neutrophils was 193 ± 2 and 200 ± 2 in the abah treatment and control groups (p=0.0973), respectively. In addition, the number of neutrophils in the hematopoietic tissue was also counted; the result was consistent with above (data not shown). In contrast, when we performed a microinjection of *S. aureus* into the muscles of the fifth somite rostral from the cloaca, the number of neutrophils accumulated at the bacterial injection site at 2 hours post infection (hpi) was similar: 25.7 ± 2 and 25 ± 2 in the abah treatment and the control group (p=0.8740), respectively (Figure 2A and 2B, in the middle of the row). In addition, sterile PBS injection induced a local sterile inflammation, causing minor accumulation of inflammatory cells at the site of injection. Similar to the tail fin injury model, in the abah treatment group the number of recruitment neutrophils was a little lower than those in the control group (p=0.0438), respectively (Figure 2A and 2B, in the right of row). These results illustrated that mpx was necessary to neutrophil migration to wounds in a mechanic injury-induced inflammatory response, whereas it was dispensable in a bacterial infection inflammation response.

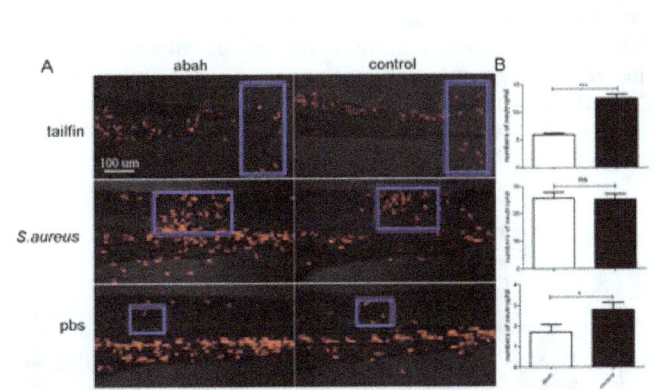

Figure 2: Inhibition of mpx activity plays a different role in injury and infection inflammation. (A) Confocal fluorescence views of the distribution of neutrophils after being subjected to tail amputation, muscle microinjection of PBS, and S. aureus, respectively. Tg (*lyz*:DsRed) transgenic fish were used. Images were taken at 2 h after injection. (B) Quantification analysis shows the neutrophil number in the wound area or *S. aureus* injection site. N=30-50.

Three hours after tail fin injury, abah treatment reduced the number of neutrophils that migrated to the wound area. To find out what happened to the neutrophil number later, whether it exist a delayed migration effect, we observed the number of neutrophils that migrated to the wound at different time points after tail fin injury. As shown in Figure 1B and 1C, from 0-6 hpa, neutrophils gradually migrated to the caudal damage area, reaching a peak at 6 hpa. After this, the number of neutrophils reduced gradually. However, at 48 hpa, a significant short-term rebound phenomenon appeared. This proved again that inhibition of mpx activity reduced the number of neutrophils recruited to the wound site. This trend of neutrophil numbers was very consistent, without any delay effect.

To verify the specific effect of mpx on neutrophil migration, tg (*coro1a*:eGFP; *lyz*:DsRed2) hkz04t; nz50 transgenic zebrafish lines were used in which macrophages and neutrophils were labeled with green and yellow fluorescent colors, respectively. As shown in Figure 3C, the number of macrophages migrating to the wound area at 3 hpa was 11 ± 2 and 10 ± 2 in the abah-treated group and the control group, respectively. This showed no significant difference in the number of macrophages (p=0.3062), whereas there were 8 ± 2 neutrophils in the abah-treated group and 17 ± 2 neutrophils in the control group (p<0.0001). This result demonstrated that mpx played a special role in neutrophil migration, but not in macrophage migration. In addition, acupuncture injury at the 4th-5th somite from the cloaca was performed for *in vivo* observation of the dynamic macrophage and neutrophil migration process (Figure 3A and 3B). Our results showed that only macrophages migrated into the wound site at 3 hpa in the abah-treated group, while almost no neutrophil came in. Interestingly, we observed a phenomenon of neutrophils coming across the vessel and maintaining a position in the lower side of the vessel (Figure 3B, top row, white arrows), but these neutrophils were not recruited to the wound site. However in the control group, consistent with the previous results, around 8 ± 2 neutrophils migrated into the wound area. The number of green macrophages recruited was 4 ± 2 (Figure 3B, in the bottom row). This result also revealed that inhibition of mpx activity specifically influenced neutrophil migration to the wound site. All of the above results demonstrated that mpx specifically regulated neutrophil migration to the wound site in injury-induced inflammation, but exerted no effect on macrophages, and was dispensable for neutrophil migration in the bacteria-infection inflammation response.

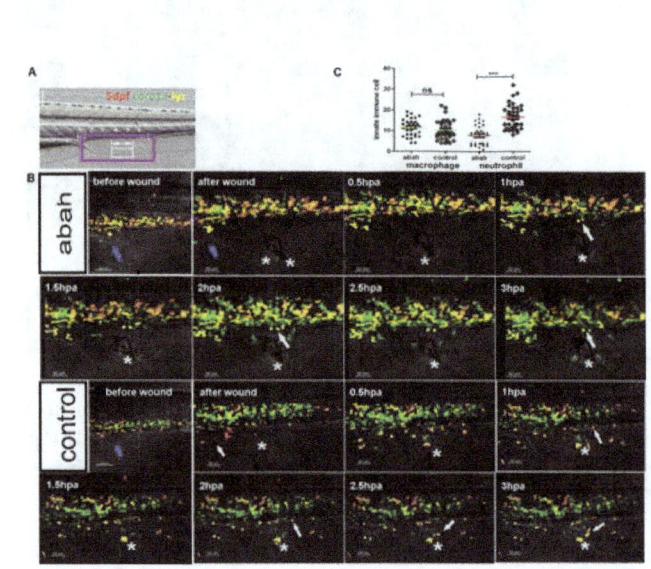

Figure 3: Inhibition of mpx activity specifically influences neutrophil migration to the wound area, but not macrophage in tg (*coro1a*:eGFP; *lyz*:DsRed2)hkz04t; nz50 transgenic zebrafish. (A) Schematic view of pelvic fin injury, about 4-5 somites away from cloaca. (B) The distribution of neutrophils and macrophages from 0-3 h after zebrafish pelvic injury in abah-treated and control groups. The white star and red arrow show the wound areas, the blue arrow indicates the position of cloaca, and the white arrows indicate neutrophil migration out of the blood vessel. (C) Quantification of neutrophils and macrophages respectively recruited to the region of 200 μm away from injury site in abah-treated and control group at 3 hpa. N=30-50.

Mpx plays different roles in basal random movement and injury-induced directional migration of neutrophils

To further characterize the effect of mpx on neutrophil responses, time-lapse analysis was performed to assess different aspects of neutrophil migration and movement behavior. Under normal conditions, the neutrophils residing in the tissues or blood vessels were constantly moving within a certain area as basal random movement, presumably patrolling the surrounding tissues to check for damaged cells and apoptosis (Figure 4A). This suggests that abah treatment had no effect on the motion behavior of the neutrophil itself. Upon tissue injury with extracellular inflammation, neutrophils migrated swiftly to the injury site, taking a relatively straight migration trajectory as a directional movement, with a change of migration velocity, speed, and forward migration distance (Figure 4B). We next assessed the migratory behavior of neutrophils in abah treatment before and after tail fin amputation. We found that in both the abah-treated and control groups, neutrophils displayed a basal migratory pattern under injury-free conditions (sham). Upon tail fin injury during 0.75 hpa-0.25 hpa or 1.5 hpa-2 hpa, neutrophils rapidly migrated toward the wound site in the control group, whereas neutrophils in abah treatment moved

randomly and failed to reach the wound site, resulting in a reduction of the number of neutrophils in the wound area. Track length, velocity, and speed of neutrophil migration to the wound site were similar to those in the control group. These data indicated that the mpx was necessary for injury-induced directional migration of neutrophils, but was dispensable for the basal random movement of neutrophils. It could be said that mpx was required for a sufficient number of neutrophils to migrate into the wound site, but not for some other neutrophil migration, which showed the same behavior as those in the control group after tail fin injury.

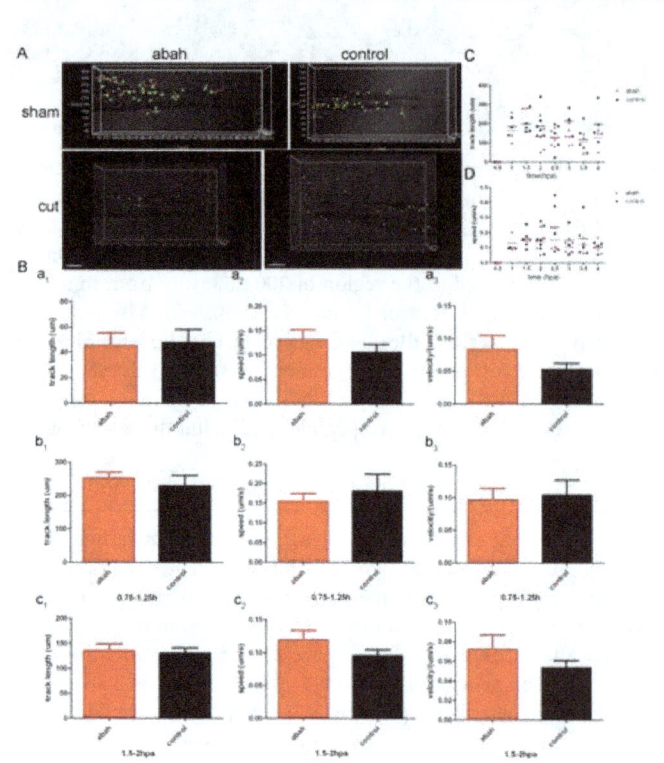

Figure 4: The injury-induced migration tracks of the neutrophils. (A) A schematic diagram of neutrophil migration tracks (B) Statistical data on the injury-induced migration tracks, speed, and velocity of neutrophils to wound area in 4 dpf tg(lyz: Dsred) were analyzed before and after tail fin injury in the abah-treated and control groups. Each panel consists of ≥ 20 neutrophils from 7 larvae. In (a), the basal random movement of neutrophils without injury is presented. The directional movement of neutrophils from 0.75 h to 1.25 h after tail fin injury is shown in (b). The directional movement of neutrophils from 1.5 h to 2 h after tail fin injury is shown in (c). A continuous real-time analysis of trends in neutrophil migration track (C) and speed (D) for each neutrophil every half an hour in the abah-treated and control groups. The sample consists of 8 cells from the same larvae.

Real-time analysis of trajectory length and speed for a single neutrophil was performed. We found that the neutrophil rapidly migrated into the wound site at 1 hpa. Later, it migrated into the wound site with the similar speed (Figure 4D). At different time points, the change in migration speed and trajectory length of neutrophil recruitment was consistent both in the abah treatment and in the control group (Figure 4C and 4D).

In order to verify whether neutrophil behavior across the vascular wall was affected by inhibition of mpx activity, we performed an acupuncture pelvic injury in 4 dpf tg (*flk*:GFP;*lyz*:DsRed) transgenic zebrafish, in which neutrophils were labelled by red fluorescent protein and blood vessels were labeled by green fluorescent protein. The result showed that neutrophils were capable of crossing the blood vessels in both the abah treatment and the control group, although they soon moved back to the blood vessels (Figure 5A, blue arrows). While in the control group, the neutrophils that migrated across the vascular wall into the wound site continued to stay in the wound area, then came into an active state, as morphological changes indicated (Figure 5A, blue arrow).

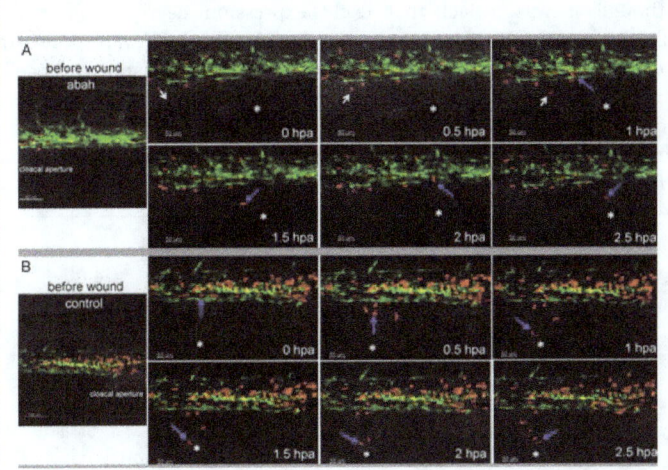

Figure 5: Abah does not affect the behavior of neutrophils across vessels in 5 dpf tg(*flk*:GFP;*lyz*:DsRed) zebrafish. The diagram of neutrophils recruited to the wound site at indicated time points before and after acupuncture injury of pelvic fin, distance from cloaca about third-fourth somite in the abah-treated group (A) and the control group (B). White asterisk indicates the wound area; scale bar 50 μm. White arrows indicate random neutrophil migration back and forth between the wound site and the vessel. Blue arrows indicate neutrophils moving normally out across the blood vessels.

Cxcl8 is involved in the regulation effect of mpx on recruitment of neutrophils to the injured site

Inhibition of mpx activity by pharmacological inhibitor (abah) decreased the migration of neutrophil to the wound site in injury-induced inflammation. Thus whether the inhibition of mpx activity affect the chemokine signaling upon inflammation needs to be verified.

After inducing inflammation in zebrafish by tail fin amputation, we first investigated genes that were considered likely to be involved in the process of inflammation. We found that cxcl8 mRNA expression at the wound sites was lower in abah treatment than those in the no treatment group at 2 h post amputation (Figure 6A and 6B), which partly explained why the abah treatment reduced the number of neutrophils recruited to the wound area. Similarly, ELISA assay showed that inhibition of mpx activity decreased the expression level of cxcl8 (Figure 6C). In addition, we also investigated it with an adult tail fin injury model by Q-RT-PCR tests (Figure 6D).

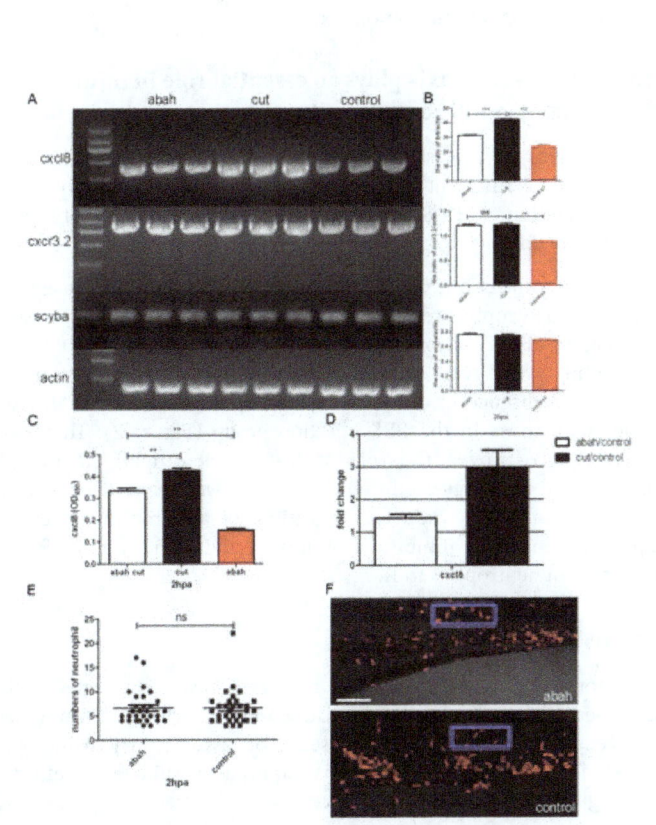

Figure 6: Mpx regulates injury-induced neutrophil directional migration via regulating cxcl8 expression. (A) The tail fins from 4 dpf zebrafish larvae were wounded and mRNA levels of indicated genes (cxcl8, macrophage-specific chemokine scyba and receptor gene cxcr3.2) were detected by RT-PCR (A), ELISA (C) at 2 hpa (80-100 tail fins per group). Quantitative RT - PCR (D) of cxcl8 in adult wound model. Control group is from tail fin tissue of unwounded zebrafish larvae. (B) Statistical data for band brightness in (A). Gene expression is normalized against actin or ef1a. (E) Counts of fluorescent neutrophils recruiting to the muscle injection area were taken at 2h after injection. Data indicate means ± SEM. N=30-50. (F) Representative wide-field fluorescence microscope micrographs of 4 dpf tg (*lyz*:DsRed) being pretreated with abah followed by muscle injection of human recombinant IL-8 into the fifth somite from the cloaca. Blue boxes indicate neutrophil number for statistical area.

It has been reported that the chemokine receptor cxcr3.2 is specifically expressed in zebrafish embryo macrophage, regulating macrophage migration to the inflammation site [22]. Scyba, a homologue of human cxcl14 in zebrafish due to its chemotactic properties, could activate macrophages and dendritic precursor cells [23]. Therefore, we needed to detect whether mpx also regulated the expression of macrophage-specific chemokine. By semi-quantitative reverse transcription we found that the mRNA expression of cxcr3.2 and scyba, after inhibition of mpx activity, was similar to those in the control group (Figure 6A and 6B). These results illustrated that abah treatment specifically reduced the expression of cxcl8 chemokine, which regulated the neutrophil migration to the wound area, but did not affect the expression of cxcr3.2 and scyba expression, which activated macrophage migration behavior.

Mpx activity–cxcl8 axis plays an essential role in injury-induced neutrophil recruitment response

To address in this study whether the mpx signal for inducing neutrophil recruitment in zebrafish was also through cxcl8, recombinant human IL-8 protein was used to detect the role of mpx on IL-8 chemotaxis in 4 dpf zebrafish.

The microinjection of IL-8 protein significantly increased the number of neutrophils recruitment in the 4 dpf tg (*lyz*:eGEP) zebrafish line (Figure 6F). Two hours post-injection, it was observed that 7 ± 2 neutrophils were recruited to IL-8 injection site whether in the presence or absence of abah treatment (Figure 6E and 6F); this was higher than those in the PBS injection group (Figure 2). This result again *in vivo* proved the role of IL-8 in neutrophil chemotactic function and also suggested that IL-8 acts downstream of mpx in the regulation of injury-induced migration of neutrophils, therefore explaining that the inhibition of mpx activity did not affect the migration of neutrophils to IL-8.

Discussion

Although many previous reports have suggested that myeloperoxidase can regulate the functional effects of immune cells involved in the inflammatory disease of tissue injury using vitro models [24-26], further evidence is required to validate the effect of mpx regulation on innate immune cells.

When applied to zebrafish, abah, identified in the zebrafish screen model, showed a dose-dependent inhibitory effect on neutrophil migration. The results showed that mpx was required for neutrophil migration to the wound area in injury-induced inflammation, but not for neutrophil migration to the infection site in infection-induced inflammation. Here, we only tested the number of neutrophil at 2 hpi, the inflammatory response at different time points after infection did not given observation. We thought our result was not contrary to previous findings in myeloperoxidase-deficient zebrafish against *C. albicans* infection showing an equal number of neutrophil at 12 hpi, but significantly higher from 24 hpi in mutant embryos than in control embryos [27]. In fact, the infection reachs to a peak at 12hpi; then the neutrophils decrease gradually later because of the resolution of inflammation or the death of immune cells. Whether treatment zebrafish with abah also affect neutrophil recruitment to *S. aureus* infection site at a more comprehensive infection time need further to verify. Here, we focused on the neutrophil recruitment in the early acute phase. Collectively, these data suggest that mpx may regulate injury-induced neutrophil directional migration, at least in part by regulating cxcl8 expression.

In addition, time-lapse analysis was performed to assess dynamic neutrophil behavior. Our results showed that the track length, speed, and velocity of neutrophil migration to the wound site in abah treatment were not different from those in control group. At 0.5 h after tissue injury, H2O2 produced by the wound regulates the initial neutrophils percepting and migrating into the wound. After 0.5 hpa, inflammatory factors released by immune cells and endothelial cells mediate a cascade inflammatory response, and induce more neutrophils to migrate to the wound area with an increase in migration velocity, speed, and forward migration distance. In this study, our results showed that the H2O2 level maintained its normal expression at

a peak of 0.5 h after zebrafish tail fin amputation, which was consistent with previous research (data not shown). However, abah treatment had no inhibitory effect on neutrophil movement characteristics. The possibilities for no differences are that our results used a time window of 0.75 hpa and 1.5 hpa for statistical analysis, which may be relatively early for observing differences in migration behavior between the abah treatment group and the control group. Meanwhile, half an hour may be a relatively short time period for statistical analysis of neutrophil migration. A longer time period may be needed to observe this process for more accurate statistical analysis. On the other hand, this study showed extensive effects by a pharmacological inhibitor of mpx activity on the entire body, especially systemic inflammatory response. Thus, we do not preclude other signal paths being induced by various factors, although it was a fact that cxcl8 expression decreases after the inhibition of mpx activity in injury-induced inflammation response. In fact, mpx has the effect of cascade amplification on inflammatory cells during inflammation. Cytokines and chemokines play a more important role. In addition, we believe that the phenotypes we observed in our study likely represent mpx inactivity without any toxic effect.

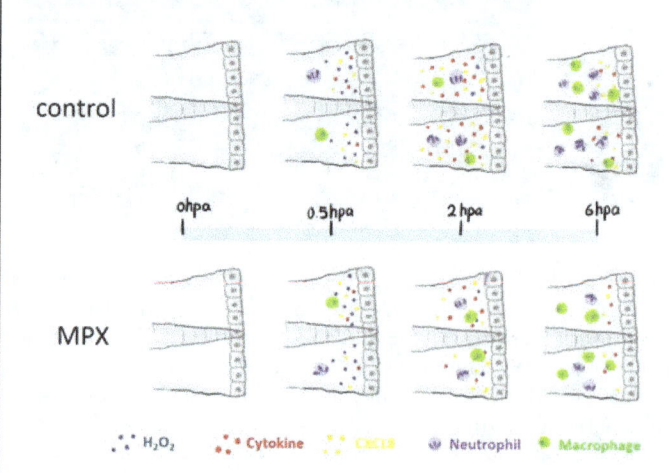

Figure 7: A schematic of regulatory mechanism of mpx on neutrophil migration. A hypothesis is that at 30 min after tail fin amputation, wound tissue produces a normal H₂O₂ signal, inducing initial neutrophils and macrophages to rapidly migrate to the wound site, releasing inflammatory cytokine and chemokine. Due to mpx inhibition, inflammatory and chemotactic signals that regulated neutrophil attenuate; subsequently, the number of neutrophils decreases.

To summarize the above results, a hypothesis is proposed in the sketch of Figure 7. Fifteen to 30 minutes after tail fin injury, a H₂O₂ signal is produced normally and induces initial neutrophils and macrophages to migrate rapidly to the wound site, releasing inflammatory cytokine and chemokine. Due to mpx inhibition, inflammatory and chemotactic signals that regulate neutrophils attenuate, and subsequently the number of neutrophils decreases. Thus functional inactivation with inhibitor after innate immune development could have great value to suppress the immune response, additional studies are required to identify the specific target of abah in neutrophils.

These results further confirm that there is a great opportunity to observe the function of mpx *in vivo* inflammation using a tg (*lyz*:GFP) transgenic zebrafish model, and that inhibitor-targeted mpx may become a powerful therapy to control inflammatory pathology. By using transgenic zebrafish lines, we can analyse the phenotypic effects of human cells to detect pathogenesis in the future. This can help screen for more effective inhibitors to treat chronic myeloid leukaemia, as well as aiding the search for new drugs to regulate neutrophil migration in inflammatory response.

Gene	Name	Nucleotide	Use
beta-actin	fwd	5'-CATTGGCAATGAGCGTTTC-3'	RT-PCR
beta-actin	rev	5'-TACTCCTGCTTGCTGATCCAC-3'	RT-PCR
Cxcl8	fwd	5'-AATGAGGGTGAAGCTCTACCTCCAC-3'	RT-PCR
Cxcl8	rev	5'-CACAGTGATAACAGTCCATTCGCAA-3'	RT-PCR
Cxcr3.2	fwd	5'-AGACGAGACGTGCCTAACATC-3'	RT-PCR
Cxcr3.2	rev	5'-ACACGATGACTAAGGAGATGA-3'	RT-PCR
Scyba	fwd	5'-TGTGGGATGACACTACCAGTGAA-3'	RT-PCR
Scyba	rev	5'-GACCGGTGTGCTTTATAAGCTTGT-3'	RT-PCR
Cxcl8	fwd	5'-CCACACACACTCCACACACA-3'	qPCR
Cxcl8	rev	5'-CCACTGAATTGTCCTTTCATCA-3'	qPCR
Ef1a	fwd	5'-CTTCTCAGGCTGACTGTGC-3'	qPCR
Ef1a	rev	5'-CCGCTAGCATTACCCTCC-3'	qPCR

Table 1: Primers.

Acknowledgements

We would like to thank Professor Zilong Wen (State Key Laboratory of Molecular Neuroscience, Department of Biochemistry, the Hong Kong University of Science and Technology, Clear Water Bay, Kowloon, Hong Kong, P.R. China), Li Li (Key Laboratory of Freshwater Fish Reproduction and Development, Ministry of Education, State Key Laboratory Breeding Base of Eco-Environments and Bio-Resources of the Three Gorges Area, School of Life Science, Southwest University, Chongqing, China), Yiyue Zhang and Wenqing Zhang (Department of Cell Biology, Southern Medical University, Guangzhou, China), and Philip S. Crosier (Department of Molecular Medicine and Pathology, School of Medical Sciences, The University of Auckland, Auckland, New Zealand) for providing Tg (*corola*: eGFP; *lyz*: Dsred) and Tg (*lyz*: Dsred) lines. This work was supported by a grant from the National Nature Science Foundation of China (NSFC U1332136, 31070950) and the National Basic Research Program of China (973) (2012CB947602, 2011CB504402) to B. Hu. Y.L. is supported by the China Postdoctoral Science Foundation (Grant No.: 2015M582007).

Disclosure

The authors declare no conflict of interest.

References

1. Soehnlein O (2012) Multiple roles for neutrophils in atherosclerosis. Circulation research 110: 875-888.

2. Ostanin DV, Kurmaeva E, Furr K, Bao R, Hoffman J, et al. (2012) Acquisition of antigen-presenting functions by neutrophils isolated from mice with chronic colitis. J Immunol 188: 1491-1502.

3. Pillay J, Kamp VM, Hoffen EV, Visser T, Tak T, et al. (2012) A subset of neutrophils in human systemic inflammation inhibits T cell responses through Mac-1. J Clin Invest 122: 327-336.

4. Ryan SO, Johnson JL, Cobb BA (2013) Neutrophils confer T cell resistance to myeloid-derived suppressor cell-mediated suppression to promote chronic inflammation. J Immunol 190: 5037-5047.

5. Deng Q, Huttenlocher A (2012) Leukocyte migration from a fish eye's view. J Cell Sci 125: 3949-3956.

6. Harvie EA, Huttenlocher A (2015) Neutrophils in host defense: new insights from zebra fish. J Leukoc Biol 4MR1114-1524R.

7. Henry KM, Loynes CA, Whyte MKB, Renshaw SA (2013) Zebra fish as a model for the study of neutrophil biology. J Leukoc Biol 94: 633-642.

8. Yoo SK, Huttenlocher A (2011) Spatiotemporal photolabeling of neutrophil trafficking during inflammation in live zebrafish. J Leukoc Biol 89: 661-667.

9. Niethammer P, Grabher C, Look AT, Mitchison TJ (2009) A tissue-scale gradient of hydrogen peroxide mediates rapid wound detection in zebra fish. Nature 459: 996-999.

10. Lindley I, Aschauer H, Seifert JM, Lam C, Brunowsky W, et al. (1988) Synthesis and expression in Escherichia coli of the gene encoding monocyte-derived neutrophil-activating factor: biological equivalence between natural and recombinant neutrophil-activating factor. Proc Natl Acad Sci USA 85: 9199-9203.

11. Kobayashi Y (2008) The role of chemokines in neutrophil biology. A journal and virtual library Front Biosci 13: 2400-2407.

12. Oliveira SD, Aldasoro CCR, Candel S, Renshaw SA, Mulero V, et al. (2013) Cxcl8 (IL-8) mediates neutrophil recruitment and behavior in the zebrafish inflammatory response. J Immunol 190: 4349-4359.

13. Chen LC, Chen JY, Hour AL, Shiau CY, Hui CF, et al. (2008) Molecular cloning and functional analysis of zebrafish (Danio rerio) chemokine genes. Comp Biochem Physiol B Biochem Mol Biol 151: 400-409.

14. Lau D, Mollnau H, Eiserich JP, Freeman BA, Daiber A, et al. (2005) Myeloperoxidase mediates neutrophil activation by association with CD11b/CD18 integrins. Proc Natl Acad Sci USA 102: 431-436.

15. Gelderman MP, Stuart R, Vigerust D, Fuhrmann S, Lefkowitz DL, et al. (1998) Perpetuation of inflammation associated with experimental arthritis: the role of macrophage activation by neutrophilic myeloperoxidase. Mediators Inflamm 7: 381-389.

16. Lieschke GJ, Oates AC, Crowhurst MO, Ward AC, Layton JE (2001) Morphologic and functional characterization of granulocytes and macrophages in embryonic and adult zebrafish. Blood 98: 3087-3096.

17. Pase L, Layton JE, Wittmann C, Ellett F, Nowell CJ, et al. (2012) Neutrophil-delivered myeloperoxidase dampens the hydrogen peroxide burst after tissue wounding in zebrafish. Curr Biol 22: 1818-1824.

18. Forghani R, Wojtkiewicz GR, Zhang Y, Seeburg D, Bautz BR, et al. (2012) Demyelinating diseases: myeloperoxidase as an imaging biomarker and therapeutic target. Radiology 263: 451-460.

19. Li L, Yan B, Shi YQ, Zhang WQ, Wen ZL (2012) Live imaging reveals differing roles of macrophages and neutrophils during zebrafish tail fin regeneration. J Biol Chem 287: 25353-25360.

20. Ren DL, Sun AA, Li YJ, Chen M, Ge SC, et al. (2015) Exogenous melatonin inhibits neutrophil migration through suppression of ERK activation. J Endocrinol 227: 49-60.

21. Ren DL, Li YJ, Hu BB, Wang H, Hu B (2015) Melatonin regulates the rhythmic migration of neutrophils in live zebrafish. J Pineal Res 58: 452-460.

22. Zakrzewska A, Cui C, Stockhammer OW, Benard EL, Spaink HP, et al. (2010) Macrophage-specific gene functions in Spi1-directed innate immunity. Blood 116: E1-E11.

23. Shurin GV, Tourkova IL, Chatta GS, Schmidt G, Wei S, et al. (2005) Small rho GTPases regulate antigen presentation in dendritic cells. J Immunol 174: 3394-3400.

24. Witkosarsat V, Descampslatscha B (1994) Neutrophil-Derived Oxidants and Proteinases as Immunomodulatory Mediators in Inflammation. Mediators Inflamm 3: 257-273.

25. Johansson MW, Patarroyo M, Oberg F, Siegbahn A, Nilsson K (1997) Myeloperoxidase mediates cell adhesion via the alpha M beta 2 integrin (Mac-1, CD11b/CD18). J Cell Sci 110: 1133-1139.

26. Klinke A, Nussbaum C, Kubala L, Friedrichs K, Rudolph TK, et al. (2011) Myeloperoxidase attracts neutrophils by physical forces. Blood 117: 1350-1358.

27. Wang K, Fang X, Ma N, Lin Q, Huang Z, et al. (2015) Myeloperoxidase-deficient zebrafish show an augmented inflammatory response to challenge with Candida albicans. Fish Shellfish Immunol 44: 109-116.

21

Population Dynamics of *Pseudotolithus Senegalensis* and *Pseudotolithus Typus* and Their Implications for Management and Conservation within the Coastal Waters of Liberia

Austin Saye Wehye[1*], Patrick K Ofori-Danson[2] and Angela Manekuor Lamptey[2]

[1]*Bureau of National Fisheries, Ministry of Agriculture, Liberia*

[2]*Department of Marine and Fisheries Sciences, University of Ghana*

[*]**Corresponding author:** Austin Saye Wehye, Bureau of National Fisheries, Ministry of Agriculture, Liberia, E-mail: austinwehye@yahoo.com

Abstract

The study evaluated some aspect of population parameters of *Pseudotolithus senegalensis* and *Pseudotolithus typus* within Liberia's coastal waters. A total of 177 and 152 samples of *P. senegalensis* and *P. typus* respectively were collected from July to December, 2016. Individual fish samples was measured for standard length and analysed using FiSAT II software. From the results, *P. senegalensis* growth parameter were estimated at asymptotic length (L_∞)=66.68 cm, growth rate (K)=0.13 yr^{-1}, the longevity (t_{max})=21.49 years, theoretical age at birth (t_0)=-1.586 years and growth performance index (φ')=2.762. While *P. typus* growth parameters asymptotic length (L_∞)=66.68 cm, growth rate (K)=0.14 yr^{-1}, the longevity (t_{max})=19.3 years, theoretical age at birth (t_0)=-2.126 years and growth performance index (φ')=2.294. Mortality parameters for *P. senegalensis* and *P. typus* were calculated as total mortality rate (Z)=0.93 yr^{-1} and 0.70 yr^{-1}, natural mortality rate (M)=0.37 yr^{-1} and 0.39 yr^{-1} and fishing mortality rate (F)=0.56 yr^{-1} and 0.31 yr^{-1} respectively. The calculated fishing mortality rates (F) compared to F_{opt}=0.4M were beyond the limit for sustainable fishing. The exploitation rate (E) of *P. senegalensis* (E=0.60) was higher than the E_{opt}=0.5 criterion. It implies that *P. senegalensis* is overexploited while *P. typus* was at the peak of exploitation (E=0.45). Results from the study revealed that the *P. senegalensis* fishery in Liberia is slightly overexploited while *P. typus* is at the optimal level of exploitation; as well as the presence of growth overfishing within the two species population within Liberian coastal waters. Thus, to avert the consequences of growth overfishing, sustainable fisheries measures including monitoring of fishing efforts, and increase in mesh size should be implemented and enforced.

Keywords: Liberia; *Pseudotolithus senegalensis*; *Pseudotolithus typus*; Growth; Mortality; Exploitation rate

Introduction

The world per capita fish consumption is reported to have increased from an average of 9.9 kg in the 1960s to 19.2 kg in 2012 [1]. On the other hand, the proportion of assessed marine fish stocks fished within biologically sustainable levels declined from 90% in 1974 to 71.2% in 2011, with 28.8% of fish stocks estimated to be overfished [1]. Although fish are renewable resources, this huge irremediable depletion of marine biodiversity by location and depth partly due to intense fishing activities has led to decline of marine capture fisheries [2,3]. The declining trend in global marine catches has led some fisheries scientists to forecast the collapse of ocean fisheries [4]. More so, the trends in catches forecast that more stocks will become overexploited and collapsed [5].

The total marine landed catch for Liberia was estimated at 1,570.82 tons in 2013 excluding artisanal catch and drop speedily to 204 tons in 2014 (BNF unpublished data, 2016). This drastic decline may be due to the reduction in fishing fleets from four (4) vessels in 2013 to two (2) vessels in 2014. The family *Sciaenidae* was among the species targeted. There was a new development in 2016 that increased the fishing intensities on this family. It was the main focus of export by local exporters to Asian markets. This has enticed local artisanal fishers to target *Pseudotolithus senegalensis* and *Pseudotolithus typus* by doubling the price per pound from one United States dollar to two United States dollars and providing outboard motors. This has exacerbated the already existing tension on these species that are mostly consumed by Liberian.

In Liberia, factors such as poor fisheries data collection, limited resources, conflicts and illegal, unregulated and unreported (IUU) do not only make it difficult to estimate the status of almost all of the marine biodiversity but also presents a great challenge to fisheries managers [6,7]. However, Togba [8] reported that Sardinella, Barracudas, Croakers (*P. senegalensis* and *P. typus*), Sharks and Ilisha africana constituted 83% and 59.06% of local fish supply in 2004 and 2005 respectively; indicating that there has been a declined in fish catches.

Furthermore, the paucity of information on population parameters and biology pertaining to commercially important fish species within Liberian coastal waters cripples any management interventions geared towards sustainable fisheries in Liberia. It is against this backdrop that the present study sought to estimate some population parameters of the family *Sciaenidae* residing in Liberian coastal waters to enhance already existing management interventions.

Methodology

Study area

Liberia is a relatively small coastal state located in West Africa with geographical coordinates as 6.4281 °N, 9.4295 °W. The coastline of Liberia is 570 kilometers comprising of relatively warm waters and low nutrient contents [9]. However, the study focused on two fish landing sampling stations (ELWA, N 06.23355° and W 010.69365°; and Marshall, N 06.13833° and W 010.38171°) within two coastal counties along the coastline of Liberia (Figure 1).

The main source of livelihood for the majority of the inhabitants residing within the selected two fish landing sampling stations is fishing and its related activities such as fish processing and fish trade. However, a few of the indigenes are engaged in alternative forms of livelihoods including farming, driving, and others.

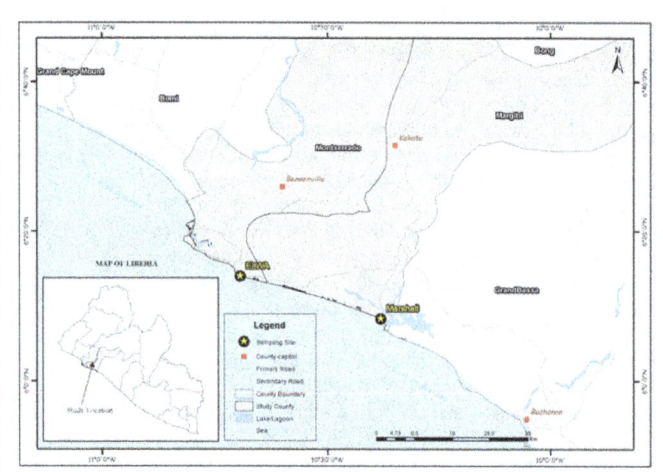

Figure 1: Map showing the sampling sites and Liberia map.

Data collection

Monthly fish samples were collected from artisanal fishermen who operated with multifilament gears from the selected fish landing stations. Fish sample collection was performed from July, 2016 to December, 2016 (six months). According to Pauly [10] analysis of the structure of fish population requires at least data collected over a period of six months. Morphometric measurements of the obtained fish samples including standard length and weights were recorded in the lab at the Bureau of National Fisheries. The standard length was measured using the 100 cm Measuring Board to the nearest 0.1 cm, whereas the weight was measured using the electronic weighing scale to the nearest 0.01 g. Fish samples were identified to the species level using fish identification keys by Schneider [11]. In all, a total of 177 and 152 specimens of *Pseudotolithus senegalensis* and *Pseudotolithus typus* respectively were sampled.

Growth and mortality parameters

The growth of the fish was assumed to follow the von Bertalanffy's growth function (VBGF) which has the basic form $L_t = L_\infty (1 - e^{-K(t-t_0)})$. Where, L∞ is the asymptotic length that is the mean length the fish of a given stock would reach if they were to grow indefinitely, K is a growth constant, t0 is the age of the fish at zero length, Lt is the length at age and t is age at length. These parameters were fitted in FISAT II [12] for estimation. Pauly's [13] empirical equation for the theoretical age at length zero (t0) was used to obtain this parameter as $\log_{10}(-t_0) = -0.392 - 0.275 \log_{10} L_\infty - 1.038 \log_{10} K$.

In order to compare different estimations of growth parameters, the empirical equation of growth performance,

$$\varphi = \log_{10} k + 2 \log_{10} (L_\infty),$$ of Pauly and Munro [14] was used.

The total instantaneous mortality rate (Z) was estimated using length converted catch curve method as implemented in FiSAT II. Natural mortality rate (M) was estimated using Pauly's empirical relationship using a mean surface temperature (T) of 27°C:

Log $M = -0.0066 - 0.279 \log L_\infty + 0.6543 \log K + 0.4634 \log T$ [15],

Where M is the instantaneous natural mortality, L∞ is the asymptotic length, T is the mean surface temperature and K refers to the growth rate coefficient of the VBGF. Fishing mortality (F) was calculated using the relationship: F=Z−M [16], where Z is the total mortality, F the fishing mortality and M is the natural mortality. The exploitation level (E) was obtained using the relationship: E=F/Z [16],

Optimum fishing (F_{opt}) which is directly related to the natural mortality (M) was calculated for the selected fish species using the expression below:

$F_{opt} = 0.4M$ ....[14].

Length at first capture (L_{C50})

The ascending left arm of the length converted catch curve incorporated in FiSAT II tool was used to estimate the probability of length at first capture (L_{c50}) in addition to the length at both 25 and 75 captures which corresponded to the cumulative probability at 25% and 75% respectively. The probability of capture gives clear idea about the estimate of the real size of the fish in the fishing area that is being caught by specific gear. It is an important tool for fisheries managers in sustainably managing a target fishery, because it helps would be managers determining the minimum mesh size of a fishing fleet.

Length at first maturity (L_m)

To estimate the length at first maturity (L_m) for the assessed species, the procedure by Hoggarth et al. [17] below was used. The input parameters for the model included asymptotic length only (L_∞).

Length at first maturity (L_m)=L_∞ * 2/3.............................[17].

Recruitment pattern and yield per recruit

The recruitment pattern of the stock was determined by backward projection on the length axis of the set of available length-frequency data as described in FiSAT. This routine reconstructs the recruitment pulse from a time series of length–frequency data to determine the number of pulses per year and the relative strength of each pulse. Input parameters included L∞ and K. Normal distribution of the recruitment pattern was determined by NORMSEP (Separation of the normally distributed components of size-frequency samples) [18] in FiSAT. The midpoint of the smallest length group in the catch was estimated as the length at recruitment (Lr) length at recruitment [19].

The relative biomass per recruit (B'/R) was estimated as B'/R=(Y'/R)/F. E_{max} which depicts exploitation rate producing maximum yield, $E_{0.1}$ highlighting exploitation rate at which the marginal increase of Y'/R is 10% of its virgin stock with $E_{0.5}$ implying exploitation rate under which the stock is reduced to half its virgin biomass were computed using the procedure incorporated using the Knife-edge option fitted in the FiSAT II Tool.

Data Analysis

The length frequency data were pooled into groups with 1cm length intervals. Then the data were analyzed using the FiSAT II (FAO-ICLARM Population Assessment Tools) software [12].

Results

Growth parameters

Figure 2 show the restructured length frequency with superimposed growth curves. The observed curves of growth portrayed the existence of six cohorts within the population of both targeted species.

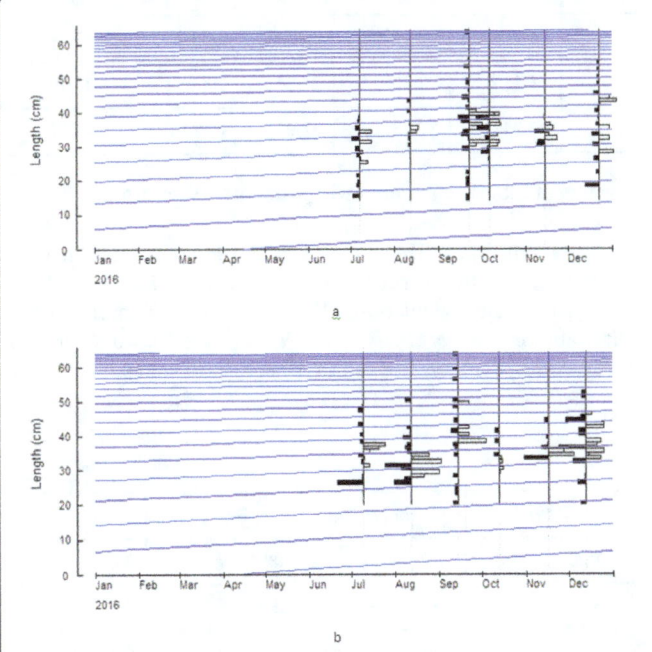

Figure 2: a. Restructured frequency distribution output from FiSAT II with superimposed growth curves for *P. senegalensis*. b. Restructured frequency distribution output from FiSAT II with superimposed growth curves for *P. typus*.

The asymptotic length (L_∞) and growth rate/ constant (K) for the two targeted species were estimated at 66.68 cm SL and 0.13 yr^{-1} respectively with its longevity (t_{max}) as 21.49 years for *P. senegalensis*, while the asymptotic length (L_∞) and growth rate/constant (K) for *P. typus* were 66.68 cm SL and 0.14 yr^{-1} correspondingly, with its longevity as 19.3 years. The growth performance index (φ') and theoretical age at birth (t_0) were estimated at 2.762 and -1.686 years for *P. senegalensis* whereas *P. typus* recorded 2.294 and -2.126 years for growth performance index (φ') and theoretical age at birth (t0) respectively. Using the estimated growth parameters (L_∞, K and t_0), the

VBGF for length at time (t) for the two targeted species were expressed as:

P. senegalensis

$$SL_t = 66.68 (1-e^{-0.13 (t-(-1.586))})$$

P. typus

$$SL_t = 66.68 (1-e^{-0.14 (t-(-2.126))})$$

Estimated growth parameters of *P. senegalensis* and *P. typus* in the current study were compared to what other authors reported in different localities (Table 1).

Species	Authors	Country	SL_∞	K	φ'	t_0
P. senegalensis	Sidibe [20]	Guinea	60.8 (TL)	0.35	3.112	-0.329
	Sossoukpe [21]	Benin (Site 1)	42.98	0.24	2.753	-0.60
	Sossoukpe [21]	Benin (site 2)	42.98	0.16	2.549	-0.90
	Current study	Liberia	66.68	0.13	2.762	-1.586
P. typus	Sidibe [20]	Guinea	73.8 (TL)	0.35	3.280	-0.149
	Sossoukpe [21]	Benin (Site 1)	48.6	0.19	2.652	-0.73
	Sossoukpe [21]	Benin (site 2)	48.6	0.15	2.549	-0.92
	Current study	Liberia	66.68	0.14	2.294	-2.126

Table 1: Estimated growth parameters of *P. senegalensis* and *P. typus* from the current study compared to other authors.

The above equations were used to estimate the lengths of the two species of *Sciaenidae* at various ages (Table 2). *P. senegalensis* attain at least 50% of the asymptotic length when at the fourth class, while *P. typus* attain at least 50% of the asymptotic length when at the third class.; both indicating less rapid growth in length at the early age class (Table 2).

Age class (yr)	Pseudotolithus senegalensis		Pseudotolithus typus	
	SL (TL) (cm)	% of L_∞	SL (TL) (cm)	% of L_∞
1	19.04 (21.14)	29	23.63 (26.24)	35
2	24.85 (27.58)	37	29.26 (32.48)	44
3	29.95 (33.25)	45	34.15 (37.91)	51
4	34.42 (38.22)	52	38.4 (42.63)	58
5	38.36 (42.58)	58	42.09 (46.73)	63

Table 2: Calculated age-length data for the two-sciaenid based on their respective von Bertalanffy growth equation.

In doing a comparative analysis of the calculated age-length data of both species for the current study, the estimated values were compared with what other authors reported in different countries (Table 3).

Species	Authors	Country	1yr	2yrs	3yrs	4yrs	5yrs
P. senegalensis	Troadec [22]	Congo	23.12	31.85	38.01	42.35	45.41
	Sun [23]	Senegal	9.77	18.01	24.97	30.84	35.79
	Coutin and Payne [24]	Sierra Leone	13.81	24.45	32.66	38.99	43.87
	Njock [25]	Cameroon	11.16	19.94	26.84	32.27	36.55
	Sidibe [20]	Guinea	23.15	36.67	44.71	49.49	52.23
	Sossoukpe [21]	Benin (Site 1)	18.3	25.68	31.34	35.72	39.13
	Sossoukpe [21]	Benin (site 2)	15.31	20.98	25.7	29.61	32.92
	Current study	Liberia	21.14	27.58	33.25	38.22	42.58
P. typus	Bayagbona [26]	Nigeria	29.21	47.79	61.68	72.09	79.87
	Poinsard [27]	Congo	26.86	36.96	45.43	52.54	58.51
	Njock [25]	Cameroon	14.26	25.24	33.69	40.19	45.19
	Sidibe [20]	Guinea	24.44	36.97	46.07	52.68	57.48
	Sossoukpe [21]	Benin (Site 1)	18.12	25.14	30.77	35.33	39.04
	Sossoukpe [21]	Benin (site 2)	17.11	22.35	27.34	31.54	35.10
	Current study [28]	Liberia	26.24	32.48	37.91	42.63	46.73

Table 3: Calculated age-length data of *P. senegalensis* and *P. typus* of the fisheries waters of Liberia compared with what others reported.

Mortality coefficients and current exploitation rate

Figure 3 shows the calculated mortalities from the length converted catch curve fitted in the FiSAT II (Table 4). From Figure 3, mortality parameters for *P. senegalensis* were estimated as: $Z=0.93$ yr^{-1}, $M=0.37$ yr^{-1} and $F=0.56$ yr^{-1} whereas the estimates for *P. typus* were $Z=0.70$ yr^{-1}, $M=0.39$ yr^{-1} and $F=0.31$ yr^{-1} (Table 5). However, the dark circles in the figure represent the points used in calculating (Z) through least squares regression lines. The yellow circles represent frequencies of fishes either not fully recruited or approaching (L_∞), and hence discarded from the calculation. The estimated optimum fishing mortality rate for *P. senegalensis* and *P. typus* were $F_{opt}=0.37$ yr^{-1} and $F_{opt}=0.28$ yr^{-1} respectively. The values of M/K ratio were 2.85 and 2.79 for *P. senegalensis* and *P. typus*, respectively (Table 6).

Species	Authors	Country	Z (yr⁻¹)	M (yr⁻¹)	F (yr⁻¹)	E
P. senegalensis	Sidibe	Guinea	1.20	0.97	0.51	0.64
	Sossoukpe (2011)	Benin (Site 1)	3.26	0.63	2.63	0.81
	Sossoukpe (2011)	Benin (site 2)	0.91	0.49	0.42	0.47
	Current study	Liberia	0.93	0.37	0.56	0.60
P. typus	Sidibe (2003)	Guinea	1.26	0.66	0.60	0.65
	Sossoukpe (2011)	Benin (Site 1)	2.65	0.52	2.13	0.80
	Sossoukpe (2011)	Benin (site 2)	1.60	0.45	1.15	0.72
	Current study	Liberia	0.70	0.39	0.31	0.45

Table 4: Estimated mortality parameters of *P. senegalensis* and *P. typus* of the fisheries waters of Liberia compared to those off other regions.

From Table 6, the current exploitation rate of the two species were estimated as follow: for *P. senegalensis* it was estimated $E_{current}=0.60$ and $E_{current}=0.45$ for *P. typus*. The Z/K ratio were estimated as 7.15 and 5.00 for *P. senegalensis* and *P. typus* respectively. The reliability of the estimated natural mortality rate, M, was ascertained using the M/K ratios because this ratio has been reported to be within the range 1.12-2.5 for most of the fish [28].

Population Dynamics of Pseudotolithus Senegalensis and Pseudotolithus Typus and Their Implications for Management...

135

Species	TL$_{m50}$ (cm)	TL$_{c50}$ (cm)	Authors	Localities
P. senegalensis	44.5 (SL)	33.84 (SL)	Current study	Liberia
	30.4	25.5	Sossoukpe [21]	Benin
	29.0	-	Sidibe [20]	Guinea
	31.0	-	Domain et al. [29]	Guinea
	26.5	-	Jock [25]	Cameroon
	26.4- 28.0	-	Troaddec [22]	Congo
P. typus	44.5 (SL)	30.23	Current study	Liberia
	30.1	22.76	Sossoukpe [21]	Benin
	40.0	-	Sidibe [20]	Guinea
	37.0	-	Domain et al [29]	Guinea
	26.5	-	Jock [25]	Cameroon
	33.0	-	Fontana [30]	Congo

Table 5: Estimated length at first capture and length at first maturity of *P. senegalensis* and *P. typus* of the fisheries waters of Liberia compared to those off other regions.

Parameters	P. senegalensis	P. typus
SL$_\infty$ (cm)	66.68 (TL$_\infty$=74.03)	66.8 (TL$_\infty$=74.03)
K (yr^{-1})	0.13	0.14
φ'	2.762	2.294
t$_0$ (yr^{-1})	-1.586	-2.126
t$_{max}$ (yrs)	21.49	19.3
Z (yr^{-1})	0.93	0.70
M (yr^{-1})	0.37	0.39
F (yr^{-1})	0.56	0.31
M/K	2.85	2.79
Z/K	7.15	5.00
F$_{opt}$ (yr^{-1})	0.37	0.28
E$_{current}$	0.60	0.45
L$_{c25}$(cm)	30.57	28.46
L$_{c50}$(cm)	33.84	30.23
L$_{c75}$(cm)	37.11	32.00
L$_{m50}$(cm)	44.45	44.45
L$_c$/L$_\infty$	0.51	0.45
E$_{0.1}$	0.807	0.704
E$_{0.5}$	0.376	0.357

E$_{max}$	0.963	0.829

Table 6: Summary of estimated growth and other derived fish population parameters for the two species of concern from July 2016 to December 2016.

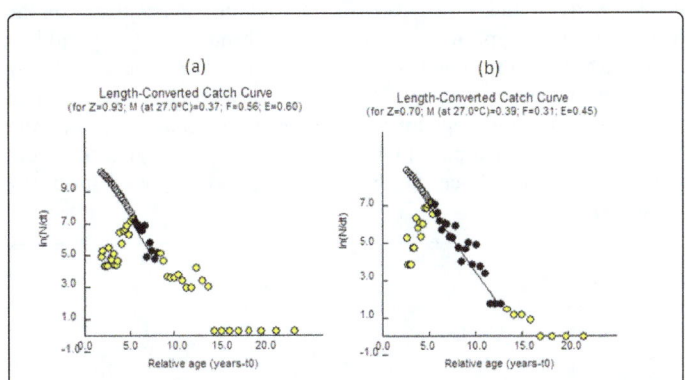

Figure 3: FISAT II output of linearized length-converted catch curve for *P. senegalensis* and *P. typus* (a for *P. senegalensis* and b for *P. typus*).

Estimated mortality parameters of both species were compared with what other authors reported from different regions in Africa (Table 4).

Length at first capture (L$_{c50}$) and Length at first maturity (L$_{m50}$)

Figure 4 shows the probability of capture and length at first maturity for the two targeted fish species. The probability of capture for *P. senegalensis* and *P. typus* at 25%, 50% and 75% were estimated as: 30.57 cm, 33.84 cm and 37.11 cm for *P. senegalensis* and 28.46 cm,

30.23 cm and 32 cm for *P. typus* respectively. The length at first maturity (Lm50) was estimated at 44.5 cm for both species (Table 6). Figure 4 below shows FiSAT II output of the probability of capture. The estimated Lc/L∞ ratios using the relationship between the Length at first maturity (L_{c50}) and the asymptotic length (L∞) for the two-treated species were 0.51 and 0.45 for *P. senegalensis* and *P. typus* correspondingly (Table 6).

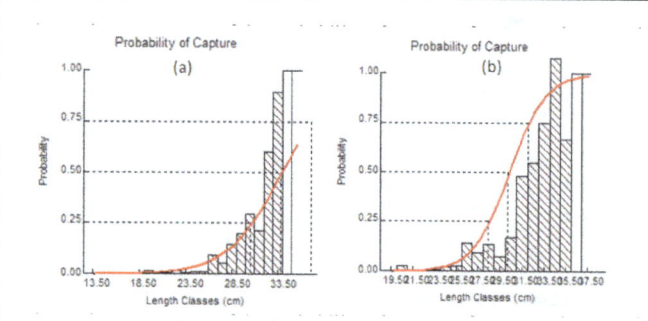

Figure 4: FiSAT II output of the probability of capture of *P. senegalensis* and *P. typus* respectively in the fisheries waters of Liberia (0.2, 0.50 and 0.75 relates to 25%, 50% and 75% respectively).

Estimated length at first capture and length at first maturity of *P. senegalensis* and *P. typus* of the present study compared with those of other authors from different localities (Table 5).

Recruitment pattern

Figure 5 shows the recruitment pattern of the two targeted fish species *P. typus* and *P. senegalensis*. The recruitment pattern for the two targeted fish species was continuous throughout the period of study with two recruitment peaks-major and minor (Figure 5). Using macro inspection, the months for the major and minor recruitment peaks were May and September for *P. senegalensis*, and October and May for *P. typus* correspondingly. The calculated length at recruitment were 8.52 cm and 9.53 cm for *P. senegalensis* and *P. typus* respectively (Figure 5).

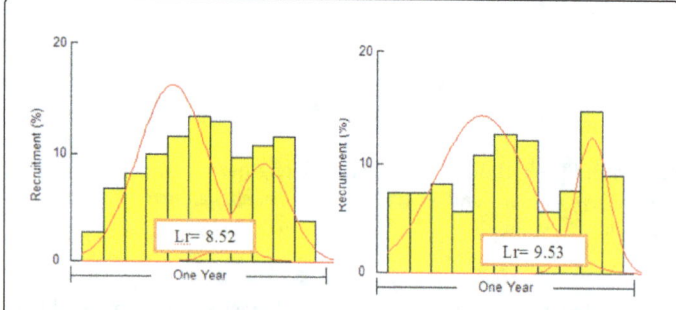

Figure 5: FiSAT II output of recruitment patterns of *P. senegalensis* and *P. typus* respectively.

Relative yield per recruit (Y'/R)

Figure 6 shows the various exploitation rates based on the Beverton and Holt relative yield per recruit model. E_{max} which implies exploitation rate producing maximum yield (yellow dashes), $E_{0.1}$

suggesting exploitation rate at which the marginal increase of Y'/R is 10% of its virgin stock (green dashes) and $E_{0.5}$ indicating exploitation rate under which the stock is reduced to half its virgin biomass (red dashes) of *P. senegalensis* and *P. typus* were estimated as $E_{0.1}=0.807$, $E_{0.5}=0.376$ and $E_{max}=0.963$; and $E_{0.1}=0.704$, $E_{0.5}=0.357$ and $E_{max}=0.829$ respectively (Figure 6). Table 5 showed the actual values of $E_{0.1}$, $E_{0.5}$ and E_{max} by sexes of the targeted fish species.

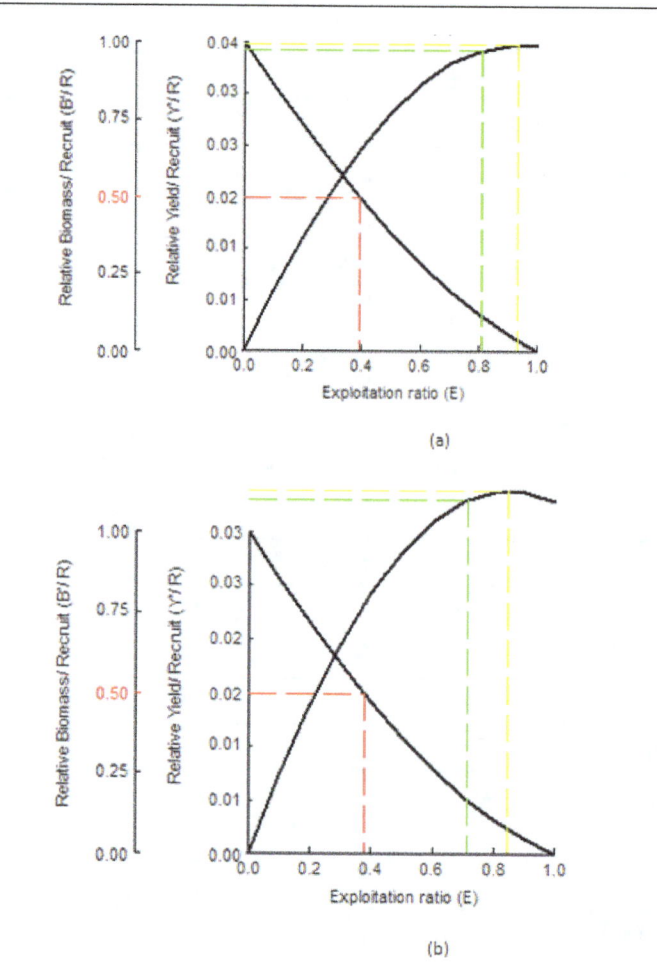

Figure 6: (a) Beverton and Holt's relative yield per recruit and average biomass per recruit models, showing levels of yield indices for *P. senegalensis* in the Coastal waters of Liberia. (b) Beverton and Holt's relative yield per recruit and average biomass per recruit models, showing levels of yield indices for *P. typus* in the Coastal waters of Liberia.

Discussion

Growth parameters

The asymptotic length (L∞) for *P. senegalensis* in the present study is higher than what were reported by other authors but similar to estimate by Sidibe (Table 1) [20]. While the value of asymptotic length for *P. typus* in the current study is close to what was reported by Sidibe [20] and higher than what were reported by other authors (Table 1). The difference in values could be linked to the diversity of methods used for the assessment of growth parameters, length of largest species, time and period of sampling as well as the nature of the length distribution The asymptotic length estimated for the two-species in the present study is lower than the highest length recorded by Sidibe [20] from Guinea (100 cm *P. senegalensis*) and 108 cm *P. typus*). This assertion suggests that the stock being exploited in Liberian fisheries waters is relatively small. The K (yr^{-1}) for the both species from the study were lower and t0 higher than what were reported from Benin and Guinea respectively, indicating that maximum size is rapidly attained for both species in Benin and Guinea than for those of Liberia (Table 1). The φ' mean values for *P. senegalensis* and *P. typus* were close to what were reported by other authors (Table 1). In this study the φ' mean values estimates of *P. senegalensis* fall within range estimated by Baijot et al. [31] for some important fishes in Africa range of 2.65-3.32 which were considered as low; moreover, *P. typus* φ' mean values estimates slightly fall below 2.65; so these species are regarded as showing slow growth performance in Liberian fisheries water. The growth performance index of the two-species in the current study appeared to be in line with estimates from other studies (Table 1). This finding demonstrates that they are of similar taxonomic family. Further, the growth performance index indicates the important availability of food and other favorable environmental conditions [32]. Moreover, the growth rate (K) was found to be lower than 0.34, demonstrating that, *P. senegalensis* and *P. typus* are slow growing fish species, evinced by the long lifespan of 21.49 and 19.3 years respectively [33]. This slow growth rate might be induced by changes in the physical and chemical characteristics of the water amplified by the persistent climate change problems [34].

The calculated age-length data result show that for *P. senegalensis* growth was similar to what were reported in Congo and Guinea; but the growth was slower in other countries (Table 3). The linear growth of *P. typus* was closer to what were reported in Congo, Guinea and Nigeria and by far higher than what was reported in other localities (Table 3). This observation show that *P. senegalensis* and *P. typus* in Liberian waters are fast growing than some countries in Table 3 and are growing almost at the same rate with others countries (Table 3). Evidently, this affirmed the first assertion that these two-species within the Liberian coastal waters are physiologically healthy.

Mortality coefficients and current exploitation rate

The fishing mortality (0.56 yr^{-1}) for *P. senegalensis* from the current study was relatively greater than the natural mortality (F=0.37 yr^{-1}). This estimate agrees with estimates from Benin from site 1, but contradicts what where reported from Guinea and Benin site 2 (Table 4). Additionally, the obtained fishing mortality was found to be higher than the optimum fishing rate (Fopt=0.37 yr^{-1}), which is an indication of over-fishing. Beverton et al. [28] suggested that when the natural mortality and fishing mortality are equal (that is; exploitation rate (E)=0.5), then the stock is in a healthy state and optimally exploited. The estimated current exploitation rate (E) was 0.60, which indicated heavy exploitation. Further, the estimated Z/K ratios for *P. senegalensis* which was greater than 1 (Z/K=7.15) as in Table 6 also strengthen the presence of heavy exploitation as a result of increased fishing mortality [35]. The estimated Ecurrent compared favourably with [28] but lower than [32] from Site 1 (Table 4). This variation could be due to the high rate of fishing pressure on stock in Benin than in Liberia. However, the calculated exploitation rate was not intense because it (Ecurrent=0.60) was much lower than the maximum allowable limit based on the yield-per-recruit calculation (Emax=0.963). Nonetheless, the exploitation of this stock could soon approach the maximum sustainable yield if the current level of exploitation is not monitored accordingly with

subsequent negative consequences on the stock biomass and food security for fishing households. Therefore, the present level of fishing mortality (in terms of number of fishing vessels, especially artisanal canoes) should be of urgent concern for fisheries managers in Liberia.

The natural mortality (M=0.39 yr^{-1}) for *P. typus* was concurrent to estimate by Sossoukpe et al. [21] from Benin site 2 (Table 6), and relatively lower than estimates reported by Sossoukpe et al. [21] from Benin site 1 and [20] from Guinea in Table 4. This observation could be due to the fact that *P. typus* stock in Liberian coastal waters is less susceptible to natural mortality conditions than the *P. typus* in Benin and Guinea waters. Further, it shows that environmental conditions in Liberian waters are more favourable than in Benin coastal waters.

Fishing mortality (F) is mostly reported to cause changes in population parameters such as size ratio, growth rate, size composition and size at first maturity [36]. The estimated fishing mortality for *P. typus* compared to other studies was relatively lower; this could be due to the low numbers of industrial fishing vessels resulting in low fishing pressure. For instance, industrial fishing vessels in Liberia in 2013 was 7 vessels and has decreased drastically to 2 vessels in 2016, although there has been an increased in artisanal fleets 2,986 canoes in 2013 to 3,250 canoes in 2016; thou their capacity still remain low. *P. senegalensis* fishing mortality in the current study was similar to estimates by Sidibe et al. [20,21] in Table 4.

The exploitation rate (Ecurrent=0.45) from the present study was lower than 0.5, depicting that the *P. typus* stock is currently underexploited. In comparison with studies done elsewhere (Table 4), the Ecurrent reported were more than 0.5, which is an indication that *P. typus* is overexploited in Guinea and Benin.

Length at first capture (Lc$_{50}$) and Length at first maturity (Lm50)

The length at first maturity (L$_{m50}$=44.5 cm for both species) was relatively higher than the length at first capture for both targeted fish species (L$_{c50}$=33.84 cm for *P. senegalensis* and L$_{c50}$=30.23 cm for *P. typus*). This signifies that the stocks of the two species are harvested before they could reach the matured stage, a characteristic feature of growth overfishing [37]. Growth overfishing is mostly characterized by small size fish species within the harvested catch. From the two-assessed fish species *P. typus* recorded L$_c$/L$_\infty$ ratios (0.45) was less than 0.5 which indicated that majority of the catch landed constituted juvenile fish species [31]. This assertion affirmed the evidence of growth overfishing. The abundance of small-sized fishes in the catches could be explained by the indiscriminate use of small mesh sized gears and the non-selectivity of fishing gears mostly deployed within the nursery zone of juvenile fishes.

In comparison of the length at first capture of the two assessed species are slightly higher than what was reported by Sossoukpe et al. [21] from Benin (Table 5). The length at first maturity in the current study of *P. typus* was close to what was reported by Sidibe et al. [20,29] both from Guinea respectively, and greater than what were reported by Sossoukpe et al. [21,25] from Cameroon (Table 5). And also the length at first maturity for *P. senegalensis* in the current study was higher than what were reported by other authors in Table 5. This confirmed the early assertion that *P. senegalensis* and *P. typus* in the Liberian fisheries waters are fast growing than those of Benin. Sossoukpe et al. [38] concluded that the size at first sexual maturity is relatively variable with the species and its bio-geographical zone.

Recruitment pattern and relative yield per recruit (Y'/R)

The population structure of *P. senegalensis* and *P. typus* was bimodal indicating probably two spawning periods for this species per year. The presence of two recruitment peaks (one major and one minor) from the study was in line with the description of the recruitment pattern for tropical fishes put forward by Pauly [39]. The observed small length at first recruitment (L_r) for *P. senegalensis* was L_r=8.52 cm and for *P. typus* was L_r=9.53 cm (Figure 5) supports the use of small mesh sized fishing gears by fishermen in Liberian coastal fishing operations with no regards to the damage they cause to the fishery [40]. The presence of recruit throughout the year indicated that recruitment within the targeted species is continuous [41]. Thus, this observation suggests the absence of recruitment overfishing within the fishery of the targeted fish species; there spite the use of small mesh sized fishing gears. For instance, the length at first recruitment was lower than the length at first capture, indicating that species get recruited before been captured by any fishing gear. This finding confirms the assertion that recruitment overfishing is absent within the population of the assessed stock.

However, the continual use of small mesh sized fishing gears in addition to high fishing effort can result in diminished economic benefits, reduced catch per effort, and the collapse of the fisheries for the current target species [42]. Therefore, it is mandatory for fisheries managers to appropriately increase the mesh sizes after careful scientific research, while ensuring that fishers comply with the use of the approved appropriate mesh sized fishing gear in order to avert the occurrence of recruitment overfishing. This is because larger mesh sized gears catch large sized fishes, while allowing juvenile fish to spawn at least once before they are harvested [43-45].

Conclusion

The study has revealed that *P. senegalensis* population within Liberia's coastal waters is currently overexploited while the population of *P. typus* is experiencing exploitation rate close to the maximum sustainable yield amidst the presence of heavy fishing pressure. However, *P. senegalensis* and *P. typus* fisheries in Liberia are currently exhibiting growth overfishing signs which could lead to severe implications on the population size and food security within vulnerable fishing households in the future. Therefore, urgent management interventions in form of monitoring fishing efforts and mesh size regulation (to increase length at first capture) are needed to safeguard these commercially important fish species from possible collapse in the future.

Acknowledgement

This is part an MPhil dissertation presented to the Department of Marine and Fisheries Sciences, University of Ghana. The authors acknowledge the financial support by the United States Agency for International Development, as part of the Feed the Future initiative, under the CGIAR Fund, award number BFSG1100002, and the predecessor fund the Food Security and Crisis Mitigation II grant, award number EEMG000400013.

References

1. FAO (2014) The State of World Fisheries and Aquaculture. Rome pp: 223.

2. Christensen V, Guenette S, Heymans JJ, Walters CJ, Watson R (2003) Hundred-year decline of North Atlantic predatory fishes. Fish and Fisheries pp: 1-24.

3. Swartz W, Sala E, Tracey S, Watson R, Pauly D (2010) The spatial expansion and ecological footprint of fisheries (1950 to present). PLoS ONE 5: 1-6.

4. Worm B, Barbier EB, Beaumont N, Duffy JE, Folke C (2006) Impacts of biodiversity loss on ocean ecosystem services. Science 314: 787-790.

5. Pitcher T, Cheung W (2013) Fisheries: Hope or despair. Mar Pollut Bull 74: 506-516.

6. MRAG (2014) Fisheries Stock Assessment. Report produced under WARFP/BNF Contract 11/001. Republic of Liberia, West Africa.

7. Sherif SA (2014) The development of fisheries management in Liberia: vessel monitoring system (vms) as enforcement and surveillance tools: national and regional perspectives. World Maritime University Dissertations.

8. Togba GB (2008) Analysis of profitability of trawl fleet investment in Liberia. University of Akureyri.

9. BNF (2014) Fisheries and Aquaculture Policy Strategy. Ministry of Agriculture. Republic of Liberia 1-73

10. Pauly D (1987) A review of the ELEFAN system for analysis of length-frequency data in fish and aquatic invertebrates. The international conference on the theory and application of length-based methods for population assessment. Italy.

11. Schneider W (1990) FAO species identification sheets for fishery purposes. Field guide to the commercial marine resources of the Gulf of Guinea. Food and Agricultural Organisation of the United nations, Rome pp: 268.

12. Gayanilo FC, Sparre P, Pauly D (2005) FAO-ICLARM Stock Assessment Tools II (FiSAT II). Revised. User's guide. Rome: FAO Computerized Information Series. pp: 168.

13. Pauly D (1979) Theory and management of tropical multispecies stocks: a review with emphasis on the Southeast Asian demersal fisheries stud.

14. Pauly D (1984) Fish population dynamics in tropical waters: a manual for use with programmable calculations. ICLARM Stud pp: 325.

15. Pauly D (1980) A selection of simple method for the assessment of tropical fish stocks. FAO pp: 54.

16. Gulland J (1971) The Fish Resources of the Oceans. FAO/Fishing News Books, Surrey pp: 255.

17. Hoggarth DD, Abeyasekera S, Arthur RI, Beddington JR, Burn RW (2006) Stock Assessment for fishery management. A framework guide to the stock assessment tools of the Fisheries Management Science Programme (FMSP). Rome pp: 261.

18. Pauly D, Caddy JF (1985) A modification of Bhattacharya's method for the analysis of mixtures of normal distributions. FAO Fish Circ pp: 16.

19. Gheshlaghi P, Vahabnezhad A, Motlagh SAT (2012) Growth parameters, mortality rates, yield per recruit, biomass, and MSY of Rutilus frisii kutum, using length frequency analysis in the Southern parts of the Caspian Sea. Iranian Journal of Fisheries Science 11 48-62.

20. Sidibe A (2003) Coastal demersal fishery resources of Guinea. Exploitation, biology and dynamics of the main species of the community at Sciaenidae. Ensar, Rennes.

21. Sossoukpe E (2011) Ecological studies on Pseudotolithus spp (Sciaenidae) in Benin (West Africa) nearshore waters: Implications for conservation and management. University of Ghana, Legon.

22. Troadec JP (1971) Biology and dynamics of an African Sciaenidae, Pseudotolithus typus. Document Scientifique Oceanographic Research Center, Abidjan 2: 1-125.

23. Sun C (1975) Study of the biology and dynamics of Pseudotolithus senegalensis V. (1833). Fish Sciaenidae on the coast of Senegal. University Doctorate, University of Western Brittany, Brest, France pp: 145.

24. Coutin PC, Payne AI (1989) The effect of long term exploitation of demersal fish populations off the coasts of Sierra Leone, West Africa. J Fish Biol 35: 163-167.

25. Jock JCN (1990) Coastal demersal resources of Cameroon: Biology and exploitation of the main fish species. University Aix-Marseille 2, Cymbium, France pp: 187.

26. Bayagbona EO (1969) Age determination and the Bertalanffy growth parameters of Pseudotolithus typus and Pseudotolithus senegalensis using the "burnt otolith technique". UNESCO 27: 349-359.

27. Poinsard F (1973) Growth of Pseudotolithus typus Blkr in the Black Point region. Doc Scient Centre Rech.

28. Beverton RJH, Holt JS (1957) On the dynamics of exploited fish populations. Fish Invest pp: 533.

29. Domain F, Chavance P, Bah A (2000) Description of the continental shelf Coastal Fisheries in Guinea. Resources and Exploitation, Paris pp: 159-171.

30. Fontana A (1979) Study of the Congolese demersal coastal stock. Biology of the main species exploited. Proposals for the management of the fishery. France pp: 300.

31. Baijot E, Moreau J (1997) Biology and demographic status of the main fish species in the reservoirs of Burkina Faso. Hydrological Aspects of Fisheries in Small Reservoirs in the Sahel Region. Technical Centre for Agricultural and Rural Cooperation, Commission of the European Communities, Wageningen, Netherlands pp: 79-109.

32. Sossoukpe E, Djidohokpin G, Fiogbe ED (2016) Demographic parameters and exploitation rate of Sardinella maderensis (Pisces: Lowe 1838) in the nearshore waters of Benin (West Africa) and their implication for management and conservation. International Journal of Fisheries and Aquatic Studies 4: 165-171.

33. Kienzle MO (2005) Estimation of the population parameters of the Von Bertalanffy Growth Function for the main commercial species of the north sea. pp: 34.

34. Ofori-Danson, PK, de Graff GJ, Vanderpuye CJ (2002) Population parameters estimates for Chrysichthys auratus and C. nigrodigitatus (Pisces: Claroteidae) in lake Volta, Ghana. Fish Res 54: 267-277.

35. Etim L, Sankare Y, Brey T, Arntz W (1998) The dynamics of unexploited population of Corbula trionga (Bivalvia: Corbulidae) in a brackisk-water lagoon. Arch Fish Mar Res 46: 253-262.

36. Chimatiro SK (2004) The biophysical dynamics of the lower shire river floodplain fisheries in Malawi. Rodes University.

37. Amponsah SKK, Ofori-Danson PK, Nunoo FKE (2016) Fishing regime, growth, mortality and exploitation status of Scomber japonicus from catches landed along the eastern coastline of Ghana. Int J Fish Aqua Res 1: 5-10.

38. Sossoukpe E, Nunoo FKE, Ofori-Danson PK, Fiogbe ED, Dankwa HR (2013) Growth and mortality parameters of P. senegalensis and P. typus (Sciaenidae) in nearshore waters of Benin (West Africa) and their implications for management and conservation. Fish Res 137: 70-80.

39. Pauly D (1982) Studying single species dynamics in a tropical multi-species context. Theory and Management of Tropical Fisheries pp: 33-70.

40. Getabu A (1992) Growth parameters and total mortality in Orochromis niloticus (Linnaeus) from Nyanza Gulf Lake Victoria. Hydrobiologia 232: 91-97.

41. Abowei J, George A, Davies O (2010) Mortality, Exploitation rate and Recruitment pattern of Callinectes amnicola from Okpoka Creek, Niger Delta, Nigeria. Asian J Agri Sci pp: 27-34.

42. Miranda L, Agostinho A, Gomes L (1999) Appraisal of the selective properties of gill nets and implications for yield and value of the fisheries at the Itaipu Reservoir, Brazil–Paraguay. Fish Res 45: 105-116.

43. Alagaraja K, Suseelan C, Muthu MS (1986) Mesh Selectivity Studies for Management of Marine Fishery Resources in India. J Mar Biol Ass India pp: 202-212.

44. Pauly D, Soriano ML (1986) Some practical extensions to Beverton and Holt's relative yield-per-recruit model. The First Asian Fisheries Forum, Asian Fisheries Society, Manila, Philippines pp: 491-496.

45. Beverton RJH Holt SJ (1966) Manual of Methods for Fish Stock Assessment. Part II. Tables of yield function. FAO 38: 67.

22

Impact Assessment on By-catch Artisanal Fisheries: Sea Turtles and Mammals in Cameroon, West Africa

Ayissi I[1,2,3,4,*] **and Jiofack TJE**[5]

[1]University of Abdelmalek Essaâdi, Department of Biology, Faculty of Science, Tetouan 2121, Morocco

[2]Cameroon Marine Biology Association, Morocco

[3]Specialized Research Center for Marine Ecosystems in Kribi-Cameroon, Cameroon

[4]Institute of Fisheries and Aquatic Sciences (ISH) at Yabassi, University of Douala, PO Box 2701, Douala, Cameroon

[5] Sub-Regional School and Postdoctoral Water Development and Integrated Management of Forests and Tropical Territories, Kinshasa, RDC, Congo

***Corresponding author:** Ayissi I, University of Abdelmalek Essaâdi, Department of Biology, Faculty of Science, Tetouan 2121, Morocco, E-mail: isidoreayissi@gmail.com

Abstract

The by-catch assessment has been carried out along Cameroon coastline to map artisanal fishing effort and quantify impact of by-catch on sea turtles and marine mammals during three months from June to September 2011 and specific objectives include:

- To interview fishermen in various fishing villages or ports in Cameroon regarding fishing effort and catch.

- To estimate fishing gears used in these fishing ports.

- To evaluate impacts of by-catch on marine mammals and sea turtles.

In total 30 fishing ports were been planned but 23 were covered with 932 files in total (245 long forms and 685 short forms). In total we have 4121boats (none motorized and motorized) and the common gears used are gillnet and surround seine.

The results reveal that, yearly around 1228 turtles with back (green, hawksbill and olive) were caught and 13 Leatherback; most not intentionally. But in Sandje port we noted the intentional catch by local fishermen with around 400 individuals per year for international commercial uses. These numbers are low according to certain data on sea turtles surveys along Cameroon coast. About cetaceans and manatee we had the following data 97 and 292 respectively for each group, but most manatees are caught intentionally for bush meat trade. The survey was limited in time and lack of baseline information on the issue but in future it could be good to involve more permanent data collectors and scientific observers. These results must be feedback to official services for good monitoring of marine faunal and their ecosystem.

Keywords: Artisanal fisheries; By-catch; Cameroon; Dolphin; Mantee; Sea turtles; Whale

Introduction

By-catch is a common threat to marine fauna such as mammals and reptiles. Recent analysis has shown that around 300,000 whales, dolphins and porpoises die each year (about one each two minutes), WWF (In Review). Our minimum by-catch estimate (~1000 loggerheads yr^{-1}) for two small-scale [artisanal] fleets [in Baja California] rivals that of North Pacific industrial-scale fisheries.

The issue of fisheries by-catch of sea turtles has been largely focused on the high-seas industrial fisheries in Alfaro Shigueto et al. [1]. By-catch in artisanal fisheries has now been recognized as a major threat Alfaro Shigueto. Nevertheless, small scale artisanal fisheries are distributed throughout the world in areas that overlap important sea turtles habitats, and are therefore significant challenge for sea turtles conservation efforts Alfaro Shigueto et al. [1].

In some areas such as Cameroon there is no information available about this problem. The few data available are from Ayissi and Moore [2,3]. Fishing is one of the most important occupations of rural population; it contributes more than half of animal protein consumed in Cameroon. This activity has undoubtedly impact on marine faunal, particularly sea turtles, manatees and cetaceans.

Goals and Objectives

The goal of this assessment is to map artisanal fishing effort and catch of sea turtles and marine mammals in a data-deficient country. Specific objectives include:

- To develop a sampling protocol specific to Cameroon that will be a representative sample of fishing in that country.

-To conduct a qualitative and quantitative analysis through questionnaires and personal interview of local fishermen in respect to fishing methods, fish catch and marketing.

- To assess the impacts of by-catch on marine mammals.

Methods

Site description

The coastal zone of Cameroon stretches over 402 km [4], from the Nigerian border in north (Akwayafe river, latitude 4°40'N) to the Equatorial Guinean border in the South (Campo river, latitude 2°20'N). In terms of longitude, it is located between 8°15'E and 9°30'E) (Figure 1).

The continental shelf is about 10,600 km^2 and gradually descends through 30, 50 and 100 m depths [5,6]. The rainfall is ordinary heavy on the coast with average at 3,000-4,000 mm with the peat of 10,160mm yearly at Debunscha around Mount Cameroon.

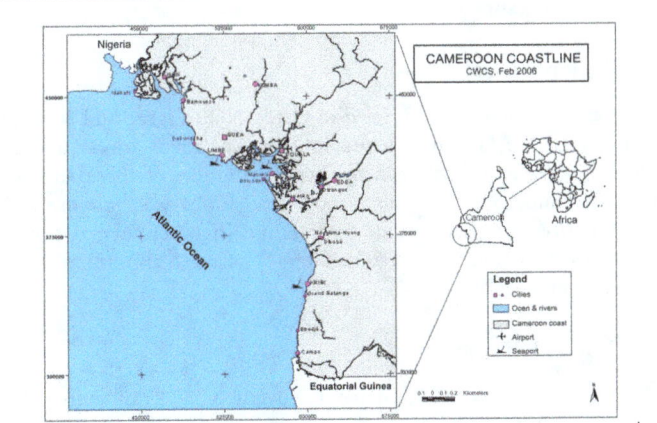

Figure 1: Map of Cameroon coastline (CWCS 2006)

The temperature is always high with average at 25°C. The main characteristic of the hydrology in the Cameroon coast is the permanent existence of a warm, low salinity surface layer cold, high salinity bottom layer [7]. The warm surface layer is 20-30 m thick and is separated from the bottom layer by a permanent thermocline whose position fluctuates with season and location, considered effect of current systems, the rainfall as well as the important water discharge from numerous coastal rivers.

Faunal

Sea turtles: The occurrence of sea turtles was first reported in Cameroon in 1902 by Tornier, Loveridge and Williams [8,9] confirmed the presence of certain species. Four species of sea turtles are common along the Cameroon coast: Leatherback turtle (*Dermochelys coriacea*), Green turtle (*Chelonia mydas*), Olive ridley (*Lepidochelys olivacea*) and Hawsksbill (*Eretmochelys imbricata*). Two of them Leatherback and Olive ridley are nesting along sanding beaches of this coast from September to April. Green turtle and Hawsbill are common in foraging habitats on rocky and on mangroves areas to feed algaes and crustaceans; the origin of these two species is unknown because they are not breeding in this zone. The nesting zones are the southern part from Campo to Kribi, the coast along Douala-Edea wildlife reserve

and the beach from Limbe to Idinaua. The foraging are the rocky sites and mangrove areas particularly in Douala estuary and Bakassi bay near Nigeria.

Manatee: The West African Manatee (*Trichechus senegalensis*) is common in all Cameroon coast in shallow and fresh water from Ntem River in southern part to Bakassi area in northern part. The specie occur particularly in the estuaries sites of main coastal Rivers as Ntem, Nyong, Sanaga, Dibamba, Wouri, Moungo, Akpa-yafe, Ndian, Manyu, Cross River and Lake Ossa which can be considered as sanctuary of Manatee. The specie is common in rainy season. Since several years the manatee is faced to many threats.

Continuing uncontrolled and likely unsustainable hunting must be considered as the major threat to the population. Despite legal protection, the manatee is still-hunted throughout its range for meat leather and oil by harpoon, trap, net, and snag line. Oil is used for medicinal and cosmetics purposes. In some areas as in Sanaga estuary hunting is highly traditional and ritualised, and the meat is consumed locally and other areas hunting are opportunistic. The bush meat trade is also the common threat because the manatee is common in local market hunted by local fishers.

Whales and dolphins: The existence of Whales and Dolphins along Cameroon coast is known, but the species distribution is unknown. The few information available are from observation to death of bycacth and collisions with ships in feeding ground, as in September 2007 and January 2008 two whales fall on Lolabe and Mombo beaches respectively. In 2003 one unknown Dolphin falls on the beach in Yoyo. Ayissi et al. reported the presence of Sousa teuszii which is endemic specie on Cameroon coast [10]. Those species are facing of many threats such as bycacth (gillnets, longlines and trawlers), habitats destruction by pollution, noise, seismic surveys by oil companies and lastly strucking by ships.

Birds: Ayissi confirmed the presence of 302 species of birds on the only coast of Kribi-Campo which according to the criteria of International Birdlife is classified like priority area for the conservation of the birds [10]. A preliminary study with the current of the months of January and March 2007, made it possible to on the whole consider the avifauna watery at a non-exhaustive 65 palearctic and afro-tropical species with a total of 18,326 individuals for 300 species.

Reptiles: With regard to the reptiles, one meets the crocodiles, in particular the crocodile with long muzzle particularly driven out for his skin and his flesh; other species of crocodiles meet in the site of Kribi - Campo are Crocodilus cataphractus, *C. niloticus* and Ostealaemus tetrapis all classified like species in danger (IUCN). With its 122 species of reptiles, the zone of Kribi-Campo is the zone richest in reptiles in the world. The saurians are represented by Rampholeum spectrum, *Chameleo quadricornis* and *C. montium*; this last species is endemic with the Mount Cameroon. The ophidiens are represented by 150 species, among which *Pithon sebae, Boulangerina annulata, Bitis gabonica*, and *Dendroaspis viridis Ayissi* [10].

Fishs and amphibians: There are 27 families and 232 species of which 18 are of major economic importance, especially *Heterotis niloticus* and *Clarias spp. Chrysichthys spp. Mormyrus spp, Synodontis spp, Labeo sp, Brycinus macrolepidotu, Lates niltoticus.* There are over 200 species of which 75 are endemic to at least the coastal forest. In the Edea region, giant frogs are found: the largest frog in the world Conrua goliath which can measure up to 30 cm and weigh more than 2.4 kg. This species is also found in the Kribi-Campo home alone more than 80 species of amphibians.

Flora

Mangroves: In Cameroon, the mangrove swamps are located in the Gulf of Guinea. They account for 30% of the 400 km coast. To Akwa-Yafe at the border with Nigeria in Rio Ntem at the border with Equatorial Guinea. They cover a total surface area of approximately 400,000 hectares (250,000 ha before the retrocession by Nigeria of the peninsula of Bakassi in Cameroon). They were left again according to three great units.

- The mangrove swamp of the Rio Del Rey who is in the area of the Western South with mouth of the rivers Akwa-Yafé, Ndian and Mémé. It has a surface of 218,000 hectares, is the larger second mangrove swamp in West Africa, one of richest in the world leaves its biodiversity.

- The mangrove swamp of the estuary of the Cameroon of 180,000 ha includes the mouths of the Rivers Moungo, Wouri, Dibamba, Sanaga and Nyong towards the South which mixes with the largest metropolis of the Douala country, which assumes the role of Economic capital through a strong establishment of industries of transformation and agro industries.

- The last block with 2000 ha, is located at the estuary of Ntem and some small islands with the mouths of the rivers Bouandjo and Lobe. It is the least important block but which under goes a strong anthropic pressure since the advent of the sea port out of deep waters.

The Cameroonian biocenose of the mangrove swamp is very diverse on the morphological level and as regards the floristic and faunal composition the flora is primarily made mangroves of the *Rhizophora* type. There are 6 indigenous species *(Rhizophora racemosa, R. harrisonii, R. mangle (Rhizophoraceae); Avicennia germinans (Avicenniaceae); Laguncularia racemosa, Conocarpus erectus (Combretaceae);* and an introduced species *Nypa fruticans* (Araceae); that is to say 7 species belonging to 4 large families of plants. Other species are associated of which most important are : *Drepanocarpus lunatus, Dalbergia ecastaphylum, Paspalum vaginatum, Hibiscus tilaceus, Phoenix reclinata, Acrostichumaureum, Pandanus candelabrun, Sesuvium portulacastrum, Alchornea cordifolia, Annonaglaba, Elaeis guinensis, Athocleista vogeli, Bambusa vulgaris, Coco nucifera, Eremospathawendlandiana, Guiborutia demensei, Raphia palma-pinus,* etc. [11,12].

Coastal forests:

Littoral forests: They are formations of low and average altitudes seasonally flooded with species like *Lophira alata (Azobé), Coula edulis (Noisettes), Saccoglottis gabonensis (Bidou),* Approximately 20 types of vegetation are identified at the level of the Kribi-Campo coast. This coast only shelters with them more than 1500 plant species divided into 640 kinds and 141 families.

Mountain forests: They are formations of altitudes on the Mt. Cameroon and the bordering formations whereas agro-forests: They result from the expansion of agricultural processing industries of the palm plantations, rubber and plantain giving the aspect of raised savannas.

Phytoplankton: They are macrophytes phanerogames, the algae abundant are formed there by the species fixed on rocks. More than 29 species of algae and 170 species of sea plants were identified in the Kribi-Campo zone.

Data's collection

This survey has been carried out by seven persons in various domains, four technician of the Ministry of fishing and livestock's, one Master's student from university of Yaounde I, one member the of National Association of sea turtle "Kuda à Tube", one conservationist biologist (Team Leader) and various guides and interpreters on the field. Three trainings were held, one per area because of the large covering distance and files were being translated in French for certain data collectors who were not able to speech English. To ensure that the collectors have understood the techniques of this survey, the Team Leader spend few days with them at the field. The sites were not covered at the same time, we started in the south Province from Campo to Lokoundjé, secondly the Douala and Tiko areas and at last in the Limbe zone. Most data collections were conducted by local guides because most fishermen were foreigners and had a poor understanding of neither French nor English and majority of the fish species and gears names were only given in local languages. In each port, we had the obligation to contact local authorities particularly from the fishing Ministry and local chiefs to explain to them the meaning of this survey.

Sampling protocol

Initially 30 ports were sampled in Cameroon (4 large and 26 small) according to the fact that in previous studies there are around 300 artisanal ports along this coast. The distribution of these sites was along the Cameroon coast landscape. The survey was suppose to be done for three months but due to certain difficulties such as rainfall, the process of disbursement and illiteracy of most fishermen it took around seven months.

Rapid by-catch survey

The survey was conducted from November 2007 to march 2008 from the Southern part of the Cameroon coast at River Ntem to the North part around Bakassi peninsula using questionnaires. The data collectors were supervised by one Biologist as Team Leader. Fishermen were selected at random and the survey was conducted particularly in the afternoon and on Sundays when fishermen were free in their homes but some were contacted at the landing site during their arrival from the fishing trip. In certain areas where data was available, we collected certain information from Government Official Service at The Ministry of Fisheries and Livestock's which the manager of fishing activities is.

The data collection took various hours to cover each port according to the size, some took few hours but others more than a week particularly in the big port. The fact that most fishermen were illiterate enabled us to use guides in certain areas to interrelate with fishermen in local languages particularly for foreign fishermen as Nigerians.

About the port description, this task was done by the Team Supervisor. When additional information was needed such the destination of fish or organization of fishing activity inside the port, the head of the village or local fisheries officer guided us to have the right information. The total number of boat and active boats was gotten from information gathered by local services.

Data analysis

The time series data were analysed using mainly simple descriptive statistics especially totals, mean, percentages, tables and charts; and inferential statistics especially correlation analysis.

Results

Landing sites characteristics

For 30 ports planned during this survey 23 were been covered (76.66%) due to the instability in the Nigerian border which host conflict in the Bakassi peninsula and one of our data collector from Nigerians (local fishermen) were out to the Cameroon during data collection. In total 3125 boats were counted with 1445 none motorize (46.24%) and 1680 motorized (53.76%). These numbers are from official documents from local fisheries services and do not reflect the reality because most boats are not registered in official documents and in certain areas, those information are not available. The landing site characteristics are reflected in the Cameroon coastline landscape which can be divided in four zones following sandy and rocky area from Campo to River Nyong, Sandy and muddy area from River Nyong to River Moungo, Rocky and sandy area from Tiko port to Idenau and lastly sandy and muddy area from Sandje to Bakassi peninsula. This geostructure has important influence on the type of gears, boats and engines used, on target species and by-catch species.

Boats and engine characteristics

The boat sizes vary from small boats (4-8 m long), large boats (8-12 m long) and extra-large boats (12-22 m long). This variation is according to the type of gear and engine. The smaller size and larger size are commonly used for Gillnet, Long line, Trawl, Beach seine and Hook line; therefore the extra-large are commonly used for surrounding seine, The distribution of engine size is also according to the type of boats, the smaller size and larger size commonly used 9.9, 15, 20 and 25 HP, but extra-large only use 40 HP to cover long distances and heavy load charge with more than 20 fishermen per boat. The average size of engine varies from 10 HP to 21.33 HP but most are between15HP.

Fishing gears distribution and target species

The Gillnet, Long line, Beach seine, Hook line are commonly used on sandy and rocky areas in Kribi, Tiko and Limbe areas but the Gillnet, Long line, Beach seine, Hook line, Trawl and Surround seine in sandy and muddy areas around Douala because of the abundance of species in Clupeidae and Peneidae families. There are more than 21 family species commonly caught on Cameroon coast, the main on caught are species in families of Clupeidae, Sciaenidae and Ariidae.

Fishing patterns

Most fishermen along Cameroon coastline are always involved in the activity, mostly from Monday to Saturday, on Sunday, they staying in their houses for cultural matters or to repair boats and nets. The peak of their activity is always with the area. The highest fishing effort is the dry season in Kribi and Douala areas but in Tiko and Limbe areas is the rainy season. The fact in Kribi and Douala areas, certain fishermen are Cameroonians and they are involve in other activities such as farming, while in Tiko and Limbe areas, most are essentially

foreigners and commonly involved in fishing activities along the week (Figures 2-4).

Number of fishing trips per month

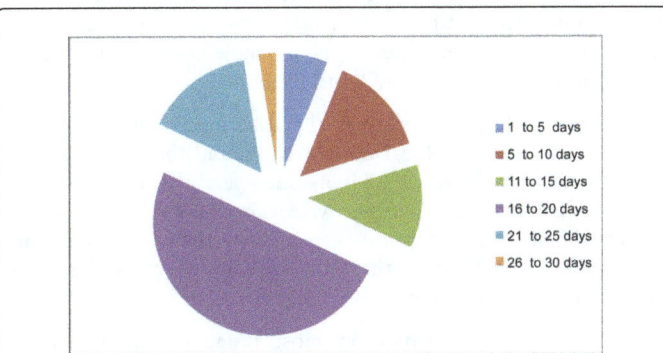

Figure 2: The distribution of the number of fishing days per month along the Cameroon Coastline

Fishing seasonal patterns

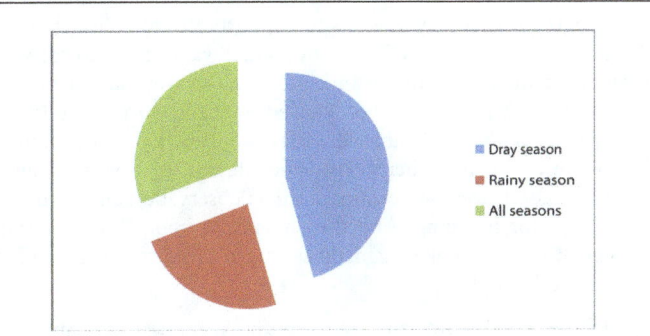

Figure 3: The seasonal distribution of fishing trips along the Cameroon Coastline

Duration of fishing trips

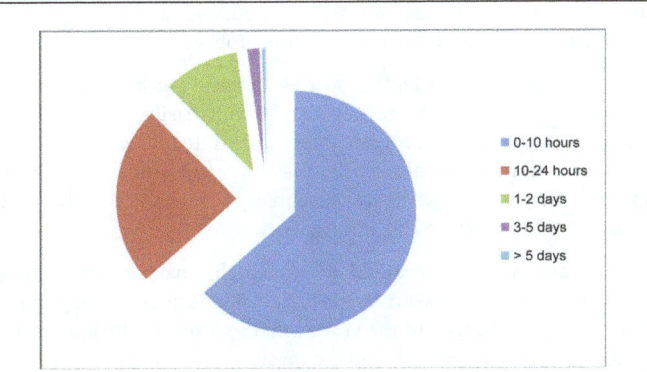

Figure 4: The timely distribution of fishing trips along Cameroon coastline

Non target species

In total we have 1241 turtles caught per year with 13 leatherback and 1228 are green, hawksbill and olive species. Turtle's meat is common in feeding habits of coastal people in Cameroon, but their catch is in majority not international. However in certain cases, these reptiles are caught internationally as in Sandje where results present 400 individuals per year by traditional fishermen.

Cetaceans are commonly caught by fishermen on the Cameroon coastline particularly in the Douala area around the Wouri estuary. But, until now, results reveal 97 individuals yearly. Most species were caught by gillnet and fishermen were not able to identify species accurately. This total may have been under estimated because most fishermen know that the caught of these species are allowed and they are afraid to give right information.

Manatee meat is common in most restaurants in Cameroon, particularly in the Sanaga estuary which can be considered as the sanctuary of this specy. This specy is either caught intentionally by nets, harpoon, trap, poison and none intentionally by gillnet around the lagoons and rovers estuaries in the Douala area.

Discussion and Recommendations

In general, this survey gave us the opportunity to collect baseline information on by-catch in Cameroon. But, it was difficult to convince certain fishermen particularly foreigners whose were afraid because most of by-catch species are not allowed by the law. In future it shall be good in each port to conduct this survey during the peak of activity, when there are many fishermen on the site and to have permanent collectors in each area that can make direct observation and involve in fishing trips for mapping. And then it shall be good before carrying this type of work to make a general census of all ports which can help to make good sampling.

Gillnet and Surround seine are common gears used in Cameroon; most national boats are none motorized while foreigners are quite organized in semi industrial companies with motorized boats. Gillnet, longline, hookline, beach seine are commonly used around sandy and rocky areas while surround seine is common in muddy areas particularly the fishing of species in Clupeidae families. It was difficult to have the exact number of boats because most are not registered in official documents, the survey was carried out at the low peak of fishing effort and also fishermen are quite mobile.

It was not easy to identify species because most fishermen are illiterates, as for the sea turtles they didn't easily identify green, hawksbill or olive, all of them were called turtle with back and leatherback turtle without back. Globally Cetaceans were rarely captured, and mortality was low essentially for whales but this study has, however, provided so far information.

- This assessment was useful to have preliminary information on by-catch in country; this survey was relevant and it could be good for the results and feedback to the Government service for instauration of scientific observers along Cameroon coast;

- In future, for good continuation of this survey it shall be good to plan activities during the fishing period and involve permanent data collectors for direct observation;

- And lastly it shall be good to involve more fishermen data collectors and to be involved in fishing trip.

Conclusion

The Impact Assessment on By-catch Artisanal Fisheries: Sea Turtles and Mammals in Cameroon, West Africa study gives us a lot of information's about the characteristics of artisanal fisheries and by catch along Cameroon coastline. This survey provides evidence of marine mammals and sea turtles by-catch in artisanal fisheries in Cameroon. But the study constitutes the basely information in the country and could be improve in future with other studies in long period according opportunity of funds available. Also other studies could be carrying out including industrial fisheries to compile more information in the thematic along the country.

Acknowledgements

Thanks are due to all who were involved in this survey particularly Dr. Minbang from Ministry of Fisheries and Livestock's in Provincial Delegation of Littoral, our data collectors Bebea Clotilde, Nking Gwendoline, Afana Dieudonne, Nga, Jiofack Bernadin, Romarius, all local Chiefs of fishing and fishers we visited. We can also thank Drs: Rebecca, Jeff Moore, Tara Cox,Sarah and Larry B. Crowder from Marine Lab in Duke University in North Carolina for the grants.

References

1. Shigueto JA, Mangel JC, Seminoff JA, Dutton PH (2008) Demography of loggerhead turtles Caretta caretta in the southeasthern Pacific Ocean: fisheries-based observations and implications for management. Endangered Species Research 5: 129-135.

2. Ayissi I, Angoni H, Amougou, et Fretey J (2007) Preliminary Assessment of the Impact of Artisanal Fishing on Sea Turtles along the Cameroon Coastline (West Africa). USA.

3. Moore JE, Cox TM, Lewison RL, Read AJ, Bjorkland R, et al. (2010) An interview-based approach to assess marine mammal and sea turtle captures in artisanal fisheries. Biological Conservation 143: 795-805.

4. Sayer JA, Harcourt CS, Collins NM (1992) The Conservation atlas of tropical forest Africa. Macmillan Publishing Ltd. London.

5. Morin MK (1989) Le Littoral Camerounais: problèmes morphologiques. Trav Labo Goegr Phys appliquée. National de Gestion de Environnement.

6. Boye M, Baltzer F, Caratini C (1974) Mangrove of the Wouri estuary. Int Symp Of biology and management of mangrove. Honolulu. 435-455.

7. Crosnier (1964) Fonds de pêche le long des côtes de la République Fédérale du Cameroun.

8. Tornier (1902) Die Crocodile Schildkroten und Eidechsen in Kamerun. Zool Jalrb 15: 163-677.

9. Loveridge A, Williams EE (1997) Revision of the African Tortoises and Turtles of the S border Cryptodira. Bull Mus Comp Zool 115: 163-557.

10. Ayissi I, van Waerebeek K, Segniagbeto G (2011) Report on the exploratory survey of cetaceans and their status in Cameroon.

11. Ajonina GN (2008) Inventory and modelling mangrove forest stand dynamics following differentlevels of wood exploitation pressures in the Douala-Edea Atlantic coast of Cameroon Central Africa. Mitteilungen der Abteilungen für Forstliche Biometrie Albert-Ludwigs- Universität Freiburg.

12. Din N (2001) Mangroves du Cameroun: statut écologique et perspectives de gestion durable.

23

Stock Assessment of Indian Scad, Decapterus Russelli in Pakistani Marine Waters and Its Impact on the National Economy

Muhammad Talib Kalhoro[1], Mu Yongtong[1]*, **Kalhoro Muhsan Ali[2], Shah Syed Babar Hussain[1], Memon Aamir Mahmood[1], Mohsin Muhammad[1]** and **Pavase Tushar Ramesh[3]**

[1]*Ocean University of China, College of fisheries, Qingdao, Shandong, China*

[2]*Faculty of marine Sciences, Lasbela University of Agriculture, Water and marine Sciences, 90150, Balochistan, Pakistan*

[3]*College of Food Science and Engineering, Seafood Safety Lab, Ocean University of China, Qingdao, 266003, China*

***Corresponding author:** Mu Yongtong, Professor, Ocean University of China, College of fisheries, Yushan road number 05 Ocean university of China Yushan Campus, Qingdao, Shandong 266003, China, E-mail: ytmu@ouc.edu.cn*

Abstract

The stock assessment, of Indian scad *Decapterus russelli* (Ruppell, 1830) from the northern Arabian Sea in Pakistan was evaluated. The samples of Indian scad (13300 specimens), ranging from 1-31 cm (FL) and 1-400 g (TW) were collected from the commercial fish landing center located at Karachi harbor. The parameters of fish length-weight relationship were calculated from the aggregated data as a=0.0323, b=2.66 with R2=0.954, indicating slightly negative allometric growth between the relationship. The length frequency samples from September 2013 to November 2014 was analyzed using FISAT II software, including the ELEFAN-I method. The growth parameters obtained using ELEFAN I was: L_∞=32.55 cm, K=0.750 per year, t_0=-0.678 with an Rn value of 0.220. Natural, total and fishing mortality M=1.42 per year, Z=3.84 per year at CI of 95% (CI=3.11-4.58) where F=2.422 per year and exploitation relation E=0.630 were obtained. Growth performance indices for L_∞ and W_∞ were performed using FiSAT-II program in order to estimate the limit and target reference points of stock exploitation were, Φ'=2.900 per year and Φ=0.170 per year, respectively. The results revealed that the natural fishing level of D. russelli (1.42 per year) was higher than the biological reference points $F_{0.1}$ (0.85) and F_{max} (0.9). Moreover the proportion of current mortality $F_{current}$ obtained was 0.630, representing that stock of *D. russelli* as highly exploited. It can be concluded from this study that the population parameters and the stock of *D. russelli* showed overexploitation in the northern parts of the Arabian Sea coast of Pakistan.

Keywords: Indian scad; Economic management; Stock assessment; Exploitation status; Pakistan

Introduction

Fish is one of the important sources of nutrition providing high quality proteins and a wide variety of essential micro nutrients, trace minerals, vitamins and fatty acids, even commonly utilized as healthy foodstuff around the globe [1]. There are more than 20,000 different fish species in the world and their utilization is more dependent on the availability in the local presence [2]. In addition, fish is also regarded as one of the widely consumed food in the coastal cities globally [3]. Seafood export play a vital role in in Pakistan's national economy. The exports of fish and fish preparations have been decreased by 7.30 percent in quantity and in value have been decreased by 5.28 percent during 2015. Gross Domestic Product (GDP) growth through this sector recorded in Pakistan was 2.9 and 2.7 during 2014 and 2015, respectively. Although Fisheries share in GDP is very little but it adds substantially to the national income through export earnings, [4]. One way to increase the role of fisheries in national GDP is to put a stop to over-exploitation of fish stocks. Thus the country has the potential to become a major producer of seafood, not only for local consumption but for the global market as well. Currently, about 400,000 people are directly engaged in fishing in Pakistan and another 600,000 in the ancillary industries, Ebrahim, [5]. Pakistan fisheries sector has an important implication towards other sectors to reduce the ongoing pressure on demand of food [6]. In the coastal regions of Sindh and

Baluchistan province, at least 90% of inhabitants are dependent on the activities related to fisheries and other fishing activities, Siddiqi [7]. Besides ocean, Pakistan is blessed with a large number of aquatic resources, including freshwater lakes, reservoirs, ponds, natural depressions, irrigation canals, waterlogged areas, rivers, and streams, contains a wide variety of commercial fish and shell-fish species [8]. The Indian scad, *Decapterus russelli* belonging to Carangidae family is a benthopelagic marine species [9]. *Decapterus russelli* is considered as a majorly important fishery resource and locally named as "seem" in the regions of Sindh and Balochistan province of Pakistan [10]. This is the most common *Decapterus* species in the western Indian Ocean. The fish forms large schools in water not exceeding 100 meters depth. *Decapterus* species reaches maturity during the first year of life, at about 10 cm total length and feeds on small planktonic invertebrates. It reaches at 35 cm (FL) and common length is 20 cm. *Decapterus* species is commonly found in large quantities in the local markets of Pakistan and is very popular in other Asian countries, Bianchi [10]. It is one of the main coastal demersal target species of commercial interest in the Northern Arabian Sea particularly in Pakistan, Bianchi [10]. This specie is most common and exclusively caught by purse seines and trawls operating in the shelf of the Arabian coast in muddy and sandy bottoms, Bianchi [10]. Generally, it is sold either fresh, or dried and sometimes salted as well as marketed as frozen and canned, Frimodt [11]. To best of our knowledge, least studies are carried out for stock assessment of this species in Pakistani marine waters. Although various studies regarding its biology, population dynamics, and stock assessment have been accounted for Indian scad fish species from all over the world mostly from Indian waters [12-14]. Despite of its economic importance, no effort was made to assess the fishery, biology and stock characteristics of Indian scad from Pakistani marine waters. Therefore, present research carried out for the first time towards evaluation of population and stock assessment parameters for *D. russelli* conducted from the Arabian Sea coast of Pakistan.

Growth, mortality, and recruitment parameters are essential for the assessment and management of fish stocks. Since these parameters determine the catch, the annual amount of fish exploited in fishery resources. Recently, numbers of research reports were conducted on the number of fish species and suggest some management steps to maintain the fish stock in northern Arabian Sea using FiSAT package [6,15-20]. This study is aimed to assess the impact of fishing pressure on fisheries resources and annual stocks of *D. russelli* along the northern Arabian Sea of Pakistan. Thus, our study is mainly focused on the annual stock, population structure and dynamics, growth, mortality, biomass and the production of Indian scads. The length frequency distribution data mainly used to determine the stock status to regulate the fishing efforts to maintain the fish stock. The output of population dynamics gives indications on the level of exploitation and the indicators of declining stocks. Hence, the generated information could be used as an input in ecosystem-based fisheries management models in Pakistani waters, which was not available previously. The present study will contribute to know the stock status of *D. russelli* fishery from Pakistani waters and suggest a strategy for better management.

Material and Methods

Data collection and sampling

Total of 13300 individual samples (both sexes combined) of *D. russelli* were obtained monthly from September 2013 to November 2014 at the Karachi fish harbor, Sindh province (Figure 1). The coastline of Pakistan is about 1120 km, which represents 772 km of Baluchistan and Makran coast to the Iranian border and 348 km measures the Sindh coast that extends to the Indian border (Figure 1) [21]. The length frequency and length weight were measured for further analysis. The total length was measured in cm and weight was taken in grams (g).

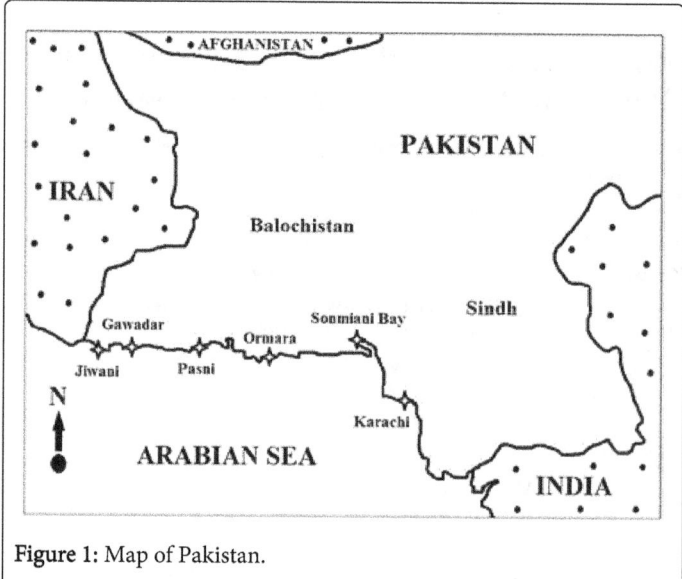

Figure 1: Map of Pakistan.

Data analysis

The pooled length frequency distribution data of both sexes combined was prepared on the monthly basis using FiSAT-II (FAO ICLARM stock assessment tool) computer software package [22]. Then the length data was merged and grouped into 1 cm class intervals in order to estimate the parameters of growth, mortality, growth performance index, yield per recruit and biological reference points, methods available in FiSAT package. Biological reference points are widely used for the management and conservation of fisheries resources nowadays, Haddon [23]. Biological reference points (BRPs) have been defined as the level of fishing mortality and/or of biomass.

Length-Weight Relationship

The length weight relationship data of *D. russelli* was calculated by the power function equation

W=aLb [24,25]

where "W" is the total weight (g),

"L" is the total length (cm),

"a" is the intercept

"b" is the slope.

Growth parameters

For the preliminary estimations of asymptotic length (L_∞) and growth constant (K) the length frequency distribution data was used in ELEFAN-I. The growth coefficient of *D. russelli* was estimated by fitting the von Bertalanffy growth function (VBGF). The van Bertalanffy growth equation was defined by Haddon [23] as:

$$L_t = L_\infty\left(1 - \exp\left(-K\left[t - t_0\right]\right)\right)$$

Where, L_t is the length at the predicted time t, L_∞ is the asymptotic length, K was the growth coefficient and t_0 is the hypothetical age or time where length was equal to zero. The value of t_0 is estimated by the empirical formula by, Pauly [26] as:

$$\log_{10}(-t_0) = -0.3922 - 0.275 \log_{10} L_\infty - 1.038 \log_{10} K$$

Mortalities rates

The length of the converted catch curve [26] was used to estimate instantaneous total mortality (Z), natural mortality (M) and fishing mortality (F) by using FiSAT package. The merged monthly data of length frequency distribution was arranged to obtain catch curve and natural logarithm (ln) of the number of individuals with respect to the age group (N) were designed against the results of their relative age (t), Pauly [26]. In order to obtain independent estimates of natural mortality (M), the subsequent formula of Pauly [27] was used as:

$$\text{Log}_{10} M = 0.0066 - 0.279 \log_{10} L_\infty + 0.654 \log_{10} K + 0.4634 \log_{10} T$$

The annual average sea surface temperature (SST) was taken as 27°C, because it was the average monthly water temperature. Fishing mortality (F) was derived by subtracting Z from M. The ratio F/Z can also be used to obtain the exploitation ratio (E).

Biological reference points

The biological reference points (BRP) was estimated by, Gulland [28] method, according to the optimum fishing mortality rate $F_{opt} = 0.5M$. The most well-known biological reference points are $F_{0.1}$ and F_{max}, they are commonly used for fisheries management [29]. The target biological reference point F_{max} is considered as a function of fishing mortality (F) for a definite exploitation pattern against the maximum value of yield per recruit (Y/P).

Beverton Holt yield recruit model

The relative yield per recruitment is analyzed by the model of Beverton-Holt yield per recruit with the knife edge selection in FiSAT-II. Yield per recruit was estimated by [30] model with the formula as under:

$$Y_w/R = FW_\infty e^{-M(t_c - t_r)} \sum_{n=0}^{3} \frac{Q_n e^{-nK(t_c - t_0)}}{F + M + nK}(1$$
$$- e^{-(F+M+nK)(t_\lambda - t_c)})$$

where, Y_W/R is yield per recruit, t_c is the average age of first capture, t_r is the age of recruitment, t_λ is the asymptotical ages, Q_n was the constant and equal to 1, -3,3 and -1 when n is 0,1, 2 and 3 correspondingly [31].

Growth performance index

The growth performance index (Φ') helps to explain the characteristics of the different ecosystems of the stock or housing of the different population of the environment [32]. Growth performance index is conducive in both movement (K and L_∞) between species and growth. To compare the growth, we used the phi prime (Φ') performance index of overall growth of Pauly and Munro [33]. In this model, the calculated values of L_∞ and K were used to estimate the asymptotic length (L_∞) and asymptotic weight (W_∞) from the routine below of the equation [33]:

$$\Phi' = \log_{10} K + 2 \log_{10} L_\infty \quad \text{and} \quad \Phi = \log_{10} K + 2 \div 3 \log_{10} W_\infty$$

Results

Length-weight relationship

Total of 997 (both sexes male and female) pairs of *D. russelli* species length and weight were observed in this study. The length range were from 1 to 31 cm (FL), the total weight ranged from 0.5 to 388 g. The foremost length ranged of *D.russelli* were from 9 to 18 cm FL (Figures 2 and 3). The average length and weight is 14.048 (± 4.775) cm (FL) and 99.082 (± 44.784) g (TW) correspondingly (Table 1). The combination of total length-weight relationship of both sexes was calculated as

$$W = 0.0323 L^{2.66} (R^2 = 0.954), n = 997$$

ML	2013			2014							
	Sep.	Oct.	Nov.	Feb.	Mar.	Apr.	June	Aug.	Oct.	Nov.
1				1		50			40	
2				1	1	102			123	3
3				10	1	100			119	41
4				8	1	80			97	113
5				20	1	65	1		85	169
6				25	3	55	1		55	157
7	5	3		22	19	3	22		3	134
8	15	34	20	15	59	1	73		1	85

9	170	488	300	8	37	3	45	1	3	90
10	380	1493	800	8	13	1	26	1	1	117
11	500	878	425	7	11	5	16	1	5	67
12	90	86	40	5	8	1	21	1	1	211
13	40	47	29	3	1		17	3		167
14	55	33	18	1		1	24	19	1	45
15	60	47	20			6	8	59	6	189
16	80	171	50			4	1	37	4	577
17	100	380	180			7	1	13	7	663
18	90	370	200			1		11	1	456
19	70	98	70					8		190
20	5	17	20			1		1	1	54
21	1	2	18							7
22	1	1								2
23										3
24										4
25										6
26										4
27										3
28										1
29										1
30										1
31										1
Σ	1662	4148	2190	134	155	486	256	155	553	3561

Table 1: Length-frequency data of Decapterus russelli from September 2013 to November 2014 in the northern part of the Arabian Sea.

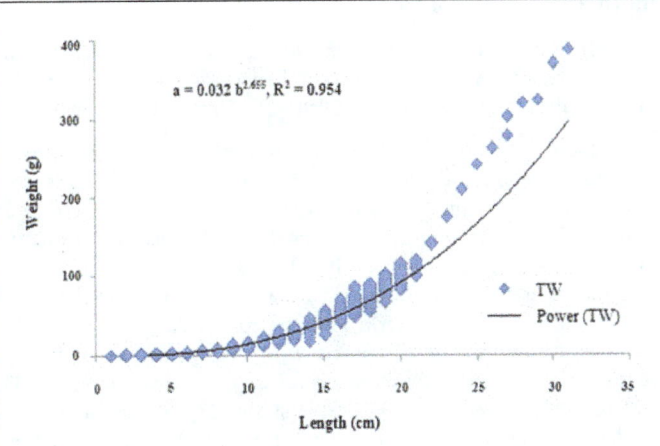

Figure 2: The Length-weight relationship of both sexes combined of *D. russelli* length and weight ranging from 1 to 31 cm (FL) and 0.5 to 400 g.

Growth parameters

A total of 13,300 length frequency distribution data was used to estimate the growth parameters by ELEFAN method. The von Bertalanffy growth parameters for *D. russelli* was estimated as $L_∞$=32.55 cm (FL) and K=0.750 per year (Figure 4). The t0 value was calculated by equation of Pauly as t_0=-0.678 per year. The R_n (goodness of fit) was estimated to be at 0.220 with ELEFAN-I method, Pauly [26].

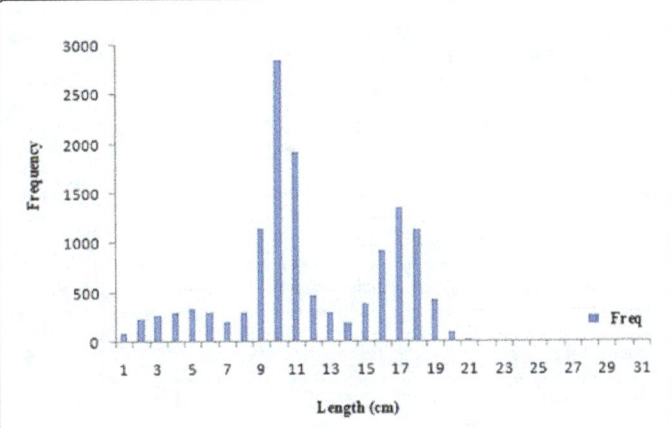

Figure 3: Length frequency distribution (n=997) ranging from 1 to 31 cm (FL) and the dominant length frequency range from 9 to 18 cm.

Mortality rate parameters

Applying the length converted catch curve analysis VBGF growth parameters ($L_∞$=32.55 cm (FL) and K=0. 750 per year) as the input value for the estimation of the mortality parameters of Z=3.84 per year of the total mortality (Z) estimates and it was estimated at 95% confidence interval (CI=3.11-4.58) (Figure 5). The value of natural mortality (M) was calculated as M=1.42 per year using annual average sea surface temperature (SST) 27°C. Thus, fishing mortality was

calculated as F=Z-M=2.422 per year and exploitation ratio (E) was selected from F/Z=0.630 per year.

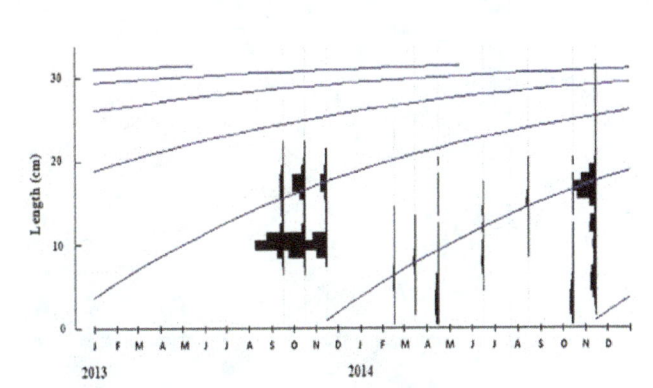

Figure 4: The total length, von Bertalanffy growth curve in this study during 2013-2014 estimated ($L_∞$=32.55 cm and K=0.750 year-1, t_0=-0.678).

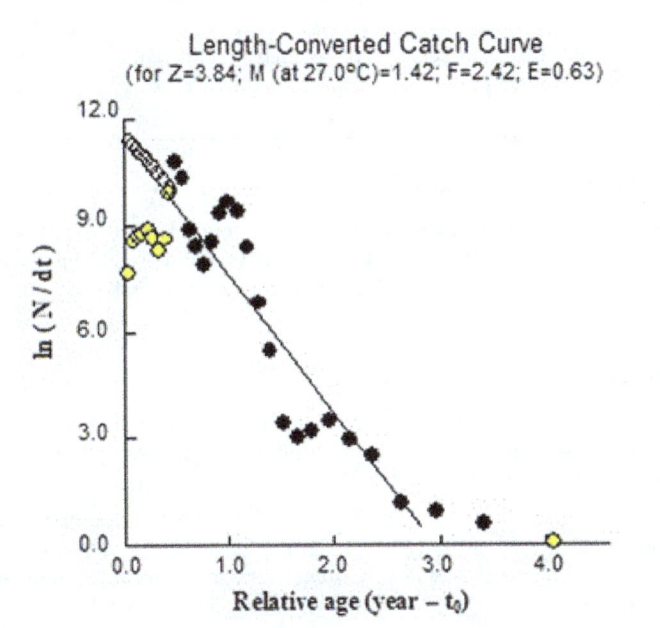

Figure 5: A Length converted catch curve analyzed for *D. russelli* using input value of VBGF growth parameters (the von Bertalanffy growth).

Biological reference points

The yield per recruit analysis representing, when the t_c was assumed to be 1, the maximum frequency F_{max} was estimated at 0.9 and $F_{0.1}$ was at 0.85 (Figure 6); therefore $F_{current}$ 2.422 per year was greater than the $F_{0.1}$ and F_{max} (Figure 1). The stock of D. russelli in marine waters of Pakistan severely overfished. Using [34] biological reference point F_{opt} M was 1.42 per year. The current fishing mortality obtained 2.422 per year was higher than the reference points obtained in Pakistan waters for the *D. russelli*.

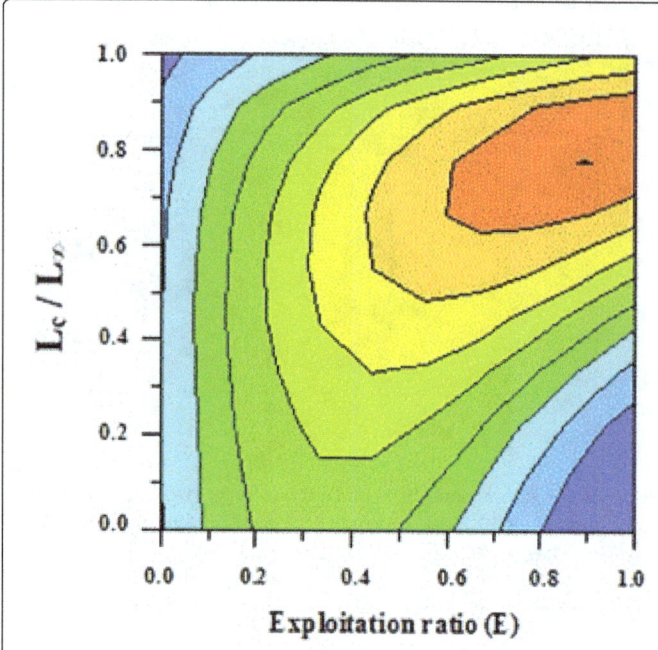

Figure 6: Yield per recruit contour map of from Pakistani waters during 2013-2014.

Growth performance index

Growth performances indices asymptotic length (L_∞) and asymptotic weight (W_∞) were $\Phi'=2.900$ per year and $\Phi=0.170$ per year were carried out for *D. russelli* from marine water of Pakistan during 2013-2014, respectively.

Beverton-Holt Y/R analysis

Von Bertalanffy growth model was used to estimate the asymptotic length growth factors such as asymptotic length and growth coefficient $L_\infty=32.55$ cm (FL) and $K=0.750$ per year by the equation of $L_t=L_\infty (1 - \exp(- k [t - t_0]))$ respectively. The value of the time t, t0 is the hypothetical age t0=- 0.678 when the length of the virtual age is considered zero. Value of t0 was estimated by the empirical formula of Pauly [26]. $\text{Log}_{10} (-t_0) = -0.3922 - 0.275 \text{Log}_{10} L_\infty - 1.038 \text{Log}_{10} K$.

Discussion

The Indian scad is one of the most important small pelagic fishes supporting the commercial fishery in Pakistan. This species has a high market demand locally due to its cheaper price relative to other pelagic fishes. Despite its significant contribution to the fishery and economic value, there are no adequate data pertaining to this species in north Arabian Sea. This study was undertaken to investigate the population dynamics and fishery of the *Decapterus russelli*. The objectives of the present study were to establish the population parameters and fishery demographics towards management practices by providing significant input in decision making for sustainable management of the fish stocks.

Length weight relationship

Length-weight relationship was mostly used for the fish growth and stock assessment [35]. The weight ratio between the lengths of the fish is very significant for the biology of fishes, Le Cren [24].

The slope values of b for *D. russelli* of both sexes combined were estimated in the present study at a=0.0323, b=2.66 with $R^2=0.954$ from Pakistani waters during 2013-2014. The slope b values range is from 2.5 to 3.5. The values higher than 2.5 shows that fish has isometric growth, whereas if fish has slop b values lower than 2.5 it may be considered that fish has allomatic growth [25,36]. Present value b show that fish have isomateric growth from Pakistani waters. The present study results were compared with previous studies in Table 2. The slope b values from Yemen, Gulf of Aden and Red Sea was 2.033 and 2.167 were lower than the present study [37]. While the b values from lagoon, New Caledonia was 2.948 [38], 2.963 in New Caledonia [39], in Indonesia Tegal's water 2.879 [40] and 2.989 in Vizhinjam, India [41], which were found to be close to the current study. On the contrary, the b value 3.000 in java Sea, Indonesia [42], 301.5 Philippines [43], and in Sofala bank, Mozambique 3.026 [44], was high than the current study However, the overall b values from different part of the world is within the range of present study (b=2.66) conducted from Pakistani waters.. Small difference in the b values, may be because of seasonal variations, environmental parameters and sample collection, or, number of individuals examined in the study, the range of the length observed to be different during the study [35].

Reference	Research area	A	b	R^2
Al Sakaff and Esseen, 1999	Gulf of Aden and Red Sea	0.11	2.033	0.99
Al Sakaff and Esseen, 1999	Gulf of Aden and Red Sea	0.08	2.167	0.97
Letourneur et al., 1998	lagoon	0.01	2.948	0.99
Kulbick et al., 2005	New Caledonian	0.01	2.963	0.99
Burhanuddin et al., 1983	Tegal	0.01	2.97	-
Sreenivasan., 1981	Vizhinjam	0.02	2.989	-
Widodo., 1988	Java Sea	0.01	3	0.96
Ronquillo., 1975	Philippines	0.01	3.015	-
Brinca et al., 1983	Sofala Bank	0.01	3.026	-
Reuben et al., 1992	east coast	0.01	3.111	-
Gjøsaeter and Sousa., 1983	Guimaras Strait	0.01	3.12	-
Gjøsaeter and Sousa., 1983	Sofala Bank	0.01	3.121	0.86
Reuben et al., 1992	south-west coast	0.01	3.136	-
Reuben et al., 1992	north-west coast	0.01	3.207	-
Present study 2015	Pakistan	0.03	2.66	0.95

Table 2: A comparative study of length weight relationship parameters of *Decapterus russelli* from the different regions of the world.

Growth parameters

Length frequency distribution data were used to evaluate VBGF parameters, namely the asymptotic length (L_∞), growth rate (K), the growth performance index (Φ ') and imaginary or hypothetical age (t_0). The present study results were compared to the previous studies from different regions (Table 3). The asymptotic length (L_∞), growth rate (K) and growth performances indices for asymptotic length (Φ ') were calculated at 19.4, 0.75 and 2.45 from northern Arabian Sea, Pakistan, respectively [45]. The values of L_∞ and Φ ' were lower than the present study while the value of K was close enough. The L_∞, K and Φ ' values from Jave Sea, Indonesia were 28.4, 0.90 and 2.86 correspondingly [42] 30.0, 0.54 and 2.69 from Manila bay, Philippines [46], 32.2, 0.86 and 2.95 from Central area, Malaysia, Bogdanov [47], were estimated by ELEFAN methods (Table 3) and were close to the present study (32.55, 0.750 and 2.900). In Palawan, Philippines the values of L_∞, K and Φ ' were 33.7, 0.36 and 2.69 [46], L_∞ was higher and values of K and Φ ' were lower Due to the correlated parameters [48], a higher K value is normally associated with a lower L_∞ value. Differences shown in Table 3 may be because of the sampling procedure, variety of data, and the differences in their lifestyle and ecological characteristics of fish [49].

Resource	Locality	l∞	K	t0	Φ'
Iqbal., 1991	Northern Arabian Sea	19.4	0.75	-	2.45
Reuben et al., 1992	east coast	22.1	0.71	-	2.54
Reuben et al., 1992	south-west coast	24.8	0.78	-	2.68
Reuben et al., 1992	Northwest coast	29.9	0.45	-	2.6
Pauly., 1978	Manila Bay	23.3	1.13	-	2.79
Murty., 1991	Kakinada	23.2	1.08	-0.08	-
Prathibha and Shanbhogue., 2005	Karnataka coast	23.2	0.7	-0.16	-
Jarzhombek., 2007	north area	23.5	1.1	-	2.78
Isa., 1987	Penang	24	0.81	-	2.67
Isa., 1987	akarta Bay (Seribu Island)	27	1.15	-	2.92
Isa., 1987	Perlis	27	1.01	-	2.87
Jaiswar et al., 2001	Mumbai (Bombay) waters	24	1.42	-	2.91
Suwarso et al, 1995	Java Sea	24.5	0.95	-	2.76
Suwarso et al., 1995	Java Sea	25.2	1.08	-	2.84
Gjøsaeter and Sousa., 1983	Sofala Bank	24.8	0.56	-0.1	2.54
Ingles and Pauly., 1984	Palawan	26	0.73	-	2.69
Ingles and Pauly., 1984	Nansha Island	26	0.52	-	2.55
Ingles and Pauly., 1984	Java Sea (Seribu Island)	26.6	0.95	-	2.83
Ingles and Pauly., 1984	Manila Bay	30	0.54	-	2.69
Ingles and Pauly., 1984	Palawan	33	0.45	-	2.69
Chen., 2003	Vizhinjam	26	0.19	-	2.1
Dwiponggo et al., 1986	Idi, Malacca Strait	26	0.9	-	2.78
Dwiponggo et al., 1986	Palawan	26.9	0.69	-	2.7
Dwiponggo et al., 1986	Manila Bay	27	0.8	-	2.77
Sreeenivasan., 1982	Vizhinjam	26	0.185	-0.5	-
Sousa., 1992	Jakarta Bay (Seribu Island)	27	1.18	-	2.93
Sousa., 1992	Sofala Bank and Boa Paz	27.3	0.68	-	2.7
Rodriguez and Sousa., 1988	Mozambique	27.8	0.57	-0.18	2.65
Rodriguez and Sousa., 1988	Mozambique	27.9	0.56	0.18	2.64
Widodo., 1988	Java Sea	28.4	0.9	-	2.86
Bogdanov and Jarzhombek., 2004	central area	28.4	0.56	-	2.65
Bogdanov and Jarzhombek., 2004	north area	28.4	1.08	-	2.94
Bogdanov and Jarzhombek., 2004	central area	32.2	0.86	-0.04	2.95
Balasubramanian and Natarajan., 2000	Vizhinjam	29	0.8	-0.04	-
Jabat and Dalzell., 1988	Camotes Sea	33.7	0.36	-	2.61
Padilla., 1991	Guimaras Strait	33.7	0.65	-	2.87
Lavapie-Gonzales et al., 1997	Camotes Sea	35.1	1.4	-	3.24
Present study., 2015	Pakistan	32.55	0.75	-0.67	2.9

Table 3: Evaluation of current growth parameters of *Decapterus russelli* from different parts of the world.

Growth parameters were estimated by using the non-parametric method ELEFAN, which is mostly used for analyzing length frequency of fish, which is essentially ad-hoc and not dependent on convention of parameters of direct subgroups. Therefore, it makes only feeble assumptions about the dissemination of the size of the cohort. The length of each cohort model is fixed lying on a curve described by the model of growth as von Bertalanffy growth model, so it makes a powerful assumption of growth, Pitcher [50].

Mortality rate

The length-converted catch curve analysis method was used with input values of VBGF growth rate parameters for *D. russelli*. These values were matched up with the earlier work in various regions of the world, respectively (Table 4). The overall mortality rate was estimated showing only the dark circles in Figure 5. The value of the mortality rate is shown in Table 5 and Figure 7, where the total mortality rate was observed as the highest mortality in March 2014, while during November 2014 showed lowest mortality.

Source	Area	Z	M	F	E
Murty., 1991	Kakinada	6.65	1.9	4.75	0.71
Jaiswar et al., 2001	Mumbai	7.75	2.63	5.1	0.66
Manoj Kumar., 2007	Malabar	3.79	2.08	1.71	0.49
Debabrata Panda et al., 2012	Mumbai	4.61	1.81	2.8	0.61
Nalini et al., 2011	Mumbai waters	6.66	2.1	4.56	0.68

Reuben et al., 1992	East coast of India	2.83	1.35	1.48	0.52
Reuben et al., 1992	N. W. coast of India	2.85	0.83	2.02	0.71
Reuben et al., 1992	S. W. coast of India	3.88	1.26	2.62	0.68
Present Investigation., 2015	Pakistan	3.84	1.42	2.42	0.63

Table 4: Compare mortality parameters from Pakistani waters in 2013-2014 with other studies from different fields and areas of the world.

Sampling month	Z	M	F	E	95% ClZ	R2
Sep-13	5.88	1.42	4.46	0.76	3.41-8.34	0.738
Oct-13	5.64	1.42	4.22	0.75	2.67-8.61	0.613
Nov-13	8.39	1.42	6.97	0.83	0.59-16.19	0.604
Feb-14	6.72	1.42	5.3	0.79	5.27-8.16	0.945
Mar-14	12.66	1.42	11.24	0.89	7.16-18.15	0.911
Apr-14	9.84	1.42	8.42	0.86	6.86-12.82	0.827
Jun-14	7.17	1.42	5.75	0.8	4.63-9.70	0.841
Aug-14	6.74	1.42	5.32	0.79	3.38-10.11	0.931
Oct-14	4.99	1.42	3.57	0.72	2.66-7.32	0.563
Nov-14	2.83	1.42	1.41	0.5	1.89-3.77	0.748

Table 5: Instantaneous rates of mortality rates based on monthly data using the length converted catch curve analysis for *Decapterus russelli.*

In general the mortality values in Table 4 were higher compare to the present study; total (Z), natural (M), fishing mortality (F) and exploitation ratio from India Mumbai were 7.75, 2.63, 5.1 and 0.66 respectively [51]. The Z, M, F, and E values were also higher in different parts of India like Kakinada, Mumbai and Malabar Table 4 [12-14,52,53], compared to present study 3.84, 1.42, 2.42 and 0.63 respectively. While lower mortality rate values were found from east coast of India and the northwest coast of India [53]. The different mortality values from different part of the world maybe different countries have different demand of this species or maybe some ecological and environmental factors effecting on the mortality of fish, Gulland [34] noted that the exploitation rate should be lower than 0.5, while Patterson [54] describe that the exploitation rate should be maintained at 0.4 it the exploitation values exceeds 0.4 than it should be assumed that stock is overexploited state. According to Gulland [34] and Petterson [54] recommendations it should be concluded that the stock of *D. russelli* from Pakistani waters is in stress and overexploited state. Because of the current fishing rate 2.422 per year revealed higher exploitation rate than the biological reference points.

Biological reference points $F_{0.1}$ and F_{max}

Characteristically $F_{0.1}$ and F_{max} are two biological reference points (BRP), used in fisheries management around the world, which is based on the information of age and the length of the structure data and depends on the executive advice for improved management [29]. $F_{0.1}$ is defined as the rate of fishing mortality on a slight increase in yield per recruit (YPR) which is 10% of that F_0 and F_{max} is the mortality of fish, which is the highest YPR to be achieved [29] (Figure 7). Output of yield per recruit (YPR) analysis pointed out that when the tc was assumed to be 1, the maximum frequency F_{max} estimated was to be at 0.9 and $F_{0.1}$ was at 0.85; therefore $F_{current}$ 2.422 per year was greater than the $F_{0.1}$ and F_{max}, respectively. The stock of *D. russelli* in marine waters of Pakistan found to be severely overfished. Using the [34] biological reference point F_{opt} M obtained was 1.42 per year. The current fishing mortality was 2.422 per year much higher than the biological reference points in Pakistan waters.

It can be concluded from this study that the population parameters and the stock of *D. russelli* showed overexploitation in the northern parts of the Arabian Sea coast of Pakistan. In the present study, the estimated value of the current fishing mortality F_c=2.422 per year was higher than the biological reference points ($F_{0.1}$=0.85 and F_{max}=0.9) [55]. Marine resources of Pakistan are fully accessible without restrictions, with lacking actual administration and planning. The current study reflected that the stock of Indian scad is overfished [56-60]; we would recommend that some management action should be taken to reduce the fishing efforts in Pakistani waters. Sustainable fisheries management measures for this species should be observed during closing season in Pakistan to protect broods stock especially during the monsoon. Thus, fisheries managers need to take rapid action on these issues, so that our fish resources can thrive with more and set unshakable advantage for the national economic growth.

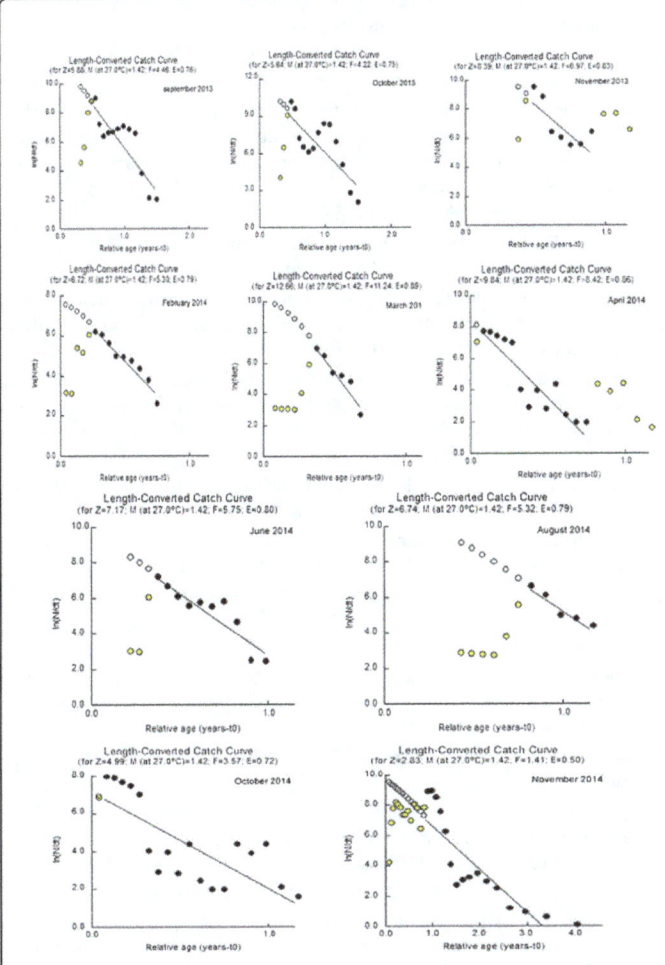

Figure 7: Length - converted catch curves for *D. russelli* from the monthly catch curve where the slope of the regression line catch curve.

Conclusion

The present study results showed that the stock of *D. russelli* fishery from Pakistani waters in overexploited state. Monitoring of the population numbers and harvest levels of this species is needed. To maintain the stock of this fishery the scientists and fishery managers have to work together for better fishery for coming future. In the light of present findings we may suggests that fishing activities must be controlled by trawl mesh size, discard of bycatch and proper check and balance of fishing and non-fishing seasons. Fishing boats must be registered under controlled authorities. Completely ban period we may suggest during peak breeding season that could save juveniles. Marine protected areas must be declared to save the fish nursery grounds. We also suggest further detailed studied maybe conducted like egg per recruit analysis for better stock management.

Acknowledgement

This work was supported by China Agriculture Research System (CARS-48-09B) Ministry of agriculture China, and the special research fund of Ocean University of China. We are very grateful to the referee for comments and suggestions, which greatly improved the manuscript. The first author would like to thank China Scholarship Council (CSC) to fund his doctoral degree.

References

1. Yoshida S, Ichimura A, Shiomi K (2008) Elucidation of a major IgE epitope of pacific mackerel paravalbumin. Food chem 111: 857-861.

2. Lopata A, Potter P (2000) Allergy and other adverse reactions to seafood. Allergy Clin Lmmunol Int 12: 271-281.

3. Sicherer SH, Sampson, HA (2010) Food allergy. J All Clin Lim 125: S116-S125.

4. Pakistan Economic Survey highlights. 2014-2015 (2015) Economic Adviser's Wing, Finance Division, Government of Pakistan: Islamabad.

5. Ebrahim Z (2015) Inside Pakistan's Untapped Fishing Industry. Inter Press Service News Agency.

6. Memon AM, Memon KH, Baloch WA, Memon A, Baset A (2015) Evaluation of the fishery status for King Soldier Bream Argyrops spinifer in Pakistan using the software CEDA and ASPIC. Chin J Ocean Limnol 33: 1-8.

7. Siddiqi AH (1992) Fishery resources and development policy in Pakistan. Geo Journal 26: 395-411.

8. Afsar N, Siddiqui G, Ayub Z (2012) Update of records of selected Prosobranch gastropod species found along the coasts of Sindh and Balochistan, Pakistan. Pak J Zol 44: 267-275.

9. Mohsin AM, Ambak MA (1996) Marine fishes and fisheries of Malaysia and neighbouring countries.

10. Bianchi G (1985) Field guide to the commercial marine and brackish-water species of Tanzania. Prepared with the support of PAK/77/033 and FAO (FIRM) Regular Programme. FAO, Rome p: 200.

11. Frimodt C (1995) Multilingual illustrated guide to the world's commercial warm water fish. Fishing News Books Ltd.

12. Murty VS (1991) Observations on some aspects of biology and population dynamics of the scad Decapterus russelli (Ruppell) (Carangidae) in the trawling grounds off Kakinada. J Mar Bio Ass Ind, 33: 396-408.

13. Panda D, Chakraborty SK, Sharma AP, Jha BC, Sawant BT, et al. (2012) Fishery and population dynamics of two species of carangids, Decapterus russelli (Ruppell, 1830) and Megalaspis cordyla (Linneaeus, 1758) from Mumbai waters. Ind J Fish 59: 53-60.

14. Manojkumar P (2007) Stock assessment of Indian scad, Decapterus russelli (Ruppell, 1830) of Malabar. J Mar Bio Ass Ind 49: 79-80.

15. Kalhoro MA, Liu Q, Memon KH, Chang MS, Jat AN (2013) Estimation of Maximum sustainable yield of Bombay duck Harpadon nehereus fishery in Pakistan using CEDA and ASPIC packages. Pakistan J Zool 45: 1757-1764.

16. Kalhoro MA, Liu Q, Waryani B, Panhwar SK, Memon KH (2014) Growth and mortality of brushtooth lizardfish Saurida undosquamis from Pakistani waters. Pakistan J Zool 46: 139-151.

17. Kalhoro MA, Liu Q, Memon KH, Chang MS, Zhang K (2014) Population dynamics of Japanese threadfin break Nemipterus japonicus from Pakistani waters. Acta Oceanol Sin 33: 49-57.

18. Kalhoro MA, Liu Q, Memon KH, Waryani B, Soomro SH (2015) Maximum sustainable yield of Greater lizardfish Saurida tumbil fishery in Pakistan using CEDA and ASPIC packages. Acta Oceanolo Sin 34: 68-73.

19. Kalhoro MA, Liu Q, Valinassab T, Waryani B, Memon KH, et al. (2015) Population dynamics of Greater lizardfish Saurida tumbil from Pakistani waters. Pakistan J Zoology. 47: 74-55.

20. Kalhoro MA, Tang D, Ye HJ, Morozov E, Liu Q, et al. (2017) Population dynamics of Randall's threadfin bream Nemipterus randalli from Pakistani waters, Northern Arabian Sea. Indian J Geo Mar Sci 46: 551-561.

21. FAO (2009) Fishery and Aquaculture Country Profile the Islamic Republic of Pakistan FAO's Fisheries Department Rome p: 1-18.

22. Gayanilo J, Sparre P, Pauly P (1996) FiSAT: FAO-ICLARM stock assessment tools. User's manual. FAO.

23. Haddon M (2010) Modelling and Quantitative Methods in Fisheries. Chapman & Hall/CRC, London.

24. LeCren E (1951) The length-weight relationship and seasonal cycle in gonad weight and condition in the perch (Perca fluviatilis). J Ani Eco 20: 201-219.

25. Froese R (2006) Cube law, condition factor and weight-length relationships: history, meta-analysis and recommendations. J Appl Ichthyol 22: 241-253.

26. Pauly D (1983) Some simple methods for the assessment of tropical fish stocks. Food and Agriculture Org.

27. Pauly D (1980) On the interrelationships between natural mortality, growth parameters, and mean environmental temperature in 175 fish stocks. J du Con 392: 175-192.

28. Gulland JA (1969) Manual of methods for fish stock assessment: Part 1 Fish population analysis. Food and Agriculture Organization of the United Nations.

29. Hilborn R, Walters CJ (1992) Quantitative fisheries stock assessment: choice, dynamics and uncertainty. Rev Fish Bio Fish 2: 177-178.

30. Beverton R, Holt S (2012) On the dynamics of exploited fish populations. Springer Science and Business Media.

31. Pitcher TJ, Hart P (1983) Fisheries ecology. Springer Science and Business Media 20: 341.

32. Baijot E, Moreau J (1997) Biology and demographic status of the main fish species in the reservoirs of Burkina Faso. Hydrobiological aspects of fisheries in small reservoirs in the Sahel region. Technical Center for Agricultural and Rural Cooperation ACP-EU, Wageningen: Netherlands pp: 79-110.

33. Pauly D, Munro J (1984) Once more on the comparison of growth in fish and invertebrates. Fishbyte 2.

34. Gulland JA (1970) The fish resources of the ocean p: 138.

35. Bagenal T, Tesch F (1978) Age and growth Methods for assessment of fish production in fresh waters.

36. Gayanilo F, Sparre P, Pauly D (2005) FAO-ICLARM stock assessment tools II: User's guide. Food and Agriculture Org.

37. Al Sakaff H Esseen M (1999) Length-weight relationship of fishes from Yemen waters (Gulf of Aden and Red Sea). Naga, the ICLARM Quarterly 22: 41-42.

38. Letourneur Y, Kulbicki M, Labrosse P (1998) Length-weight relationship of fishes from coral reefs and lagoons of New Caledonia: An update. Naga, the ICLARM Quarterly 21: 39-46.

39. Kulbicki M, Guillemot N, Amand M (2005) A general approach to length-weight relationships for New Caledonian lagoon fishes. Cybium, 29: 235-252.

40. Burhanuddin A, Martosewojo S, Mulyanto R (1983) Evaluation of potential and business of fish resource management (Decapterus spp.). National Oceanographic Institute-LIPI, Jakarta.

41. Sreenivasan P (1981) Length-weight relationship in Decapterus dayi Wakiya. Ind J Fish 28: 283-286.

42. Widodo J (1988) Population biology of Russell's scad (Decapterus russelli) in the Java Sea, Indonesia. FAO Fisheries Report (FAO).

43. Ronquillo I (1974) A review of the roundscad fishery in the Philippines. Proceedings Indo-Pacific Fisheries Council 15: 351-375.

44. Brinca L, Silva J, Sousa L, Sousa I, Saetre R (1983) A survey on the fish resources at Sofala Bank, Mozambique [Indian Ocean].

45. Iqbal M (1990) Length-based fish stock assessment method: Their application in estimation of the growth and mortality rates of Indian scad Decapterus russelli(Carangidae) in Pakistani waters: Northern Arabian Sea. Zoological Society of Pakistan, c/o Dr. A. R. Shakoori, Punjab University, Lahore (Pakistan) pp: 5-8.

46. Ingles JD, Pauly D (1984) An Atlas of the Growth, Mortality, and Recruitment of Philippine Fishes. World Fish.

47. Bogdanov G (2004) Compilation of studies on the growth of scombrids, horse mackerels, swordfishes and sailfishes. Russian Federal Institute of Fisheries and Oceanography, (in Russian).

48. Pauly D, Morgan G (1987) Length-based methods in fisheries research. ICLARM Conference Proceedings 13. International Center for Living Aquatic Resources Management, Manila, and Kuwait Institute for Scientific Research, Safat.

49. Adams P (1980) Life history patterns in marine fishes and their consequences for fisheries management. Fis Bul 78: 1-12.

50. Pitcher TJ (2002) A Bumpy Old Road: Size-Based Methods in Fisheries Assessment. Handbook of Fish Biology and Fisheries 2: 189-210.

51. Jaiswar A, Chakraborty S, Swamy R (2001) Studies on the age, growth and mortality rates of Indian scad Decapterus russelli (Ruppell) from Mumbai waters. Fish Res 53: 303-308.

52. Poojary N, Tiwari L, Chakraborty S (2011) Stock assessment of the Indian Scad, Decapterus russelli (Ruppell, 1830) from Mumbai waters. Ind J Mar Sci 40: 680.

53. Reuben S, Kasim H, Sivakami S, Nair P, Kurup K, et al. (1992) Fishery, biology and stock assessment of carangid resources from the Indian seas. Ind J Fis 39: 195-234.

54. Patterson K (1992) Fisheries for small pelagic species: an empirical approach to management targets Rev Fish Biol Fisheries 2: 321.

55. Al-Barwani, SM, Arshad A, Amin SN, Japar S, Siraj S, et al. (2007) Population dynamics of the green mussel Perna viridis from the high spat-fall coastal water of Malacca, Peninsular Malaysia. Fish Res 84: 147-152.

56. Fischer W, Bianchi G (1984) FAO species identification sheets for fishery purposes: Western Indian Ocean (Fishing Area 51) 5.

57. Prager M (2005) A stock-Production model incorporating covariates (version 5) and auxiliary programs. CCFHR (NOAA) Miami Laboratory Document MIA-92/93-55, Beaufort Laboratory Document BL-2004-01.

58. Riede K (2004) The Global Register of Migratory Species-First Results of Global GIS Analysis. In: Werner D (eds) Biological Resources and Migration. Springer, Berlin: Heidelberg.

59. Smith-Vaniz WF, Kent EC (2007) Review of the crevalle jacks, Caranx hippos complex (Teleostei: Carangidae), with a description of a new species from West Africa. Fish Bul 105: 2.

60. Tah L, Joanny TN, Douba V, Kouassi J, Moreau J (2010) Preliminary estimates of the population parameters of major fish species in Lake Ayamé I (Bia basin; Côte d'Ivoire). J Apl Icht 26: 57-63.

24

Pathogenic Bacteria in *Oreochromis Niloticus* Var. Stirling Tilapia Culture

Huicab-Pech ZG, Castaneda-Chavez MR* and Lango-Reynoso F

National Technological Institute of Mexico/Technological Institute of Boca del Rio Veracruz, Mexico

*Corresponding author: Castaneda-Chavez MR, National Technological Institute of Mexico/Technological Institute of Boca del Rio Veracruz, Mexico,
E-mail: castanedaitboca@yahoo.com.mx

Abstract

Tilapia *Oreochromis niloticus* is an aquaculture resource that represents one of the most popular crops in the world. However, species cultivation presents health problems, which are associated with the presence of pathogenic bacteria and causes high economic losses. The aim of this study was to determine the diversity of these bacteria at the genus level in the species *O. niloticus var.* Styrling during growing stage in the fattening and pre-fattening phases. Tilapia samples were collected and analyzed; each sample was subjected to a macroscopic external and internal observation of organs and tissues. Subsequently, samples were evaluated by microbiological tests using Trypticase Soy Agar (TSA), Thiosulfate Citrate Bile Salts Sucrose Agar (TCBS) and selective media (Pseudomonas sp. Group), and conventional biochemical tests aimed at the production of glucose, sucrose, lactose, oxidase, catalase, indole, ornithine and Gram staining. External analysis revealed clinical signs of disease such as skin bleeding, body ulceration, corneal opacity, and intestine and vesicle inflammation. Microbiological and biochemical analysis showed the presence of eleven bacterial genera known as *Arthrobacter sp., Enterococcus sp., Staphylococcus sp., Micrococcus sp., Streptococcus sp., Aeromonas sp., Pseudomonas sp., Edwardsiella sp., Flexibacter sp.* and *Flavobacterium sp.*, with a predominance of 55% Gram-negative bacilli in tilapia crops. According to the results, it is necessary to take preventive and corrective measures in order to avoid possible risks during production cycles, mainly when handling organisms. It is also important to promote good crop management practices and quality systems in production units to benefit the aquaculture sector.

Keywords: Diversity; Bacteria; Disease; Microbiological; Biochemical

Introduction

Aquaculture is considered an activity that contributes to the production of foods of high nutritional value, generating employment and economic income for the world population. In addition, it strengthens the source of inputs for the food industry and foreign exchange for the country [1]. During 2014, aquaculture reached 73.8 million tons with an estimated value of USD 160.2 billion [1]. According to the statistics, from 1970 to 2008 aquaculture production presented an annual growth of 8.3%, with a per capita consumption of 0.7 to 7.8 kg per person, which meant a 6.6% increase of the annual average. However the apparent fish consumption per capita will be 21.8 kg in 2025, this consumption is associated with the improvement of distribution channels, increase in fish production and urbanization [1]. At present, the aquaculture productive sector is made up of species that are grown in fresh water reaching a world production of 59.9%. Tilapia cultivation, as an introduced species, has favoured aquaculture development in recent years, because its cultivation shows high resistance in the chain production [1]. Among other characteristics, 99% of this species is grown outside its natural habitat and is positioned as the second species cultivated worldwide, mainly in countries such as Philippines, Indonesia, Thailand, Malaysia, China, Chile, Mexico, Ecuador, Brazil, and Colombia, with productions that exceed the cultivation of salmonids and carps [2]. Among commercial species are *O. niloticus, O. mossambicus, O. aureus, O. hornorum, Tilapia rendalli and Tilapia zilli* [1]. Tilapia cultivation in Mexico has shown an increase in production, favouring the country's economy with yields of 12,529 t [3]. Mexico reaches development levels ranging from the experimental scale as it happens with white fish, native mojarra, abalone, scallops, mussel, lobster and snail, to industrial-commercial production of species such as catfish, carp, tilapia, trout, oysters, shrimp and prawns. The state of Veracruz reports a production of tilapia *O. aureus and O. niloticus* of 74,659 kg and 121,459 kg, respectively [4].

The success of tilapia production is due to the fact that the species exhibits rapid growth, ease of propagation, tolerance to environmental conditions, easy acceptance of natural foods and dietary supplements, and resistance to diseases. *Oreochromis sp.* is resistant to bacterial diseases caused by stress as a result of high planting densities and minimal control in crop management. The state of disease in fish is completed from the appearance of clinical signs by bodily alterations and the behaviour of the organism. According to Rodriguez et al. [5], diseases are classified by physical, chemical and biological risks, and the disturbing result is due to the association of one or two factors, which causes a physiological alteration of the organism [6]. There are two categories of disease called infectious and non-infectious diseases. The first is related to the pathogenic organisms present in aquatic environment and in fish, which causes contagious diseases, therefore treatments to control bacterial outbreak are required. While non-infectious diseases are related to environment, for example those problems associated with biotic and abiotic factors, minimum management of good practices, deficits in the nutrition of the organisms, and genetic abnormalities in relation to the quality of progeny. MSD Animal Health says that the mortalities correspond to the increase of seed densities (number of fish/m^3) and species introduction are the result of the appearance of emerging diseases [7,8]. Snieszko [9] mentions that disease is related to the interaction between fish, the pathogen and the aquatic environment as habitat, that is, when an organism is exposed to pathogenic bacteria in unfavourable environments, where poor quality of water or excess of organic matter prevails. The incidence of disease is higher because balance between host, guest and aquatic environment is broken [6]. However, fish exhibit a high bacterial diversity, a symbiotic effect among bacteria, which protects them to adapt to nutritional changes and assimilation of food in the digestive tract [10]. Among bacteria that cause mortality in tilapia pathogens of Flavobacterium columnare stand out, *Edwardsiella tarda, Aeromonas sp., Vibrio sp., Francisella sp., Streptococcus iniae, Streptococcus agalactiae, Vibrio anguillarum, V. harveyi, Photobacterium damsele subsp.* [11]. Some of these pathogens have a geographical distribution in tropical and temperate regions where warm water species, such as Nile tilapia, are commonly grown. Among the most important bacteria, the species Aeromonas hydrophila is the causal agent of hemorrhagic septicemia syndrome or red pest on skin [12], it is also considered as an opportunistic and contaminating pathogen in aquaculture environments. Aeromonas hydrophila has a prevalence of 10% and 85% at any stage of culture and is considered the most important bacterial agent in freshwater fish, especially rainbow trout, tilapia, including ornamental and marine fish [13-15]. Another pathogen is *Streptococcus agalactiae*, which is characterized by septicemia and meningoencephalitis in fish. It has a prevalence of 40% and 70% in stages of fry, juvenile and fattening. It is also reported in several species of fish around the world, due to its geographic distribution includes regions with temperate and tropical climate, as presented by Brazil, China, Malaysia and the United States [16-18]. Some diseases are caused by not applying sanitary protocols on growing farms, as well as the use of antibiotics as an almost mandatory preventive or corrective practice. In this context, the main objective was to identify bacterial diversity with pathogen potential in *O. niloticus var. Styrling* during pre-fattening and fattening phases, to suggest management alternatives related to the implementation of quality systems in the production processes of aquatic organisms.

Materials and Methods

Study area

Aquaculture production units are located between 18°53' 35.64" north latitude and 95°56' 40.89" west longitude, in the municipality of Alvarado, Veracruz, Mexico. The production units consist of 15 rustic ponds of 0.5 ha in average and 12 circular concrete ponds of 12 m in diameter with a capacity of 200 m^3. Aquaculture farms also have their own water system and a sedimentation lagoon with groundwater supply.

Collection of organisms

Organisms were collected in aquaculture production units in order to perform analyzes corresponding to the identification of pathogenic bacteria. The weight of each organism was taken into consideration to choose the ones with a final weight between 350 gr and 450 gr. The whole process was carried out under aseptic conditions. Each sample of organism was protected following the chain of custody, placed in polyethylene bags with water at 40% capacity. They were transported in containers at 4°C (NOM-109-SSA1-1994; [19]) to the Research and Aquatic Resources Laboratory of the Technological Institute of Boca del Río, Veracruz.

Taking physicochemical parameters of water in the in the culture ponds

Water parameters were measured monthly at the sampling points. Oxygen (mg/L), temperature (°C), salinity (ppm) and pH were recorded with a multi-parameter probe YSI 556 MPS. Nitrites (mg/L) and nitrates (mg/L) were determined with the CHEMets® Colourimetric Test Kit.

Bacterial identification and processing

Each organism was observed in the laboratory by macroscopic external for the identification of lesions in skin and fins. A ventral section was then made to observe internal organs such as gills, intestine, spleen, liver, gallbladder and kidney.

These organs were selected to determine bacterial genus and isolate pathogens. A minimum portion of the sample was taken with a platinum handle and seeded in duplicate by cross-streaking in boxes of Trypticase Soy Agar (TSA), Thiosulfate Citrate Bile Salts Sucrose Agar (TCBS) and selective media for the Pseudomonas group (Pseudomonas F Agar) The incubation time was 24-48 hrs at 34°C. After bacterial growth, some colonies were selected to perform Gram staining and biochemical and presumptive tests, mainly motility, indole, catalase, ornithine, triple sugar iron agar, methyl red, and Voges-Proskauer tests [20]. These tests are explained below:

For the Gram staining test a small sample was taken with a platinum handle and spread over the slide. The extension was fixed by heat, passing it gently over the flame of the burner until dry. Staining was performed with crystal violet or gentian violet, covering the sample homogeneously, this solution is allowed to work for one minute at 25°C. Briefly rinsed with water, air dried and Lugol was added for one minute, and then continue washing it with water. The discolouration was performed with alcohol for about 15 to 20 seconds and subsequently washed with water. Safranite was used, as contrasting staining, for 15 seconds, followed immediately with a water wash and allowed to dry. Direct observation, under a microscope, continued where Gram (+) bacteria showed a purple colour and Gram (-) bacteria a pink or red colour.

The catalase test was carried out with a sterile straight bacteriological loop, taken from the center of a pure colony of 18 to 48 hours and placed on a clean glass slide. Then, a drop of Pasteur 30% hydrogen peroxide (H_2O_2) was added. A positive result is evidenced by bubbling and no bubbling as negative.

Motility test, indole and ornithine (MIO) was performed in 3 ml tubes with MIO agar, in each tube the microorganism strains were seeded in the culture medium and incubated for 24 hours at 37°C. After 24 hours of incubation, the mobility test took as positive those that generated turbidity or growth beyond the line of sowing, while the negative ones were those that showed growth only in the line of seed. The indole test was determined by the Erlinch or Kovacs Reagent, the medium was allowed to warm to room temperature prior to inoculation. The tubes were inoculated with a platinum handle, where a portion of pure culture was transferred and incubated at 37°C for 40 to 48 hours. Then 0.5 mL of the Kovac Reagent was added and gently shaken to determine the production of Indol. It was taken as positive when adding the reagent presented the formation of a red colour band at the top of the medium. A yellow colour denotes a negative indole after addition of Kovacs Reagent.

A KIA agar was used for the Kligler test. By tilting, a loop of pure culture was taken from a well-isolated colony of the bacteria and in the culture medium agar it was stabbed to the bottom for later to perform a seeding striation on the bezel. They were incubated for 24 hrs. The results were observed to determine the different characteristics that this test indicated. It was also observed if it had bubble formation, which would indicate the production of gas and the blackening indicated the production of hydrogen sulphide.

For the methyl red tests 3 ml tubes and MR-VP broth were used as the culture medium. A bacterial colony was taken and seeded by shaking, with 24 hours incubation. After incubation, 5 drops of the red methyl indicator were added, the bright red colour indicates a positive reaction and the yellow colour indicates a negative one.

For the Voges-Proskauer (VP) test tubes were used with MR/VP broth where an aliquot of the colony was placed and seeded by shaking in the broth, with an incubation of 24 hrs. After the incubation, 0.2 ml of potassium hydroxide (KOH) was added; the positive result is indicated by a pink-red colour and the negative result by a yellow or copper colour.

As a presumptive test the presence of the cytochrome oxidase complex was determined with Bactident oxidase indicator strips (Merck, Merck KGaA Germany®) for each bacterial colony. The results scored positive showing a violet colour and negative when there were no changes in colouration [20].

Statistical analysis

Statistical analysis was performed using Excel and presence/absence of pathogens. The physicochemical parameters are presented as mean ± SE.

Results

Physicochemical parameters of crop water

The physicochemical parameters of water in pre-fattening and fattening ponds were found within the limits for the cultivation of Nile Tilapia *O. niloticus var. Styrling* (Tables 1 and 2).

Physicochemical parameters of the culture water						
	Temperature (°C)	Oxygen (mg/L)	pH	Salinity (ppm)	Nitrites (mg/L)	Nitrates (mg/L)
March	22.5 ± 1.12	7.5 ± 1.9	9.1	0.28	0.25	0.25
April	22.9 ± 0.91	7.7 ± 1.83	8.9	0.27	0.5	0.5
May	23.1 ± 0.77	7.4 ± 1.93	9.2	0.29	0.25	0.25

Table 1: Physicochemical parameters of water in pre-fattening ponds.

Physicochemical parameters of culture water						
	Temperature (°C)	Oxygen (mg/L)	pH	Salinity (ppm)	Nitrites (mg/L)	Nitrates (mg/L)
March	23 ± 2.34	7 ± 2.26	8.07	0.31	0.25	0.25
April	24 ± 1.14	6.9 ± 2.21	8.5	0.3	0.5	0.5
May	24.8 ± 0.97	7 ± 2.20	8.3	0.32	0.25	0.25

Table 2: Physicochemical parameters of water in fattening ponds.

External analysis of Tilapia *Oreochromis niloticus*

External analysis showed cynical signs of disease caused by handling in production units. The signs commonly found in the two phases of culture are related to bleeding skin, body depigmentation, frayed fins, distended gallbladder, liver discolouration, body ulceration, corneal opacity and intestine inflammation (Figure 1).

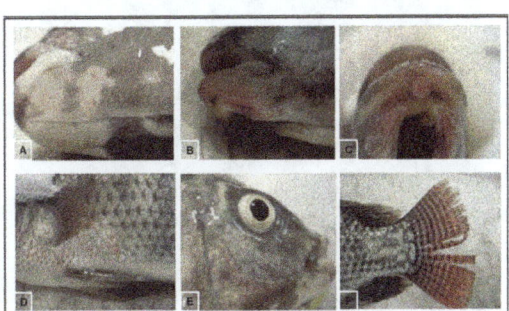

Figure 1: External signs of damage and anomalies caused by mismanagement and handling of organisms. A, B and C: Skin discolouration, D: Skin bleeding, E: Corneal Opacity, F: Frayed fins.

Taxonomic composition of bacteria found in species of Tilapia *Oreochromis niloticus*

Eleven bacterial genera were identified in pre-fattening and fattening tilapia, foremost among them *Arthrobacter sp., Enterococcus sp., Staphylococcus sp., Vibrios sp., Micrococcus sp., Streptococcus sp., Aeromonas sp., Pseudomonas sp., Edwardsiella sp., Flexibacter sp.* and *Flavobacterium sp.* According to the biochemical tests, Gram staining verified the presence of 55% Gram-negative organisms predominating *Aeromonas sp., Pseudomonas sp., Edwardsiella sp., Flexibacter sp.,* and *Flavobacterium sp.;* and 45% Gram-positive organisms *Arthrobacter sp., Enterococcus sp., Staphylococcus sp., Micrococcus sp.* and *Streptococcus sp.* (Figures 2 and 3, Tables 3 and 4).

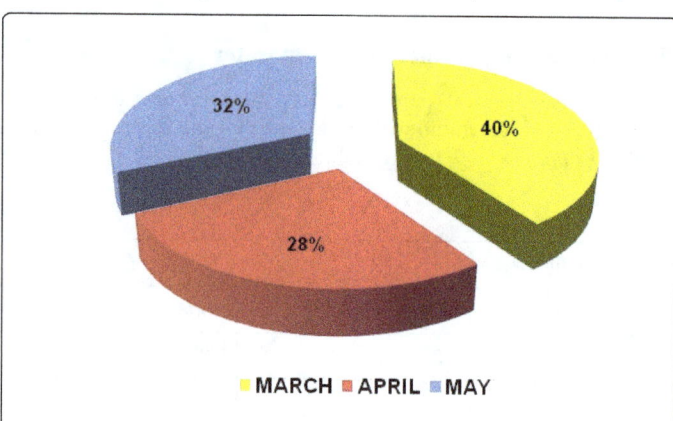

Figure 2: Distribution of bacterial pathogens in *O. niloticus* Var. Stirling.

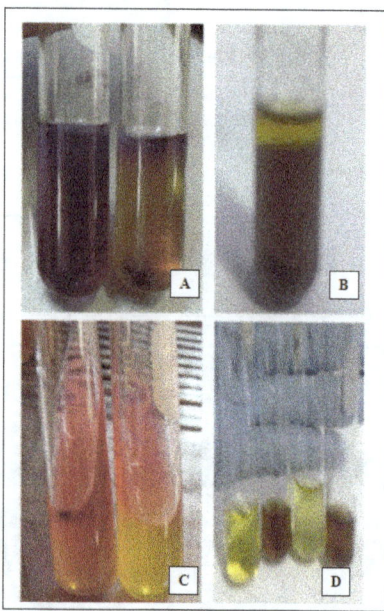

Figure 3: Reaction of biochemical tests. A: Ornithine test; B: Indole test; C: Triple sugar test; D: MR-VP test.

Conforming to presence/absence of pathogens, it is observed that in the sampling months of March, April and May, where *Aeromonas sp.,*

Pseudomonas sp., Vibrios sp., and *Enterococcus sp.* were detected; March turned out to be the month with the highest incidence of bacterial genera compared to April and May (Tables 3 and 4).

The presence of five bacterial genera was registered, during the fattening stage, *Aeromonas sp., Pseudomonas sp., Vibrios sp., Enterococcus sp.,* and *Micrococcus sp.,* with a higher incidence in March, April and May. The highest presence of pathogens, during this stage, was determined mainly in March and May (Tables 3 and 4).

Months	Pre-fattening	Fattening
	Bacteria	
March	*Aeromonas sp.*	*Aeromonas sp.*
	Edwardsiella sp.	*Streptococcus sp.*
	Streptococcus sp.	*Staphylococcus sp.*
	Staphylococcus sp.	*Pseudomonas sp.*
	Pseudomonas sp.	*Vibrios sp.*
	Vibrios sp.	*Flexibacter sp.*
	Flexibacter sp.	*Enterococcus sp.*
	Arthrobacter sp.	*Micrococcus sp.*
	Enterococcus sp.	
April	*Aeromonas sp.*	*Aeromonas sp.*
	Pseudomonas sp.	*Edwardsiella sp.*
	Vibrios sp.	*Pseudomonas sp.*
	Arthrobacter sp.	*Vibrios sp.*
	Enterococcus sp.	*Enterococcus sp.*
		Micrococcus sp.
May	*Aeromonas sp.*	*Aeromonas sp.*

Table 3: Presence-Absence of pathogens in *O. niloticus* var. Styrling, Pre-fattening and fattening phases.

Bacteria	Gram		O	C	Mot	Ind	Orn	Lys	MR	VP
Arthrobacter sp.	+	C	+	-	-	-	-	-	-	-
Enterococcus sp.	+	C	+	-	-	-	-	+	-	-
Staphylococcus sp.	+	C	+	+	-	-	-	-	-	-
Micrococcus sp.	+	C	+	+	-	-	-	+	-	-
Streptococcus sp.	+	C	+	-	-	-	+	-	-	-
Aeromonas sp.	-	B	+	+	-	-	-	+	-	-
Vibrio sp.	-	B	-	+	+	+	+	+	+	-
Pseudomonas flourescens	-	B	+	+	-	-	-	+	-	-
Edwardsiella sp.	-	B	-	+	-	-	-	-	-	-
Flexibacter sp.	-	B	+	-	-	-	-	-	-	-

APRIL										
Enterococcus sp.	+	C	+	-	-	-	-	+	-	-
Staphylococcus sp.	+	C	+	+	-	-	-	-	-	-
Micrococcus sp.	+	C	+	+	-	-	-	+	-	-
Aeromonas sp.	-	B	+	+	-	-	-	+	-	-
Vibrio sp.	-	B	-	+	+	-	-	+	-	-
Pseudomonas sp.	-	B	+	+	-	-	-	-	-	-
Edwardsiella sp.	-	B	-	+	-	-	-	-	-	-
MAY										
Streptococcus sp.	+	C	+	-	-	-	+	-	-	-
Enterococcus sp.	+	C	+	-	-	-	-	+	-	-
Micrococcus sp.	+	C	+	+	-	-	-	+	-	-
Aeromonas sobria	-	B	+	+	+	-	+	+	-	-
Flavobacterium columnare	-	B	+	+	-	-	-	-	-	-
Pseudomonas sp.	-	B	+	+	-	-	-	+	-	-
Vibrios sp.	-	B	-	+	+	+	+	+	+	-
Edwardsiella sp.	-	B	-	+	+	+	+	+	-	-

Table 4: B: Bacillus: Coconuts, O: Oxidase, Ca: Catalase, Mot: Motility, Ind: Indol, Orn: Ornithine, Lys: Lysine, MR: Methyl Red, VP: Vogues-Proskeaur.

Discussion

During the study, it was shown that the presence of pathogenic bacteria is not significantly related to the physicochemical parameters (PQ) of water, i.e., the PQ parameters are not considered as indicative of disease in the production farm. In addition, they are within the optimal values for tilapia cultivation [21,22]. In contrast, *Aeromonas hydrophila* is related to sudden changes in temperature, dissolved oxygen and inadequate nutrition, as pointed out by Conroy [23-25] who indicate that the constant variation of physicochemical parameters is a stress factor that benefits the outbreak of disease caused by opportunistic bacteria. However, in the study there was no direct relationship with physicochemical parameters and the presence of bacterial genera.

External analysis showed clinical signs of disease, foremost among them skin hemorrhages, corneal opacity, body ulceration, liver discolouration, frayed fins and intestine and vesicle inflammation, these abnormalities are considered as the main symptoms of infection reported by Giordano, et al. [17,26,27], who observed signs such as skin alterations, anorexia, exophthalmia, corneal opacity, extension of the visceral cavity, bleeding and abdominal inflammation, hepatomegaly and splenomegaly. According to Soto [28] it is proven that bacteria are the cause of epithelial hyperplasia in gills, splenomegaly, renomegaly and necrosis in internal organs, mainly in spleen, heart, liver, kidney, brain and musculature. For example Yardimci, et al. [29-31] indicate that species *Aeromonas hydrophila* is associated with hemorrhages in gills and skin, weakness and anorexia, as well as vision loss by breaking' orbital eyes, the above agrees with

those reported by Clavijo et al. [32], who reported that the presence of genus *Edwardsiella sp.,* causes septicemia in internal and external organs, including kidney, liver and spleen, skin, rectum, fins, abdominal inflammation, and opaque eyes. Some pathogens are transmitted horizontally as indicated by Mauel et al. [33], who pointed out that bacteria use water as a precursor vehicle, causing fish-to-fish outbreaks by direct contact. Newman [34] also mentions that the degree of pathogenicity depends on species resistance and environmental conditions. Because some pathogens are present in environments with high temperatures, poor water quality and accumulation of organic matter, these conditions allow bacterial adhesion and replication in host cells.

In this sense, the management and handling of organisms in this study is considered inadequate, since the activities carried out in the production phases do not comply with the Manual of Good Practices established by the National Service of Health, Safety and Agro-Food Quality [35], which demand the references established by the United Nations Organization (UNO), Food and Agriculture Organization (FAO), and the World Health Organization (WHO), through the Codex Alimentarius Commission; concentrated on water management, food handling, handling of chemicals and drugs, and product safety during harvesting. These safety measures reduce risks of biological, physical and chemical contamination and avoid possible losses caused by diseases, in this case losses by opportunistic bacteria.

In the pre-fattening and fattening phases, Gram-positive and Gram-negative bacteria predominated, which agrees with [36], which determine a prevalence of 77% of Gram negative bacilli in the intestinal flora of tilapia and a prevalence >10% of bacteria such as *Aeromonas hydrophila, Swewanella putrefaciens, Corynebacterium urealyticum, Escherichia coli* and *Vibrio Cholerae.* Some of the bacteria found in the present study are considered native to aquatic environment, such as *Aeromonas hydrophila* and *Vibrios sp* [36]. The presence of various bacterial genera varies according to the growth of organisms. However, the reason for its diversification is due to food consumption and water quality. Gram negative bacteria is the main cause of bacterial disease, for example Austin et al. [37] mention that genus *Aeromonas sp* causes *furunculosis* and *hemorrhagic septicemia* in skin, this coincides with Calvo et al. [38], who consider that Gram negative bacteria is listed at the margins of public health as well as its high impact by antibiotic resistance.

Pseudomonas sp., Staphylococcus sp. and *Aeromonas sp.* were found in pre-fattening and fattening phases. The incidence is related to tilapias grown in floating cages [28] also points out that these genera can provoke an epidemic outbreak due to its potential bacterial pathogen. Some reports by Al-Harbi et al. [39] indicate that the presence of bacteria in fish's digestive flora is normal. However, the outbreak of disease is related to the existence of a stress factor based on the interaction between fish, pathogens and aquatic environment as a natural habitat of the organism, as well as poor water quality or excess of organic matter factors which allows the incidence of disease to be greater, as mentioned by Huicab-Pech et al. [6]. Although fish exhibit high bacterial diversity, [9] points out that there is a symbiotic effect among bacteria, that is, the host adapts to nutritional changes and food assimilation in the digestive tract through bacterial balance.

Bacterial genera, found in the present work, belong to the lineage of species such as *Aeromonas hydrophila, Vibrio parahaemolyticus, Vibrio vulnificus, Vibrio cholerae, Streptococcus iniae, Pseudomonas spp., Edwardsiella sp., Flexibacter sp.* and *Flavobacterium sp.,* which present risk for human health; bacteria are autochthonous organisms of the aquatic environment, including water and sediment. The presence, as an opportunistic pathogen, is due to the conditions of the aquatic environment and stress, as reported by Burr et al. [14,15,40-42] who consider that organisms under stress conditions are susceptible to the presence of opportunistic pathogens. These pathogens cause *hemorrhagic septicemia* and clinical signs of erratic or circling swimming, uncoordinated movements, anorexia or decreased appetite, exophthalmia, corneal opacity, visceral cavity extension, bleeding and abdominal inflammation, softening of the brain and liver, hepatomegaly and pallor in the organ, as well as splenomegaly and visceral adhesion commonly found in crops and studies at experimental level.

The genus *Sthapylococcus sp.* and *Micrococcus sp.* were presented in fattening organisms, however, according to Mhango, et al. [43,44] who relate lack of hygiene in the management of fish, since bacteria is common in human skin and in normal bacterial flora of freshwater fish. The culture of tilapia presented those bacteria that are in the NOM-115-SSA1-1994, reason why its infective activity is related to alimentary intoxications in human beings. In addition to that, some are classified like pathogens indicative of fecal contamination; such is the case of *Enterococcus sp., Vibrio sp.* and *Sthapylococcus* sp., as indicated by Soto [28], so that temperature variation, human settlements, amount of food supplied, nutritional quality and harvesting methods are related to the incidence of opportunistic bacteria in crops.

Conclusion

The occurrence of diseases is due mainly to several factors that act individually or jointly in an aquaculture crop, therefore, it is considered necessary the implementation of strategies for its optimal management, with the objective of achieving a sustainable production under safety and good management practices programs, as well as an alternative of effective vaccination as a preventive and corrective treatment. Having also the support of friendly treatments for crops and human being, in order to avoid direct and indirect losses, and to assure the production and success of *O. niloticus* crops.

References

1. FAO (2016) Food and Agriculture Organization of the United Nations. The State of World Fisheries and Aquaculture. Contribution to food security and nutrition for all pp: 23.

2. FAO (2010) Food and Agriculture Organization of the United Nations World Review of Fisheries and Aquaculture. In: FAO's State of the World Fisheries and Aquaculture pp: 197.

3. Official Journal of the Federation (2008) Statistical Yearbook of Fisheries and Aquaculture. Official Journal of the Federation of Mexico.

4. Ministry of Agriculture (2009) Diagnosis of the Primary Sector in Veracruz. State Technical Evaluation Committee pp: 106.

5. Rodríguez GM, Rodríguez DG, Monroy Y, Mata JA (2001) Manual of Fish Diseases. Bulletin of the National Aquaculture Health Program and the Diagnostic Network.

6. Huicab-Pech ZG, Landeros-Sánchez C, Castañeda-Chávez MR, Lango-Reynoso F, López-Collado CJ, et al. (2016) Current State of Bacteria Pathogenicity and their Relationship with Host and Environment in Tilapia Oreochromis niloticus. Journal Aquaculture Research & Development 7: 428.

7. MSD Animal Health (2011) Bacterial Disease in Warm water Fish: New Strategies for Sustainable Control. Natal, Brazil. Held in conjunction with the World Aquaculture Society Conference.

8. Bondad-Reantaso MG, Subasinghe RP, Arthur JR, Ogawa K, Chinabut S, et al. (2005) Disease and health management in Asian aquaculture. Vet Parasitol 132: 249-272.

9. Snieszko SF (1975) History and present status of fish diseases. J Wildlife Dis 11: 446-459.

10. Al-Harbi H, Uddin MN (2005) Bacterial diversity of tilapia (Oreochromis niloticus) cultured in brackish water in Saudi Arabia. Aquaculture 250: 566-572.

11. Wang X, Li H, Zhang X, Li Y, Ji W, et al. (2000) Microbial flora in the digestive tract of adult penaeid shrimp (Penaeus chinensis), Journal Ocean University Qingdao 30: 493-498.

12. Austin B, Austin DA (1999) Bacterial fish pathogens diseased of farmed and wild fish. Praxis publishing Ltd.

13. Shao J, Liu J, Xiang L (2004) Aeromonas hydrophila induces apoptosis in Carassius auratus lymphocytes in vitro. Aquaculture 229: 11-23.

14. Burr SE, Goldschmidt-Clermont E, Kuhnert P, Frey J (2012) Heterogeneity of Aeromonas populations in wild and farmed perch. Perca fluviatilis. L. J Fish Dis 35: 607-613.

15. Soto-Rodríguez SA, Cabanillas-Ramos J, Alcaraz U, Gómez-Gil B, Romalde JL (2013) Identification and virulence of Aeromonas dhakensis, Pseudomonas mosselii and Microbacterium paraoxydans isolated from Nile tilapia. Oreochromis niloticus, cultivated in Mexico. J Appl Microbiol 115: 654-662.

16. Ye X, Li J, Lu M, Deng G, Jiang X, et al. (2011) Identification and molecular typing of Streptococcus agalactiae isolated from pond-cultured tilapia in China. Fisheries Sci 77: 623-632.

17. Pretto-Giordano LG, Eckehard-Müller E, Freitas JCD, da Silva VG (2010) Evaluation on the Pathogenesis of Streptococcus agalactiae in Nile Tilapia (Oreochromis niloticus). Braz Arch Biol Techn 53: 87-92.

18. Mian GF, Godoy DT, Leal CAG, Yuhara TY, Costa GM, et al. (2009) Aspects of the natural history and virulence of S. agalactidae infection in Nile tilapia. Vet Microbiol 136: 180-183.

19. Official Journal of the Federation (1994) Mexican Official Standard NOM-109-SSA1-1994. Procedure for taking and handling and transporting food samples for microbiological analysis. Health Secretary Clarification. Official Journal of the Federation.

20. Whitman KA, MacNair NG (2004) Finfish and Shellfish. Bacteriology Manual Techniques and Procedures. Iowa State Press. A Blackwell publishing Company pp: 258.

21. Saavedra M (2006) Text of Subject Agriculture and Aquaculture Production.. Department of Technology and Architecture. Faculty of Science, Technology and Environment Centroamerican University. Managua Nicaragua.

22. Meyer D (2007) Introduction to aquaculture. Pan American Agricultural School, Zamorano, Honduras pp: 159.

23. Conroy G (2014) Most frequent tilapia diseases in Latin America and the Caribbean. 9th International Aquaculture Forum. World Aquaculture Society.

24. Conroy G (2007) New advances in warm water fish farming. Consulted.

25. Li Y, Cai SH (2011) Identification and pathogenicity of Aeromonas sobria on Tail-rot disease in juvenile tilapia Oreochromis niloticus. Curr Microbiol 62: 623-627.

26. Eldar A, Bejerano Y, Livoff A, Horovitcz A, Bercovier H (1995) Experimental streptococcal meningo-encephalitis in cultured fish. Vet Microbiol 43: 33-40.

27. Figueiredo HPC, Carneiro DO, Faria FC, Costa GM (2006) Streptococcus agalactiae associated with meningoencephalitis and systemic infection in Nile tilapia (Oreochromis niloticus) in Brazil. Brazilian Archive of Veterinary Medicine and Zootechnics 58: 678-680.

28. Soto SA (2009) Water quality and bacteria present in cultivated tilapia. Foundation produces Tilapia A.C pp: 7-19.

29. Yardimci B, Aydin Y (2011) Pathological findings of experimental Aeromonas hydrophila infection in Nile Tilapia (Oreochromis niloticus), Ankara Univ Vet Fak 58: 47-54.

30. Penagos G, Barato P, Iregui C (2009) Immune system and vaccination of fish. Acta Biologica Colombiana 14: 3-24.

31. Suchanit N, Kunihiko F, Masato E, Masashi M, Takayuki K (2010) Immunological effects of glucan and Lactobacillus rhamnosus GG. A probiotic bacterium, on Nile tilapia Oreochromis niloticus intestine with oral Aeromonas challenges Fisheries Sci 76: 833-840.

32. Clavijo AM, Conroy G, Conroy DA, Santander J, Aponte F (2002) First report of Edwarsiella tarda from tilapias in Venezuela. B Eur Assoc Fish Pat 22: 280-282.

33. Mauel MJ, Soto E, Moralis JA, Hawke J (2007) A piscirickettsiosis-like Syndrome in Cultured Nile Tilapia in Latin America with Francisella spp. As the Pathogenic Agent, J Aquat Anim Health 19: 27-34.

34. Newman SG (1993) Bacterial vaccines for fish. Annu Rev Fish Dis 3: 145-185.

35. SENASICA (2009) Manual of Good Practices of aquaculture production of tilapia for food safety. 1st eds pp: 155.

36. Al-Harbi A, Uddin MN (2004) Seasonal variation in the intestinal bacterial flora of hybrid tilapia (Oreochromis niloticus x Oreochromis aureus) cultured in earthen ponds in Saudi Arabia. Aquaculture 229: 37-44.

37. Austin B, Austin DA (2007b) Aeromonadaceae Representatives (Motile Aeromonads). Disease of Farmed and Wild Fish. Bacterial Fish Pathogens 119-146.

38. Calvo J, Martínez-Martínez L (2009) Mechanisms of action of antimicrobials. Infectious Diseases and Clinical Microbiology 27: 44-57.

39. Al-Harbi H, Uddin MN (2005) Bacterial diversity of tilapia (Oreochromis niloticus) cultured in brackish water in Saudi Arabia. Aquaculture 250: 566-572.

40. Palumbo S, Abeya C, Stelma G (1992) Aeromonas hydrophila group. In: Compendium of methods for the microbiological examination of food. Washington: Asian Pacific American Heritage Association pp: 497-515.

41. Woo PTK, Bruno DW (2010) Fish diseases and disorders. Volume 3: viral, bacterial and fungal infections. In Edwardsiella septicaemias. 2nd eds. Wallingford: CABI International, pp: 512-534.

42. Johri AK, Paoletti LC, Glaser P, Dua M, Sharma PK, et al. (2006) Group B Streptococcus: global incidence and vaccine development. Nat. Revista Microbiología 4: 932-942.

43. Mhango M, Mpuchane SF, Gashe BA (2010) Incidence of indicator organisms, Opportunistic and Pathogenic bacteria in fish, African journal of food agriculture nutrition and development 10: 4202-4218.

44. González-Rodríguez CT, López-Díaz M, Gracia-López M, Prieto, Otero A (1999) Bacterial microflora of wild trout (Salmo trutta). wild pike (Esox lucius) and aquacultured rainbow trout (Oncorhynchus mykiss). J Food Prot 62: 1270-1277.

25

Survey on Penaeidae Shrimp Diversity and Exploitation in South East Coast of India

Perumal Rajakumaran and Baskralingam Vaseeharan[*]

Department of Animal Health and Management, Alagappa University, Karaikudi 630003, Tamil Nadu, India

[*]**Corresponding author:** Baskralingam Vaseeharan, Crustacean Molecular Biology & Genomics lab, Department of Animal Health and Management, Alagappa University, Karaikudi 630003, Tamil Nadu, India, E-mail: vaseeharanb@gmail.com

Abstract

The assessment of Penaeidae species diversity in a particular region is very important in formulating conservation strategies. In the present study, the survey on diversity of Penaeidae species in south east coast of India has been assessed on the basis of landing of variety of species in this group. Penaeidae species were collected from various main landing centers of south east coast of India for three years. Identification and nomenclature was done based on previously published literature. Among the 59 species observed, the *Penaeus semisulcatus, Penaeus monodon* and *Fenneropenaeus indicus* were found mostly in all landing centers. As first and foremost, the *Metapenaeus papuensis, Metapenaeus anchistus, Metapenaeopsis wellsi, Parapenaeopsis sinica (Kishinouyepenaeopsis amicus), Parapenaeopsis hungerfordi, Parapenaeopsis venusta, Parapenaeopsis coromandelica, Parapenaeopsis gracillima, Trachysalambria longipes* and *Parapenaeus lanceolatus* landed in south east coast of India. As far as Penaeidae shrimp diversity is good in south east coast region of India, and needed the fishing site and mesh size regulation to protect the juvenile and adult of Penaeidae from inshore and offshore catching.

Keywords: Penaeidae; Diversity; Nomenclature; Exploitation; Conservation

Introduction

Biodiversity performs a number of ecological services for mankind that have commercial and recreational or resources management purpose. Globally, more than 30, 000 marine crustacean species have been reported. Crustacean fishery is one of the major resources of India that includes the commercially important shrimps, prawns, lobsters and crab which are important in the tropical food chain of marine ecosystem. Penaeidae, a family of marine crustacean in the suborder Dendrobranchiata, often referred to as penaeid shrimp or penaeid prawn with 48 recognized genera, 23 of them is known only from the fossil record. Total averaged Penaeidae catching at world level was 1.21 million tons per annum for the year 2008-10. The Penaeidae shrimp constitute the backbone of Indian seafood export industry as the major foreign exchange earner as well as source of livelihood for millions of fishermen in the country. India exported US $ 2.8 billion worth marine products in 2010-11, of which shrimp contributed 3.09% in volume and 69.5% in value of the total export [1].

South east coast mainly is situated on the south east of Peninsular India covers an area of 1,30,058 Sq.km. The length of its coastline is about 1050 km with its significant portion on the east coast bordering Bay of Bengal. Increase in human population and demand for shrimp in the world market has resulted in over exploitation of shrimp from Indian coastal waters. This is believed to have caused over fishing of all stock and population of shrimp by the use of banned gears and methods [2,3]. In general decline in resource availability as evidenced by decline in catches and catch rate and incidence of large proportion of juveniles and young fish in the landings and decrease in average length at capture of many of the targeted species. Considerable volume of discards of non-target edible fishes by the multiday trawlers is also a serious concern. The intensive fishing of prawns at 50 m depth line persistently over the past several years and the destruction of habitats occur during the process of exploitation caused by various kinds of human activities adversely affecting the crustacean fauna and their resources [2,4-8]. The substantial portion of the penaeid catches by indigenous gears such as fixed bag nets ('Dol'), seines, gill nets etc. which operate in the inshore areas. A number of innovative gears such as ring seines, trammel nets and minitrawls operated by motorized country crafts are being increasingly employed along the coasts of India which state led to state of over exploitation.

Many researchers have been surveyed on penaeid species exploitation by using variety of gear in India [9,10-22]. The juvenile destruction in the marine environment in the Palk Bay where young ones of *P. semisulcatus* are indiscriminately captured by 'Thalluvalai' a kind of small conical bagnet draged along the shallow near-shore areas, if allow fully grown to them could get crores of money [23,24]. Damaging practice commonly found in the estuarine systems is the widespread removal of young shrimp for aquaculture purpose. 97% of the shrimp fry are destroyed or thrown on the land during the collection of only 3% seed of tiger shrimp for culture. During wild collection of 1 million *P. monodon*, an estimated annual loss of 75 million non-target fin and shellfish larvae occurs [3].

The maintenance and management of our rich biodiversity requires accurate and continuous updating of data, identification of biological organism and documentation of biological diversity is a primary step towards any research work, management and conservation [1,25]. Sudarsan stated that specimen of prawns had often catches and as such a special survey for prawn resources have to be undertaken [26]. The pioneering survey work on penaeid species diversity in north east coast of India such as Andaman and Nicobar Islands 12 littoral species

[17]. Chanda and Bhattacharya described three new species of shrimp from Indian waters [27-29]. The survey of trawl fishing of Penaeidae shrimp species in the Northern Mandapam Coast of Palk Bay has been described by Siva Rama Krishnan [30]. Radhakrishnan reported annotated checklist of the penaeoid, Sergestoid, Stenopodid and Caridean prawn fauna of India [1]. Not much inventory work has been carried of this penaeidae diversity and details of species exploitation in south east coast of India. Conservation of penaeidae diversity is the urgent need of the hour in order to maintain the balance of nature and support the availability of natural resources for future generation. Assessment of biodiversity of a particular region is very important to formulate conservation strategies [31]. Therefore, the present study on the Penaeidae diversity and exploitation of the east coast on the basis of landing from inshore and offshore water. The objectives of present study are to report the diversity of penaeidae, new species availability in the south east coast of India for a period from April 2010 to April 2013. In addition report the pattern of exploitation and preventive strategies to protect Penaeidae diversity in the study area.

Materials and Methods

Different kinds of Penaeidae shrimp were collected from the landing centers of Chennai, Nagapattinam, Pudukkottai, Ramanathapuram, and Tuticorin of south east coast of India in Figure 1. The collection of species for three years from April 2010- April 2013. Collected shrimps were kept in ice pack, brought to the laboratory. All species were identified and grouped according to published literature [32-35] and nomenclature of Penaeidae based on the availability of published literature present in the form of research articles, monographs, books, species checklist and technical reports. The WoRMS Register, ITIS Standard Search and the Carideorum catalogus also have been referred for confirmation of the genera and species [36].

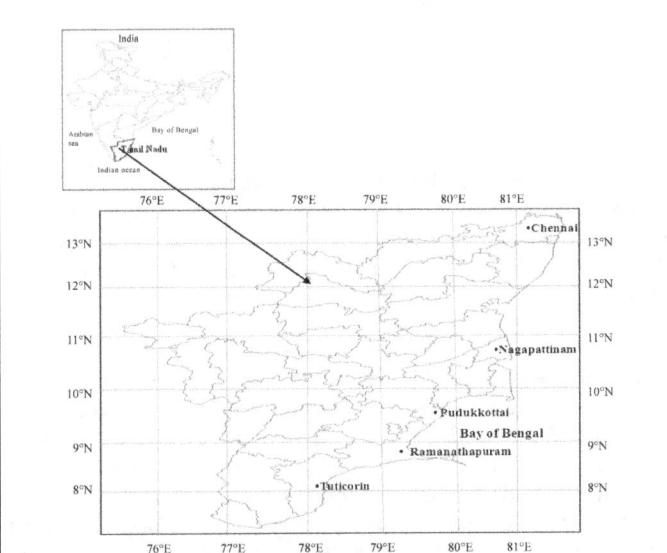

Figure 1: Different kinds of Penaeidae shrimp collected from the provinces of east coast of India.

General diagnostic characters for identification of penaeidae shrimp (FAO Species identification guide for fishery purposes, 1998)

The major criteria used to identify penaeidae shrimp are as follows:- Penaeidae rostrum is well developed and generally extending beyond eyes, always bearing more than 3 upper teeth. No styliform projection at base of eyestalk and no tubercle on its inner border. Both upper and lower antennular flagella of similar length, attached to tip of antennular peduncle. Carapaces lacking both post orbital and post antennal spines. Generally cervical groove are short, always with distance from dorsal carapace. All 5 pairs of legs are well developed, fourth leg bearing a single well-developed arthrobranch (hidden beneath carapace, occasionally accompanied by a second, rudimentary arthrobranch). In males, endopod of second pair of pleopods (abdominal appendages) with appendix masculine only. Third and fourth pleopods divided into 2 branches. Telson sharply pointed, with or without fixed and/or movable lateral spines. Colour: body colour varies from semi-translucent to dark grayish green or reddish, often with distinct spots, cross bands and/or other markings on the abdomen and uropods; live or fresh specimens, particularly those of the genus *Penaeus*, can often be easily distinguished by their coloration. General diagnostic characters for identification of penaeidae shrimp is given in Figure 2.

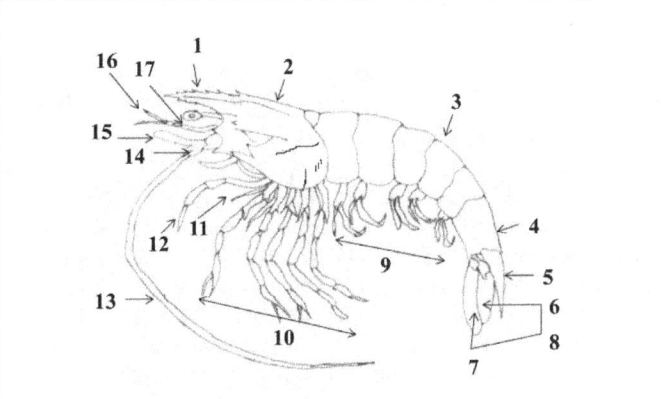

Figure 2: General diagnostic characters for identification of penaeidae shrimp. 1. Rostrum, 2. Carapace, 3.Abdomen, 4.Dorsal crest, 5.Telson, 6. Exopod of uropod, 7, Endopod of uropod, 8. Uropod, 9. Pleopods, 10. Periopods, 11. Exopod, 12. Third maxilliped,13. Antennal flagellum, 14. Antenna, 15. Antennal scale, 16. Antennular flagella, 17. Antennule

Genus wise diagnostic characters followed for penaeidae shrimp identification (FAO Species identification guide for fishery purposes, 1998)

The major criteria used to identify penaeidae shrimp genus wise are as follows: Fifth leg with exopod, carapace without longitudinal or vertical sutures; second leg with ischial spine; eyes small: *Atypopenaeus* (Figure 3a). Third maxilliped with epipod; male petasma asymmetrical: *Funchalia* (Figure 3b). Telson without large subapical fixed lateral spines, body densely covered with short hairs, with grooves and crests on carapace obscure; petasma asymmetrical: *Metapenaeopsis* (Figure 3c). Rostrum without lower teeth, third maxilliped without epipod; male petasma symmetrical, benthic, fifth leg without exopod (carapace without longitudinal or vertical sutures), third maxilliped without epipod; male petasma symmetrical benthic: *Metapenaeus* (Figure 3d and e). Rostrum with lower teeth, abdomen glabrous and smooth: *Penaeus* (Figure 3d), grooved carapace *(Penaeus-Melicertus)*, non – grooved carapace *(Penaeus- non Melicertus)* (Figure 3f). Body almost naked, with crests and grooves on carapace distinct; petasma, symmetrical, Carapace with longitudinal and vertical sutures telson without movable lateral spines: *Parapenaeus* (Figure 3g and h). Carapace with both longitudinal and vertical sutures second leg without ischial spine; eyes large body naked, with crests and grooves on carapace distinct; longitudinal suture usually long third leg without epipod: *Parapenaeopsis* (Figure 3j). Carapace lacking longitudinal and vertical sutures telson with movable lateral spines, rostrum extending far beyond eye; pterygostomian spine present, deep water: *Penaeopsis* (Figure 3c and i). Carapace with both longitudinal and vertical sutures, second leg without ischial spine; eyes large body usually hairy, with crests and grooves on carapace obscure, longitudinal suture short; third leg generally with epipod: *Megokris, Trachysalambria* (Figure 3k). The coloration of the uropods of Megokris are yellowish with grey or brown margins and centre is red or reddish brown with golden margin but in Trachysalambria the colouration of uropods are red or reddish brown, with conspicuous white margin.

Results

Morphological and seasonal variation of Penaeidae species

In all penaeidae species collected in the present study, the female specimens were large size than male specimens. As far as penaeidae species sexual variation, a large copulatory organ on first pair of pleopods in males (petasma), and on posterior thoracic sternites in females (thelycum). In males, endopod of second pair of pleopods was with appendix masculina only. In *Metapenaeus* species in adult male, merus of fifth leg (pereiopod) with basal notch followed by prominent keel was present which was absent in females. In *Fenneropenaeus* species dactyl of third maxillipeds had half as long as protopodus. In *Parapenaeopsis hardiwickii*, the female rostrum very long with sigmoidal shape with distal half is toothless, extending beyond antennular peduncle, in adult male tooth less portion is absent and slightly curved downward, only reaching middle of the second antennular segment. The specimen colour was varied from place to place, depended the residence environmental condition such as sea weeds and sedimentation. *Penaeus monodon, Fenneropenaeus indicus, F. merguiensis, F. penicillatus, Parapenaeopsis, Metapenaeus, Megokris, Trachysalambria* species were obtained highly in the North east monsoon season (October-November), *Penaeus semisulcatus* were obtained throughout the year. The more variety of *Parapenaeus* species were obtained in the summer rainy season (April –May) from the south east coast of India. In the present study, Gender number of Penaeidae species were somewhat equal only.

Distribution of Penaeidae species

In the present study totally fifty nine Penaeidae species were obtained from all landing centers from south east coast of India. The *Penaeus* species like *P. monodon, F. indicus,* and *P. semisulcatus* were obtained mostly in all landing centers rather than other Penaeidae species. The species under genus *Penaeus* senu lato (old *Penaeus), Melicertus latisulcatus* was available only in southern part of east coast

of India (Ramanthapuram and Tuticorin), and species under genus *Penaeopsis, P. jerryi* and *P. rectaculata* also were only in southern part (Tuticorin). The more variety of *Parapenaeus* and *Parapeneopsis* species were landed as by-catch of other Penaeidae species from offshore water in the southern region of south east coast of India (Ramanathapuram).

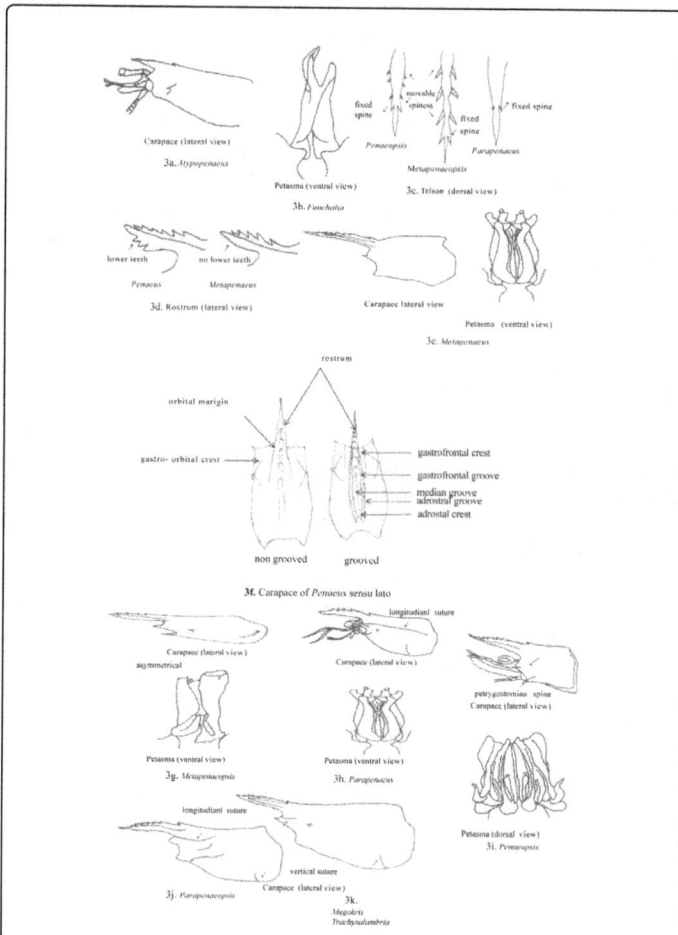

Figure 3: Genus wise diagnostic characters followed for penaeidae shrimp identification. a). *Atypopenaeus,* b). *Funchalia,* c). Telson, d). Rostrum, e). *Metapenaeus,* f). *of Penaeus sensu,* g). *Metapenaeopsis,* h). *Parapenaeus,* i). *Penaeopsis,* j). *Parapenaeopsis,* k). *Megokris, Trachysalambria*

More number of Penaeidae species diversity was seen in central part of south east coast of India (Ramantahpuram and Nagapattinam). *Trachysalambria* and *Megokris* species were mostly landed in southern and northern part of east coast of India (Ramanathapruam, Nagapattinam and Chennai). *P. semisulcatus* landed predominantly in southern regions of southeast coast of India (Tuticorin, Ramanathapuram and Pudukkottai); whereas, *F. indicus* was predominant in northern regions (Nagapattinam and Chennai).

All over south east coast of India, the most dominant species were *Atypopenaeus stenodactylus, Fenneropenaeus indicus, Fenneropenaeus merguiensis, Penaeus monodon, Penaeus semisulcatus, Melicertus latisulcatus, Metapenaeus dobsoni, Metapenaeus monoceros, Metapenaeus affinis, Metapenaeus brevicornis, Metapenaeus lysianassa, Parapenaeopsis stylifera,*

Parapenaeopsis hardwickii, Megokris granulosus, Megokris sedili, Megokris pescadoreensis, Penaeopsis jerryi, Penaeopsis rectaculata, Parapenaeus fissures, Parapenaeus investigatoris, Metapenaeopsis stridulans, Metapenaopsis palmensis, Metapenaeopsis moegensis, Metapenaeopsis barbata, Trachysalambria curvirostris. Lesser number of species landed were *Fenneropenaeus penicillatus, Marsupenaeus japonicus, Metapenaeus ensis, Metapenaeus stebbingi, Funchalia woodwardi, Parapenaeopsis cornuta, Parapenaeopsis tenella, Parapenaeopsis nana, Parapenaeopsis sulptilus, Parapenaeopsis hungerfordi, Parapenaeopsis venusta, Parapenaeopsis coromandelica, Parapenaeopsis gracillima, Kishinoyepenaeopsis maxillipedo (Parapenaeopsis maxillipedo), Ganjampenaeopsis uncta (old name Parapenaeopsis uncta), Parapenaeus longipes, Parapenaeus fissuroides indicus, Trachysalambria aspera, Trachysalambria fulva, Trachysalambria longipes.* Details of variety and distribution of Penaeidae species in the period 2010 -2013 were collected from south east coast of India are given in Table 1. In the present study, new candidate shrimp in Indian water like *Metapenaeopsis wellsi, Kishinouyepenaeopsis amicus* (old name *Parapenaeopsis sinica*), *Parapenaeus lanceolatus [32], Metapenaeus papuensis, P. gracillima, Metapenaeus anchistus, Parapenaeopsis venusta, Trachysalambria longipes, Parapenaeopsis coromandelica, Parapenaeopsis hungerfordi* were landed in south east coast of India in Figure 4. The special morphological characters of new species obtained from south east coast of India distinguished from other Penaeidae species are given in Table 2.

Figure 4: Details of variety of Penaeidae species collected in the period 2010-2013.

Genus	Species	Common Name (English name)
Atypopenaeus	*A. stenodactylus* (Stimpson, 1860)	Periscope shrimp
Funchalia	*F. woodwardi* (Johnson, 1868)	-
Penaeus sensu lato	*Penaeus semisulcatus* (De Haan, 1844)	Green Tiger shrimp
	Penaeus monodon (Fabricius, 1798)	Jumbo Tiger
	Fenneropenaeus indicus (H. Milne Edwards, 1837)	Indian white shrimp
	Fenneropenaeus penicillatus (Alcock, 1905)	Red tail prawn
	Fenneropenaeus Konkani [28]	-
	Marsupenaeus japonicus (Bate, 1888)	Kuruma prawn
	Melicertus latisulcatus (Kishinouye, 1896)	Western king prawn
	Fenneropenaeus merguiensis (De Man, 1888)	Banana prawn
Metapenaeus	*M. brevicornis* (H. Milne Edwards, 1837)	Yellow shrimp
	M. dobsoni (Miers, 1878)	Kadal shrimp
	M. affinis (H. Milne Edwards, 1837)	Jinga prawn
	M. lysiannassa (De Man, 1888)	Bird shrimp
	M. ensis (De Haan, 1844)	Greasy back shrimp
	M. monoceros (Fabricius, 1798)	Speckled shrimp
	M. joyneri (Miers, 1880)	Shiba shrimp
	M. endeavouri (Schmitt, 1926)	Endeavour shrimp
	M. moybei (Kishinouye, 1896)	Moybei shrimp
	M. elagans (De Man, 1907)	Fine shrimp
	M. papuensis (Racek and Dall, 1965)*	Papua shrimp
	M. anchistus (De Man, 1920)*	Spiny greasy shrimp back shrimp
	M. stebbingi (Nobili, 1904)	Peregrine Shrimp
Metapenaeopsis	*M. stridulans* (Alcock, 1905)	Fiddler shrimp
	M. barbata (De Hann, 1844)	Whisked velvet shrimp
	M. palmensis (Haswell, 1879)	Southern velvet shrimp
	M. megoiensis (Rathbun, 1902)	Mogi velvet shrimp
	M. wellsi (Racek, 1967)*	-
	M. toloensis (Hall, 1962)	Tolo velvet shrimp

	M. novaeguineae (Haswell, 1879)	Northern velvet shrimp
Parapenaeopsis	*P. hardwickii* (Miers, 1878)	Spear shrimp
	P. stylifera (H. Milne Edwards, 1837)	Kiddy shrimp
	P. cornuta (Kishinouye,1900)	Coral shrimp
	P. tenella (Spence Bate, 1888)	Smooth shell shrimp
	P. acclivirostris (Alcock, 1905)	Hawknose shrimp
	P. nana (Alcock, 1905)	Dwarf shrimp
	*P. venusta (De Man, 1907)**	Adonis shrimp
	P. coromandelica (Alcock, 1906)*	Coromandel shrimp
	P. gracillima (Nobili, 1903)*	Thin shrimp
	P. sculptilis (Heller, 1862 a)	Rainbow shrimp
	P. hungerfordi (Alcock, 1905)*	Dog shrimp
	P. uncta (New name *Gangampenaeopsis uncta* Alcock, 1905)	Uncta shrimp
	P. maxillipedo (Alcock,1905) New name *Kishinoyepenaeopsis maxillipedo*	Torpedo shrimp
	P. sinica (Liu and Wang, 1987) New name *Kishinoyepenaeopsis amicus* (V.C.Nguyên, 1971)*	
Megokris	*M. sedili* (Hall, 1961)	Malayan rough shrimp
	M. granulosus (Haswell, 1879)	Coarse shrimp
	M. pescadoreensis (Schmitt, 1931a)	Big head Sand Prawn
Trachysalambria	*T. curvirostris* (Stimpson, 1860)	Southern rough shrimp
	T. aspera (Alcock, 1905)	-
	T. fulva (Dall, 1957)	-
	*T.longipes** (Paul'son, 1875)	Long legged rough shrimp
Penaeopsis	*P. jerryi* (Perez Farfante, 1979)	Gondwana shrimp
	P. rectaculata (Spence Bate, 1881)	Needle shrimp
Parapenaeus	*P. investigatoris* (Alcock & Anderson, 1899)	Explorer rose shrimp
	P. longipes (Alcock, 1905)	Flamingo Shrimp
	P. fissuroides indicus (Crosnier, 1986a)	False rose shrimp.
	P. fissurus (Spence Bate, 1881)	Neptune Rose Shrimp
	P. sextuberculatus (Kubo, 1949)	Domino shrimp

*P. lanceolatus** (Kubo, 1949)*	Lancer rose shrimp

Table 1: List of species availability in south east coast of India 2010-2013. *New species in the south east coast of water.

Penaeidae species	Morphological features
Metapenaeus anchistus	Rostrum distinctly directed upward, bearing 10 to 12 teeth along entire upper margin, almost straight and slightly curved downward at tip; rostrum extending to about distal segment of antennular peduncle. Postrostral crest is low. Branchiocardiac crest, distinct. First leg with distinct ischial spine. Telson with 3 pairs of large movable spine, Body covered with fine pubescence. In males, a narrow space between distomedian projections of petasma; in females,lateral plates of thelycum without raised posterior edge and continuous to posterior transverse ridge.
Metapenaeus pauensis	Look like a Metapenaeus elegans, rostrum armed with teeth along entire upper border, armed 8-12 upper teeth, in males, ditomedian projection of petasma directed forward, their inner margin almost parallel, tubercle on merus of fifth leg slightly bent inwards, in female, ridges on lateral plates of thelycum curved outward posteriorly.
Metapenaeopsis wellsi	Posteriolateral carapace no stidulating ridges, pterygostomian spine very strong, rostrum not forming crest.
Parapenaeopsis gracillima	Rostrum short, not extending beyond eyes, first leg without basial spine.
Parapenaeopsis coromandelica	Rostrum sigmoid shape, half-length toothless, telson armed with 1 or 2 pairs of fixed lateral spines.
Parapenaeopsis hungerfordii	Rostrum long and exceeding antennular peduncle, longitudinal suture extending almost to posterior carapace.
Parapenaeopsis venusta	Rostrum short and extending just beyond eyes, longitudinal suture only reaching as far as level of hepatic spine.
Kishinouyepenaeopsis amicus (*Parapenaeopsis sinica*)	Rostrum usually with 9 or 10 upper teeth, third leg with a basial spine. The end of the rostrum upward with black dot, absence of dark band in the last abdominal segment as seen in Kishinouyepenaeopsis maxillipedo.
Parapenaeus lanceolatus	Rostrum extending beyond second segment of antennular peduncle, branchiostegal spine present.
Trachysalambria longipes	Look like Trachysalambria curvirostris, rostrum straight and armed with 8 to 11upper teeth, fourth and fifth abdominal segments without posteromedian incisions fifth leg extending beyond antennal scale, posterior plate of female thelycum without distinct notch.

Table 2: Morphological characters of new species obtained from south east coast of India during 2010-2013

Exploited Penaeidae species

Marsupenaeus japonicus, Melicertus latisulcatus, P. monodon, P. semisulcatus, F. indicus, F. merguiensis, F. penicillatus in *Penaeus* genus, *M. dobsoni. M. monoceros, M. brevicornis*, and *M. ensis* in *Metapenaeus* genus, and *P. stylifera, P. hardwickii* in *Parapenaeopsis* genus were commercially exploited along the south east coast of India. Also the by-catch species like *Parapenaeopsis, Parapenaeus, Penaeopsis, Metapenaeopsis, Atyopenaeus, Megokris, Trachysalambria, Trachypenaeus*, both juvenile and adult from offshore water were thrown as waste are shown in Figure 5. Particularly *Penaeus semisulcatus* juvenile populations were exploited by fishing from inshore water are given in Figure 6.

Figure 5: Exploited Penaeidae species

Figure 6: *Penaeus semisulcatus* juvenile populations were exploited by fishing from inshore water.

Discussion

Diversity of Penaeidae species

The structure of decapod cructacean assemblages on the continental regions is different by spatial differences in environmental and oceanographic conditions particularly by depth, bottom type and characteristic of the water masses [37-43]. In the present study, the south east coast of India from Tuticorin to Chennai, 59 species of Penaeidae were landed, this high number of species availability shows that the good environmental and oceanographic conditions for living of these species. The main species landed were *P. semisulcatus F. indicus, P. monodon, M. latisulcatus, P. stylifera, P. hardiwickii, M. dobsoni*, and *M. brevicornis*. In addition the following deep sea shrimps observed were *Penaeopsis jerryi, Penaeopsis rectaculata, Parapenaeus investigatoris, Parapenaeus fissures, Atyopenaeus stenodactylus*, and *M. stridulensis* were landed in south east coast of India. According to the Central Marine Fisheries Research Institute [44], in south east cost particularly in Tamilnadu, the inshore Penaeidae shrimp comprised of 25 species, of which *P. semisulcatus, F. indicus, Melicertus latisulcatus, Parapenaeopsis maxillipedo, Ganjampenaeopsis uncta* (Old name *Parapenaeopsis uncta*) and *Metapenaeus dobsoni* are predominant and other species of deep sea prawns were *Parapenaeus fissuroides, Penaeopsis jerryi*, and *P. investigatoris*.

According to Suseelan [21], predominant Penaeidae species in the Indian coast are, *F. indicus, P. monodon, P. semisulcatus, F. merguiensis, F. penicillatus, M. dobsoni, M. monoceros, M. affinis, M. brevicornis, P. stylifera, Metapenaeus moyebi, Metapenaeus kutchensis P. hardwickii*, and *P. sculptilis*. In the present study, *F. indicus, P. monodon, P. semisulcatus, F. merguiensis, F. penicillatus, M. dobsoni, M. monoceros, M. affinis, M. brevicornis, P. stylifera, Metapenaeus moyebi*, and *P. sculptilis* was landed. *P. semisulcatus, P. monodon* and *F. indicus* were obtained mostly all landing centre, which indicates the three species are the major commercial species in south east coast of India. The *M. latisulcatus* landed mainly in southern most regions (Tuticorin and Ramanathapuram) which consistent with geographical location of this species as reported by Rao et al. and indicate geographic specific distribution of *Melicertus latisulcatus* [45].

In the present study, *Penaeopsis jerryi* and *Penaeopsis rectaculata*, landed southern region only (Tuticorin), which could be the environmental condition favors such as temperature and substratum sand with mud in deeper region for these two species as reported by John and Kurien, and Radhika Rajasree [46], that the *Penaeopsis* species, *Metapenaeopsis andamanensis* to be showing strong preference towards slightly higher water temperature and a substrate demarcated by mixture of sand and mud.

Kurian and Sebastian reported that the *Parapenaeus longipes, Parapenaeus fissures*, and *Parapenaeus investigatoris* in Indian water landings were by the long trip deep sea trawls from the depth of 70-90m [47]. In present study, the *P. fissures, P. investigatoris, P. longipes, Parapenaeus sextuberculatus, P. fissuroides indicus, Metaepaneopsis barbata, Metapenaeopsis stridulans*, and *Atyopenaeus stenodactylus*, were catched in the south east coast water by long trip deep trawels catching. This observation indicate that the depth profoundly influences the assemblage structure of deep sea prawn and the hydrographic features and the fishing intensity can affect the distribution and abundance of marine species such effect on the species diversity and species richness as stated by Radhika Rajasree [46].

Previously, many new Penaeidae species has been reported like *Parapenaeopsis hardwickii*, *Atypopenaeus compressipes*, *Parapenaeopsis acclivirostris*, *Metapenaeopsis novae-guineae*, and *Trachypenaeus curvirostris* from west cost of India [48], *Penaeopsis eduardoi* from Indo-west pacific region, *Penaeopsis jerryi* from Indian Ocean [49], and *Parapenaeus fissuroides indicus* from west cost of India [50]. In the present study, new Penaeidae species of *Metapenaeopsis wellsi*, *Kishinouyepenaeopsis amicus* (old name *Parapenaeopsis sinica*), *Parapenaeus lanceolatus*, *M. papuensis*, *P. gracillima*, *Metapenaeus anchistus*, *Parapenaeopsis venusta*, *T. longipes*, *P. coromandelica*, and *Parapenaeopsis hungerfordi* were landed in south east coast of India. These Penaediae species were not reported previously in Indian coasts; hence the present report would give the novel insight on Penaeidae diversity in the east coast of India.

Exploitation of Penaeidae species

The total landing of Penaeidae prawn in India was 2,72,969 tons in 2011 in which trawlers account for about 60% and the indigenous gears 40%. The annual report for 2011-12 meant the Penaeidae species participated 7.1% in total marine fish catches, and catches (35,200 t) declined by 1.3% as compared to 2010 [44]. As far as south east coast of India (Tamilnadu) about 20,163 t of penaeid prawns were landed, accounting for 54.8% of the crustacean landings in 2011. About 85.4% of this was landed by trawl nets. The catch of non-penaeid prawns was relatively meager, accounting for 3.4% of the prawn landings. Prawn fishery along south Tamil Nadu coast (off Tuticorin) is done by mechanized trawl, indigenous trawl and gillnet (mainly in the estuarine areas). Mechanized trawls landed 119 t of prawns from inshore waters and 468 t of deep sea prawns. Indigenous trawl landed 46t of prawns while gillnets landed 13t [44]. In the present study, *Penaeus* species *Marsupenaeus japonicus*, *Melicertus latisulcatus*, *P. monodon*, *P. semisulcatus*, *F. indicus*, *F. merguiensis*, *P. penicillatus* species and among the *Metapenaeus* species, *M. dobsoni*, *M. monoceros*, *M. affinis*, *M. brevicornis*, *M. ensis*, *M. endovori*, *M. lysianassa*. *Parapenaeopsis* species *P. stylifera*, and *P. hardwikii*, which grow to a large size, were commercially exploited by Mechanized large-scale operation of ring seines, mini-trawls, trammel net and indigenous gears in south east coast of India.

In deep shrimp trawling, the juvenile and seldom adult of *G. uncta*, *P. cornuta*, *K. maxillipedo*, *P. sulptilis*, *P. tenella*, *M. granulosus*, *M. sedili*, *T. curvirostris*, *Parapenaeus*, *Metapaneaopsis*, *Penaeopsis* species are exploited as by-caught, and are considered less ecomomic value and thrown as waste. Thus deep shrimp exploitation affects the penaeidae and also marine species biodiversity. Limiting the operation of fixed nets like stake nets, dip net etc. together with appropriate mesh size restrictions, and a ban of export of count sizes of shrimps below a fixed minimum level would be the methods for conserving the penaeidae shrimp.

As far as the *Penaeus semisulcatus* juvenile stage in the shallow inshore water seagrass ecosystem is the nursery ground for *P. semisulcatus* which have restricted distribution and are facing depletion. It is estimated that about 2500 indigenous fishing units are engaged in this type of fishing in 0-4 m depth and over 3000 t of the juveniles of the species are exploited every year [20]. In the present observation of *P. semisulcatus* highly landed in southern region, in the inshore water at juvenile stage is major exploitation of this species. In order to protect the juvenile population of these valuable species,

minimum legal size may also be fixed for the capture fishery. A restriction on the export of undersized prawns will discourage capture of smaller size groups of these resources by commercial nets and this will go a long way in improving their fishery.

The fast developing brackish water prawn farming in the country depends on the nature of seeds of fast growing species like *P. monodon* and *F. indicus*. Therefore it is essential to establish adequate number of shrimp hatcheries and legally prohibit seed collection from estuaries. It may be seen from the foregoing account, the penaeidae species which brings considerable foreign exchange to the country are threatened by several ways. For their sustained survival and productivity in the natural habitats. The activities of man and appropriate methods to conservation measures should be implemented before further damage is inflicted on the fauna and its environment.

Conclusion

In present study reports 59 different Penaeidae species landed showed that the healthy diversity of Penaeidae species in south east coast of Indian water. The indiscriminate exploitation of juvenile at inshore areas and deep sea Penaeidae species like *Parapenaeus*, *Metapenaeopsis*, *Penaeopsis*, *Atyopenaeus*, and selected species of *Parapenaeopsis*, *Megokris*, and *Trachysalambria* are thrown as waste which reducing the diversity of Penaeidae. Therefore, the conservation measures generally adopted for this Penaeidae species include restriction of fishing effort, imposition of closed seasons for fishing, allotment of catch quotas, cod-end mesh regulations for fishing nets, and restriction on capturing juveniles from nursery grounds of entire coastline of south east coast of India to protect the prevailing Penaeidae diversity in south east coast of India.

Acknowledgement

The authors express their special thanks to the Department of Science and Technology, New- Delhi, India, for financial assistance from DST–PURSE.

References

1.   Radhakrishnan EV, Deshmukh VD, Maheswarudu G, Josileen J, Dineshbabu AP, et al. (2012) Prawn fauna (Crustacea: Decapoda) of India - An annotated checklist of the Penaeoid, Sergestoid, Stenopodid and Caridean prawns. Journal of Marine Biological Association of India 54: 50-72.

2.   Suseelan C, Pillai NN (1993) Crustacean fishery resources of India - An overview. Indian Journal of Fisheries 40: 104-111.

3.   Quader O (2010) Coastal and marine biodiversity of Bangladesh (Bay of Bengal) Proc. of International Conference on Environmental Aspects of Bangladesh (ICEAB10), Japan.

4.   Gopalan UK, Vengayil D, Udayavarma P, Krishnankutty M (1983) The Shrinking Backwaters of Kerala . Journal of Marine Biological Association of India 25: 131-141.

5.   Suseelan C (1987) Impact of environmental changes and human interference on the prawn fishery resources. The marine fisheries information service Technical and Extension Series 73: 1-5.

6.   Kathirvel M, Suseelan C, Rao PV (1988) Biology population and exploitation of the Indian deep-sea spiny lobster Pueriiliis sewelli Ramadan. Fishing Chimes 8: 16-25.

7.   Suseelan C, Rajan KN, Nandakumar G (1989) The Karikadi fishery of kerala. The marine fisheries information service Technical and Extension Series: 102: 4-8.

8. Suseelan C, Nair KP (1994) Endangered vulnerable and rare estuarine shellfishes of India. In: Threatened fishes of India. Natcon Publication 4: 237-251.

9. Mohamed KH, Suseelan C (1973) Deep-sea prawn resources of the South-West Coast of India. Proc Symp. Living Resources of the seas around India. Spl Publ Cochin : 614-633.

10. Muthu MS (1971) On some new records of penaeid prawns from the East Coast of India. Indian Journal of Fisheries 15: 145-154.

11. Muthu MS (1972a) Parapenaeopsis indica sp. now. (Decapoda, Penaeidae) from the Indian waters. Indian Journal of Fisheries 16: 174-180.

12. Muthu MS (1972b) Taxonomic notes on the penaeid prawn Metapenaeopsis gallensis (Pearson, 1905). Journal of Marine Biological Association of India 14: 564-567.

13. Rao GS (1984) On collection of two species of pelagic penaeids (Crustacea : Decapoda) from the oceanic waters of the South-West Arabian sea. Journal of Marine Biological Association of India 26: 165-166.

14. Ravindranath K (1989) Taxonomic status of the Coromandel shrimp Parapenaeopsis stylifera coromandelica Alcock (Decapoda, Penaeidae). Crustaceana 57: 258-262.

15. Silas EG, Muthu MS (1976a) On a new species of penaeid prawn of the genus Metapenaeus Wood-Mason and Alcock from the Andamans. Journal of Marine Biological Association of India 16: 645-648.

16. Silas EG, Muthu MS (1976b) Notes on a collection of penaeid prawns from the Andamans. Journal of Marine Biological Association of India18: 78-90.

17. Thomas MM (1977) Decapod crustaceans new to Andaman and Nicobar Islands. Indian Journal of Fisheries 24: 56-61.

18. Thomas MM (1979) On a collection of deep sea decapod crustaceans from the Gulf of Mannar. Journal of Marine Biological Association of India 21: 41-44.

19. Thomas MM (1986) Decapod crustaceans from Palk Bay and Gulf of Mannar. In: Recent Advances in Marine Biology. New Delhi: 405-435.

20. George MJ (1979) Taxonomy of Indian prawns (Penaeidae, Crustacea, Decapoda) In: Contributions to Marine Sciences. 21-59.

21. Suseelan C (1996) Crustacean biodiversity, conservation and management. In: Marine Biodiversity: Conservation and management 41-65.

22. Kathirvel M, Thirumilu P, Gokul A (2007) Indian penaeid shrimps their biodiversity and economical values. National symposium on Conservation and Valuation of Marine Biodiversity. Zoological Survey of India: 161-176.

23. Manickam SPE, Arputharaj MR, Rao PV (1989) Exploitation of juveniles of green tiger prawn, Penaeus semisulcatus along Palk Bay and its impact on the tiger prawn fishery of the region. Bulletin of the central marine fisheries Research institutes 44:137-45.

24. Rajamani M, Palanichamy S (2009) Need to regulate the thalluvalai fishery along Palk Bay, southeast coast of India. Journal of the Marine Biological Association of India 51: 223-226.

25. Pathak AK, Singh SP, Kumar R, Chaturvedi R (2012) Glimpses on Marine ornamental and shell fish resources of India. International day for biological diversity Marine Biodiversity 152-154.

26. Sudarsan D (1978) Fish trawl catches of Shoal Bay Port Blair (Andamans) in relation to hydrology and plankton. Matsya 3: 83-86.

27. Chanda A, Bhattacharya T (2002) Melicertus similis a new species of prawn Decapoda: Penaeidae from India. Bombay Natural History Society 99: 495-498.

28. Chanda A, Bhattacharya T (2003) Fenneropenaeus konkani a new species of prawn Decapoda: Penaedae from Indian coast. Science and Culture 69: 229-230.

29. Chanda A, Bhattacharya T (2004) A new species of the genus Parapenaeopsis Alcock, 1900 (Penaeoidea: Penaeidae) from Orissa India. Proceedings of the Zoological Society. 57: 23-27.

30. Sivaramakrishnan T, Rajesh S, Patterson J (2012) Trawl Fishing of Penaeid Prawn in the Northern Mandapam Coast of Palk Bay. World Journal of Fish and Marine Sciences 4: 278-283.

31. Basha SKM, Rajya Laksmi RE, Ratneswara Rao B, Murthy CVN, Savithramma N (2012) Biodiversity and conservation of Pulicat lake-Andhra Pradesh. International Journal of Geology, Earth and Environmental Sciences 2: 129-135.

32. Kubo I (1949) Studies on the penaeids of Japanese and its adjacent waters. Journal of the Toyko University Fisheries 20: 870-872.

33. Holthuis LB (1980) FAO species catalogue. Shrimps and prawns of the world. An annotated catalogue of species of interest to fisheries. FAO Fisheries Synopsis 1: 1-261.

34. Farfante PI, Kensley B (1997) Penaeoid and Sergestoid shrimp and Prawns of the world. Paris: Museum National d'Histoire Naturelle.

35. Chan YT (1988) Shrimps and prawns. FAO Species Identification Guide for Fishery purposes. The living Marine Resources of the western central pacific Rome: FAO. 851-971.

36. De Grave S, Fransen CHJM (2011) Carideorum Catalogus: The Recent Species of the Dendrobranchiate, Stenopodidean, Procaridididean and Caridean Shrimps (Crustacea: Decapoda), Zool. Med Leiden 85: 195-589.

37. Abello P, Valladares FJ, Castellón A (1988) Analysis of the structure of crustacean assemblages off the Catalan coast (North-West Mediterranean). Marine Biology 98: 39-49.

38. Markle DF, Dadswell MJ, Halliday RG (1988) Demersal fish and decapod crustacean fauna of the upper continental slope off Nova Scotia from LaHavre to St. Pierre Banks. Canadian Journal of Zoology 66: 1952-1960.

39. Basford DJ, Eleftheriou A, Raffaelli D (1989) The epifauna of the northern North Sea (56°-61°N). Journal Marine Biological Association UK 69: 387-407.

40. Olaso I (1990) Distribución y abundancia del megabentos invertebrado en fondos de la plataforma Cantábrica. Publicaciones Especiales Instituto Espanol de Oceanografia,1.

41. Macpherson E (1991) Biogeography and community structure of the decapod crustaceanfauna off Namibia (southeast Atlantic). Journal of Crustacean Biology 11: 401- 415.

42. Sarda F (1993) Bio-ecological aspects of the decapod crustacean fisheries in the Western Mediterranean. Aquatic living Resources 6: 299-306.

43. Sarda F, Cartes J, Company J (1994) Spatio-temporal variations in megabenthos abundance in three different habitats on the Catalan deep-sea (Western Mediterranean). Marine Biology 120: 211-219.

44. CMFRI (2012) Annual Report 2011-12. Central Marine Fisheries Research Institute Cochin 186.

45. Rao SG, Subramaniam VT, Rajamani M, Manickam PES, Maheswarudu G (1993) Stock assesement of Penaeus spp. off the east coast of India. Indian journal of Fisheries 40: 1-19.

46. Rajasree RSR (2011) Biodiversity of Deep Sea Prawns in the Upper Continental Slope of Arabian Sea, off Kerala (South West India): A Comparison between Depths and Years Turkish Journal of Fisheries and Aquatic Sciences 11: 291-302.

47. Kurian CV, Sebastian VO (1976) Prawn and Prawn fisheries of India. Hindustan publishing corporation (India) 280.

48. Kunju MM (1960) On new records of five species of Penaeidae Decopoda Macrura :Penaeidae) on the west coast of India. J Mar boil Ass India 2: 82-84

49. Farfante IP (1980) Penaeopsis jerryi new species from the Indian ocean Crustacea Penaeoidea. Proceedings of the biological society of Washington 92: 208-215.

50. Dineshbabu AP (2004) An account on the fishery and biology of Parapenaeus fissuroidesindicus Crosnier, 1985 recorded for the first time from Indian waters. Journal of Marine Biological Association of India 46: 215-219.

26

Marketing and Livelihood Contribution of Fishermen in Lake Tana, North Western Part of Ethiopia

Kidanie Misganaw and Addis Getu*

Department of Animal Production and Extension, University of Gondar, P.O. Box: 196, Gondar, Ethiopia

*Corresponding author: Addis Getu, Faculty of Veterinary Medcine, Department of Animal Production and Extension, University of Gondar, P.O. Box: 196, Gondar, Ethiopia, E-mail: addisgetu2002@yahoo.com

Abstract

The study area was conducted in the North Western part of Lake Tana which are three commercially fish species are found (Tilapia, Catfish and Barbus species). The study was focused on fish production and marketing system. Three landing sites were selected purposively for the survey based on the experience of fishing practices. A total of 95 fishers were interviewed: from each landing site ("Delgie 27", "Goregora 35" and "Infranze 33."). The data collection was conducted from October 2012-June 2013. This consists of both form primary and secondary source. A simple random sampling technique was employed covering fishers. Descriptive and statistical package for social sciences (SPSS V-17) was used in analysing. From sample respondents, 100% were reed boat owners. All sampled fishers from the three fish landing sites were used to catch Nile tilapia (*Oreochromis niloticus*), African catfish (*Clarias gariepinus*) and large barbs (*Labeobarbus* spp.). Fishing, crop production, animal husbandry, petty trade and causal labourer contributed 60%, 21%, 12%, 2% and 5% of fisher's livelihood, respectively. Fisheries development interventions should be aimed at addressing both fish production and marketing problems. The study further suggested that fish quality, fish supply, education and training, licensing of the fishers and improving access to services should receive due attention to improve fish marketing and production system.

Keywords: Fish; Fishers; Lake Tana; Marketing; Production

Introduction

Fishery industry provides a vital source of food, employment, recreation, trade and economic well-being for people throughout the world. Fishery is the most important livelihood option for the inhabitants. It is threatened with problems of overexploitation, environmental degradation and consequently unrecovered resources resulting in loss of its potentials. These resources, although renewable, are not infinite and need to be properly managed, if their contribution to the nutritional, economic and social well-being of the growing world population is to be sustained [1]. Ethiopia depends on the inland waters for the supply of fish as a cheap source of animal protein. It has a number of lakes and rivers with substantial quantity of fish stocks. The total area of the lakes and reservoirs stands at about 7000 to 8000 km^2 and the important rivers stretch over 7000 km in the country [2,3].

Lake Tana contains 28 types of fish species, from this 21 are endemic [4-6] and it is the home of a variety Labeobarbus species [7,8]. But, still there are only three types of fish species that joined the market. These are the Nile tilapia, *Oreochromis niloticus*, the African catfish, *Clarias gariepinus* and *Labeobarbus* species. The other fish species, Beso (*Varicorhinus beso*) contribute very low percentage and mainly found in rivers [9]. The fishery of the Lake Tana expanded through the introduction of modern technology such as the modernized boats with 100 m gillnet. Traditional fishing activity is also growing as provision of modern gillnets is increasing for small-scale fisheries. Thus, fishing is becoming more important both economically and socially for the low-income rural population around the Lake.

At present, the Lake Tana fishery is open access in practice and it is with no well-developed of the fish market system. Besides, illegal trading, absence of clear market policies, fishing during peak spawning and breeding time, lack of standard and grades of fish and poor supply of export market are common problems in the region. Furthermore, weak legal systems, lack of institutions to plan, implement and evaluate fish market development, presence of too many intermediaries, lack of vertical and horizontal coordination on fish production and the market centres are some of the core problems to be mentioned. Despite the significance of fish in the livelihood of many fishers and income generation, fish in the study area has not been given due attention. In Ethiopia, the fishery research activities mainly focused on biological related studies and lesser extent to fishing technology and marketing issues [9]. Systematic and adequate information on the process of market competition and on market structure, conduct and performance was not well identified. Furthermore, fish marketing channels and their characteristics livelihood have not yet been studied. Hence, this study attempts to fill in these gaps. The objective of the study was fish production and marketing system in North Western part of Lake Tana.

Materials and Methods

Study site

The study site was conducted in three fishing landing sites along North Western part of Lake Tana. It is the Largest Lake in Ethiopia. The sites were Enferanz, Delgie and Goregora. The fish market was selected on the basis of the size of operation, number of forward and backward linkages and volume of daily trade on site selection criteria.

Data collection

The study was conducted from October 2012-June 2013. Both primary and secondary data sources were used. In this study, random sampling techniques were employed to select fishers. A total of 95 fishers were interviewed: from each landing site ('Infranze 33", "Delgie 27" and "Goregora 35").

Data analysis

In this study, both descriptive and econometric methods were used to analyse the data. To address the household level marketable supply determinants, a Regression Model was used. Linear Ordinary Least Squares (OLS) Regression has been fitted to analyse and estimate supply of fish in North Western gulf of Lake Tana. According to the model OLS regression [10] can be specified as in the following.

$$y_i = b_o + b_i x_i + u_i$$

where, y_i=Market supply of fish, b_o=Intercept, b_i=Coefficient of ith explanatory variables, x_i=A vector of explanatory variables and u_i =Disturbance term.

The hypothesized independent variables for supply function include:

Y=f (price, inputs, education level, fishing trip, gillnet number, family size, non-fishing income source, fishing ground distance, boat type, distance to market, age and credit services).

Results and Discussion

Demographic characteristics of sample fishers

Fishers were diversified in their demographic aspect (sex, age, marital status, education level and family size). All respondents were male and 81.1% were married and the rest 18.9% were not married. According to Demessie in Lake Tana fishery, fishing is observed totally as men's duty; while the post-harvest activity and marketing seem to be almost women [9]. Most of those married respondents bear responsibility to one or more dependants under them. The overall mean family size was 3.63 with a minimum of 1 and maximum of 8 (Table 1).

Variable	Enferanze		Delgie		Goregora		Total		
	Mean	SD	Mean	SD	Mean	SD	Mean	SD	F-test
Age	35.56	4.04	33.56	10.09	32.56	7.2	33.9	7.11	1.233
Family size	3.9	1.54	3.43	2.54	3.5	2.1	3.63	2.06	0.571
Fishing experience	14.8	7.38	11.46	8.08	14.08	6.32	13.45	7.26	2.084

Table 1: Demographic characteristics of sample fishers (N=95). Source: Own survey result 2012/2013.

Education facilitates the person with ability to do basic communications for business purpose. Most of the respondents involved in this could at least read and write (Table 2). This indicates that the fishing activity is open not only for poor and illiterate communities but also for students and others to get job opportunity after completing secondary school. These groups could be able to interpret market and other information better than those who have less or no education.

Livelihood contribution of fishers

The majority of respondents in the three districts ranked fishing as first priority source of income for the family and followed by income generated from animals and animal products. Trading and labour although are less important as source of income, trading appears relatively important. Analysing the livelihood of the respondents gave an overview of the status and the relevance of fisheries in contributing to the livelihood of the fishers. Of the sampled fishers (Figure 1) 60%, 21%, 12%, 2% and 5% were engaged in fishing, animal husbandry, crop production, petty trade and causal labourer, respectively. Fishing is indicated as the first major source of livelihood for 60% of the fishers and Animal husbandry the second source of livelihood for 21% of the fishers and the third was crop production 6%. Fishing contributes 64.6% [9] and 70% [11] of income among the sampled respondents. This shows fishing involved around the Lake Tana to be one of the major livelihood contribution of among landing sites.

Fishing practices and fish production

All fishers from the three fish landing sites were used to catch tilapia, catfish and *labeobarbus* species. 100% fishers were associated with reed boats for fishing. From the total respondents, fishers working with the reed boats have full owner ship rights over both the small vessels and gears. Fishers who are working with reed boat were estimated to have average of 6.5 gillnet in all fishing landing sites i.e., mean of 6.6 Inferanze and 7.7 for Delgie and 5.3 for Goregora. It was found that fishers spend 18 to 25 days (22 days, on average) per month or 9.09 month per year on fishing (Table 3). According to Demessie fishing was about 257 and 300 days, but the survey result it was about 272.7 days per year for reed boats [9].

Fishers working with reed boat were estimated to produce tilapia, catfish and labeobarbus species rank first to third, respectively. The fish species composition of sample fishers supplied to market in North Western part of Lake Tana was 50%, 30% and 20% of Nile tilapia, *Labeobarbus* species and catfish respectively (Figure 2). According to Abebe Getahun and Tewabe the species composition of Lake Tana is found to be 64%, 21% and 15% for Nile tilapia, catfish and *Labeobarbus* species, respectively [12,13]. Both results indicates sharp decline of the endemic *Labeobarbus* species (reed boat fishers) which needs serious attention for the sustainable development of Lake Tana fishery.

		Sampled fishers				
		Enferanze	Delgie	Goregora	Total	%
Sex	Male	33	27	35	95	100
	Female	0	0	0	0	0
	Total	33	27	35	95	100
Education level	Illiterate	4	4	13	21	22.1
	Read and write	10	2	8	20	21.1
	Primary school education	14	18	10	42	44.2
	High school and preparatory	5	3	4	12	12.6
	Total	33	27	35	95	100
Marital status	Single	7	6	5	18	18.9
	Married	26	21	30	77	81.1
	Total	33	27	35	95	100

Table 2: Socio-economic characteristics of sample fishers.

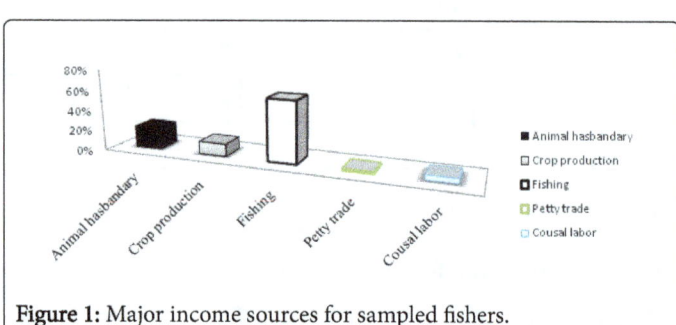

Figure 1: Major income sources for sampled fishers.

Fishery	Name of landing site			
	Inferanze	Delgie	Goregora	Total
Average monthly trip/year for reed boat	9.31	7.1	10.86	9.09
Reed boat average gillnet number/fishers	6.6	7.7	5.3	6.5

Table 3: Average monthly fishing trip per year and average gillnet number/fishers in different landing sites.

Marketable supply of fish and its determinants

Among the 12 predictor variables, eight were found to significantly affect the fisher's level marketable supply (Table 4). Of the predictors variables; non-fishing activity sources (petty trade and casual labour), educational level, fishing trip per year, distance from fishing ground in hours, credit supply for fishing, average selling price of fish (tilapia, catfish and *Labeobarbus*) and number of gillnets were significant variables.

Distance from fishing ground (Fish-ground): A continuous variable that was found significant (P<0.01) is distance from fish landing site that came up with positive signs against the expectation. As distance from landing centre to fishing ground increased by 5.6 hours, fish production increased by 0.003 (or 0.3%) due to less gear competition in the fishing ground.

Age: Age of the fishers, a continuous variable showed no significance (P>0.05), was taken as one of the explanatory variables came up with positive signs against the expectation. As an individual stays long, he will have better knowledge and will decide to allocate more size mesh, produce good quality fish and supply more.

Education level (Edu-level): This variable that was found significant (P<0.01) measured using formal schooling of the fishers and predictors to affect marketable supply positively. This is due to the fact that fishers with good knowledge can adopt better practices than illiterates that would increase marketable supply. Holloway argued that education had positive significant effect on quantity of milk marketed in Ethiopian highlands [14].

Income from the non-fishing sources (Income-Source): It is continuous variable measured in percentage. Petty trade was one of the significant variables (P<0.05) but in contrary to the proposed direction of influence and causal labour was one of the significant variables (P<0.01) affect marketable supply positively.

Fishers own consumption (Consumed-year): A continuous variable that was found no significant (P>0.05) contrary to the proposed direction of influence. As fishers own consumption in kg per year increased by 0.042 kg the probability to participate in fish supply market decrease by 0.093 (9.3%).

Variable name	Unstandardized Coefficients		Standardized Coefficients	t	Sig.
	B	Std. Error	Beta		
Fishing-ground	5.7	1.9	0.2	3.0	0.003**
Age	0.3	0.3	0.1	1.3	0.209
Educational level	4.0	1.1	0.3	3.7	0.000***
Land-owned	9.2	5.7	0.1	1.6	0.108
Animal husbandry	0.2	0.3	0.1	0.8	0.432
Petty trade	-0.7	0.3	-0.2	-2.3	0.026*
Casual-labor	1.1	0.3	0.3	3.5	0.001**
Consumed –year	-0.04	0.0	-0.2	-1.7	0.093
Avg-selling price	-6.001	2.691	-0.237	-2.230	0.029*
Credit-supply	0.001	0.000	0.378	4.714	0.000***
Fishing – trip	-3.032	1.132	-0.221	-2.679	0.009**
Gillnet – No	1.494	0.299	0.628	5.002	0.000***
Constant	7.081	14.694		0.482	0.631
R2 /R Square/	0.664				
Adjusted R Square	0.611				
F Value	12.67				

Table 4: Determinants of fish marketed surplus (OLS result). Dependent Variable: Volume of fish production for market supply. **Note:** *** represents significance at 0.1%, ** at 1% and * at 5%.

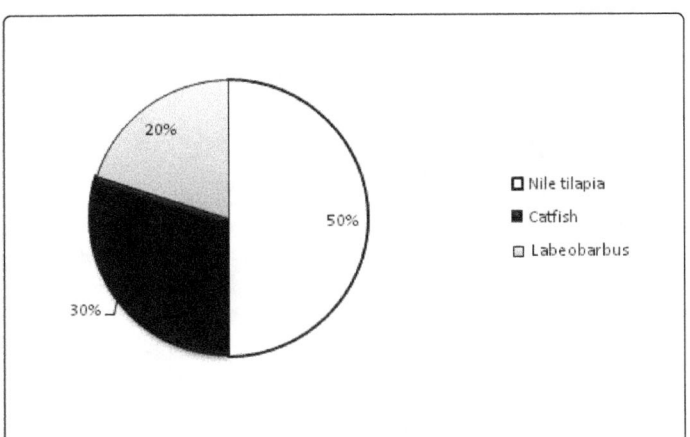

Figure 2: Fish species composition from the sampled reed boat fishers.

Average selling price of fish (Avg-sell): It is one of the continuous variables that was found to be significant (P<0.05). Tomek argued that the product price has direct relations with marketable supply and hence it was expected to affect the fisher's marketable supply of fish [15]. The average selling price decrease by 6 Birr fish production supply to the market decrease by 0.029 (2.9%).

Access to credit (Credit-Supp): Access to credit is measured as continuous significant variables (P<0.01) and is expected to influence the marketable supply of fish positively on the assumption that access to credit improves the financial capacity of fishers to buy more motorized and improved fishing gear, thereby increasing fish production. As the access to credit increased by one, fish supplied to market increased by one quintal.

Working month for fishing (fish-trip): Month for fishing was one of the significant variables (P<0.01) but in contrary to the proposed direction of influence. As can be understood from the results to participate in fish production increased by 0.009 (or 0.9%) when the working month for fishing decreased by three month. This indicates as fisher's increased working month due to traditional reed boat has less operational cost than commercial motorized boat. Fisher's tendency to use reed boat, resulted in less supply.

Number of Gillnet (Gillnet-No): Gillnet number was also another variable found highly significant (P<0.01) to influence the volume of fish supplied to market. It came up with positive sign and it was as expected. As the number of gillnet increased, fish supplied to market increased.

Conclusions

In the North Western part of Lake Tana significance of fish in the livelihood of many fishers and income generating fish in the study area. All sample fishers from the three fish landing sites are used to catch tilapia, catfish and *Labeobarbus* species. The fishers respondents used for fishing with 100% had been using the reed boats for fishing. Fishers said that fish production decrease from year to year due to the use of illegal fishing practice (catch undersized fish by using narrow mesh size), fishing during peak spawning and breeding season, deforestation of vegetation around the Lake and lack of attention. Policy implications are suggested so as to be considered in the future intervention strategies which are aimed at the promotion of fish production and marketing in the study area in particular and in the country in general.

Acknowledgement

The authors would like to thank the University of Gondar, Research and community core process for providing fund.

References

1. FOA (1995) Review of the Fisheries and Aquaculture sector, Ethiopia, FAO Rome.

2. Mebrat A (1993) Overview of the fishery sector in Ethiopia. In: Proceedings of the National Seminar on Fisheries Policy and Strategy. Addis Ababa, Ethiopia pp: 45-53.

3. Wood R, Talling J (1988) Chemical and algal relationship in salinity series of Ethiopian inland waters. Hydrobiologia 158: 29-67.

4. Dejen E (2008) Endowments, potentials and constraints on fisheries and aquaculture in the Tana and Beles sub-basins.

5. Vijverberg J, Sibbing FA, Dejen E (2009) Lake Tana: Source of the Blue Nile. The Nile: Origin, Environments, Limnology and Human Use, Monographiae Biologicae, Springer pp: 163-192.

6. Getahun A, Dejen E, Anteneh W (2008) Fishery studies of Rib River, Lake Tana Basin, Ethiopia. Presented to the World Bank-financed Ethiopian-Nile irrigation and drainage project coordination Office, Ministry of Water Resources Addis Ababa, Ethiopia.

7. de Graaf M (2003) Lake Tana's piscivorous barbus (Cyprinidae) Ethiopia: Ecology; Evolution; Exploitation. Wageningen Institute of Animal Sciences, Wageningen University, The Netherlands.

8. Dejen E (2003) Ecology and potential for fishery of the small barbus (Cyprinidae, Teolesti) of Lake Tana, Ethiopia. The Netherland, Wageningen University.

9. Demessie S (2003) Socio-economic study on Lake Tana Fishery: Its role in the livelihood of one fishing community and local people in the region. Norwegian Fisheries College of Science, University of Troms, Norway.

10. Gujarat DN (2004) Basic of econometrics, 4thEd. McGraw Hill Company, in United States Military Academy, West Point.

11. Tefera B, Tessema A, Dejen E (2009) Dry fish market assessment from Lake Tana to Metema. Amhara Regional Agricultural Research Institute (ARARI), Bahir Dar Fishery and Other Aquatic Life Research Center.

12. Getahun A, Dejen E, Anteneh W (2008) Fishery studies of Rib River, Lake Tana Basin, Ethiopia. Presented to the World Bank-financed Ethiopian-Nile Irrigation and drainage project coordination Office, Ministry of Water Resources Addis Ababa, Ethiopia.

13. Tewabe D, Goshu G (2010) Status of Lake Tana commercial fishery, Ethiopia. Amhara Regional Agricultural Research Institute (ARARI), Bahir Dar Fishery and Other Aquatic Life Research.

14. Holloway G, Nicholson C, Delgado C (1999) Agro industrialization through Institutional Innovation: Transactions Costs, cooperatives and milk-market Development in the Ethiopian Highlands. Missed Discussion p: 35.

15. Tomek WG, Robinson KL (1990) Agricultural products prices. 3rd Ed Cornel University Press. London p: 360.

27

Histopathological Study in Stomach and Intestine of *Anabas testudineus* (Bloch, 1792) under Almix Exposure

Palas Samanta[1,2], **Sandipan Pal**[3], **Aloke Kumar Mukherjee**[4], **Debraj Kole**[1] **and Apurba Ratan Ghosh**[1*]

[1]*Ecotoxicology Lab, Department of Environmental Science, The University of Burdwan, Golapbag, Burdwan 713104, West Bengal, India*

[2]*Division of Environmental Science and Ecological Engineering, Korea University, Anam-dong, Sungbuk-gu, Seoul 02841, Republic of Korea*

[3]*Department of Environmental Science, Aghorekamini Prakashchandra Mahavidyalaya, Subhasnagar, Bengai, Hooghly 712611, West Bengal, India*

[4]*Department of Conservation Biology, Durgapur Government College, Durgapur 713214, West Bengal, India*

*Corresponding author: Apurba Ratan Ghosh, Ecotoxicology Lab, Department of Environmental Science, The University of Burdwan, Golapbag, Burdwan 713104, West Bengal, India, E-mail: apurbaghosh2010@gmail.com

Abstract

The aim of the present study was to investigate the histopathological alterations in the stomach and intestine of Indian freshwater teleost, *Anabas testudineus* (Bloch, 1792) after Almix® exposure both under laboratory and field conditions. The field (dose 8 g/acre) and laboratory (dose 66.67 mg/l) experiments was carried out for 30 days. Special type of cage was prepared and installed in the pond for the field experiment. Pathological alterations in the concerned fish organs namely stomach and intestine were assessed through light microscopy, scanning and transmission electron microscopy. Lesions observed under light microscopy also endorsed the findings of ultrastructural observations both under laboratory and field conditions. Cytopathological alterations observed under light and electron microscopy revealed that the degree of responses were different in different fish tissues as well as under conditions, here in particular effects in stomach were more prominent in laboratory condition. The overall responses registered in the fish tissues under laboratory condition were more pronounced than field condition. Therefore, these symptoms and/or alterations in the present study due to almix intoxication could be considered as biomarkers in toxicity study in aquatic ecosystem.

Keywords: Cytopathological; Stomach; Intestine; *Anabas testudineus*; Almix

Introduction

In the agricultural fields, the use of herbicides to protect the crops from the attack of pests and unwanted plants has been considered as an integral part of the modern agricultural practices worldwide. But, the indiscriminate use of it might endanger the aquatic ecosystems and fish farms close to the agricultural fields, as they ultimately reach to these aquatic bodies as runoff and caused harmful effects to the natural inhabitant reside in water especially non-target aquatic organisms such as aquatic insects, molluscs and fish. Almix is one of the most widely used herbicide in Indian agricultural fields in recent times. It is sulphonyl urea group type of herbicide, and is composed of 10.1% metsulfuron methyl, 10.1% chlorimuron ethyl and remaining 79.80% adjuvants [1]. It is used for controlling the broad leaf weeds and sedges such as such as *Cyperus iria* (Linnaeus, 1753), *Cyperus defformis* (Linnaeus, 1836), *Frimbristylis* sp., *Eclipta alba* (Linn), *Ludwigia parviflora* (Roxb, 1820), *Cyanotis axillaris* (Don, 1826), *Monochoria vaginalis* (Presl, 1827), *Marsilea quadrifoliata* (Linn), etc., both in the terrestrial and aquatic system. It is a selective, both pre-emergent and post-emergent herbicide and destroy the unwanted plants both through contact and systematic pathway. It was applied in the field at a very low use rate i.e., 8 g per acre and did not show any volatilization property; therefore do not affect the adjacent crops [1].

Sentinel organisms play an important role in the assessment of environmental quality and simultaneously provides a sensitive as well as reliable approach to evaluate the contamination level caused by xenobiotic substances in aquatic bodies [2]. Fish, among them considered as an excellent experimental aliquot for toxicity studies because they are the best understood organisms in the aquatic environment, held at the top of the trophic level and finally, they are directly exposed to these xenobiotic substances directly via surface run-off or indirectly through food chain [3,4]. Therefore, the use of fish for better understanding the pollution-induced environmental conditions in the aquatic environment have gained much more importance worldwide in last few decades and helps to monitor the health status of the entire aquatic environment [4,5]. In the present study, *Anabas testudineus* (Anabantidae) was selected as experimental model for toxicity study. Some of the characteristics of this fish species make them as excellent experimental model such as wide distribution in aquatic environment, non-invasive property, wide availability throughout the year, economic importance and ease acclimatization e.t.c.

A number of studies demonstrated the histopathological alterations including ultra structural observations (scanning electron microscopy and transmission electron microscopy) which is considered as an efficient and extensively used methods to evaluate the health status of the organisms exposed to a complex mixture of environmental contaminants both in the laboratory and field conditions [6-8]. One of the most important advantage of using histopathological biomarkers in monitoring the environmental quality is that it allows only the examination in the specific target organ toxicity, in particular, stomach and intestine. In addition, histopathological biomarkers also play a pivotal role in assessing the overall health status of the entire population in the aquatic ecosystem. Furthermore, the alterations observed in these target organs are more easier and reliable to identify specifically than the functional ones [9], and ultimately serve as warning signal of deterioration in animal health [10,11]. These biomarkers, in last few decades, have opened up a new vista in assessing the aquatic ecosystem toxicology as the fish alimentary canal are continuously being exposed during the digestion of ingested food stuff contaminated with xenobiotic substances directly through primary producer organisms. In last few years, a number of studies are available on biochemical, physiological and metabolic alterations of this herbicide in different fish species [12-20]. Regarding the pathological alterations through histological and ultrastructural observations of this herbicide on various organs in different fish species are scanty [19,20] as major advancement in science has been made in recent years. Therefore, considering this scarce information of this agrochemical, the objectives of the present investigation was to characterize and compare the histological and ultrastructural alterations induced by Almix, with particular emphasis on stomach and intestine of *Anabas testudineus*.

Materials and Methods

Fish

Indian Freshwater teleost, *Anabas testudineus* (Bloch, 1792) with an average weight of 23.58 ± 2.05 g and total length of 11.15 ± 0.548 cm, respectively were purchased from the local fish farm and were acclimatized for 15 days. During acclimatization, fish were kept in continuously aerated water (250 L capacity) with a static system, and at natural photoperiod of 12 h light/12 h dark. Average value of water parameters during the acclimatization period were as follows: temperature, 18.61 ± 0.81°C; pH, 7.23 ± 0.082; electrical conductivity, 413.67 ± 0.90 µS/cm; total dissolved solids, 295.11 ± 1.16 mg/l; dissolved oxygen, 6.46 ± 0.22 mg/l; total alkalinity, 260.00 ± 16.90 mg/l as $CaCO_3$; total hardness, 177.33 ± 5.50 mg/l as $CaCO_3$; sodium, 19.20 ± 0.36 mg/l; potassium, 2.45 ± 0.22 mg/l; orthophosphate, 0.02 ± 0.002 mg/l; ammoniacal-nitrogen, 2.31 ± 0.43 mg/l and nitrate-nitrogen, 0.30 ± 0.06 mg/l. After completion of acclimatization, fish were separated into two parts: one group of fish was transferred to field ponds situated at Crop Research and Seed Multiplication Farm (CRSMF) premises of the University of Burdwan, and remaining fishes were brought to the laboratory aquarium. Fish were fed commercial fish pellets (32% crude protein, Tokyu) once a day during acclimatization and experimentation. The experiments were conducted in accordance to the guidelines of the University of Burdwan for animal experiment and were authorized by the Ethical Committee of this University.

Field experimental design

After transferring the fish to the field ponds, fish were divided into two groups: control group with 10 fish species in each cage (triplicate), and exposure group also with 10 fish species in each cage (triplicate). The recommended dose (8 g/acre) used for rice cultivation was dissolved in water and sprayed on the surface of each Almix-treated plots on the first day. Almix® was purchased from the local market (DuPont India Pvt. Ltd., Gurgaon, Haryana, India). The duration of the experiment was 30 days. Special type of cage was prepared based on the method described by Chattopadhyay with slight modifications and installed at the ponds of CRSMF premises [21]. Cages were square in shape with the dimension of 2.5 m × 1.22 m × 1.83 m (submerged height was 0.83 m) and were structured by light strong bamboo. Four-sided wall of the cage, floor and cage cover was fabricated with nylon net, which was embraced by two PVC (poly vinyl chloride) nets: the inner mesh (1.0 × 1.0 mm²) and outer mesh (3.0 × 3.0 mm²). The average of pond water, during the experimentation period were as follows: temperature, 15.67 ± 0.15°C; pH, 7.89 ± 0.03; electrical conductivity, 390.33 ± 2.19 µS/cm; total dissolved solids, 276.33 ± 1.45 mg/l; dissolved oxygen, 7.47 ± 0.09 mg/l; total alkalinity, 101.33 ± 0.68 mg/l as $CaCO_3$; total hardness, 152.0 ± 2.32 mg/l as $CaCO_3$; sodium, 20.56 ± 0.29 mg/l; potassium, 2.89 ± 0.11 mg/l; orthophosphate, 0.12 ± 0.01 mg/l; ammoniacal-nitrogen, 6.06 ± 0.88 mg/l and nitrate-nitrogen, 0.58 ± 0.02 mg/l.

Laboratory experimental design

After acclimatization, fish were transferred to laboratory aquarium and maintained in six aquariums, three for control and three for treatment in the Ecotoxicology Lab, Department of Environmental Science, The University of Burdwan. Each aquarium contains 10 fish (40 L capacity). Treated aquarium exposed to single sub-lethal concentration of Almix i.e., 66.67 mg/l for a period of 30 days [13-18]. On every alternate day water was replaced and after water replacement dose was applied. During experimentation almix-treated and control were subjected to same environmental conditions. Average water parameters, during the experimentation period, were as follows: temperature, 19.67 ± 0.29°C; pH, 7.48 ± 0.05; electrical conductivity, 478.33 ± 9.70 µS/cm; total dissolved solids, 341.44 ± 6.56 mg/l; dissolved oxygen, 5.82 ± 0.39 mg/l; total alkalinity, 317.30 ± 15.60 mg/l as $CaCO_3$; total hardness, 188.89 ± 8.58 mg/l as $CaCO_3$; sodium, 21.36 ± 0.76 mg/l; potassium, 2.80 ± 0.29 mg/l; orthophosphate, 0.02 ± 0.001 mg/l; ammoniacal-nitrogen, 6.63 ± 1.16 mg/l and nitrate-nitrogen, 0.46 ± 0.11 mg/l.

Sampling

Water quality during acclimatization and experimentation was assessed as per APHA [22]. At the end of the experiment (i.e., 30 days), fish were collected from both conditions using hand net and were anesthetized with tricaine methanesulphonate (@ 100 mg/l). After anesthetization, fish were dissected and desired organs namely stomach and intestine were taken immediately and fixed in respective fixatives prescribed for histological, scanning and transmission electron microscopic study.

Histopathological analysis

Stomach and intestine after dissection were fixed in aqueous Bouin's solution for overnight. Then dehydrated through graded series of ethanol (70%, 90% and 100%) and finally embedded in paraffin for preparing the paraffin block. Tissue sections were cut at 3-4 µ using Leica RM2125 microtome and stained with haematoxylin-eosin (H&E). Finally stained sections were examined under Leica DM2000 light microscope and photographs were taken by Leica Image Organizer software to examine the pathological alterations.

Ultra structural analysis

For electron microscopic (SEM) study, stomach and intestine were fixed in 2.5% glutaraldehyde solution prepared in phosphate buffer (0.2 M and pH 7.4) for 24 h at 4°C and then post-fixed with 1% osmium tetra oxide solution prepared in same phosphate buffer for 2 h at 4°C. After fixation tissues were washed with phosphate buffer and dehydrated through graded series of acetone, followed by amyl acetate and finally subjected to critical point drying with liquid carbon dioxide in CPD (critical point drying) machine. After drying tissues were mounted on metal stub and sputter-coated with gold (thickness 20 nm) and examined under scanning electron microscope (Hitachi S-530) at University Science Instrumentation Centre of the University of Burdwan, Burdwan, West Bengal, India and photographs were taken for analysis by Image Organizer software.

For transmission electron microscopic (TEM) study, stomach and intestine (2 × 2 mm in size) were fixed in Karnovsky fixative prepared in 0.1 M phosphate buffer for 12 h at 4°C and then post-fixed with 1% osmium tetra oxide prepared in phosphate buffer (0.2 M and pH 7.4) for 2 h at 4°C. After fixation tissues were washed with phosphate buffer and then dehydrated through graded series of acetone, infiltrated and finally embedded in epoxy resin (araldite CY212). Ultrathin sections of the respective tissues were then cut by using a glass knife on "Ultracut E Reichart-Jung" machine (thickness 70 nm). Sections were then collected on naked copper-meshed grids, and stained with uranyl acetate and lead citrate. Finally, tissues were examined under TECHNAI G2 high resolution transmission electron microscope at Electron Microscope Facility, Department of Anatomy, AIIMS, New Delhi, India and photographs were captured by Image Organizer software.

Results

Stomach

Histologically, stomach is made up of as usual of four layers viz., mucosa, submucosa, strong muscularis, and serosa. The gastric mucosa is lined with a single layer of compactly arranged columnar epithelial cells (CEC) with centrally placed nuclei. The tubular gastric glands are present at basal portion of gastric mucosa. In gastric gland, the gastric cells with centrally placed nucleus are present such as encircling the central lumen. Gastric glands are simple, tubular along with either rounded or elongated in shape. Sub mucosa is well vascularised with thick layer of loose connective tissue (Figure 1.1). Most notable changes observed under light microscopy in the laboratory condition were degenerative changes in columnar epithelial cells, fatty deposition in the basal region, brush border disappearance, top plate thinning, damage in gastric glands and mucosal folds in stomach of A. testudineus (Figure 1.2), while under field condition no such prominent changes were observed (Figure 1.3).

SEM study also confirmed the damages observed under light microscopy such as severe degeneration in CEC such as fragmented CEC, severe mucus secretion over epithelial surface and damage in the microridge structures (Figures 1.4 and 1.5), while under field condition damages were comparatively less than laboratory condition (Figure 1.6). Transmission electron microscopic observation showed deformation in nucleus and mitochondria (Figure 1.7), damage in rough endoplasmic reticulum, and vacuolations in stomach of A. testudineus (Figure 1.8), but only deformed mitochondria and vacuolations were observed under field condition and damages were less than laboratory condition (Figure 1.9).

Intestine

Intestine, histologically, also possesses prominent four histological layers like stomach. The intestinal villi are narrow and slender. The mucosa of intestine is made up of simple, long absorptive columnar epithelial cells each with basally and centrally placed nucleus. Mucous cells are present scatteredly throughout the intestinal mucosa. The

loose connective tissue fibres of submucosa projected into the mucosal folds forming the lamina propria. The lamina propria is narrow, long vascular and mucous cells are dispersed. Columnar epithelial cell are prominent and nucleus are centrally placed and deep stained (Figure 2.1). The most conspicuous changes in intestine under laboratory condition were severe damage in CEC, distortion in connective tissues of lamina propria, detachment of epithelial layer from lamina propria and severe mucus secretion (Figure 2.2), while under field condition intestine showed almost normal appearance but in some places mucus secretion was prominent (Figure 2.3).

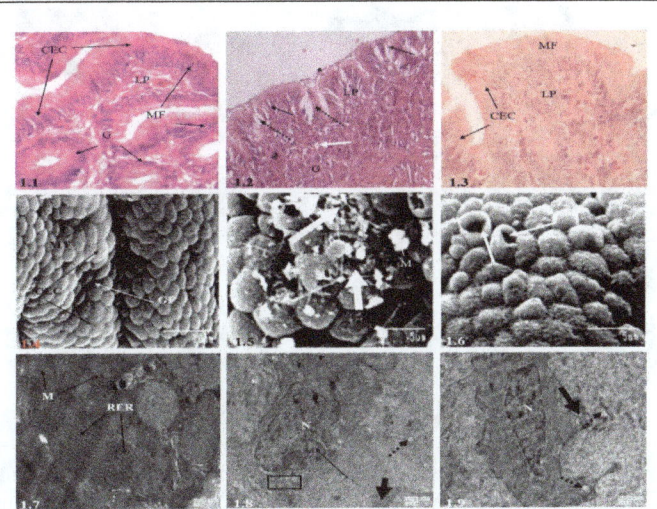

Figure 1: Photomicrographs of stomach in *A. testudineus* showing control condition (C), laboratory condition (AL) and field condition (AF).

where 1.1: Normal and compact arrangement of columnar epithelial cells (CEC) with distinct nucleus under light microscopy and presence of rounded gastric glands (G) separated by lamina propria (LP) (C × 1000), 1.2: Showing degenerated CEC (arrow), vacuolization (broken arrow), thin top plate (arrow head) and lesions in gastric gland (white arrow) under normal microscopy (AL × 400), 1.3: Compact CEC with distinct nucleus under light microscopy (AF × 1000), 1.4: SEM observation showing normal mucosal folds (MF) surrounded by oval or round shaped CEC and stubby microvilli (MV). Note presence of gastric pits (GP) (C × 3000), 1.5: Showing degeneration in columnar epithelial cells (bold arrow) and mucus secretion (M) under scanning electron microscopy (AL × 6000), 1.6: Showing damage on tip of CEC (arrow) under SEM (AF × 6000), 1.7: Showing normal gastric glands with mitochondria (M) and rough endoplasmic reticulum (RER) under TEM observation (C × 7000), 1.8: Deformed nucleus (arrow) and mitochondria (bold arrow), damage in RER (square) and vacuolization (broken arrow) under TEM study (AL × 4000), and 1.9: Presence of deformed mitochondria (bold arrow) and vacuolization (broken arrow) under TEM obsevation (AF × 5000).

Ultra structural lesions displayed severe mucus secretion over epithelial surface and necrosis under laboratory condition (Figures 2.4-2.9). While mucosal folds and CEC showed less damage in comparison to laboratory condition but in-between the primary mucosal folds debris of the fragmented secondary mucosal folds was observed under field condition under SEM study (Figure 1.6). TEM study also showed fatty deposition and vacuolations, damage in the glycocalyx structure, dilated mitochondria, and damage in the tubular network under laboratory condition (Figure 1.8), while mitochondrial deformation and vacuolations were prominent under field condition (Figure 1.9).

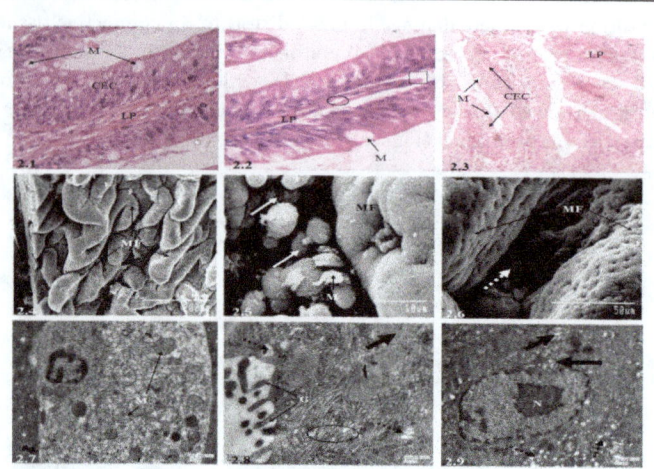

Figure 2: Photomicrographs of intestine in A. testudineus showing control condition (C), laboratory condition (AL), field condition (AF).

where 2.1: Normal lamina propria (LP), columnar epithelial cells (CEC) under light microscopy (C × 1000), 2.2: Damage in lamina propria (oval), detachment of epithelium layer from lamina propria (square) under light microscopy (AL × 1000), 2.3: Light microscopy showed almost normal CEC with distinct nucleus (AF × 400), 2.4: Distinct mucosal folds (MF) with oval or round shaped CEC and supported by microvilli (MV) under SEM observation (C × 200), 2.5: Showing deformed MV (arrow) and mucin droplets (M) under SEM observation (AL × 3000), 2.6: Showing normal mucosal folds (MF) and mucin droplets between MF (broken arrow) (AF × 1000), 2.7: TEM observation showing normal appearance of columnar epithelial cells with abundant mitochondria (M) (C × 5000), 2.8: Showing severe vacuolization (broken arrow), dilation in mitochondria (bold arrow), damaged tubular network (oval) and glycocalyx (G) under TEM (AL × 9900), and 2.9: TEM observation showed deformation in mitochondria (bold arrow) and vacuolization (broken arrow) (AF × 4000).

Discussion

Present study is reporting first time the toxicity of the sulfonylurea-based commercial agrochemical, Almix with regard to histological and ultrastructural observations through scanning and transmission electron microscopy in A. testudineus under field and laboratory conditions on comparative basis, although Senapati reported histopathological alterations under laboratory condition in oesophagus, buccopharynx, stomach and intestine of A. testudineus and Samanta on some biochemical parameters in different fish species including A. testudineus but only under laboratory condition [13-20].

Stomach is one of the prime organs of fish alimentary canal and plays an important role in the digestion of ingested food stuffs for the growth and development of fish species. Histopathological study under light microscopy showed marked differences in gill epithelium between two conditions. In the present study, degenerative changes in columnar epithelial cells, vacuolated basal region, brush border disappearance,

top plate thinning, distortion in gastric glands and fusion of submucosa with mucosal folds were frequently observed pathological alterations in stomach of A. testudineus. Damage in gastric glands observed under present study was also reported by Crespo and indicating lower production of mucin which ultimately reduced the protection ability of the gastric epithelium against the chemical injuries [23]. Distortion in the digestive glands also hampers the digestive enzymes production which ultimately leads to reduced absorption of food materials by intestinal part. Loss of structural integrity and vacuolation seen under this study were also reported by Establier, Sastry and Gupta [24,25]. Swelling, distortion and/or vacuolation in the mucosal epithelial cells of stomach also reported after chronic exposure of endosulfan and methyl ethyl mercurial in *Gymnocorymbus ternitzi* by Amminikutty and Rege [26]. Vacuolization in sub mucosa, shrinkage of mucosal folds observed under present study was also reported by Ghanbahadur in *Rasbora daniconius* after endosulfan exposure [27]. Prominent ultrastructural alterations under laboratory condition include severe degeneration in CEC such as fragmentation, severe mucus secretion and damage in microridge structures observed under scanning electron microscopy. Severe mucus secretion, damage in the microridge structures and CEC observed under SEM observation can also be corroborated with the findings of light microscopic observations and this might be due to herbicidal action which ultimately reduces the protection ability of gastric epithelium and triggers the activity of the gastric epithelium. The results were also agreement with the findings reported by Senapati [19]. Haque also observed damages in columnar epithelial cells which include fragmentation, profound mucus secretion and loss of microridge structure in stomach of *Channa punctatus* exposed to fluoride [28]. At the transmission electron microscopic level, gastric epithelium of stomach showed deformation in nucleus and mitochondria, damage in rough endoplasmic reticulum, and vacuolation under both conditions but the degree of changes was comparatively less in field condition. Similar result as observed under present study was also reported by Rebolledo and Vial [29]. On the other hand, Carrassón demonstrated abundance of rough endoplasmic reticulum and mitochondria, damage in tubule-vascular network, and heterochromatinic nucleus in stomach of *Dentex dentex* [30]. Comparatively less pathological alterations under field condition might be due to self-perpetuating mechanism as fish are in natural habitat and taking almost natural food.

Intestine is the next most important part of the fish alimentary canal after stomach and plays an important role in digestion and absorption of food materials as well as considered as a sensitive organ for toxicity assessment of xenobiotic substances in fish species as they are directly exposed to complex mixture of toxic substances via ingestion of contaminated food stuffs or indirectly via blood and/or lymph [31]. A number of studies on histopathological effects of different pesticides on fish intestine have been reported by several authors but histological and ultrastructural studies related to intestinal epithelium due to Almix exposure, are relatively scanty [32-35]. Walsh and Velmurugan noticed degeneration in the tip of villi, loss of structural integrity in mucosal folds, hypertrophy, vacuolation and necrosis in *Cyprinus carpio* and *Cirrhinus mrigala* exposed to atrazine and fenvalerate, respectively [35,36]. These pathological alterations can also be resembled with our findings observed under present study such as damage in CEC, distortion of connective tissues in lamina propria, detachment of epithelial layer from lamina propria and excessive mucus secretion. Similar type of pathological alterations as observed in the present investigation was also reported in *C. batrachus* and *C.*

mrigala after pesticidal exposure by Mandal and Sharma [37,38]. Ravanaiah and Narasimha Murthy also noticed vacuolations, lesions in villi and serosa layer, necrosis, congestion in blood capillaries and severe mucus secretion in *Tilapia mossambica* exposed to industrial pollutants [39]. Damage in brush border and blood vessels was also supported by Ghosh, which indicated reduction in the absorption of various macromolecules from the intestinal lumen [40]. Scanning electron microscopic observations depicted severe secretion of mucus and necrosis in laboratory condition. These results are also in agreement with the findings of Senapati who reported damage in mucosal folds and CEC, and degeneration in microvilli structure with profound secretion of mucus in intestine of A. testudineus afetr Almix exposure under laboratory condition [19]. Similar observations were also described by Ghosh in intestine of *Notopterus notopterus* after arsenic exposure and by Bose in A. testudineus after lead and cadmium exposure [41,42]. Severe secretion of mucus under present study indicated that fish were under stress and trying to overcome these stress as compensatory response. Debris of the fragmented secondary mucosal folds in between the primary mucosal folds was observed under SEM study in field condition. Transmission electron micrographic observation depicted severe vacuolations, damage in glycocalyx structure, dilated mitochondria and damage in the tubular network under both conditions, indicating fish were in stress and approached to protect the imposed stress. Comparatively less pathological responses in field condition might be due to dilution capability and self-regulating mechanism of the natural environment. Therefore, the alterations impaired the intestinal transportation process as well as absorption of food materials.

In summary, the present study revealed that Almix exposure caused severe pathological alterations in stomach and intestine of A. testudineus under laboratory condition. Pathological lesions displayed stronger responses under laboratory condition compared to field study. Finally, these pathological alterations to this herbicide exposure could be considered as indicators to evaluate fish health status under stressed conditions in freshwater ecosystem, and careful handling and monitoring should be taken before application of this herbicide in agricultural farms or aquatic bodies for controlling weeds.

Acknowledgements

The authors like to thank the INSPIRE Program Division, Department of Science & Technology, Govt. of India (DST/INSPIRE Fellowship/2011/164, Dt. 29.09.2011) for the financial assistance. We also like to thank the Head, Department of Environmental Science, the University of Burdwan, Burdwan, West Bengal, India for providing the laboratory facilities and library facilities during the course of research. We are also thankful to the reviewers for improving our manuscript.

References

1. DuPont Safety Data Sheet (2012) DuPont™ Almix® 20 WP. Version: 2.1.

2. Farrington JW, Tripp BW (1995) International Mussel Watch Project: Initial Implementation Phase. Final Report, NOAA Tech Memorandum NOS ORCA 95. US Department of Commerce, National Oceanic and Atmospheric Administration, Silver Spring, MD, USA pp: 55-59.

3. Goksoyr A, Husoy A, Larsen H, Klungsoyr J, Wilhelmsen S, et al. (1991) Environmental contaminants and biochemical responses in flatfish from the Hvaler Archipelago in Norway. Arch Environ Contam Toxicol 21: 486-496.

4. Lopez-Barea J (1996) Biomarkers to detect environmental pollution. Toxicol Lett 88: 77-79.

5. Teh SJ, Adams SM, Hinton DE (1997) Histopathological biomarkers in feral freshwater fish populations exposed to different types of contaminant stress. Aquat Toxicol 37: 51-70.

6. Thophon S, Kruatrachue M, Upatham ES, Pokethitiyook P, Sahaphong S, et al. (2003) Histopathological alterations of white seabass, Lates calcarifer, in acute and sub chronic cadmium exposure. Environ Pollut 121: 307-320.

7. van der Oost R, Beyer J, Vermeulen NPE (2003) Fish bioaccumulation and biomarkers in environmental risk assessment: A review. Environ Toxicol Pharmacol 13: 57-149.

8. Kasherwani D, Lodhi HS, Tiwari KJ, Shukla S, Sharma UD (2009) Cadmium toxicity to freshwater Catfish, Heteropneustes fossilis (Bloch). Asian J Exp Sci 23: 149-156.

9. Fanta E, Rios FS, Romão S, Vianna ACC, Freiberger S (2003) Histopathology of the fish Corydoras paleatus contaminated with sublethal levels of organophosphorus in water and food. Ecotoxicol Environ Saf 54: 119-130.

10. Hinton DE, Lauren DJ (1990) Liver structural alterations accompanying chronic toxicity in fishes: potential biomarkers of exposure. In: Biomarkers of environmental contamination. Boca Raton, Lewis pp: 17-57.

11. Sorour J (2001) Ultrastructural variations in Lethocerus niloticum (Insecta: Hemiptera) caused by pollution in Lake Mariut, Alexandria, Egypt. Ecotoxicol Environ Saf 48: 268-274.

12. Samanta P, Senapati T, Mukherjee AK, Mondal S, Haque S, et al. (2010) Effectiveness of Almix in controlling of aquatic weeds and fish growth and its consequent influence on water and sediment quality of a pond. The Bioscan 3: 691-700.

13. Samanta P, Pal S, Mukherjee AK, Senapati T, Ghosh AR (2013) Evaluation of enzymatic activities in liver of three teleostean fishes exposed to commercial herbicide, Almix 20 WP. Proc Zool Soc pp: 9-13.

14. Samanta P, Pal S, Mukherjee AK, Senapati T, Ghosh AR (2014a) Effects of Almix herbicide on metabolic enzymes in different tissues of three teleostean fishes Anabas testudineus, Heteropneustes fossilis and Oreochromis niloticus. Int J Sci Res Environ Sci 2: 156-163.

15. Samanta P, Pal S, Mukherjee AK, Senapati T, Kole D, et al. (2014b) Effects of Almix herbicide on alanine aminotransferase (ALT), aspartate aminotransferase (AST) and alkaline phosphatase (ALP) of three teleostean fishes in rice field condition. Global J Environ Sci Res 1: 1-9.

16. Samanta P, Pal S, Mukherjee AK, Ghosh AR (2014c) Biochemical effects of almix herbicide in three freshwater teleostean fishes. HydroMedit, 1st International Congress of Applied Ichthyology & Aquatic Environment.

17. Samanta P, Pal S, Mukherjee AK, Senapati T, Ghosh AR (2014d) Alterations in digestive enzymes of three freshwater teleostean fishes by Almix herbicide: A comparative study. Proc Zool Soc.

18. Samanta P, Pal S, Mukherjee AK, Senapati T, Kole D, et al. (2014e) Effects of Almix herbicide on profile of digestive enzymes of three freshwater teleostean fishes in rice field condition. Toxicol Rep 1: 379-384.

19. Senapati T, Mukherjee AK, Ghosh AR (2012) Observations on the effect of Almix 20WP herbicide on ultrastructure (SEM) in different regions of alimentary canal of Anabas testudineus (Cuvier). Int J Food Agri Vet Sci 2: 32-39.

20. Senapati T, Samanta P, Mandal S, Ghosh AR (2013) Study on histopathological, histochemical and enzymological alterations in stomach and intestine of Anabas testudineus (Cuvier) exposed to Almix 20WP herbicide. Int J Food Agri Vet Sci 3: 100-111.

21. Chattopadhyay DN, Mohapatra BC, Adhikari S, Pani PC, Jena JK, et al. (2013) Effects of stocking density of Labeo rohita on survival, growth and production in cages. Aquacult Int 21: 19-29.

22. APHA (2005) Standard methods for the examination of water and wastewater. APHA, AWWA, WPCF, Washington DC.

23. Crespo SG, Nannotte G, Colin DA, Leray C, Nonnotte L, et al. (1986) Morphological and functional alterations induced in trout intestine by dietary cadmium and lead. J Fish Biol 28: 69-80.

24. Establier R, Gutierrez M, Arias A (1978) Accumalation and histopathological effect of inorganic and organic mercury in the lisa Mugil auratus. Investigación Pesquera 42: 65-80.

25. Sastry KV, Gupta PK (1978) Histopathological and enzymological studies on the effects of chronic lead nitrate intoxication in the digestive system of fresh water teleost Channa punctatus. Environ Res 17: 472-479.

26. Amminikutty CK, Rege MS (1977) Effects of acute and chronic exposure to pesticides Thiodan E.C. 35 and Agallot 3 on the liver of widow tetra Gymnococrymbus ternetzi (Boulenger). Indian J Exp Biol 15: 197-200.

27. Ghanbahadur A, Ghanbahadur G (2012) Histopathological effect of organochloride endosulfan on intestine and stomach of larvivorous fish Rarbora daniconius. DAV Int J Sci 1: 126-127.

28. Haque S, Pal S, Mukherjee AK, Ghosh AR (2012) Histopathological and ultramicroscopic changes induced by fluoride in soft tissue organs of the air-breathing teleost, Channa punctatus (Bloch). Fluoride 43: 263-273.

29. Rebolledo IM, Vial JD (1979) Fine structure of the oxynticopeptic cell in the gastric glands of Elasmobranch species (Halaelurus chilensis). Anat Rec 193: 805-822.

30. Carrassón M, Grau A, Dopazo LR, Crespo S (2006) A histological, histochemical and ultrastructural study of the digestive tract of Dentex dentex (Pisces, Sparidae). Histol Histopathol 21: 579-593.

31. Muniyan M (1999) Effect of ethofennprox (trebon) on the biochemical and histological changes in the selected tissues of the freshwater fish, Oreochromis mossambicus (Peters). Ph.D. Thesis, Annamalai University, India.

32. Das BK, Mukherjee SC (2000) A histopathological study of carp (Labeo rohita) exposed to hexachlorocyclohexane. Vet Arhiv 70: 169-180.

33. Cengiz EL, Unlü E, Balci K (2001) The histopathological effects of Thiodan* on the liver and gut of mosquitofish, Gambusia affinis. J Environ Sci Health 36: 75-85.

34. Yildirim MZ, Benli KC, Selvi M, Ozkul A, Erko F, et al. (2006) Acute toxicity behavioural changes and histopathological effects of deltamethrin on tissues (gills, liver, brain, spleen, kidney, muscle, skin) of Nile Tilapia (Oreochromis niloticus) fingerlings. Environ Toxicol 21: 614-620.

35. Velmurugan B, Selvanayagam M, Cengiz EI, Unlu E (2007) Histopathology of lambda-cyhalothrin on tissues (gill, kidney, liver and intestine) of Cirrhinus mrigala. Environ Toxicol Pharmacol 24: 286-291.

36. Walsh AH, Ribelin WE (1975) The Pathology of Pecticide Poisoning. In: The Pathology of Fishes. University of Wisconsin Press, Madison, Wisconsin pp: 515-558.

37. Mandal PK, Kulshrestha H (1980) Histopathological changes induced by the sublethal sumithion in Clarias batrachus (Linn.). Indian J Exp Biol 18: 547-552.

38. Sharma RR, Pandey AK, Shukla GR (2001) Histopathological alterations in fish tissues induced by pesticides toxicity. Aquaculture 2: 31-43.

39. Ravanaiah G, Murthy CVN (2010) Impact of aquaculture and industrial pollutants of Nellore district on histopathological changes in the liver and intestine tissues of fish, Tilapia mossambica. National J Life Sci 7: 110-115.

40. Ghosh AR (1991) Arsenic and cadmium toxicity in the alimentary canal and digestion of two Indian air-breathing teleosts Notopterus notopterus (Pallas) and Heteropneustes fossilis (Bloch). Ph.D. Thesis, The University of Burdwan, West Bengal, India.

41. Ghosh AR, Chakrabarti P (2001) Impact of inorganic arsenic on mucosal surface of stomach, intestine and intestinal caeca of Notopterus notopterus (Pallas). ICIPACT pp: 469-473.

42. Bose R (2005) Effects of lead and cadmium on the digestive system and kidney of Indian fresh water perch, Anabas testudineus (Cuvier) and subsequent recovery by EDTA. Ph.D. Thesis, The University of Burdwan, West Bengal, India.

28

The Production of Catfish and Vegetables in an Aquaponic System

Nawwar Zawani Mamat*, **Mohd Idrus Shaari and Nur Amirul Anas Abdul Wahab**

Marine Technology Programme, Universiti Teknologi MARA (Perlis), 02600 Arau, Perlis, Malaysia

*Corresponding author: Nawwar Zawani Mamat, Faculty of Applied Sciences, Marine Technology Programme, Universiti Teknologi MARA (Perlis), 02600 Arau, Perlis, Malaysia, E-mail: nawwar353@perlis.uitm.edu.my

Abstract

Aquaponic is a system that mutually integrates aquaculture and plant cultivation (by means of hydroponic). Both crops are combined in a recirculating system that utilizes less water than the traditional farming. Nutrients contained in fish tanks are recycled into plant biomass with the presence of nitrifying bacteria that convert the excreted ammonia to nitrite and then to nitrate. In this study, fifteen sets of aquaponic system were developed to study the growth of African catfish (*Clarias gariepinus*) and three types of plants; the red and green-red amaranth (*Amaranthus spp.*) and water spinach (*Ipomoea aquatica*). The combination of aquaculture and hydroponic gives a new insight into increasing the efficiency of food production which respects principles of sustainable agriculture.

Keywords: Aquaponic; Aquaculture; Integration; Sustainable; Integrated aquaculture

Introduction

The development of aquaculture industry has become a major economic importance worldwide. Aquaculture continues to show increasing production at an average annual growth rate of 6.1% between 2002 and 2012. The production increased from 36.8 million tons in 2002 to 66.6 million tons in 2012. Major aquaculture producers in 2012 were China (41.1 million tons), India (4.2 million tons), Vietnam (3.1 million tons), Indonesia (3.1 million tons), Bangladesh, Norway, Thailand, Chile, Egypt and Myanmar. These producers contributed 88% of the total aquaculture production worldwide [1].

Aquaculture in Malaysia started in the 1920's with multiple carp species reared in ex-mining pools. Since then, the industry has developed into a lucrative and sustainable industry. Freshwater pond culture has been the greatest contributor to local aquaculture production. A total production of 14,162 tons with estimated value of MYR 97.6 million was achieved in 1992 [2]. The main fish species cultured are red tilapia hybrid (*Oreochromis sp.*), catfish (*Clarias sp.*) and climbing perch (*Anabas testudineus*) [3]. However, it is noticeable that the development of aquaculture worldwide is slowing. The growth of aquaculture for land-based and near-shore systems has peaked due to political, environmental, economic and resource constraints. Therefore, the development of aquaculture is currently on going, driven by new ideas and innovations [4].

Integrated aquaculture has gained attention as an innovated system to add value to water, recycle nutrients and wastes in the system to produce more crops. Integration of crops is also regarded as an environmentally friendly practice which intensifies the use of land [5,6]. Integration of aquaculture and hydroponic is known as aquaponic. It combines the rearing of aquatic organisms (mainly fish) and plant production in a recirculating water system [7,8].

The concept of aquaponic is to reuse the nutrient-enriched water from the fish rearing tank for the growth of plants in hydroponic system. In addition, aquaponic is a productive method of producing fish and vegetables in a green, sustainable and energy efficient system. Aquaponic presents opportunities to increase economical operations because spaces, nutrients and water are optimized. This contributes to lower infrastructural costs, production of inexpensive food and consequently poverty reduction.

The objective of the present study is to assess the growth of African catfish (*Clarias gariepinus*) and three vegetable types; red amaranth, green-red amaranth and water spinach in an aquaponic system.

Materials and Methods

Fifteen aquaponic sets were installed in an aquaculture setting at Kuala Sungai Baru, Perlis, Malaysia. Each set consisted of a 150 gallon polyethylene tank and four rows of hydroponic trays. Water was circulated from the rearing tank to the trays with a 15 watt submersible water pump. Fifty juveniles of African catfish (*C. gariepinus*) were assigned to each tank filled with 80 gallons of water. They were fed twice daily (at 0830 and 1600 hrs.) with commercial pellets at 6% of their total body weight (wet weight).

Seeds of the vegetables (red amaranth, green-red amaranth and water spinach) were sown and sprouted in the sowing trays before being transferred to the hydroponic system. The fish were reared for 60 days while the vegetables were grown twice within the period. The fish were weighed weekly and the vegetables were weighed at the end of the cultivation period.

Data on average length and weight of the fish were assessed at the end of the experimental period. In addition, plants were weighed and number of leaves were determined in this study. One-way analysis of variance (ANOVA) tests were conducted to determine significant difference in growth of fish with different plant types.

Student's T-tests were carried out to assess significant difference of plant growth between the two cycles. All significant differences were accepted at $p < 0.05$. All statistical tests were determined using SPSS Statistics (version 17.0).

Results

Results on fish growth performance (total length and weight) are presented in Figures 1 and 2. The highest average length and weight of catfish was obtained with the green-red amaranth (20.22 ± 0.19 cm/fish; 55.42 ± 1.34 g/fish), followed by fish co-cultured with red amaranth and water spinach (19.58 ± 0.95 cm/fish; 49.17 ± 5.45 g/fish and 19.39 ± 0.17 cm/fish; 48.13 ± 1.17 g/fish, accordingly). However, there was no significant difference in average length and weight of catfish co-cultured with any of the three plant types (one-way ANOVA, $p > 0.05$).

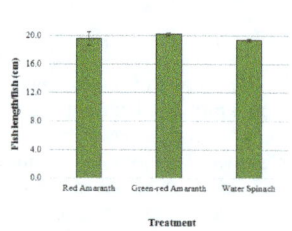

Figure 1: Average length per fish (cm) integrated with hydroponic cultivations of red amaranth, green-red amaranth and water spinach. No significant difference in fish length with different plant types ($p > 0.05$).

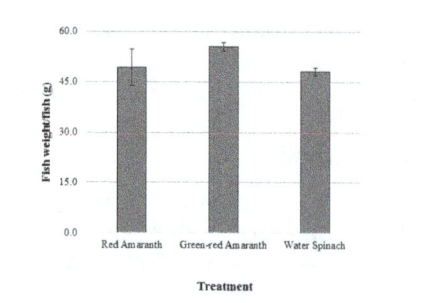

Figure 2: Average weight per fish (g) integrated with hydroponic cultivations of red amaranth, green-red amaranth and water spinach. No significant difference in fish weight with different plant types ($p > 0.05$).

The plants were grown in two cycles within the eight weeks of culture period (Figure 3). The wet weight (g) and number of leaves of the plants are shown in Figure 4. Results show that both amaranth plant types grew better in the second cycle. The red amaranth produced an average wet weight of only 43.67 g/plant in the first cycle. However, the growth reached 92.38 g/plant in the second cycle.

There was a significant difference in the wet weight of red amaranth between the first and the second cycle (T-test, $p < 0.05$). Similarly, growth of the green-red amaranth showed significant improvement in the later cycle. The plant weighed 72.63 g/plant in the first cycle and it grew to 103.71 g/plant in the second cycle. However, the growth of water spinach in the first and second cycles was significantly similar (72.01 g/plant and 70.75 g/plant, accordingly) (T-test, $p > 0.05$).

Figure 3: Cultivation of plants with African catfish in an aquaponic system. (A) the green-red amaranth (B) the red amaranth (C) the water spinach (D) the aquaponic system.

Discussion

Results of this study suggest that none of the three types of plants had a deleterious effect on the growth of catfish. The catfish is a hardy, robust fish species. They can thrive in turbid and low-oxygen water, and sometimes be found in drying rivers [9].

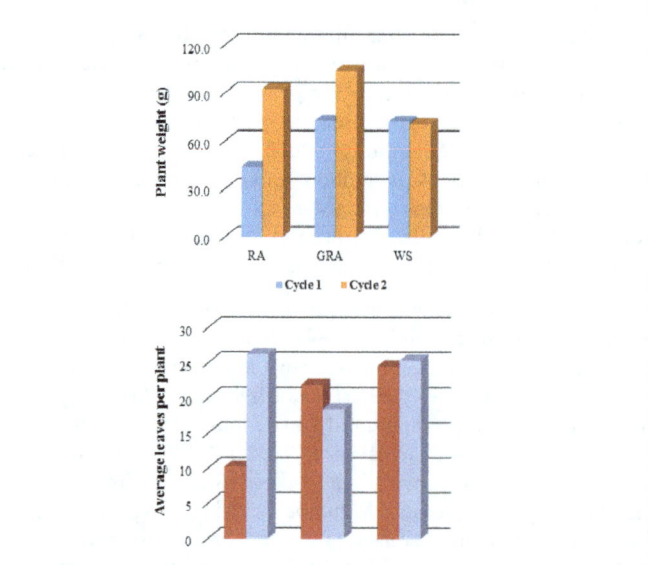

Figure 4: Wet weight (left) and average number of leaves (right) of plants cultivated in the aquaponic system (RA: Red Amaranth; GRA: Green-red Amaranth; WS: Water Spinach).

Therefore, catfish has been considered as a good candidate for aquaculture [10,11]. The present study also shows significant differences in growth of the red and green-red amaranth between the two crop cycles. Vegetable amaranth require high medium fertility, particularly potassium and nitrogen [12,13]. Previous studies have indicated that levels of nutrients in water increased with culture time [14,15]. Therefore, the lower plant weight in cycle 1 for both amaranth types could be associated with the lower nitrogen content in the water.

The two most common species of nitrifying bacteria are *Nitrosomonas sp.* and *Nitrobacter sp. Nitrosomonas sp.* converts ammonia (NH_3) to nitrite (NO_2) while *Nitrobacter sp.* use nitrites for their energy source during its conversion to nitrate (NO_3). Nitrogen in nitrate form is absorbed and used as a nutrient by plants. It can take a significant amount of time for these bacteria to grow in a system. This might have contributed to lower nitrogen content in the first few weeks.

Integration of fish farming (aquaculture) and plants has the potential to be environmentally friendly and sound because water use is minimized, less water discharge and the water form fish tank is reused to produce vegetables. In addition, the recirculating water keeps the water quality adequately safe for the fish. In the common aquaculture practices, water is changed regularly to reduce the accumulation of nitrogenous compounds [16]. These compounds mainly come from the faecal materials of the fish and uneaten feeds. The discharge of nutrient-enriched water creates environmental problems such as sedimentation and eutrophication [17,18]. This eventually will pollute the waterways and we may run out of clean water we depend on. It is very imperative to sustain the production of food sources without increasing pressure on the environment. Advances in knowledge can have significant benefits towards improving the living of people and protecting the natural environment [19].

This study shows that aquaponic practice is an effective way to raise fish and vegetables in one system. It utilizes the existing water containing nutrient from the fish tanks to water plants in the hydroponic system. The culture of catfish seemed to be positively co-exist with amaranth plants and water spinach. The aquaponic system is an efficient way to not only produce food crops, but also a successful system to recycle wastewater in aquaculture.

Acknowledgements

The authors would like to thank the Department of Higher Education, Ministry of Education, and Malaysia for financially supporting this study.

References

1. Food and Agriculture Organization of the United Nations (2015) In: Aquaculture topics and activities. Aquaculture: FAO Fisheries and Aquaculture Department, Rome.

2. Kechik IA (1995) Aquaculture in Malaysia. In: Bagarinao TU, Flores EEC (ed.) Towards Sustainable Aquaculture in Southeast Asia and Japan: Proceedings of the Seminar-Workshop on Aquaculture Development in Southeast Asia, Iloilo City, Philippines.

3. Food and Agriculture Organization of the United Nations (2008) In: National Aquaculture Sector Overview. Malaysia.

4. National Aquaculture Sector Overview Fact Sheets. In: FAO Fisheries and Aquaculture Department, Rome.

5. Food and Agriculture Organization of the United Nations (2006) State of World Aquaculture: 2006. FAO Fisheries Technical Paper No: 500, Rome.

6. Blidariu F, Grozea A (2011) Increasing the economic efficiency and sustainability of indoor fish farming by means of aquaponic-review. Scientific Papers Anim Sci and Biotech 44: 1-8.

7. Klinger D, Naylor R (2012) Searching for solutions in aquaculture: Charting a sustainable course. Annu Rev Environ Resour 37: 247-276.

8. Graber A, Junge R (2009) Aquaponic systems: Nutrient recycling from fish wastewater by vegetable production. Desal 246: 147-156.

9. Rakocy JE (2012) Aquaponics-Integrating Fish and Plant Culture. Aquaculture production systems pp: 344-386.

10. Van der Waal BCW (1998) Survival strategies of sharptooth catfish *Clarias gariepinus* in desiccating pans in the northern Kruger National Park. Koedoe 41: 131-138.

11. Copp GH, Britton JR, Cucherousset J, García-Berthou E, Kirk R, et al. (2009) Voracious invader or benign feline? A review of the environmental biology of European catfish Silurus glanis in its native and introduced ranges. Fish 10: 252-282.

12. Isyagi AN (2007) The aquaculture potential of indigenous catfish (*Clarias gariepinus*) in the Lake Victoria Basin, Uganda.

13. National Research Council (1984) National Research Council Amaranth: Modern Prospects for an Ancient Crop. National Academy Press, Washington DC.

14. Gopalakrishnan TR (2007) Vegetable crops. New India Publishing.

15. Li W, Li Z (2009) In situ nutrient removal from aquaculture wastewater by aquatic vegetable *Ipomoea aquatica* on floating beds. Water Sci Technol 59: 1937-1943.

16. Endut A, Jusoh A, Ali N, Wan Nik WB (2011) Nutrient removal from aquaculture wastewater by vegetable production in aquaponics recirculation system. Desalin Water Treat 32: 422-430.

17. Pillay TVR (2008) Aquaculture and the Environment. John Wiley & Sons.

18. Gabric AJ, Bell PR (1993) Review of the effects of non-point nutrient loading on coastal ecosystems. Mar Freshwater Res 44: 261-283.

19. Meyer-Reil LA, Köster M (2000) Eutrophication of marine waters: effects on benthic microbial communities. Marine Poll Bull 41: 255-263.

29

Microbial Evaluation of Selected Post Harvest Processing Techniques for Quality Fish Product at Bahir Dar Town, Ethiopia

Adamu Yimer[1*], Minwyelet Mingist[1] and Behailu Bekele[2]

[1]Department of Fisheries, Wetlands and Wildlife Management, College of Agriculture and Environmental Sciences, Bahir Dar University, P.O. Box 5501, Bahir Dar, Ethiopia

[2]School of Food and Chemical Engineering, Bahir Dar Institute of Technolgy, Bahir Dar University, P.O. Box 79, Bahir Dar, Ethiopia

*Corresponding author: Adamu Yimer, Department of Fisheries, Wetlands and Wildlife Management, College of Agriculture and Environmental Sciences, Bahir Dar University, P.O. Box 5501, Bahir Dar, Ethiopia, E-mail: zyamhel@gmail.com

Abstract

The objective of this study was to evaluate the quality of fish processed by open air rack, solar tent and smoking methods. African catfish (*Clarias gariepinus*) was filleted, washed, sliced, brine salted and processed by the selected methods, packed in plastic bags and stored at room temperature. Abalo (*Brucea antidysenterica*) and Olic tree (*Olea europaea*) were used as smoking wood. For moisture content test, 25 g of processed fish was put in an oven at 105°C and weight change of the samples was measured until the change become constant. It was calculated as the difference between the initial and final weight. Twenty-five gram of processed fish was taken aseptically and standard procedures of dilution and spread plating were done based on the type of microorganism to be identified. Then, the number of colonies were counted and changed into log 10 cfu/g. Solar tent reduced the moisture content to 20% and 23% for Nile tilapia and African catfish, respectively. Microbial load of solar tent dried fish samples was below the standard norm than open air rack and smoking methods. There was statistical difference between treatments (p=0.05). Solar tent drier produced better quality of fish product.

Keywords: *Clarias gariepinus*; Solar tent; Microbial count; Moisture content; Quality

Introduction

Fish is a highly nutritious food for providing high quality protein and income to many people in the developing world [1]. In Africa, 5% of the population (35 million people) depends on the fisheries sector for their livelihood [2]. However, fish is one of the most perishable of all the foods because it is a suitable medium for growth of microorganisms after death [1]. In the tropics at ambient temperature fish will spoil within 12-20 hrs depending on species and method of capture [3].

Lake Tana, the largest lake in Ethiopia, creates job opportunity for 3,514 fishers [4]. Most communities engaged in fisheries of *L. Tana* have been experiencing significant loss. Fish postharvest loss in Lake Tana is more than 30%, excluding low value fish parts [5]. Losses occur as a result of flaws in the handling, storage, distribution, processing and marketing techniques. Traditional fish processing and preservation method is the only method used to dry fish in the study area where the price of the such fish is very low due to quality problems affecting the fishers' livelihood and nutrient loss for the consumers. Hence, it is important to identify appropriate improved fish processing methods to reduce postharvest fish loss, increase quality, and market value and their income. This could be by upgrading the traditional fish processing technology and adoption of solar dryer [6]. Therefore, the main purpose of this study was to evaluate the quality of fish processed by solar tent, open air rack and smoking methods.

Materials and Methods

Description of the study area

The study was conducted in Bahir Dar town, capital of Amhara National Regional State. It is located 565 km North-Western from the capital city of Ethiopia, Addis Ababa, at 11° 4'N and 37° 3'E. The altitude is 1800 masl. The minimum and maximum temperature is 9°C and 34°C, respectively, with annual rainfall of 1300 mm (National Meteorology Agency Bahir Dar Branch, 2011).

Sample preparation and processing

Sample of African Catfish (Clarias gariepinus) was used. For the drying method, the sample was gutted, filleted and sliced using knife and washed thoroughly; 3 kg of fillet was used for each treatment. Then it was soaked in 16% brine solution for 30 minutes, drained and arranged on rack. For smoking method, the fish was only gutted and washed thoroughly. Then it was soaked in 16% brine solution until the colour of the eye become white. Immediately after the completion of fish sample preparation, it was processed by using open air rack, solar tent and smoking methods. After the processing was completed, the sample was packed in plastic bags and stored at ambient temperature with no preservative.

Open air rack drying

For open air drying, rack was used. The rack, bed like rectangular structure, was made up of wood frames and plastic ropes to avoid rusting of wire mesh (Figure 1). The fish was arranged on the rack and placed on the open air to be dried by the direct sunlight and the flow of air.

Figure 1: Open air/sun drying rack.

Solar tent drying

Solar tent was made up of wood frame and transparent polythene sheet which was 1.70 m length, width and height (Figure 2A and 2B) with constructed ventilation located at the top of the tent made up of wire mesh which was vermin proof. Black plastic sheet was put on the ground inside the tent to increase heat trapping capacity of the tent. Fish sample was arranged on the rack and put inside the drying tent.

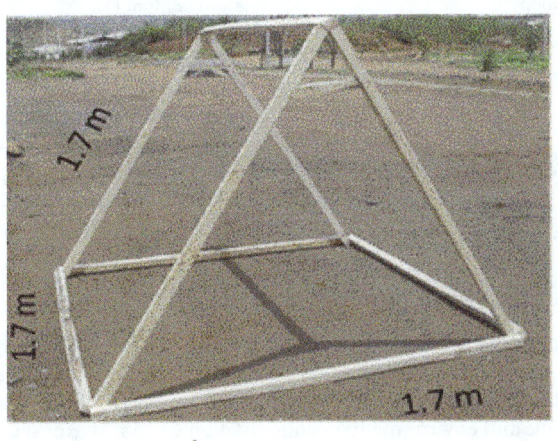

Figure 2: Solar tent frame (a) and solar tent covered with polythene sheet (b).

Smoking

Smoking was done by using rectangular iron sheet having container like structure (Figure 3). It had two doors; lower opening to supply and control the fire for smoke and the upper for inserting and checking of fish samples.

Figure 3: Smoking unit.

The sample was hanged over by metal rods arranged on top of the smoking unit. It had also three small opening at the top to escape the smoke. Abalo (*Brucea antidysenterica*) and Olic tree (*Olea europaea*) woods were used to produce smoke. The fish was turned at intervals and smoked for 8 hrs and cooled. The smoking process followed surface drying, smoking and cooling procedures. Surface drying is the removal of surface moisture leaving a protein coating (pellicle) on each piece of fish so that it accepts an even smoke deposit and the smoked product will not be fragile soon [7].

Data Collection Methods

Moisture content test

A sample of 25 g of processed fish from each treatment was taken and put in an oven at 105°C. The weight change of the samples was measured until the change becomes constant. The moisture content was calculated as the difference between the initial and final weight.

Microbiological test

Twenty-five gram of each processed fish was taken aseptically and individually transferred to sterile plastic bags. The sample was placed in a sterile standard stomacher bag containing distilled water and peptone water with a dilution ratio of 1:100 and was blended and homogenized for 1 to 2 minutes in a stomacher. Samples were, then, serially diluted (1:10) and spread plated onto various media depending on the type of microorganism to be tested. The microorganisms isolated, media used, incubation and indicator of its presence is shown by Table 1. Then, the number of colonies were counted and changed into log10 cfu/g of sample.

Microorganism	Media used	Incubation		Indicator
		Temperature (°C)	Time (Days)	
Staphylococcus species	Mannitol Salt Agar	37	01-Feb	round white colonies
Enterobacteriaceae	Violet Red Bile Dextrose Agar	37	01-Feb	blue to red-blue colonies
Aerobic mesophilic count	Plate Count Agar	35	01-Feb	
Yeasts and moulds	Potato Dextrose Agar	35	01-Feb	development of black and white colonies

Table 1: Microbiological isolation and identification.

Data analysis

Descriptive statistical analysis was used to analyse the mean and standard deviations of microbial load of samples with respect to the processing methods. One-way ANOVA was used to determine statistical significant difference among fish processing methods. Statistical significance was done at the p=0.05 value. Statistical package SPSS version 18.0 software was used for the analysis.

Results

Drying duration

Fish samples dried with solar tent took shorter period of time, 2.5 days. However, rack drying took 4 days. This can be attributed by the capacity of trapping heat by the tent. Solar tent drier had higher internal temperature, 56°C, than atmospheric temperature which was 27°C.

Moisture content

The moisture content of fish samples processed by solar tent was lower than those processed by open-air rack drying and smoking methods (Table 2). Solar tent reduced the moisture content to 23%. Solar tent reduced the moisture content to below 25% where the growth of bacteria and moulds is suppressed.

Treatment	Moisture content (%)
Solar tent dried	23
Open air rack dried	31
Smoked	41

Table 2: Moisture content of African catfish sample treated by different processing technologies.

Microbial count analysis

High yeast and mould, total plate count (TPC), *S. aureus* and *Enterobacteriaceae* count was found in open air dried catfish samples (Table 3). This high load may come from higher load from raw materials and environmental contamination since it is exposed to open atmosphere. Lower yeast and mould, total plate count (TPC), *S. aureus* and *Enterobacteriaceae* counts were found at the fish sample processed by solar tent drying (Table 3). Smoked fish samples had lowest yeast and mould counts, but, it was only for 15 days of storage that the sample had low yeast and mould count (Table 3). There was significant difference (p=0.05) between treatments.

Storage time (days)	Yeast and mould			Total plate count			*Staphylococcus* **species**			*Enterobacteriaceae*		
	Solar dried	Rack dried	Smoked	Solar dried	Rack dried	Smoked	Solar dried	Rack dried	Smoked	Solar dried	Rack dried	Smoked
0	3.54 ± 0.01^d	8.39 ± 0.02^e	3.24 ± 0.00	3.69 ± 0.06^e	7.69 ± 0.04^d	3.72 ± 0.02	3.16 ± 0.04^f	7.55 ± 0.02^c	ND	ND	5.81 ± 0.03^f	ND
15	4.55 ± 0.03^a	8.63 ± 0.04^d	5.26 ± 0.11	3.90 ± 0.00^d	7.93 ± 0.02^c	6.16 ± 0.00	3.45 ± 0.07^e	8.08 ± 0.05^b	4.14 ± 0.08	1.30 ± 1.84^b	5.90 ± 0.00^e	2.60 ± 0.00
30	3.56 ± 0.03^d	7.15 ± 0.17^f		4.35 ± 0.06^c	8.37 ± 0.06^b		3.63 ± 0.04^d	7.93 ± 0.01^b		3.46 ± 0.085^a	5.57 ± 0.04^g	
45	4.00 ± 0.00^c	9.72 ± 0.03^a		4.80 ± 0.03^a	8.48 ± 0.04^b		3.58 ± 0.06^d	8.07 ± 0.04^b		3.32 ± 0.02^a	6.90 ± 0.00^d	
60	4.27 ± 0.10^b	9.24 ± 0.09^b		4.60 ± 0.02^b	8.45 ± 0.05^b		4.33 ± 0.04^c	8.93 ± 0.00^a		3.45 ± 0.06^a	7.81 ± 0.01^c	
75	4.48 ± 0.04^a	8.87 ± 0.01^c		4.79 ± 0.02^a	9.18 ± 0.12^a		3.71 ± 0.01^a	8.82 ± 0.22^a		3.64 ± 0.04^a	8.76 ± 0.01^b	

Microbial Evaluation of Selected Post Harvest Processing Techniques for Quality Fish Product at Bahir Dar Town, Ethiopia

187

90	4.28 ± 0.03b	9.59 ± 0.03a		4.68 ± 0.01b	9.10 ± 0.06a		4.59 ± 0.04b	9.00 ± 0.01a		3.62 ± 0.02a	8.87 ± 0.01a	

Table 3: Microbial load of African catfish (*C. gariepinus*) processed by solar tent, open air rack and smoking methods (log10 of cfu/g). All values are means of duplicate ± standard deviation. Means with the same superscript letters within a column are not significantly different (p=0.05).

Discussion

Drying duration

The drying duration of solar tent was shorter than open air rack drying method. Hot air might be trapped by a black plastic sheet put on the floor inside the tent. The opening designed by a wire mesh would reduce the humidity inside the tent. These conditions caused the sample to be dried faster than rack open air/sun drying methods. The result agreed with the studies conducted by different authors. Assefa et al. [8,9] reported that the time taken to dry tilapia with solar tent was one and half days and African Catfish was dried one day later.

Moisture content

Moisture loss rate of solar tent dryer was higher than open air rack dryer and smoking methods. The black plastic sheet put on the ground inside solar tent trapped heat and the tent maintained high heat inside the tent resulting higher temperature, hence, increased moisture loss rate and reduced moisture content. This result is agreed with [8,9] in which solar drying resulted in lower moisture content than rack drying and rock drying. Solar tent reduced moisture content of tilapia (*O. nilaticus*) to 18.51% [10]. If the moisture content of fish is reduced to 25%, spoilage bacteria can't survive as lower moisture content inhibits growth of bacteria and moulds [11]. The result obtained from this study showed that the final moisture content of solar tent dried fish samples was below the specified critical value (25%). This can be explained by the capacity of the solar tent to maintain high internal heat, which had reached up to 56°C.

Microbial count

Yeast and mould count: High yeast and mould count was found in open air dried tilapia and catfish samples (Table 3). Open air dried fish samples were significantly different (p=0.05) and highest yeast and mould count from other treatments 7.82 ± 0.47 and 8.80 ± 0.84 for tilapia and catfish, respectively. This highest yeast and mould count might come from high yeast and mould load from raw materials and environmental contamination since it is exposed to open atmosphere. Fungi generally prefer substrate with low water activity and usually very high on dry fish samples [12]. The water activity in dried products is low and in favor moulds which spore are spread by air since fish samples are exposed to the ambient atmosphere.

Lower yeast and mould count was found on fish samples dried by solar tent (Tables 3). The reason for this may go to the processing condition (Table 4). It was dried under relatively higher temperature (560°C) with constructed ventilation and protected from environmental contamination by the tent. The higher internal temperature enabled the sample to have relatively higher moisture loss rate. Dehydration preserves fish by destroying enzymes and removing the moisture necessary for bacterial and mold growth [7].

Smoked fish samples had lowest yeast and mould counts, but, it was only for 15 days of storage that the sample had low yeast and mould count. Smoked fish are preserved primarily by control of salt and moisture content. However, smoke deposition is effective only in controlling surface spoilage [13].

Total plate count: Total plate count (TPC) indicates the level of microorganisms in a product [14]. Higher TPC was found on open air rack dried fish samples (Tables 3 and 4). The TPC of open air dried fish samples was significantly different (p=0.05). Rahman et al. [7] reported that the drying temperature of 50°C or below has no lethal effect on the microflora. Therefore, relatively lower atmospheric temperature during drying, exposure towards the open atmosphere, post-processing contamination during packaging and/or higher microbial load of raw material contributed to higher TPC for open air dried tilapia and catfish samples. The logic behind this mainly focused on the exposure of the fish samples to the open environment. This reason is because fish products treated at the open environment with low level of temperature in which microorganisms are found anywhere is expected to have high microbial load specifically aerobic plate count.

Fish samples processed by solar tent drying method had lower TPC (Tables 3 and 4). This is because of simultaneous reaction of brining, relatively higher temperature and protection of the samples from environmental contamination. According to Hardy [15] salting converts fresh fish into shelf-stable products by reducing the moisture content and acting as a preservative. Moreover, the bacterial load of dried fish decreased due to removal of water level below that needed for microbial growth and enzyme activity [16]. Drying process remove enough moisture from fish to a limit that greatly decreases these destructive effect [17]. This microbial load is even lower than the study conducted by Ahmed [17] where the total number of bacterial count for dried African catfish (*Clarias gariepinus*) was 5.75 ± 5.5 (log10 cfu/g).

The TPC of smoked fish samples was initially low because of brining, relatively high temperature treatment and anti-microbial effect of smoke, however, the lower total count was only for 15 days of storage. This can be attributed by the higher moisture content (Catfish (44%) and Tilapia (35%) of smoked fish. The level of growth of microorganisms on the smoked fish depends on the amount of water which has been expelled from them [18] and smoke deposition is effective only in controlling surface spoilage [13]. This result is agreed with Getachew [19] in which the microbial load of smoked tilapia samples stored at room temperature reached maximum level after 15 days of storage.

Staphylococcus aureus count: Higher S. aureus count was found on open air rack dried fish samples; 7.56 ± 0.31 and 8.34 ± 0.55 for tilapia and catfish, respectively (Tables 3 and 4). Relatively lower temperature during processing, exposure to the open atmosphere, post-processing contamination during packaging and/or higher microbial load of raw material contributed to higher staphylococcus count for open air dried tilapia and catfish samples (Tables 3 and 4). The staphylococcus load of open air dried fish samples was significantly different (p=0.05). The reason behind might be the exposure of the fish samples to the open environment. Samples processed by sun drying appeared to be of poor microbiological quality since *E. coli, S. aureus* and moulds were detected at concentrations above recommended norms [12] sun drying reduces microbial loads but did not eliminate completely contaminant in most samples.

Lower *S. aureus* count was scored for fish samples processed by solar tent drying method (Tables 3 and 4). In smoke dried fish samples *E. coli, S. aureus* and enterococci were totally absent [12]. Since the processing temperature was relatively higher the vegetative microbes will be destroyed.

Enterobacteriaceae count: *Enterobacteriaceae* is a large family of bacteria including many of the more familiar pathogens, such as *Salmonella, Escherichia coli, Shigella* species etc. Usually, its occurrence in a final product is due to contamination during processing by an infected asymptomatic carrier with poor personal hygiene [20].

Enterobacteriaceae count in Nile tilapia and African catfish processed by smoking and solar tent drying was not detected until 15 days of storage for Nile tilapia and at zero-time for African catfish. But very low count was found after the later days (Tables 3 and 4). This was because mainly smoking process was gone for high temperature for longer time duration (8 hours).

After the treatment, if there is no post contamination due to poor hygiene of the processors the smoked fish (especially hot smoked) is free from pathogenic micro-organisms. *Enterobacteriaceae* load was high and significantly different (p=0.05) from other processing techniques for fish samples processed by open air drying.

Open air rack drying	Solar tent drying
Not covered/protected	Covered/protected by polythene sheet
Environmental factor is very high	Environmental factor is low
Less temperature value	High temperature value
High environmental contamination	Low environmental condition
High incidence of flies	No incidence of flies
More responsive to changes in atmospheric condition	Less responsive to changes in atmospheric condition
Long time to dry	Short time to dry

Table 4: Comparison of open air rack and solar tent drying methods.

References

1. Ojutiku RO, Kolo RJ, Mohammed ML (2009) Comparative study of sun drying and solar tent drying of Hyperopisus bebe occidentalis. Pak J Nutr 8: 955-957.

2. Davies OA, Davies RM (2009) Traditional and improved fish processing technologies in Bayelsa State, Nigeria. European Journal of Scientific Research pp: 539-548.

3. Igene JO (1983) Drying of fish factors to consider. Proceedings of 3rd Annual Conference on Fisheries Society of Nigeria (FISON) pp: 123-131.

4. ANRSLRDPA (2011) Lake Tana fisheries management plan and processing manual. Bahir Dar, Ethiopia pp: 107.

5. Mohammed B (2011) Assessment of motorized commercial gill net fishery of the three commercially important fish species in Lake Tana, Ethiopia. MSc thesis, Bahir Dar University.

6. Akinola OA, Akinyemi AA, Bolaji BO (2006) Evaluation of Traditional and solar drying systems towards enhancing fish storage and preservation in Nigeria (Abeokuta Local Government as a case study). Journal of Fisheries International 1: 44-49.

7. Rahman MS, Al-Amri OS, Al-Bulushi IM (2002) Pores and physico-chemical characteristics of dried tuna produced by different methods of drying. J Food Eng 53: 301-313.

8. Tessema A, Demissie S, Goshu G, Bekele B, Fentahun A, et al. (2008) Evaluation of solar tent and drying rack methods for the production of quality dried fish in Lake Tana area. Proceedings of the 3rd Annual regional Conference on Completed Livestock Research Activities (CLRA). Bahir Dar, Ethiopia pp: 15-26.

9. Degebassa A (2010) A comparative study on the effect of three drying methods for better preservation of fish: Management of shallow water bodies, EFASA.

10. Mohamed F, Hegazy M, Abdellatef M (2011) Physico-chemical properties and mycotoxins contents of tilapia fish fillets after solar drying and storage. Global Veterinaria, Dokki, Giza, Egypt pp: 138-148.

11. Waterman JJI (1976) The production of dried fish. FAO Fisheries technical paper, No.16 Rome pp: 52.

12. Ahmed A, Ahmedou D, Mohamadou BA, Saidou C, Tenin D (2011) Influence of traditional drying and smoke drying on the quality of three fish species (Tilapia nilotica, Silurus glanis and Arius parkii) from Lagdo Lake, Cameroon. In: Journal of animal and veterinary advances 10: 301-306.

13. Hilderbrand KS (1992) Fish smoking procedures for forced convection smokehouse. Oregon State University Extension Service. Oregon pp: 1-41.

14. Maturin L, Peeler J (1998) Aerobic plate count. In: Food and Drug Administration Bacteriological Analytical Manual, AOAC International, Gaithersburg, MD.

15. Hardy R (1980) Fish lipids. In: Advances in fish science and technology. England, Fishing News Books pp: 103.

16. Doe PE, Curran CA, Poulte RRG (1983) Determination of the water activity and shelf life of dried fish products, FAO pp: 202-208.

17. Ahmed SH, Eltegani IE (2012) Effect of drying on microbial load of Clarias Sp. Fish Meat In: International Journal of Biology, Pharmacy and Applied Sciences (IJBPAS).

18. Oyewole BA, Agun BJ, Omotayo KF (2006) Effects of different sources of heat on the quality of smoked fish. J. Food and Agric. Environ 4: 95-97.

19. Getachew E (2012) Effect of hot smoking on the quality and shelf stability of Nile tilapia (Oreochromis niloticus) fillets. MSc thesis in Addid Ababa Institute of Technology, Department of Chemical Engineering, Addis Ababa University.

20. Huss HH (1994) Assurance of seafood quality. FAO Fisheries Technical pp: 334.

30

Investigations on Mass Mortalities among *Oreochromis Niloticus* at Mariotteya Stream, Egypt: Parasitic Infestation and Environmental Pollution Impacts

Nisreen E Mahmoud[1*], **MM Fahmy**[1] **and Mohga FM Badawy**[2]

1Department of Parasitology Faculty Of Veterinary Medicine, Cairo University, Egypt

2Department of veterinary hygiene and management, Faculty Of Veterinary Medicine, Cairo University, Giza 11221, Egypt

*Corresponding author:** Nisreen E Mahmoud, Department of Parasitology Faculty Of Veterinary Medicine, Cairo University, Egypt, E-mail: drnisreene@hotmail.com

Abstract

The present study was carried out to determine the possible causes of an emergent event of respiratory distress with consequent mass mortalities among Nile fish, *Oreochromis niloticus* (*O. niloticus*) at Mariotteya stream, an intrastate tributary of River Nile, Egypt. The area of incident extended from Shabramant till Abouseer city (along 4 km distance) with the direction of water current. Field visits have recorded thousands of dead large sized fish on both sides of the stream while huge numbers of fish of different size accumulated on the water surface showing typical signs of asphyxia. It was also noted that Mariotteya water body is being subjected to multiple sources of pollution through the dumping of improperly treated organic and inorganic chemical wastes in addition to sewage materials. Results of field and laboratory investigations have revealed that, all the examined 60 fish samples were heavily infested with different types of parasites including zoonotic species: *Cichlidogyrus arthracanthus* (*monogenea*), *Lamproglena monody* and *Ergasilus sarsi* (*Copepod*), *Myxobolus dermatobia*, *Chilodonella hexastica*, *Trichodina truttae*, *Trichodina fultoni*, *Cryptosporidium spp.* and *Balantidium spp.* (*Protozoa*), *Acanthosentis tilapae.* (*Acanthocephala*), *Clinostomum spp.*, *Euclinostomum spp.*, Heterophid and Prohemistomatid metacercarae (larval of Trematodes). The physical and chemical examinations of water samples and the analysis of heavy metals concentration indicated marked abnormal water quality parameters and environmental pollution which might be incriminated as a primary stress factor that promoted the invasion of parasites as a secondary stress factor. The study concluded that both factors would have interacted to produce this catastrophic intense case of respiratory distress and mass mortalities. The impact of the recorded parasitic infestation and environmental pollution was briefly discussed.

Key words

Egypt; *Oreochromis niloticus*; Pollution; Mariotteya

Introduction

River Nile is the main fresh water resource in Egypt, meeting all demands for drinking water, irrigation and industry [1]. It is also the major source of many aquatic food organisms. Anthropogenic sources play a role in introducing several pollutants into the River Nile either directly or indirectly through different drains. Egyptian drains receive large quantities of untreated or partially treated waste water which in turn discharge into River Nile with the amount exceeds its natural ability to attenuate it. Mariotteya stream (El Moheet drain) is an interstate tributary of River Nile that extends through Giza and October governorates and is such a receiving environment for agricultural, industrial and domestic wastes through Sakkara 7 drainage where Al-Hawamdia sugar factory and Industrial and sewage treatment plants outlets are dumped. These types of pollutants reported to create both localized and regional problems in nearly every country around the world [2], and in some cases, it has been extensive enough to lead to environmental disasters such as mass mortalities of many aquatic species [3].

Numerous causes of fish mass mortalities in natural water resources were reported in many parts of the world and mostly related to number of environmental problems such as acute toxicity of some pollutants [4] also physico-chemical factors such as low dissolved oxygen, high level of un-ionized ammonia and high concentration of heavy metal could be potential causes [5-7]. Mass mortality might be also attributed to pathogen invasion as potential primary causes among cultured and wild fish population particularly viral [8]; bacterial [9]; fungal [10] and parasitic infection [11].

In Egypt, and as a result of pollution, aquatic ecosystems were subjected to repeated cases of mass mortalities among fish population in the past few year; in 2008, mass mortality was recorded among grouper fish along the Mediterranean coast at Marsa Matrouh area, and in 2009 catastrophic mass mortalities occurred among cultured Tilapia fish in King Mariot at Alexandria province, also in Kanater Edfina at El- Behera province. Another cases of mass mortalities erupted among many species of fish through the period of 2010 and 2012 at different localities along River Nile branches in Dakhlia and Behera provinces [12,13].

In January 2010, a disaster of mass mortality and respiratory distress among fish population has been erupted in Mariotteya stream at Giza province leading to great economic loss, and consequently, the aim of the current study was to estimate the consequences of environmental back ground, and the subsequent invasion of the affected fish with parasites in order to estimate the causes that lead to this catastrophic mass mortality.

Materials and Methods

Field visit

On the 4th of January 2010, the Egyptian media announced catastrophic mass mortalities among fish in Mariotteya water stream and denoted that, several thousands of fish were lost. An emergent visit to the location of incident was performed by the investigation team for reporting the case history and recording the mortality patterns, the abnormal behavioural change and clinical finding among the affected fish species and also for sampling in order to estimate the potential causes of this mass mortality problem.

Water samples

Five points along the affected area representing the outlet of Al-Hawamdea sugar factory (site 1), the drainage of industrial and sewage treatment plants (site 2), and the outlet of the drainage Sakara 7 to the Mariotteya water stream (site 3). Mariotteya water stream at Shubramant, north to the drainage Sakara7 (site 4), and Mariotteya water stream south to the drainage Sakara7 (site 5).

Pre-cleaned polyethylene sampling bottles (1.5 liter capacity) were immersed about 50 cm below the water surface, 2 samples each of 500 ml of water were taken at each sampling site and acidified on spot with 10 % nitric acid then transported immediately to the laboratory in an ice bath and prepared for analysis as described by APHA [14]. The water samples were physically examined for colour and temperature (°C) which are measured by a dry mercury thermometer. For Chemical parameters, hydrogen ion concentration was measured by Orion Research Ion Analyser 399A PH meter, dissolved oxygen (DO) was measured using the five days incubation method, where one water sample of each site was analyzed immediately for dissolved oxygen according to APHA, 1998, and the second is incubated in the dark at 20 C for 5 days and then tested for the amount of dissolved oxygen remaining. The difference in oxygen levels between the first test and the second test, in milligrams per liter (mg/L) represents the amount of oxygen consumed by microorganisms to break down the organic matter present in the sample during the incubation period which in turn indicated the level of organic pollution. The concentration of ammonia was determined by using the colorimetric techniques according to APHA [15]. Heavy metals ; Copper (Cu), Cadmium (Cd) and Lead (Pb) in water samples were determined using atomic absorption spectrophotometer (Model 3100, Perkin-Elma, Norwalk, Conn, USA) according to standard methods described by APHA [16]. Depending on data during field visit, the water samples were tested for Phenol and Polycyclic aromatic hydrocarbons (PAHs) using High Performance Liquid Chromatography (HPLC, Jasco, Model UV-2076 plus) (Figures 1–5, Mape 1).

Fish samples

A total of 60 O.niloticus(average weight 50-100 g) showing typical signs of asphyxia manifested by gasping, rapid opercular movement were collected from the affected Mariotteya stream water and were transported alive to the laboratory of Parasitology department, Faculty of Veterinary Medicine, Cairo University. The fish were kept in aerated glass aquaria at 25°C and subjected to parasitological examination.

Figure 1: Al-Hawamdeya outlet (Site 1).

Figure 2: Drainage of sewage treatment plants (Site 2).

Figure 3: The outlet of the drainage Sakara 7 to Mariotteya water stream (Site 3).

Figure 4: Mariotteya water stream north to Sakara 7 (Site 4).

Figure 5: Mariotteya water stream south to Sakara 7 (site 5).

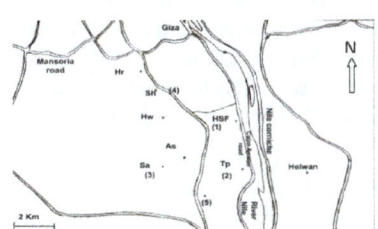

Mape (1): Hr: Haram area Sh: Shabramant. Hw: Hawamdia. As: Aboseer. Sa: Sakara 7. HSF: Hawamdia Sugar Factory. Tp: Treatment plant. 1,2,3,4 and5: Sampling sites. N: North direction.

Parasitological examination

Fish samples were examined macroscopically and microscopically for parasites infection. For ectoparasite examination, mucous was scraped with spatula from skin surface, fins, and oral cavity. Samples were examined with fresh preparation method. Fixed preparation was created from the preparation in which parasites were detected. For endoparasite examination, body cavity, muscles and internal organs (livers, spleens, stomach, gall bladder, heart, brain and intestine contents) were examined. Samples were separated according to their species. All the isolated parasites were prepared for permanent mounting followed the standard protocol after Prichard and Kruse, Lom and Dyková, Kabata, Paperna and Yamaguti [17-21] Prevalence of the detected parasites was calculated according to Bush et al. [22].

Statistical analysis

Data were presented as Mean ± standard deviation for numerical variables. To compare mean between groups, student's t test was used. The Statistical analysis was performed using SPSS© version 16.

Results

Field visits

The emergent visit to the site of incident revealed that, fish mortalities were confined to the area extended from Shabramant till Abouseer (distance of about 4 km). Thousands of dead and dying large sized Oreochromis niloticus (200-250 g) was recorded along both sides of Mariottya stream. Huge numbers of small sized (50-100 g) were aggregated on the water surface with typical signs of asphixia manifested by gasping and rapid opercular movement. Tanked cars and large numbers of pipes were seen dumping huge quantities of sewage collected from houses into the water stream in addition to the large quantities of garbage accumulated on both sides of Mariotteya

water. The mortality and signs of respiratory distress were not detected among fish in mariotteya stream south to sakara 7. The case history revealed that, the same case of mass mortalities was previously occurred 3 years ago but in limited scale also refers to the role of Alhawamdia sugar factory in polluting Mariotteya stream by pouring its industrial effluents to the water through Sakara 7 drainage (Figures 6-10).

Figure 6: Thousands of dead and dying fish along Mariotteya stream.

Figure 7: Tanked car dumping sewage material.

Figure 8: large quantities of garbage accumulated on both sides of Mariotteya water.

Water sample examination

Pysico-chemical parameters:

Table (1) shows the mean values of physicochemical parameters of the sampling sites along Mariotteya stream. It is obvious that, the mean values of the different parameters of water in sites 1, 2, 3, and 4 were very high with the exception of temperature and PH while the parameters at site 5 South to Sakara 7 were within the permissible limits except for PAHs. The result also showed depletion in oxygen content in all sampling sites.

Figure 9: Pipes dumping wastes from houses into Mariotteya water sream.

Figure 10: The polluted water is used in irrigation.

Figure 11: Nodule of Myxobolus dermatobia in skin.

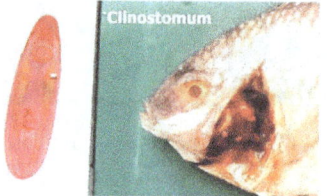

Figure 12: Clinostomum spp metacercarae in branchial cavity

Heavy metals

Data reported in Table 2 showed the values of the detected heavy metals from different sites along Mariotteya strem and indicated that lead is the most abundant element in Mariotteya stream followed by cadmium. Parasitological examination of fish samples. All the examined fish samples were found parasitized (100%) with different species of parasites belonging to Trematodes (Monogenea and Digenean metacercarae), Protozoa and Copepods. Protozoan species showed higher prevalence among the examined fish (tables 3 and 4). Macroscopic examination revealed presence of skin nodules of Myxobolus spp. (Figure 11) and large numbers of Clinostomum spp. metacercarae in the branchial cavities of the examined *O. niloticus* fish (Figure 12).

Discussion

In the present investigation, data of field visit revealed that the mass mortalities and signs of respiratory distress among *O. niloticus* were restricted only along the distance of about 4 km starting from the outlet of Sakara 7 drainage into mariotteya stream at Abuseer at the south and extend to Shabramant at the North with the direction of water current. This restricted area indicated the role of Sakara 7 as source of pollution by receiving the discharge of different sources of pollutants including Al-Hawamdea sugar factory and Industrial and sewage treatment plants.

Results of water analysis revealed that, Mariotteya stream was subjected to high levels of pollutants that introduced to its water through improperly treated industrial, agricultural and sewage wastes. Among the marked abnormal water quality that was detected in the present study, the increase in un-ionized form of ammonia level (reached 55 mg/L). According to Elghobashy [23], the high un-ionized ammonia level is an indicator of the presence of sewage discharge, agriculture-runoff and industrial effluents and it might be also attributed to the increase in oxygen consumption of the decomposing organic matter and oxidation of chemical constituents. Toxicty due to un-ionized form of ammonia may predispose to asphyxia and mortalities as it is easily spread through the gills causing behavioural, physiological and histological changes [6]. In addition, Prolonged exposure to concentration of un-ionized ammonia higher than 1 mg/L causes losses especially among fries and juveniles in water with low dissolved oxygen [24].

The reduction in dissolved oxygen content in Mariotteya water samples may be due to decomposition of suspended organic matter of sewage [25]. According to Adeogun [3], prolonged exposure of fish to low dissolved oxygen level (>5 mg/L) has direct consequences for the survival of fish and other aquatic animals as DO reduction elicit physiological regulatory mechanisms involved in the maintenance of oxygen gradient from water to tissues which is essential to maintain the metabolic aerobic pathways. Also low DO concentration leads to an increase of organic matter in the water system and considered as a factor that predisposing fish diseases carried by pathogens in water [26].

Al-Hawamdea sugar factoy was incriminated as the main source of the detected phenol which was mostly detected in sugar factories as an end product of hydrolysis of molasses [27]. Phenol toxic action is due to its effect on the nervous system also has the same toxic effect of ammonia on fish respiratory function resulting in asphyxia [28].

Poly cyclic aromatic hydrocarbons are the most widespread organic pollutants that known to be carcinogenic and have damaged effects on epithelial tissues of fish body [29]. The detected high level PAHs in Mariotteya water samples (2.56 mg/L) might be resulted from throwing the untreated industrial wastes of many factories located along the Mariotteya road.

Site No	color	Temp°C	PH	Ammonia mg/L	DO mg/L	Phenol mg/L	PAHs mg/L
1	Redish brown with foams	21.3	8.7	17	3.5	5	60.8
2	Redish brown with foams	22	8.2	15.5	3.5	1.5	4.3
3	Dark brown	20.7	8.6	55	3	4.5	25.7
4	Redish brown	20	8.4	10.5	4	1.8	13.3
5	Yellowish	21	7.5	0.02	4.5	0	8.5
Mean ±SD		21±0.73	8.28±0.47	19.60**±0.87	3.70±0.57	2.56**±0.11	22.52**±0.85
P.I		Over 5	6.5-9	0.5	> 5	(-)	(-)
P.I=Permissible limits				**Higher than P.I			

Table 1: Physicochemical parameters of the collected water samples from Mariotteya stream.

Site No	Cu mg/L	Cd mg/L	Pb mg/L
1-	0.0399	0.06	0.83
2	0.05	0.05	0.3
3	0.035	0.045	0
4	0.034	0.14	0.65
5	0	0.162	0.435
Mean ±SD	0.031±0.01	0.091**±0.05	0.443**±0.31
P.I	1	0.01	0.05
P.I=Permissible limits		**Higher than P.I	

Table 2: Concentrations of heavy metals in the collected water samples from Mariotteya stream.

Examined number	Infected number	Infection rate	Protozoa		Crustacea		Helminthes	
			Number	%	Number	%	Number	%
60	60	100	42	70*	37	61.66	27	45

Table 3: Prevalence of parasitic infection.

Parasites	Site of Infection	Number of infected fish	Prevalence %
Monogenea:			
Family Cichlidogyridae Cichlidogyrus arthracanthus	Gills	25	41.67
Digenea (metacercarae):			
Family Clinostomatidae Clinostomum sp.metacercarae	Branchial cavities	19	31.66
Euclinostomum sp.metacercarae	Branchial cavities	3	5

Family Heterophyidae	Muscles and skin	21	35
Heterophid sp. metacercarae			
Family Prohemistomatidae Prohemistomid sp.metecercarae	Muscles	20	33.33
Acanthocephala:			
Quadrigyridae Acanthosentis tilapea	Intestine	11	55
Protozoa :			
Family Chilodonelidae Chilodonella sp.	Skin	12	20
Family Trichodinidae Trichodina truttae. Trichodinafultoni	Skin and gills	32	53.33
Family Balantididae Balantidium sp.	Intestine	10	16.67
Family Myxobolidae Myxobolus dermatobia.	Skin and eyes	13	21.67
Family Cryptosporididae Cryptosporidium sp.	Intestine	24	40
Copepodes:			
Familiy Lernaeidae Lambrogena monody	Gills	27	45
Family Ergasilidae Ergasilus sarsi	Gills	16	26.67

Table 4: Prevalence and site of infection of isolated parasite species.

Concerning heavy metals, data of analysis revealed that, Cu level in Mariotteya water were within the permissible limits indicted by E.O.S [30] and Egyptian Governmental Law No. 48, [31] while Cd and Pb levels were higher than the permissible limit. Kock and Hofer [32] reported that even low concentration of heavy metals in the water may results in high concentration of them in fish flesh the point which is of great concern regarding to fish health directly and human health indirectly [33]. Prolonged exposure of fish to heavy metals causes several histopathological changes especially in the fish gills [34-36] and since gills are the respiratory and osmo-regulatory organ of fish, these histopathological changes might impair the gill respiratory function by reducing respiratory surface area resulted in hypoxia, respiratory failure problems [37-39] which may lead to fish death [40].

Results of water analysis of the five sampled sites, revealed low quality and detection of multiple types of pollutants Sakara 7 showed the worst results, that is might be attributed to the excessive shooting levels of different types of pollutants particularly through Sakara 7 drainage especially with the marked low water column and the stagnant nature of the drainage during that time period of the year (January). On the other side, results indicated that site 5 showed lower level of pollution this is could be due to the flow direction of water current which diluted and so decreased the concentration of pollutants along the stream.

It could be concluded that the recorded low quality with the detection of multiple pollutants may contribute to the respiratory distress and mass mortalities event among O. niloticus in Mariotteya stream.

Parasitic species can be found on every living organism, their presence in their hosts is generally at equilibrium in aquatic ecosystem [41]. When natural or anthropogenic changes either environmental such as temperature, climate, or anthropogenic such as pollution and urbanization occur it can change the state of balance of the parasite between host and nature, thus resulting in disease or mortality in fish population. Parasites can cause mechanical damage (fusion of gill lamellae, tissue replacement), physiological damage (cell proliferation, immunomodulation, detrimental behavioral responses, alteration of growth) and also reproductive damage [42]. In addition, zoonotic parasites have considerable human hazard impacts resulted in pathological lesions and diseases [43,44].

Regarding parasitic infestation, results revealed that all (100%) of the examined O. niloticus sampled from Mariotteya stream were heavily infested with different species of parasites, the result which might be attributed to that Heavy pollution increases the susceptibility of the exposed fish to parasitic infection [45] and also causes biochemical and behavioural changes of the host that can influence the prevalence of parasitism by impairing the host's immune responses [46]. These data could support the present record concerning the high prevalence of parasitic infestation (100%) among the examined O. niloticus sampled from Mariotteya stream. It is important to mention that the detected ectoparasite species including protozoan and crustacean were recorded to cause severe pathological changes and tissue damages in the affected hosts [47] also to keep consideration, the zoonotic importance of the isolated heterophiid and prohemistomatid metacercariae.

In the present investigation, it could be concluded that mass mortalities and asphyxia among O.niloticus population along Mariotteya stream might be attributes to multi-factorial synergism of environmental chemical and biological pollution. Depending on data

of field visit and results of investigations it could be concluded that the synergism of the marked abnormal water quality and environmental pollution that was indicated by the high levels of Un-ionized ammonia, phenol, PAHCs and heavy metals, sever reduction of DO, in addition to the deleterious impacts of the high prevalence of parasitic infection could constitute a complex problem that might lead to the eruption of this catastrophic event of mass mortalities and respiratory distress among O. niloticus at Mariotteya stream on 4th of January 2010. Releasing new water into Mariotteya stream through other adjacent tributaries of River Nile for three successive days lead to

gradual correction of the condition and consequently cessation of signs of asphyxia and mortalities.

Acknowledgment

The authors thank Dr. Eva Sharaby, researcher at the Egyptian Central Laboratories of the Ministry of Health for the logistic support in field work.

References

1. Mohamed MAM. Osman MA, Potter TL, Levin RE (1998) Lead and Cadmium in Nile River water and finished drinking water in greater Cairo Egypt. Environment International 24: 767-772.

2. Adeogun AO, Chukwuka AV, Ibor OR (2011) Impact of abattoir and saw-mill effluent on water quality of upper Ogun River (Abeokuta). American Journal of Environmental Sciences 7: 525-530.

3. Adeogun AO (2012) Impact of industrial effluent on water quality and gill pathology of Clarias gariepinus from Alaro stream Ibadan Southwest Nigeria. Europian Journal of Scientific Research 76: 83-94.

4. Sawako H, Diasuke H, Shin T, Toshiyuki I, Michio K, et al. (2009) Mass mortality and trace element residues in Isaza (Gymnogobitus isaza) collected from lake Biwa Japan. Interdisplinary studies on environmental chemistry- Env Research in Asia 177-183.

5. Grib V, Goncharenko N, Voytyshina D (2006) Saponin as factor of mass fish mortality in the Rivers of Ukraine. Hydrobiol 4: 61-71.

6. Evans JJ, Pasnik DJ, Brill GG, Kleisius PH (2006) Un-ionized ammonia exposure in Nile Tilapia: Toxicity Stress response and susceptibility to Streptococcus agalactia. North American Journal of Aquacultural 68: 23-33.

7. Lasheen MR, Abdel-Gawad F, Alaneny AA, Abd El Bary HMH (2012) Fish as Bio Indicators in Aquatic Environmental Pollution Assessment A case study in Abu-Rawash Area Egypt. World Applied Sciences Journal 19: 265-275.

8. Herdick RP, Gilad O, Yun S, Spangenberg JV, Marty GD, et al. (2000) A Herpes virus associated with mass mortality of juvenile and adult Koi a strain of common Carp. J Aquatic Animal Health 12: 44-57.

9. Subramanian, Purushothamann A (1985) Mass mortality of fish and invertebrates associated with a bloom of Hemidiscus hardmannianus (Bacillariophyceae) in Parangi-Pettai (South India). Limnology and Oceanography 30: 910-911.

10. Mian GF1, Godoy DT, Leal CA, Yuhara TY, Costa GM, et al. (2009) Aspects of the natural history and virulence of S. agalactiae infection in Nile tilapia. Vet Microbiol 136: 180-183.

11. Elsayed E, Mahmoud A (2006) Icthyophthiriasis various fish susceptibility or presence of more than one strain of the parasites. Nature and Science 4: 5-13.

12. Ismael AA (2012) Benthic bloom of cyanobacteria associated with fish mortality in Alexandria water. Egyptian Journal of Aquatic Research 38: 241-247.

13. Abou El-Gheit, EN Abdo MH, Mahmoud SA (2012) Impacts of Blooming Phenomenon on Water Quality and Fishes in Quarun Lake, Egypt. International Journal of Environmental Science and Engineering. 3:11-23.

14. APHA (1989) Standard methods of the examination of water and wastewater. American Public Health Association Washington DC: 1268.

15. APHA (1989): Standard methods of the examination of water and wastewater. Washington DC USA 1193.

16. APHA (1995) Standard methods for examination of water and wastewater analysis. (19 edtn) Washington DC.

17. Pritchard MH, Kruse GOW (1982) The collection and preservation of animal parasites. Univ Nebrasca Lincoln London 144.

18. Lom J, Dyková I (1992) Protozoan parasites of fishes. Elsevier Science publishers Amsterdam 315.

19. Kabata Z (1985) Parasites and diseases of fish cultured in the Tropics. Taylor and Evanics London and Philladelphia Chapter 10 injuries caused by Crustacean Parasites: 121-154.

20. Paperna I (1996) Parasites Infections and Disease of Fishes in Africa. FAO CIFA Technical Paper No. 31 Food and Agriculture Organization, Rome.

21. Yamaguti S (1963) Parasitic Copepoda and Branchiura of fishes. Interscience Publishers New York.

22. Bush AO, Lafferty KD, Lotz JM, Shostak AW (1997) Parasitology meets ecology on its own terms. Margolis et al. Revisited J Parasitol 83: 575-583.

23. Elghobashy HA, Zaghlul KH, Metwally MAA (2001) Effect of some water pollutants on the Nile tilapia Oreochromis niloticus collected from the River Nile and some Egyptian Lakes. Egyptian Journal of Aquatic Biology and Fisheries 5: 251-279.

24. Harris JO, Greg BM, Stephen E, Stephen MH (1998) Effect of ammonia on the growth rate and oxygen consumption of juvenile greenlip abalone Haliotis laevigata Donovan. Aquaculture 3: 259-272.

25. Tayel SI, Ibrahim SA, Authman MMN, El-Kashef MA (2007) Assessment of Sabal drainage canal water quality and its effect on blood and spleen histology of Oreochromis niloticus. African Journal of Biological Sciences 3: 97-107.

26. Suhet MI, Schocken-Itturrino RP (2013) Physical and chemical water parameters and Streptococcus species occurance in intensive Tilapia farming in the state of Espirito Santo Brazil. Acta Scientiarum. Biological Science Maringa 35: 29-35.

27. Klinke HB, Thomsen AB, Ahring BK (2004) Inhibition of ethanol-producing yeast and bacteria by degradation products produced during pre-treatment of biomass. Appl Microbiol Biotechnol 66: 10-26.

28. Saha NC, Bhunia F, Kaviraj A (1999) Toxicity of phenol to fish and aquatic ecosystems. Bull Environ Contam Toxicol 63: 195-202.

29. De Maagd PGJ, Vethaak AD (1998) Biotransformation of PAHs and their carcinogenic effects in fish. Nelson. Berlin 265-309.

30. Egyptian Organization of Standardization (1993) Egyptian Standard Maximum level for heavy metals concentration in food ES. 546-815.

31. Egyptian Governmental Law No. 48 (1982) The implementer regulations for law 48/1982 regarding the protection of the River Nile and water ways from pollution. Map Periodical Buul, 3-4: 12-35.

32. Köck G, Hofer R (1998) Origin of cadmium and lead in clear softwater lakes of high-altitude and high-latitude, and their bioavailability and toxicity to fish. EXS 86: 225-257.

33. Olaifa FE, Olaifa AK, Adelaja AA, Owolabi AG (2004) Heavy metal contamination of Clarias gariepinus from a lake and fish farm in Ibadan Nigeria. African Journal of Biomedical Research 7: 145-148.

34. Mohamed FAES (2003) Histopathological studies on some organs of Oreochromis niloticus, Tilapia zilli and synodontis schal from El-Salam canal Egypt. Egyptian journal of Aquatic Biology and Fisheries 7: 99-138.

35. Balah AM, El-Bouhy ZM, Easa ME (1993) Histologic and Histopathologic studies on the gills of Tilapia nilotica (Oreochromis niloticus) under the effect of some heavy metals. Zagazig Veterinary Journal 21: 351-364.

36. Rajeshkumar S1, Munuswamy N (2011) Impact of metals on histopathology and expression of HSP 70 in different tissues of Milk fish (Chanos chanos) of Kaattuppalli Island, South East Coast, India. Chemosphere 83: 415-421.

37. Agatha AN (2010) Levels of Some Heavy Metals in Tissues of Bonga Fish Ethmallosa fimbriata from Forcados River. Journal of Applied Environmental and Biological Sciences 1: 44-47.

38. Javed M, Usmani N (2011) Accumulation of heavy metals in fishes: A human health concern. International Journal of Environmental Sciences. 2: 659-670.

39. Yasser AG1, Naser MD (2011) Impact of pollutants on fish collected from different parts of Shatt Al-Arab River: a histopathological study. Environ Monit Assess 181: 175-182.

40. Abdel-Baki AS, Dkhil MA, Al-Quraishy S (2011) Bioaccumulation of some heavy metals in tilapia fish relevant to their concentration in water and sediment of Wadi Hanifa Saudi Arabia. African Journal of Biotechnology 10: 2541-2547.

41. Marcogliese DJ (2005) Parasites of the superorganism: are they indicators of ecosystem health? Int J Parasitol 35: 705-716.

42. Lafferty KD, Kuris AM (1999) How environmental stress affects the impacts of parasites. Limnology and Oceanography 44:925-931.

43. Scholz T (1999) Parasites in cultured and feral fish. Vet Parasitol 84: 317-335.

44. Al-Jahdali MO, El-S Hassanine RM (2010) Ovarian abnormality in a pathological case caused by Myxidium sp. (Myxozoa, Myxosporea) in onespot snapper fish Lutjanus monostigma (Teleostei, Lutjanidae) from the Red Sea. Acta Parasitologica 55:1-7.

45. El-Seify MA, Zaki MS, Razek A, Desouky Y, Abbas HH, et al. (2011) Study on Clinopathological and Biochemical Changes in Some Freshwater Fishes Infected With External Parasites and Subjected to Heavy Metals Pollution in Egypt. Life Science J 8: 401-405.

46. Khan RA, Thulin J (1991) Influence of pollution on parasites of aquatic animals. Adv Parasitol 30: 201-238.

47. Yildiz K, Kabackci N, Yarim M (2004) Pathological changes of tench intestines infected with Pomphorhynchus laevis. Revista de Medicina Veterinaria 155: 71–73.

31

Price Modulation Policy of Federal Government of Nigeria: Effects on Fish Production

Ayeloja AA[1*], George F[2], Sodeeq E[3] and Adebisi GL[1]

[1]Fisheries Technology Department, Federal College of Animal Health and Production Technology Moor Plantation,

PMB 5029, Ibadan, Nigeria

[2]Department of Aquaculture and Fisheries Management, Federal University of Agriculture, Abeokuta (FUNAAB)

PO Box 2240, Abeokuta, Nigeria

[3]Department of Agric. Extension and Management, Federal College of Animal Health and Production Technology

Moor Plantation, PMB 5029 Ibadan, Nigeria

*Corresponding author: Ayeloja AA, Department of Fisheries Technology, Federal College of Animal Health and Production Technology Moor Plantation, PMB 5029, Ibadan, E-mail: ayeloja2@gmail.com

Abstract

The study investigated the effect of price modulation policy of Federal Government of Nigeria on fish production in Oyo State, South West Nigeria. A multistage sampling technique was used to select 150 respondents from 9 wards within Oyo State using well-structured questionnaire to obtain information on socio-economic characteristics of the respondents, types of fish cultured and the impact of price modulation policy on profitability of fish production. Data were analyzed using descriptive statistics, regression and t-test analysis. There is influx of young men (57.5%) in fish production in the study area with 95% of them having between 1-5 years of experience. *Clarias spp* is the most cultured fish species in the study area and cost per kg of fish produced after price modulation had a negative and significant ($p<0.05$) effect on profitability of fish production during the period of this study. T-test further showed that there was significant difference between production cost before price modulation (PCBF) and production cost after price modulation (PCAF). This could have negative health and welfare implications on the citizenry. Policy makers should therefore put in place welfare packages that will ameliorate the effect of this policy.

Keywords: Price modulation; Policy; Fish production

Introduction

Fish is an important source of good quality protein required in human diets, it has the higher level of easily metabolisable protein, fats, vitamins, calcium, iron, and essential amino acids when compared to other sources of animal protein such as poultry and beef [1-2]. Aquaculture provides nearly 50% of the annual world fisheries production with 110 million tonnes of food fish in 2006. Half of all aquaculture production is finfish, a quarter is aquatic plants and the remaining quarter is made up of crustacean (such as shrimp, prawn, crabs, oyster and mussels) [3]. Although aquaculture activity in Nigeria started about 50 years ago [4], aquaculture production in Nigeria is currently about 40,000 metric tonnes contributing only 6% of domestic fish production [5]. Nigeria has become one of the largest fish importers in the developing world, importing about 600,000 metric tonnes annually [4]. Ifejika et al. [6] reported that in spite of the fact that over 1.5 million hectares of surface water area is available for fish culture; no appreciable result had been recorded in the aquaculture sub-sector due to instability in Nigerian policies which affect all aspect of her economy. One of such policies is subsidy reform or price modulation policy. According to Ebewore et al. [7], Subsidy is a price intervention policy measure whereby financial assistance is granted by a government for the purpose of promoting public welfare. The adjustment of price shocks along the chain of fish producers is an important characteristic of the functioning of fish markets. As such, the process of price transmission through the supply chain has long attracted the attention of agricultural economists as well as policy makers. Due to subsidy reform introduced in Nigeria within 2015/2016 fiscal year and other monitory policies of the Federal Government of Nigeria, the cost of performing production function of fish is assumed to have been affected through increase in transportation cost and cost of fish production inputs. It is therefore imperative to study the effect of these policies which was termed price modulation by the Government on fish production in Oyo state South-West Nigeria. The introduction of price modulation policy by the Federal Government of Nigeria during the same period had also brought about controversy of what the overall effect of such policy on agricultural production and products will be most especially fish production and products. It was assumed that the policy will increase the cost of production, reduce farmers gain and reduce purchasing power of consumers thus the need to carry out this research. This study therefore aimed at describing the socio-economic characteristics of fish producers in Oyo State, identify the types of fish they culture as well as determining the impact of price modulation policy of Federal Government of Nigeria on their production of fish.

Methodology

Study area

The study was carried out in Oyo state, South-West Nigeria. The study area has heterogeneous population of Yoruba, Igbo and Hausa. Oyo state is located in the south west geopolitical zone of Nigeria. Oyo state consist of 33 local government areas which are Akinyele, Afijio,

Egbeda, Ibadan north east, Ibadan north west, Ibadan south east, Ibadan south west, Ibarapa central, Ibarapa east, Ibarapa north, Ido, Irepo, Iseyin, Kajola, Lagelu, Ogbomosho south, Oyo west, Oyo east, Atiba, Atigbo, Saki west, Saki east, Itesiwaju, Iwajowa, Olorunsogo, Oluyole, Ogo-oluwa, Surulere, Ori ire and Ona ara. The state covers a total of 27,249 square kilometers of land mass and it is bounded in the south by Ogun state, in the north by Kwara state, in the west partly by Ogun state and partly by Benin republic, and in the east by Osun state. The state has four political zones namely-Saki, Ogbomosho, Oyo, and Ibadan/ibarapa out of which one zone was studied which was Ibadan/Ibarapa. The study population comprises of different fish species processors in Oyo state metropolis.

Source of data and data collection

Primary data were used for this study. Data on socio-economic characteristics of the respondents, types of fish cultured and the impact of price modulation policy on profitability of fish production were collected using a well-structured questionnaire.

Sampling technique and sample size

Multistage sampling techniques were employed in the selection of respondents. The first stage involved the selection of three (3) local government areas out of the thirty-three (33) local government areas of Oyo state using simple random technique. The local government visited included, Akinyele, Lagelu and Oyo East. The second stage involved the selection of three wards under each local government selected which give the total sum of nine (9) wards. The third stage involved the selection of five (5) communities under each wards using simple random technique, which give the total sum of forty-five (45) communities. The third stage involved the questioning of 10 processed fish mongers from each community selected using structured questionnaire which makes the total sum of 150 respondents used as the sample size.

Data Analysis

The data were analysed using appropriate statistical tools. Data on personal characteristics and type of fish cultured were analysed using descriptive statistics (Percentage, frequency and mean) and bar chart. Multiple regressions were used to determine the impact of price modulation policy of Federal Government of Nigeria on production of fish and T-test was used to test the stated hypothesis.

Result and Discussion

Simple descriptive statistics such as percentage, mean and frequency were used to analyze the objectives while chi-square analysis was used for the hypothesis testing.

This study (Table 1) indicated that out of 150 respondents, 62.5% falls within the age group of 30 years or less, 25.0% were between 31-40 years, 7.5% were 41-50 years and only 5.0% were 51-60 years of age. This indicates that majority of fish farmers in Oyo State are in their middle active age (<40). Similar finding was reported by Akinbile et al. [8] that majorly of fish farmers in Oyo state and Nigeria generally are in their active age, while people of age 50 years and above are most likely to have retired. Akinpelu et al. [9] also reported similar finding in their study of the gender differentials in knowledge and utilization of ICTS among fish farmers in Ido Local Government area of Oyo state South-Western Nigeria where it was reported that that majority of the

respondents fall within the age group less than 40 years. Yisa et al. [10] also observed similar result for rural women marketing fish in Niger State Nigeria.

Variable	Freq	%
Age		
≤ 30	75	62.5
31-40	30	25
41-50	9	7.5
51-60	6	5
Total	120	100
Mean	33	
Sex		
Male	69	57.5
Female	51	42.5
Total	120	100
Marital status		
Single	69	57.5
Married	51	42.5
Total	120	100
Experience		
01-May	114	95
06-Oct	6	5
Total	120	100
Mean		
Education		
Primary	111	92.5
Secondary	-	-
Tertiary	9	7.5
Total	120	100

Table 1: Socio-economic characteristics of respondent.

The result of this also shows that 57.5% of the respondents were male while 42.5% were female and 57.5% of them were single. This means that unmarried men dominate fish production in the study area, this is contrast with the report of many authors like Akinpelu et al. [11-13] were they reported that fish processing and marketing in Nigeria is dominated by women. This could be attributed to the influx of young people (mostly male) in the sector as indicated by the result of this study which show that most of the respondents had low experience (95% of the farmers had 1-5years of experience while 5% had 6-10 years' experience) on the job. This study also revealed that majority (92.5%) of the respondents had primary education while only 7.5% had tertiary education which shows that most of the farmers were not well educated and the government needs to improve on seminars, orientation for the farmers. Similar opinion was expressed by Ayo et al. [10,11,14]. More than half (55%) were Christians 40% were Muslims and 2.5% were either pagans or worshippers of other religions showing that these religion do not prohibit fish production. Seventy five per cent (75%) of the respondents in the study area had household size between 1-5 members and 25% had household size between 6-10 members. Yisa et al. [10] reported similar result that thirty five percent (35%) of the fish marketers from Katcha Local Government Area of Niger State, Nigeria had large household size with 4 children and above. Furthermore 45% of the respondents engaged in fish production as a primary occupation while less than 50% took fish marketing as secondary occupation, Ayeloja et al. [1] gave similar report in their study of the effect of insect infestation on the economic value of smoked fish sold in selected markets within Oyo State, South West Nigeria where it was reported that 98% of the respondents took fish marketing as secondary occupation.

Types of fish cultured by the respondents

Table 2 shows the types of fish cultured in the study area, *Clarias* spp was the most commonly cultured spp in the study area with 82.5% followed by Tilapia spp with 42.5% while Carp was the least cultured fish in the study area with 7.5%. This is in line with the report of Ayeloja et al. [15,16] as well as Adewumi et al. [17] who stated that *Clarias* spp is the most cultivated fish spp because it enjoys wide acceptability in most parts of the country for its unique taste, flavour and texture (Figure 1).

Fish Type	Yes	No
Clarias spp	99 (82.5%)	21 (17.5%)
Tilapia spp	51 (42.5%)	69 (57.5%)
Heterotis spp	24 (20.0%)	96 (80%)
Heterobranchus spp	18 (15.0%)	102 (85%)
Heteroclarias	24 (20.0%)	96 (80%)
Carp	9 (7.5%)	111 (92.5%)

Table 2: Types of fish cultured in the study area.

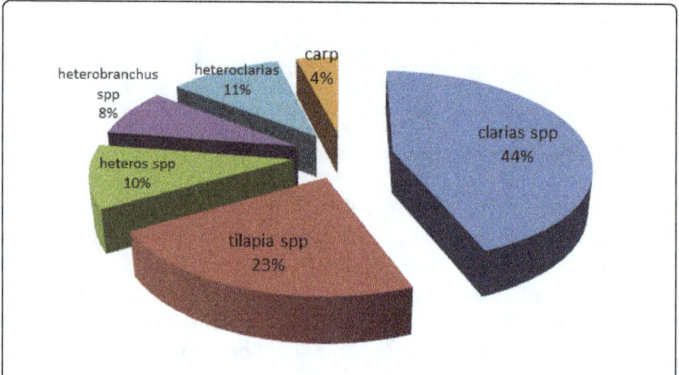

Figure 1: Types of fish cultured in the study area.

Impact of price modulation policy on profitability of fish production

The effect or impact of price modulation policy on profitability of fish production is shown in equation 1 while equation 2 shows the causality effect of production cost per kg of fish before and after price modulation on profitability of fish production.

The result of the study showed that the production cost before the price modulation policy of the Federal Government of Nigeria had a positive significant (p<0.05) effect on profitability of fish production in the study area. This implies that a unit increase (naira) of production cost per kg of fish produced before price modulation will increase the profitability per kg of fish production by 4.3%. However, the case was opposite after price modulation policy of the Federal Government of Nigeria as production cost after price modulation had a negative and significant (p<0.05) effect on profitability of fish production. This implies that a unit (naira) of production cost per kg of fish produced after price modulation will reduce profitability per kg of fish produced by 4.6% indicating that arbitrary change of Government policy do have direct impact on cost of fish production just as the price modulation policy have reduced the expected gains on fish farming in Nigeria during the period of this study. This opinion was further justified by the result presented on Table 3 where T-test was used to test for significant difference (p<0.05) in the cost of fish production before and after price modulation policy of the Federal Government of Nigeria. The result indicated that the mean cost of 1kg of fish produced was N318.38 before price modulation policy, however after price modulation policy; the production cost rose to N411.25. The percentage effect of price modulation on the production cost of fish was therefore estimated to have increased by 29.17% indicating that there is linear correlation between Government price policy and asymmetric price transmission. This is in line with the report of [18-19] who stated that asymmetric price transmission does not only respond to market power and market structure alone but there is also linear correlation between Government price policy and price transmission in the market. Price policy makers should therefore be cautious of the overall implications of their policy especially those policy reforms that could have inflationary implications on the citizenry and put necessary economic welfare measures that will cushion the adverse effect of this price modulation policy so as to improve living standard of Nigerians. Similar study that will compare the effect of price modulation policy of the Federal Government of Nigeria as well as similar Governmental reforms on the production and marketing of other products in Nigeria should be conducted. It is also important for Nigeria Economists to proffer indigenous economic models that can move the economy forward for the Government.

Variable	N	Mean	Std. dev	t	df	sig
PCBF	120	318.38	106.03861	32.89	119	0
PCAF	120	411.25	122.60178	36.745	119	0

Table 3: T-Test showing the result of hypothesis.

References

1. Ayeloja AA, George FOA, Awobifa OM, Sodeeq AE, Jimoh WA, et al. (2015) Effect of insect infestation on the economic value of smoked fish sold in selected markets within Oyo State, South West Nigeria. In the 30th annual proceeding of Fisheries Society of Nigeria (FISON) held at

ETF Lecture Theatre, Delta State University, Asaba Campus, Delta State pp: 558-561.

2. Ayoola SO (2010) Sustainable fish production in Africa. African Journal of Food Agriculture Nutrition and Development 10: 1-5.

3. FAO (2001). Pests and insects infesting smoked dried fish FAO fish technical paper p: 87.

4. Olagunju FI, Adesiyan IO, Ezekiel AA (2007) Economic viability of Cat fish production in Oyo state Nigeria. J Hum Ecol 21: 121-124.

5. Adeogun OA, Ogunbadejo HK, Ayinla OA, Oresegun A, Oguntade OR, et al. (2007) Urban Aquaculture: Producer Perceptions and Practices in Lagos State, Nigeria. Middle-East Journal of Sci Res 2: 21-27.

6. Ifejika KI, Okunade EO, Ifejika LI, Asadu AN (2011) Physical assets ownership of fisherfolk in fishing communities of Kainji Lake Nigeria: Implications for Climate change. Journal of Agricultural Extension 16: 92-103.

7. Ebewore SO (2012) Policy Instruments. Department of Agricultural Economics and Extension, Delta State University, Asaba Campus (Unpublished lecture) pp: 23.

8. Akinbile LA, Alabi OE (2010) Use of ICTs among Fish Farmers in Oyo State. Journal of Agricultural Extension 14: 1.

9. Akinpelu OM, Akinbile LA, Ayeloja AA, Akinosho GA, George FOA, et al. (2013) Gender differentials in knowledge and utilization of ICTs among fish farmers in Ido Local Government Area of Oyo State South-western Nigeria. Journal of Agricultural Economics and Development 2: 255-263.

10. Yisa TA, Tsadu SM, Mohammed I (2011) Socioeconomic evaluation of rural women and the estimation of profitability of fish marketing in four markets in Nigeria. Int J Fish Aquaculture 3: 180-183.

11. Akinpelu OM, Ayeloja AA, George FOA, Adebisi GL, Jimoh WA, et al. (2013) Gender Analysis of Processing Activities among Commercial Catfish Processors within Ibadan Metropolis, Oyo State South-Western Nigeria. J Aquac Res Development 4: 176.

12. Oluwatoyin DK, Stella BW, Awujola AF (2010) Indigenous fish processing and preservation practices amongst women in Southwestern Nigeria. Indian Journal of Traditional Knowledge 9: 668-672.

13. Nwabueze AA (2010) The role of women in sustainable aquacultural development in delta state. Journal of Sustainable Development in Africa 12: 284-293.

14. Ayo-Olalusi CI, Anyanwu PE, Ayorinde F, Aboyweyere PO (2010) The Liverpool fish market in Lagos State, Nigeria. African Journal of Agricultural Research 5: 2611-2616.

15. Ayeloja AA, George FOA, Obasa SO, Sanni LO (2011) Effect of post-slaughter time intervals on the quality of the African catfish, Clarias gariepinus (Burchell, 1822). American Journal of Food Technology 6: 790-797.

16. Kumolu-Johnson CA, Aladetohun NS, Nolimele PE (2010) The effects of smoking on the nutritional qualities and shelf-life of Clarias gariepinus (LACEPEDE). African Afr J Biotechnol 9: 073-076.

17. Adewumi AA, Olaleye VF (2011) Catfish culture in Nigeria: Progress, prospects and problems Agricultural and Fisheries Working p: 3.

18. Pavel VBKG. (2005) "Analysis of Price Transmission along the Food Chain", OECD problems. African Journal of Agricultural Research 6: 1281-1285.

19. Aguero JM (2004) "Asymmetric Price Adjustments and Behavior Under Risk: Evidence from Peruvian Agricultural Markets".

32

Iranian Fisheries Status: An Update (2004-2014)

Harlioglu MM[*] and Farhadi A

Department of Fisheries, Fırat University, Elazig, Turkey

[*]**Corresponding author:** Harlioglu MM, Fisheries Faculty, Firat University, Elazig, Turkey, E-mail: mharlioglu@firat.edu.tr

Abstract

Iran's appropriate geographical location (i.e., large brackish water source in the north, Caspian Sea, salt water source in the south, Persian Gulf and Gulf of Oman) and a wide range of brackish, freshwater and marine species (i.e., trout, carp, sturgeon, sea bass, sea bream, turbot, mackerel, sardine, tuna, sea cucumber, marine shrimp, crayfish) provides Iran to be a great fish producer country. The total fishery production was 947,352 tons in 2014. In this year, 575,512 tons (60.74% of total fishery production) of fish production were obtained from the capture fisheries and 371,840 tons (39.26% of total fishery production) of production was obtained from the aquaculture production. Fisheries in the Persian Gulf and Gulf of Oman are the most important fishery (93% of total fishery) in Iran. There has been a fast increase in the aquaculture production in Iran. For example, total aquaculture production for 2004 and 2014 was 124,560 and 349,365 tons, respectively. Therefore, the percentage of aquaculture in total fish production has been rising every year. The ratio of aquaculture production to total fish production was 26.26% in 2004, 32.65% in 2008 and 39.26% in 2014. Rainbow trout and carps are the main cultured freshwater fish species. In recent years, fisheries production export of Iran has been increased from US$ 85 million in 2004 to US$ 300 million in 2014. In conclusion, despite Iran's long coastline, fishery has not been developed completely and has the potential to be developed more by enhancing aquaculture and fish cage culture.

Keywords: Aquaculture; Export; Statistics; Fisheries; Iran

Introduction

Iran is surrounded by three seas: the Caspian Sea at the north, the Persian Gulf and Gulf of Oman at the south. It has a long coastline exceeding 5,800 km, about 890 km in the north (Caspian Sea coast) and 4,900 km in the south (Persian Gulf and Gulf of Oman coast including coastline around the islands) [1]. In addition, Iran has a great variety of marine and freshwater resources. Table 1 shows important marine and freshwater resources, and their surface area in Iran. There are differences between biological contents and climatic conditions among Iran's sea. These difference water conditions result in diversity of fish species and provides many resources for fisheries activity in Iran. The fisheries are one of the most important agriculture industries in Iran (REF). Providing human nutrition and raw material for industrial sectors, creating employment possibilities and generating high potential for export earnings.

Since 1996, fisheries production statistics have been collected every year by Iran Fisheries Organization. Marine fisheries is divided into two sectors north water (Caspian Sea) and south water fisheries (Persian Gulf and Gulf of Oman coast), while for inland aquaculture, Iran is divided into thirty-one provinces.

There has been a recent increase in the fishery production of Iran. Total harvest increased from 474,500 tons in 2004 to 838,892 tons in 2012 (Table 2). In 2014, total fisheries production of Iran peaked at 947,352 tons, with 575,512 tons (60.74%) of the total production were obtained from capture fisheries [2,3]. In 2014, Iran was 20th place in world aquaculture production, 28th place in world fish capture production, and 27th in overall production [4]. After Egypt (1,481,882 tons), Iran (947,354 tons) is the biggest fish producer in the Middle East and western Asia. Turkey (536,516 tons), Oman (211,319 tons) and Yemen (190,000 tons) are other big producers [4]. The percentage contribution of aquaculture production to total harvest increased from 26.26% in 2004 to 39.26% in 2014.

Although the bulk (about 93%) of capture production (535,865 tons) was obtained from the south, 48% of the aquaculture production came from three provinces; Mazandaran (71,784 tons) and Guilan (46,802 tons) provinces in the north and Khuzestan (60,172 tons) in the southwest [2,3].

Seas and Lakes	Dam Lakes	Rivers
Caspian sea (370,987)	Aras (145)	Helmand (1,150)
Gulf of Oman (903,000)	Shahyun (65)	Hari (1,100)
Persian Gulf (251,000)	Amir Kabir	Aras (1,072)
Urmia (5,200)	Latyan	Karun (950)
Hamoun (3,820)	Sivand (11)	Karkheh (900)
Bakhtegan (3,500)	Mulla Sadra	Sefid Rud (670)
Namak (647)	Upper Gotvand	Zayanderud (400)
Maharloo (600)	Golpayegan	Zarrineh (302)

Table 1: Important marine and freshwater resources of Iran and their surface area.

Years	Aquaculture				Capture fisheries				Aquaculture + Capture fisheries
	Fresh water	Sea	Fresh water + sea	% Ratio of aquaculture in total production	Fresh water	Sea	Fresh water + sea	% Ratio of capture fisheries in total production	Total production
2004	115657	8903	124560	26.26	35775	314165	349940	73.74	474500
2005	130603	3577	134180	25.68	44887	343492	388379	74.32	522559
2006	148974	5700	154678	26.88	46435	374447	420882	73.12	575560
2007	191169	2508	193677	34.44	39174	329571	368745	65.56	562422
2008	179275	4372	183647	32.65	36967	341980	378947	67.35	562594
2009	202225	5128	207353	34.58	44279	348122	392401	65.42	599754
2010	245015	6359	251374	37.88	43805	368505	412310	62.12	663684
2011	277325	8026	285351	38.82	37831	411897	449729	61.18	735079
2012	328725	10152	338877	40.4	40314	459701	500015	59.6	838892
2013	358178	12698	370876	41.91	40423	473658	514081	58.09	884957
2014	349365	22475	371840	39.26	39647	535865	575512	60.74	947352

Table 2: Aquaculture and capture fisheries productions (tons/year) obtained from seas and freshwaters in Iran between 2004 and 2014 [2,3].

A review on the aquaculture development in Iran until 2008 was primarily published by Kalbassi et al. [5]. They mainly reported carp, rainbow trout, sturgeon and marine shrimp aquaculture status in Iran. In addition, in another study, a review on the status of fisheries in Iran was primarily published by Karimpour et al. [6]. They presented and discussed the fishery, aquaculture, importance aquaculture species and aquaculture industry in Iran between 1997 and 2008. On the other hand, no studies have been published on the status of fisheries in Iran in recent years. This review presents and discusses the notable expansion of fisheries and aquaculture in Iran between 2004 and 2014 [7].

Fishery production

Marine fishery: Over 60% of marine fishery products is from Persian Gulf and Gulf of Oman. The most abundant species were *Eleutheronema tetradactylum* (fourfinger threadfin), *Otolithes ruber* (tigertooth croaker), *Pampus argentus* (silver pomfret), *Scomberomorus commerson* (narrow-barred Spanish mackerel), *Scomberomorus guttatus* (Indo-Pacific king mackerel), *Pomadasys kaakan* (javelin grunter), *Epinephelus coioides* (orange-spotted grouper), *Thunnus tonggol* (longtail tuna), *Dussumieria* (rainbow sardines), *Coryphaena hippurus* (mahi-mahi), *Acanthopagrus latus* (yellowfin seabream) and *Cynoglossus arel* (largescale tonguesole) [2,3] (Figure 1). There is no notable difference in abundant of each species caught from the Persian Gulf and Gulf of Oman.

Regard into marine crustaceans; three shrimp species are caught in the south of Iran. *Penaeus indicus* (Indian white shrimp), *Penaeus merguiensis* (banana shrimp) and *Penaeus semisulcatus* (green tiger shrimp). Annual harvest in 2013 and 2014 was 8,789 and 8,567 tons shrimp were from Persian Gulf and Gulf of Oman, respectively [3].

Iran's marine fishery has increased steady in the last decade. It raised to 535,865 tons in 2014 from 314,165 tons in 2004. Although there are no statistics to show exact portion of fish species in Iran's marine fishery, about 50% of harvest is large pelagic species. In addition, catching of tuna and tuna-like (*Auxis rochei, Auxis thazard, Euthynnus affinis, Katsuwonus palamis, Rastrelliger kanagurta, Scomber japonicas, Scomberomorus commerson, Scomberomorus guttatus, Thunnus albacares* and *Thunnus tonggol*) species is a major component in large pelagic fisheries in Iran [2,3].

After large pelagic fish species, demersal fish species with 32-35% constitute the highest rate in the catch caught from the seas. *P. kaakan, O. ruber, C. arel* and *P. argenteus* are the most important demersal fish species [2,3].

Inland fishery: Iran has two major inland basins in the north and south and several smaller basins in center and east. Taking into account the newly described species of cyprinids and loaches, freshwater and brackish water fish of Iran exceed 200 species. Inland waters contain 163 of these species including mainly cyprinids with 87 species, balitorids with 22 species and gobiids with 10 species [8].

The main inland fishing area is the Caspian Sea the largest inland body water in the world with salinity around 12 ppt. In 2014, the total fish catch from the Caspian Sea was 39,647 tons. The most important commercial fish species in the Caspian Sea are divided into three groups; bony fish, Caspian Sea sprat and sturgeon fish species.

The most important commercial bony fish of the Caspian Sea are *Rutilus frisii kutum* (Caspian kutum), *Liza aurata* (Golden grey mullet), *Liza saliens* (Leaping grey mullet), *Sander lucioperca* (Pikeperch), *Cyprinus carpio* (European carp), *Rutilus rutilus* (Roach), species of genus *Alosa* (Caspian shads), *Abramis brama* (bream), *Chalcalburnus chalcoides* (Caspian shamaya) and *Vimba vimba* (Caspian vimba). Three species of Caspian Sea sprat live in the Caspian

Sea: *Clupeonella engrauliformis* (Anchovy kilka), *Clupeonella grimmi* (Big eyed kilka) and *Clupeonella cultriventris* (Common kilka). The Caspian Sea is inhabitant of five species of Sturgeon fish: *Huso huso* (Great sturgeon), *Acipenser gueldenstaedtii* (Russian sturgeon), Acipenser persicus (Persian sturgeon), Acipenser stellatus (Stellate sturgeon) and *Acipenser nudiventris* (Spiny sturgeon) [6]. Annual average catch of these three fish groups between 2004-2014 is given in Figure 2.

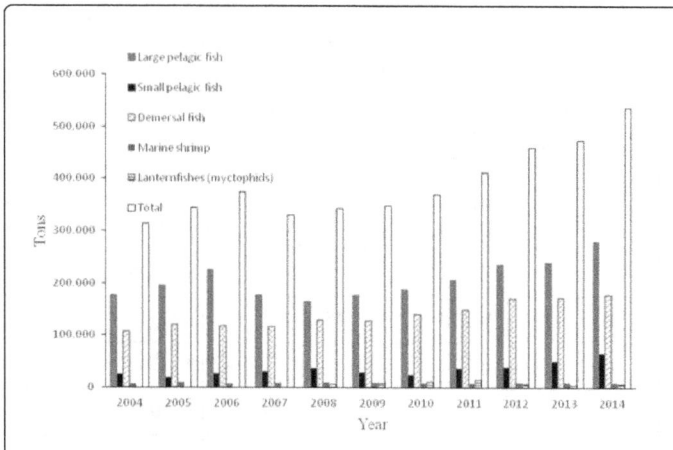

Figure 1: The most abundance caught fish types in the marine water of Iran between 2004-2014 [2,3].

Caspian kutum, grey mullet and European carp are the most important caught fish species of fishermen in the Caspian Sea. Grey mullet is the most abundant species and contributes in 70% of the total bony fish fishery in the Caspian Sea.

Sturgeon fishery: There are 27 sturgeon species living in the seas and rivers of the Northern hemisphere [9]. The maximum sturgeon catch

in the world was 32,078 tons in 1977 [10]. Iran is the biggest exporter caviar and sturgeon fish meat in the world. However, sturgeon fisheries reduced in the last two decades. For example, sturgeon fisheries from Iranian waters dropped to 41 tons in 2014 from 500 tons in 2004 [2,3]. Bronzi et al. [10] suggested that variety of reasons contributed to the sturgeon fishery decline:

- River fragmentation and channelization with subsequent changes in hydrology and hydrodynamics.
- Overharvest by legal and illegal fisheries.
- Increasing pollution, from agricultural practices, urban growth and industrial developments.

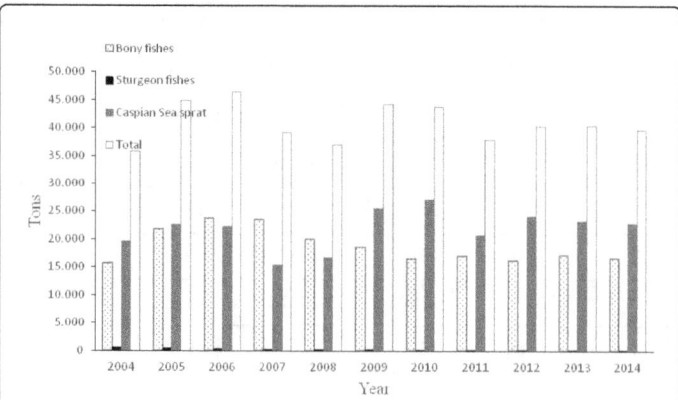

Figure 2: The amount of captured bony fish, Caspian Sea sprat and sturgeon fish species from the Caspian Sea between 2004-2014 [2,3].

	2004	2005	2006	2007	2008	2009	2010	2011	2012	2013	2014
Common carp	65400	73396	77463	97262	87748	100430	121608	132177	154565	167883	170341
Rainbow trout	30000	34760	46275	58761	62630	73642	91519	106409	131000	143917	126515
Sturgeon	-	-	-	-	-	363	251	312	456	564	650
Marine shrimp	8903	3577	5700	2508	4372	5128	6359	8026	10152	12698	22475
Narrow-clawed crayfish	27	268	270	258	275	287	298	338	341	263	70
Harvesting from natural water resources	20230	22179	24970	34888	28622	27503	31339	38089	42363	45551	51666
Marine fish in cage	-	-	-	-	-	-	-	-	-	-	123
Total	124560	134180	154678	193677	183647	207353	251374	285351	338877	370876	371840

Ornamental fish	30	31	35	54	79	93	107	132	148	186	204

Table 3: Name and quantity (tons) of important aquacultured and ornamental fish (million) products between 2004 and 2014 [2,3].

Aquaculture production

In recent years, there has been a fast increase in the aquaculture production of Iran. It contributed approximately 40% of fisheries production of Iran in 2014 (Table 3).

Aquaculture production in Iran increased about 11.5% per year from 2004 (124,560 tons, 26% of fisheries production) to 2014 (349,365 tons, 40%). This rapid increase was higher than the global average of 8% per year and resulted from production of carp species and rainbow trout (carp culture increased from 54,801 tons in 2002 to 170,341 tons in 2014). This increment was more rapid for rainbow trout. Rainbow trout culture in Iran enhanced from 16,026 tons in 2002 to maximum amount of 143,917 tons in 2013. After Chile, Iran was the biggest rainbow trout producers in the world.

Most aquaculture production in Iran is comprised of freshwater species except for marine shrimp production (about 6% of the total). However, there is also a very small cage culture industry in the south of Iran.

Iranian Fisheries Organization (IFO) divided aquacultured species into six groups 1) cyprinid species, 2) trout species, 3) sturgeon species, 4) marine shrimp species, 5) crayfish (besides it is not cultured) and 6) marine fish species. The most important aquaculture species are listed in Table 4.

No	Group of species	Species
1	Cyprinid	*Cyprinus carpio, Hypophthalmichthys molitrix, Hypophthalmichthys nobilis, Ctenopharyngodon idella*
2	Trout	*Oncorhynchus mykiss*
3	Sturgeon	*Huso huso, Acipenser baerii, Acipenser persicus, Acipenser ruthenus, Acipenser stellatus*
4	Marine shrimp	*Litopenaeus vannamei, Penaeus merguensis, Penaeus monodon, Fenneropenaeus indicus, Penaeus semisulcatus*
5	Crayfish	*Astacus leptodactylus*
6	Marine fish	*Lates calcarifer, Acanthopagrus latus, Sparidentex hasta, Sparus aurata*

Table 4: Important aquaculture species in Iran.

Cobia (*Rachycentron canadum*), silver pomfret (*Pampus argenteus*), fourfinger threadfin (*Eleutheronema tetradactylum*), Asian sea bass (*Lates calcarifer*), sobaity seabream (*Sparidentex hasta*), grouper (*Epinephelus coioides*) and rabbit fish (*Siganus canaliculatus*) (11-12) are produced in experimental or pilot scales.

One of the important freshwater crustacean species in Iran is the narrow clawed crayfish (*Astacus leptodactylus*). It is the only freshwater crayfish species in Iran. The commercial value of exported *A. leptodactylus* between 2000 and 2009 varied from 1.5-2.5 million US$ annually [13]. Iranian crayfish production reached to maximum 341 tons at 2012.

Future species: The candidate species for mariculture development include groupers (*Serranidae*), cobia (*Rachycentron canadum*), silver pomferet (*Pampus argenteus*) and fourfinger threadfin (*Eleutheronema tetradactylum*) [5]. Also recently, the Caspian salmon, *Salmo trutta caspius*, has attracted interest for aquaculture in cages and raceways in Iran, with emphasis on using triploid populations to omit problems associated with sexual maturation, which can reduce commercial benefits of salmonid culture, especially beyond the maturation phase [5-14].

There are more than 130 species of seaweed found in the Iranian marine waters. *Gracilaria spp., Sargassum spp.* and *Eucheuma spp.* are some of the commercial seaweed species. Over the past four years several trials have been carried out on the farming of Gracilaria in ponds and the open sea and a pilot project has been initiated to develop commercial seaweed farming. Persian Gulf pearls are well-known on the international markets, however, due to over fishing, oyster stocks have been reduced dramatically. Iranian Fisheries Research Organization (IFRO) has conducted various research projects for seed production and in 2004 successfully produced seed. Access to seed production technology could lead, in the future, to pearl culture activity [5].

In addition, a recent development occurred in Iran in the culture of two sea cucumber species (*Holothuria lecospliota* and *Holothuria scabra*) and black-lip pearl oyster (*Pinctada margaritifera*).

Fish processing industry: IFO has divided fish processing industry into four sections:

- Fish canning factory.
- Fish meal factory.
- Freezer and refrigerator units in shoreline.
- Fish processing unit.

At present, there are 134 fish canning factories, 46 fish meal factories, 122 freezer and refrigerator units in shoreline and 143 fish processing units. Number and production capacity of fishery processing industry of Iran between 2004 and 2014 are presented in Table 5.

In the last decade, Iranian fishery processing industry increased in all sections except fish meal production. Although the number of fish meal factory increased the fish meal production decreased in recent years. The reason of this decrease is due to the fluctuations in Caspian Sea sprat catch. For example, fish meal production reached to the maximum in 2005 and 2006 when the Caspian Sea sprat catch reached to the maximum.

	2004	2005	2006	2007	2008	2009	2010	2011	2012	2013	2014
Fish canning factory	113	118	127	134	134	134	134	134	134	134	134
	458.95	491	543	569	569	569	569	569	569	569	569
Fish meal factory	38	44	44	36	36	36	46	46	46	46	46
	960	1100	1100	910	921	921	921	921	921	921	921
Freezer and refrigerator unit	133	126	109	112	113	113	114	114	120	120	122
	86.9	101	91	96	116	126	126	126	151	151	159
Fish processing unit	116	119	125	125	129	132	135	135	142	142	143
	1891	1940	2038	2038	2067	2135	2179	2179	2255	2255	2266

Table 5: Number (up lines) and production capacity (down lines) of fishery processing industry in Iran between 2004 and 2014 [2,3].

Exports

Strong infrastructure (such as access to open waters) and high volume product (marine shrimp, marine fish, carp species and rainbow trout) of Iran have caused a continuous increase in fisheries exports. Iran's exports worth and amount displayed a steady increase in the last decade. The value of aquatic products export in 2014 was more than 300 million US$. The worth of export between 2004 and 2014 is presented in Figure 3.

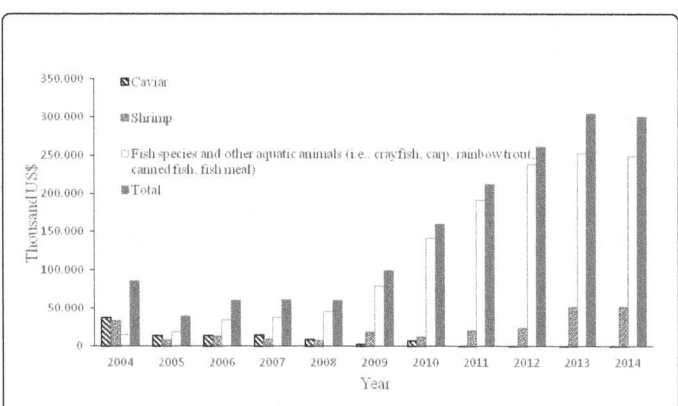

Figure 3: Worth (thousand US$) of aquatic products export of Iran between 2004 and 2014 [2,3].

Caviar is one of the most valuable export products of Iran. In 2014, the price of caviar increased 94.1% in comparison to that of 2013. The price of one kg Iranian caviar reached to 1,652 US$ in 2014. In 2015, Iran exported 1,029 kg caviar to Japan, Germany, England, Italy, Belgium, the South Korea, Norway and United Arab Emirates.

Other important product is shrimp. Iran exports shrimp to 40 courtiers in the world. The worth of shrimp exported from Iran in 2014 was more than 56 million US$. Countries in the Southeast Asia, Iraq, Kuwait, Lebanon and United Arab Emirates are the major export markets for Iranian shrimp. Iran also exports shrimps to some European countries such as Spain, Italy, Norway and Turkey. Fisheries export amount (ton) between 2004 and 2014 is presented in Table 6.

	2004	2005	2006	2007	2008	2009	2010	2011	2012	2013	2014
Caviar	38.4	9.1	9.98	6.6	2.3	0.4	4	0.3	0.4	1.2	0.8
Shrimp	7681	1918.7	2986	2289.1	1346.2	3801.1	2602	4141	4904	11585	11610
Fish and other aquatic animals	12610	14856	27312	31102	23028	29633	41894	52673	59096	60219	60182
Total	20329	16784	30308	33398	24376	33435	44500	56814	64000	71805	71793

Table 6: Amount of fisheries export (ton) from Iran between 2004 and 2014 [2,3].

Fish consumption in Iran

In Iran, fish consumption per person was 4.5 kg in 1997. However, fish consumption in Iran has increased from 5.2 kg in 2002 to 9.2 kg in 2014. Global per capita fish consumption has risen to above 20 kg.

Although fish consumption is increasing from year to year, on the other hand, fish consumption in Iran is still behind the world average. Fish consumption per capita between 2004 and 2014 in Iran is presented in Table 7.

	2004	2005	2006	2007	2008	2009	2010	2011	2012	2013	2014
Fish consumption	6.7	7.03	7.7	7.35	7.32	7.51	8.5	9.1	10.2	8.5	9.2

Table 7: Fish consumption (kg per capita) between 2004 and 2014 in Iran [2,3].

Fisheries and aquaculture facilities

Fish farm numbers and fish farm areas: Fish farm numbers and their areas raised very fast in recent years. For example, fish farm number increased from 4,859 in 2002 to 18,795 in 2014. Fish farm area for cyprinids increased from 25,890.6 hectares in 2004 to 50,853

hectares in 2014 (approximately 96% increment). This increment was 0.4% for rainbow trout farms. Fish farm area for rainbow trout increased from 104.6 hectares in 2004 to 225 hectares in 2014. Number and area of fish farms between 2004 and 2014 are presented in Table 8.

	2004	2005	2006	2007	2008	2009	2010	2011	2012	2013	2014
Cyprinid	6084	6319	6863	7261	7923	8362	10527	11968	14295	14615	16254
	25891	28332	29836.7	33793	31892	34504	40261	43722	46587	48697	50853
Rainbow trout	662	698	750	1200	1085	1180	1387	1607	1907	1923	1595
	104.6	132	111.4	162.6	157	169	230	236.5	258	230	225
Harvesting from natural water resources	220	240	356	307	283	351	332	296	367	412	428
	450000	848500	570183	545287	455709	499117	496579	485259	555515	562227	746096
Shrimp	310	298	189	208	219	145	214	209	320	313	518
	4272	3641	2625.7	1207	2481	2148	2873	3220	4427	4779	7053
Total	7276	7555	8158	8976	9510	10038	12460	14080	16889	17285	18795
	480267	880605	602756	580449	490239	535938	539943	532437	606787	615933	804227

Table 8: Number (up) and area (hectare) of fish farms (down) in Iran between 2004 and 2014 [2,3].

The number of employees in the fisheries sector: Number of employees in the fisheries sector increased as fisheries industry enlarged after 2004. For example, the number of employees in the fisheries sector increased from 144,584 persons in 2002 to 208,472 persons in 2014. Number of employees in the different parts of fisheries sector between 2004 and 2014 are presented in Figure 4. The number of fishermen in the north and south waters did not increase during last decade. In contrast, the number of fish farmers increased dramatically from 16,894 in 2004 to 68,287 in 2014 (approximately 24% increment).

Number of fishing fleets: The number of fishing fleets (boat, doha dhow, fishing ship) does not show any significant differences between 2004 and 2014. Number of fishing fleets between 2004 and 2014 is presented in Table 9.

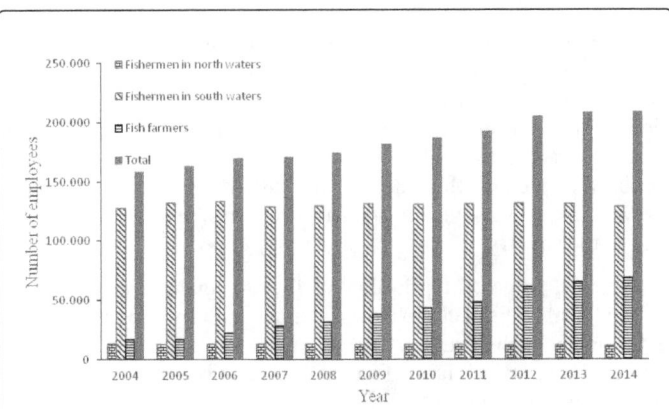

Figure 4: Number of employees in the different parts of fisheries sectors in Iran between 2004 and 2014 [2,3].

		2004	2005	2006	2007	2008	2009	2010	2011	2012	2013	2014
Caspian Sea	Boat	582	713	1007	980	980	890	827	811	804	865	825
	Doha Dhow	152	145	146	135	115	81	73	73	73	73	73
	fishing Ship	-	-	-	-	-	-	-	-	-	-	-
	Total	734	858	1,153	1,115	1,095	971	900	884	877	938	898
Persian Gulf and Gulf of Oman	Boat	7,496	7,563	7,663	7,847	7,970	7,932	7,855	7,689	7,520	7,423	7,385
	Doha Dhow	3,210	3,250	3,257	2,999	3,033	3,066	3,087	3,090	3,135	3,151	3,165
	Fishing Ship	77	78	47	45	44	47	51	54	54	51	50
	Total	10,783	10,891	10,967	10,891	11,047	11,045	10,993	10,833	10,709	10,625	10,600

Table 9: Number of fishing fleets in Iran between 2004 and 2014 [2,3].

Discussion and Conclusion

Iran has a great potential for fisheries and aquaculture production in its both freshwater and marine resources. It seems that potential for mariculture will also significantly enlarge with the completion of the cage aquaculture projects in the north and south of Iran. In addition, Although Iran has the potential to produce 900,000 tons fish in cages according to the Development Strategy Department of IFO the total production from cage aquaculture will increase up to 400,000 tons in 2025. The cage culture potential for Caspian Sea, Persian Gulf and Gulf of Oman are 300,000, 150,000 and 450,000 tons respectively.

Statistics in Iran show that fishery productions between 2004 and 2014 have significantly been developed and freshwater aquaculture has achieved remarkable attention due to high demand for aquatic products because of the fact that fast growing rate of human population and increase in fish consumption of per person in Iran. Therefore, fishery sector is considered as one of the most important promising industries of Iran economy.

However, although the suitable aquatic production in Iran is about 947,352 tons this industry is still far away from the production aims of the Iranian government. There is no statistics available to see the contribution of fisheries in Gross Domestic Products (GDP) of Iran. Nevertheless, the agriculture sector is not an important part of the economy at present; the whole sector (fisheries, agriculture, poultry and animal husbandry) was estimated to contribute 13% of GDP in 2014 in the country.

The most important priorities of IFO to financially support are shrimp farming, fish culture in cages, lantern fish (myctophids) fishery and sturgeon fish farming. For example, Iranian Government has started to support fish farms financially especially for cage culture and sturgeon fish farming.

Conversely, there are difficulties in Iranian fisheries, such as shortage of experts, qualified technical stuff and lack of high technology in cage culture (i,e., importing cages, automatic feeding machinery, water quality monitoring, etc.). Despite to having high potential for doing mariculture in cages, fish cage culture has not developed well and needs future investments and supports.

Furthermore, aquaculture in Iran is associated with other problems such as low stocking density large earthen ponds for shrimp farming, lack of technical knowledge among farmers, diseases especially white spot disease in shrimp culture, unsuitable feed quality especially for juvenile diets, improper feeding management, low water quality in some of aquaculture sites, low hatching and survival rate in larval production units, low quality seed production, improper brood stock production especially in shrimp aquaculture industry, financial problem, and low cultural species diversity [5]. Therefore, it can be concluded that although Iran has a good potential to improve fish production it is necessary to overcome above problems.

References

1. Mousavi A, Karimi J, Mohammadi AA, Vafayi F (2008) Determination the length of coast line north and south Iran. 8th International Conference of Coasts, Ports and marine structures, Ports and Shipping Organization, Tehran, Iran.

2. Iran Fisheries Organization (IFO) (2013) Annual Iranian Fisheries Statistics between 2002 and 2012. Fisheries Design and Program Office, Tehran, Iran pp: 64.

3. Iran Fisheries Organization (IFO) (2015) Annual Iranian Fisheries Statistics 2013 and 2014. Fisheries Design and Program Office, Tehran, Iran pp: 33.

4. FAO (2016) The State of World Fisheries and Aquaculture 2016, Contributing to food security and nutrition for all, Italy, Rome pp: 200.

5. Kalbassi MR, Abdollahzadeh E, Salari-Joo H (2013) A review on aquaculture development in Iran. Ecopersia 1: 159-178.

6. Karimpour M, Harlıoğlu MM, Khanipour AA, Abdolmalaki S, Aksu O (2013) Present status of fisheries in Iran. Journal of Fisheries Sciences 7: 161-177.

7. Kardavani P (2012) Iran Water Resources. Tehran University Press, Tehran, Iran pp: 420.

8. Keivany Y, Nasri M, Abbasi K, A Abdoli (2015) Atlas of inland water fishes of Iran. Iran Department of Environment Press, Tehran.

9. Birstein VJ, Waldman JR, Bemis WE (2006) Sturgeon biodiversity and conservation. Envir Biol of Fis 48: 13.

10. Bronzi P, Rosenthal H, Gessner J (2011) Global sturgeon aquaculture production: an overview. J Appl Ichthyol 27: 169-175.

11. Regunathan C, Kitto MR (2005) Persian Gulf fish culture in Iran-pointers for success. Aquaculture Asia 10: 40-42.

12. Hajirezaee S, Ajdari D, Matinfar A, Aghuzbeni SHH, Rafiee GRA (2015) preliminary study on marine culture of Asian sea bass, Lates calcarifer in the coastal earthen ponds of Gwadar region, Iran: an assessment of growth parameters, feed intake efficiency and survival rate. J Appl Anim Res 43: 309-313.

13. Karimpour M, Harlıoğlu MM, Aksu Ö (2011) Status of freshwater crayfish (Astacus leptodactylus) in Iran. Knowl Manag Aquat Ec 401: 18.

14. Dorafshan S, Kalbassi MR, Pourkazemi M, Amiri BM, Karimi SS (2008) Effects of triploidy on the Caspian salmon Salmo trutta caspius hematology. Fis phy and bio 34: 195-200.

33

To Reduce Mortality of Fry Fish (*Oncorhynchus mykiss*) Caused with Viral Infection (IPNV and VHSV) by Water Treatment with Chloramin-T as Disinfectant

Saeed Ganjoor M*

Genetic and Breeding Research Centre for Cold Water Fishes (Shahid-motahari Centre), Iranian Fisheries Science Research Institute, agricultural Research education and extension Organisation (areeo)Yasuj, Iran

*Corresponding author: Saeed Ganjoor M, Microbiologist, Genetic and Breeding Research Centre for Cold Water Fishes (Shahid-motahari Centre), P.O Box: 75914-358, Yasuj, Kohgiloye-va-Boyerahmad Province, Iran, E-mail: msg_isrc@yahoo.com

Abstract

In winter of 2015 we observed gross mortality of fry fish in some tanks of a hatchery in Iran. They had dissonant swimming, spiral swimming, skin darkness, abdominal distension, and anorexia. At the beginning, mortality was low but it increased more and more during the several days. The fry were in fiberglass tanks with 1000 litres of water. It was about 20000 fry in each tank at the beginning. Two activities did synchronously while mortality observation. At the first, some fry sampled from each tanks and sent to laboratory for pathogen detection. The next, 9 tanks selected and grouped as 3 treatments (control, treatment-1 and treatment-2). Control treatment was consisting of 3 tanks that they had not mortality. Treatment-1 and Treatment-2 have the highest mortality and each of them was consisting of 3 tanks. Then, 10 ppm chloramin-T as disinfectant compound added to each tanks of treatment-1 during 1 hour in 3 continuous days (3 times). The tanks of treatment-2 added no drug. After 7 days mortality of fry in each tank estimated and compared with each other. Survival in tanks of treatment-1 was about 76% while survival in tanks of treatment-2 was about 27% while survival in control tanks was about 98%. One month later, results of laboratory tests reported. We found that fish of control tanks were safe (without pathogen) while fish of treatment-1 and treatment-2 were infected with IPN-virus and VHS-virus based on RT-PCR test. Totally 36 fry had been examined by RT-PCR. We founded that 10 fry were IPN+ and 2 of them were VHS+. Clearly; results showed that chloramin-T is able to control the viral infection of *Oncorhynchus mykiss* ($p < 0.05$) in statistical comparison, it confirmed with SPSS software by using Anova-test. Chloramine-T increased surveillance from 27% to 76% while viral contamination had been confirmed.

Keywords: Chloramin-T; *Oncorhynchus mykiss*; Virus; Control; Infection

Introduction

Aquaculture is developing in our era. Totally aquaculture productions have been increased from 41 million tons in 2004 to 55.1 million tons in 2009, so it increased 2.63 million tons per year. There are conflicts over aquaculture development. Aquaculture is going to produce more food for the human while disease out breaks in aquatic animals that makes food losing and economic loss [1]. Infectious diseases in farmed fish such as viral diseases are a significant economic problem for aquaculture producers. There are several important viral diseases in rainbow trout fish (*Oncorhynchus mykiss*). IPNV (Infectious Pancreatic Necrosis Virus) and VHSV (Viral Haemorrhagic Septicemia Virus) are major viral pathogen for the fish which causes high mortality especially in small fish. They reported from Iran [2-4].

VHS is pathogen for a broad range of aquatic animals. It causes disease at 48 fish species for which there is conclusive evidence of susceptibility with VHSV. VHSV belongs to the genus *Novirhabdovirus*, within the family Rhabdoviridae. Virions are bullet-shaped (approximately 70×180 nm in size), contain a negative-sense, single-stranded RNA genome of approximately 11,000 nucleotides. Diseased fish may display nonspecific clinical signs in the early stages of infection, including rapid onset of mortality (which can reach up to 100% in fry), lethargy, darkening of the skin, exophthalmia, anaemia (pale gills), haemorrhages at the base of the fins, gills, eyes and skin, and a distended abdomen due to oedema in the peritoneal cavity. VHS can also occur in a nervous form, characterized by severe abnormal swimming behaviour, such as constant flashing and/or spiralling. Mortality varies, depending on many environmental and physiological conditions, most of which have not been fully determined. In generally, the disease is a cool or cold water disease with highest mortality at temperatures around 9-12°C. Small rainbow trout fry (0.3-3 g) are most susceptible with the virus (genotype Ia) while mortalities close to 100%, but all sizes of rainbow trout can be affected with mortalities ranging from 5 to 90%. Although research on vaccine development for VHS has been continued for more than three decades, a commercial vaccine is not yet available. No therapies are currently available. Several immunostimulants, such as yeast-derived beta-glucans, IL-1β-derived peptides, and probiotics have been assessed for enhancing protection against VHS. Several authors reported positive effects, but no immune-stimulant directed specifically at enhanced resistance to VHS is available [5,6].

IPNV is a highly contagious viral disease of young fish of salmonid species held under intensive rearing conditions. The disease most characteristically occurs in rainbow trout (*Oncorhynchus mykiss*), brook trout (*Salvelinus fontinalis*), brown trout (*Salmo trutta*), Atlantic salmon (*Salmo salar*), and several Pacific salmon species (*Oncorhynchus* spp.). Fish susceptibility generally decreases with age increasing. Clinical signs include darkening pigmentation, a pronounced distended abdomen and a corkscrewing-spiral swimming motion. Cumulative mortalities may vary from less than 10% to more than 90% depending on the combination of several factors, such as virus strain and quantity, host and environmental conditions. The disease is transmitted both horizontally via the water route and vertically the egg. The causative agent, IPNV, is a bi-segmented double-stranded RNA virus belonging to the family Birnaviridae. Control methods currently rely on the implementation of control policies and hygiene practices in salmonid husbandry, through the avoidance of the introduction in fertilized eggs originating from IPNV-carrier brood stock, and the use of a protected water supply. In outbreaks, in the population density a reduction may help to reduce the overall mortality. No treatment or entirely effective vaccine is available at present [7].

Chloramin-T is a disinfectant compound which applying to inactivate some pathogens. Linear formula of chloramin-T is $C_7H_7ClNO_2S.Na$ ($3H_2O$) and its IUPAC name is: N-chloro 4-methylbenzenesulfonamide, sodium salt. Chloramin-T is white powder. Its solubility in water is 150 g/L (25°). It is well known antimicrobial agent especially as antiviral agent for sanitation.

However, there are several strategies to control of disease in aquaculture such as sanitation, antibiotic and drug recommendation, vaccination, probiotics recommendation and use of disinfectant compounds. We use of a disinfectant to control mortality in fry fish successfully. It was Chloramin-T.

Material and Methods

Sampling

About 6 fry fish gathered from each tank. Moribund fry which showed sign of disease (bad swimming) gathered from tanks which have the highest mortality rate. The fry of other tanks (Control) had not sign of disease. Each samples (each fry) immersed in a sterile-plastic tube which contained about 5 ml VTM (Virus transport medium). The tubes accommodated in a special-transport tank (nitrogen tank) that which semi filled with liquid nitrogen. It transferred to a laboratory to do viral detection.

Virus detection method

It consists of two steps. At the first step, samples cultured on cell lines. Second step was PCR-test which done based on RT-PCR (Reverse Transcriptase-PCR) method.

Culture on cell line: EPC and BF-2 cell lines had growth at EMEM (Eagle's minimum essential media) 24 hours earlier. Then, Each sample (fry fish) homogenized with about 2 ml (PBS, pH: 7) and filtrated with 0.45 µ membrane filter. About 500 µl of filtrate added to EPC cell line as much as BF-2 cell line. Cell culture plates incubate at 15°C for 7 days and evaluated for CPE effect of virus. Samples which had been CPE positive selected and they evaluated with RT-PCR (reverse transcriptase-PCR) method by means of viral detection [5,8].

RNA extraction and PCR method: One-step RT-PCR method done. Qiagen OneStep RT-PCR System applied according to the manufacturer's instructions.

Primers for PCR:

- Primer which handled for VHS-virus detection [8].
- VHS3: CGGCCAGCTCAACTCAGGTGTCC,

- VHS4: CCAGGTCGGTCCTGATCCATTCTGTC. Primer which handled for IPN-virus detection [8].
- WB1:CCGCAACTTACTTGAGATCCATTATGC, and
- WB2: TCTGGTTCAGATTCCACCTGTAGTG.

Antiseptic agent: It was chloramin®-T. It was made by BOCHEMIE Company.

Method of antiseptic application

10 grams of chloramin-T dissolved in 20 litres of water and added to each tank of treatment-1 slowly (During 20 minutes). A plastic barrel with a tap used to do it. Water of the each tank was about 1000 L and its draining was about 30 L/min. However, 10 ppm of chloramin-T added to each tank while water flow of each tank was about 30 L/min, thus water of each tank was refreshing during 33 minutes. Drug concentration increased slowly during of 20 minutes then it decreased slowly by water flow of tanks. So, during the day, concentration of antiviral agent had a dynamic rule. The procedure was done for 3 days frequently and 30 grams drug used totally for a 1000 L tank. The tanks were in shade not under the sunlight. They were in a hall.

Statistical method

One-way Anova test used as statistical method and SPSS (ver: 21) software applied to do it [9].

Results

In this study, there were three treatments; each of them had been 3 replications. Fry fish of control treatment had very low mortality. They were active without sign of disease such as abnormal swimming, so we believe they were free of pathogen. It confirmed by laboratory test. Fry fish of control tanks show surveillance 97.36% to 97.77% at the end of study (Table 1). It was acceptable as a normal condition. Fry fish of other tanks show mortality which increased during the time. It believed that an exotic agent caused the mortality. Pattern of mortality was as like as infectious disease. Some fry fish show sign such as lethargy, skin darkness, exophthalmia, abdomen distended and abnormal swimming. They were swimming without harmony with flock of fish. At the next days they settled at bottom of tank and were deceased a few hours later. Some moribund fry fish of the tanks gathered and send to laboratory for diagnostic test such as virological tests by PCR method. The tests were time consuming and needed to 7-10 days for reporting. On the other hand; mortality was increased, so it was necessary to do something rapidly. We decided to do research on it. Tanks with mortality arranged in two treatments, each of them with 3 replication and called treatment-1 and treatment-2. We use chloramin-T as disinfectant for inactivation of unknown infectious agent while did nothing to control of mortality for tanks of treatment-2. Chloramin-T applied for 3 days by tanks of treatment-1. After 3 days we found rapturous results. Fry fish of tanks of treatment-1 show mortality less than treatment-2. They show decreasing of mortality during the next days. Mortality was controlled in tanks of treatment-1. Fry fish of treatment-2 show increasing in mortality unlike of treatment-1. They show mortality more and more, it continued after the study. One week after that chloramin-T recommendation, surveillance of fry fish of each tank estimated (Table 1). It analysed by SPSS software with ANOVA-test. Total surveillance of fry fish of Control was 97.56% while surveillance of Treatment-1 and Treatment-2 were 76.47% and 26.66% orderly. Difference between surveillance of treatments with control confirmed by statistical test

(P<0.05) (Tables 2 and 3). On the other hand, difference between surveillance of treatment-1 with treatment-2 confirmed by statistical test (P<0.05). Laboratory results show that fry fish of treatment-1 and treatment-2 were infected by IPN-Virus or VHS-Virus. Frequency of each viral infection mentioned in Table 4.

Treatment	Drug application	Tank[1] number	No of sampled fry	Laboratory test result (Detected virus)[2]	Population of fry at stacking time (Estimated)		Population of fry 7 days after drug recommendation while signs of disease had started (Estimated)		Surveillance (%): (7 days after drug recommendation)	
Control	No	1	6 (Fry fish)	Negative (Virus didn't detect). VHS-and IPN-	19000	Mean: 20500	18500	Mean: 20000	97.36	Mean: 97.56% =100(20000/20500)
		2	6	VHS- and IPN-	22500		22000		97.77	
		3	6	VHS- and IPN-	20000		19500		97.50	
Treatment-1	Yes (10 ppm of Chloramin-T used once a day for 3 days repeatedly).	4	6	IPN+(2 of 6 sample) VHS-(0 of 6 sample)	19500	Mean: 19833	14500	Mean: 15166	74.35	Mean: 76.47%
		5	6	IPN+(1 of 6 sample) VHS-(0 of 6 sample)	22000		17500		79.54	
		6	6	IPN-(0 of 6 sample) VHS+ (1 of 6 sample)	18000		13500		75.00	
Treatment-2	NO	7	6	IPN+(1 of 6 sample) VHS-(0 of 6 sample)	18500	Mean: 20000	6000	Mean: 5333	32.43	Mean: 26.66%
		8	6	IPN+(1 of 6 sample) VHS-(0 of 6 sample)	20500		4500		21.95	
		9	6	IPN-(0 of 6 sample) VHS+(1 of 6 sample)	21000		5500		26.19	

[1]-Tank wall was made from a part of a large pipe which made of fiberglass and plastic while its bottom was made of concrete. Each tank contains 1000 liters of fresh water. Fresh water added to each tank with a valve by 30 L/min and drained by the same rate.

[2]-Viral detection done by RT-PCR method following cell culturing.

Table 1: The properties of tanks and treatments.

Group (Treatment)	Mean of surveillance	Std. error of mean	F	P
Control	20000	1040.83		
Treatment-1	15166	1201.85	61.56	0.000
Treatment-2	5333	440.95		

Table 2: Variance analysis.

Group	Group	Mean difference	P
Control	Treatment-1	4833	0.032
	Treatment-2	14666	0.000
Treatment-1	Control	-4833	0.032
	Treatment-2	9833	0.001
Treatment-2	Control	-14666	0.000
	Treatment-1	-9833	0.001

Table 3: ANOVA result and multiple comparisons by scheffe-test.

Treatment	Tank number	Frequency of infected fry with IPN-virus (%)	Frequency of infected fry with VHS-virus (%)	Total frequency of infected fry with viruses (%)
Control	1	0	0	0
	2	0	0	
	3	0	0	
Treatment-1	4	33.33	0	22.22
	5	16.66	0	
	6	0	16.66	
Treatment-2	7	16.66	0	16.66
	8	16.66	0	
	9	0	16.66	

Table 4: Frequncy (%) of infected fry fish in each tank based on laboratory test.

Discussion

Surveillance of fry fish of 3 groups (Control, Treatment-1 and Treatment-2) was difference based on statistical test (P<0.05). Fry fish of control were safe so they have high surveillance. Fry fish of Treatment-1 and Treatment-2 were infected with the pathogens so their surveillance could be equal but statistical test show difference between them. Total frequency of infected fry of Treatment-1 was 22.22% while it was 16.66% for Treatment-2 (Table 4). Therefore, fry fish of Treatment-1 were more illness than Treatment-2. Based on laboratory test predicted surveillance of Treatment-2 was more than Treatment-1; it not happened. Surveillance of fry fish of Treatment-1 was 49.81% more than Treatment-2. As a result, the surveillance of fry fish of Treatment-1 not only didn't less than Treatment-2 but also it was more. We ascribe to antiviral effect of chloramin-T the higher surveillance rate of Treatment-1 compare with Treatment-2. Chloramin-T recommended only for tanks of Treatment-1. Based on this study, chloramin-T inhibit mortality of fry fish which caused by IPN-virus and VHS-virus. If chloramin-T didn't recommended, mortality will be about 73.33% while it decreased as much as 23.53% by application of chloramin-T.

Conclusion

While mortality is going to start, it is important that a farmer be able to control of disease before 10th day, otherwise mortality increase day to day and cause high financial damage. Laboratory tests are time consuming; they need 3-10 days to get a result. But, sometimes, you must choice the best activity before getting the laboratory results and use an antiseptic. Chloramin-T is economic, because 30 grams of chloramin-T is sufficient for 20000 fry fish to be safe. So, a breeder can apply it while think a viral disease is going to emerge. Samples (for laboratory) must gather before applying of antiseptics or antibiotics, otherwise laboratory can't detect pathogenic agents.

Chloramin-T isn't an antiviral drug so you can't use it as food additive for antiviral therapy of fry fish. I think that chloramin-T inhibit virus transfer in water so it controlled the mortality. On the other hand, chloramin-T deactivated virus in water and blocked its spreading rule. Therefore, some fry fish survive.

In future; it is require a study to evaluate antiviral effect of Chloramin-T against VHS-virus and IPN-virus on molecular level in an equipped laboratory.

References

1. FAO (2010) The State of World Fisheries and Aquaculture. Food and Agriculture Organization of the United Nations p: 197.
2. Soltani M, Rouholahi S, Ebrahimzadeh-mousavi HA, Abdi K, Zargar A, et al. (2014) Genetic diversity of infectious pancreatic necrosis virus (IPNV) in farmed rainbow trout (Oncorhynchus mykiss) in Iran. Bull Eur Ass Fish Pathol 34: 155.

3.   Raissy M, Momtaz H, Ansari M, Moumeni M, Hosseinifard M (2010) Distribution of Infectious Pancreatic Necrosis Virus (IPNV) in two major rainbow trout fry producing provinces of Iran with respect to clinically infected farms. J Food Agric Environ 8: 614-615.

4.   Stickney RR (2000) Encyclopedia of Aquaculture. John Wiley & Sons, Inc p: 1068.

5.   OIE (2016) Manual of Diagnostic Tests for Aquatic Animals. Viral Haemorrhagic Septicaemia p: 24.

6.   King AMQ, Adams MJ, Carstens EB, Lefkowitz EJ (2012) Virus Taxonomy. Classification and Nomenclature of Viruses. Ninth Report of the International Committee on Taxonomy of Viruses. International Union of Microbiological Societies, Virology Division. Elsevier academic press.

7.   OIE (2003) Manual of Diagnostic Tests for Aquatic Animals, 4th edition p: 142-151.

8.   Williams K, Blake S, Sweeney A, Singer JT, Nicholson BL (1999) Multiplex Reverse Transcriptase PCR Assay for Simultaneous Detection of Three Fish Viruses. J Clin Microbiol 37: 4139-4141.

9.   IBM (2010) IBM SPSS Statistics Base 19: 316.

PERMISSIONS

LIST OF CONTRIBUTORS

Yun-Guo Liu
College of Life Sciences, Yantai University, Yantai 264005, China

Ling-Xiao Liu
College of Life Sciences, Yantai University, Yantai 264005, China
Linyi Academy of Agricultural Sciences, Linyi 276012, China

Shi-Chao Xing
Gout laboratory, The Affiliated Hospital of Qingdao University, Qingdao 266003, China

Wehye AS and Jueseah AS
Bureau of National Fisheries, Ministry of Agriculture, Liberia

Amponsah SKK
Food Research Institute, Box M20, Accra, Ghana

Rahmi Can Ozdemir
Department of Fisheries, Kastamonu University, Kastamonu, Turkey

Aygül Ekici
Department of Fisheries, Istanbul University, Turkey

Ahmed Mohammed Musa Ahmed
Department of Fish Sciences, Neelain University, Khartoum, Sudan

Dipak Pandey
The United Graduate School of Agricultural Sciences, Bioresource Production Science, Ehime University, Japan
South Ehime Fisheries Research Center, Ehime University, Japan

Yong-Woon Ryu and Takahiro Matsubara
South Ehime Fisheries Research Center, Ehime University, Japan

Mosepele K
Senior Research Scholar- Fisheries Biologist, Research Services and Training, Botswana

Kareem OK and Osho EF
Department of Aquaculture and Fisheries Management, University of Ibadan, Nigeria

Olanrewaju AN
Federal College of Freshwater Fisheries Technology, P.M.B 1060, Maiduguri, Nigeria

Orisasona O
Department of Wildlife and Fisheries Management, Osun State University, Nigeria

Akintunde MA
National University of Lesotho, Department of Agriculture, Roma 120, Kingdom of Lesotho, Southern Africa

Jette Jakobsen
National Food Institute, Technical University of Denmark, Kemitorvet, DK-2800, Lyngby, Denmark

Cat Smith
Bantry Marine Research Station, Gearhies, Bantry, Co. Cork, Ireland

Vincent Oké, Youssouf Abou and Alphonse Adité
Laboratory of Ecology and Management of Aquatic Ecosystems (LEMEA), Department of Zoology, University Of Abomey-Calavi, PO Box 526, Republic of Benin

Jean-André T Kabré
Laboratory of Research and Training in Fishing and Wildlife, Institute of Rural Development, Polytechnic University of Bobo-Dioulasso, BP. 1091 Bobo 01, Burkina Faso

Hazem S Abedalhammed and Haitham L Sadik
Department of Animal Production, College of Agriculture, University of Al- Anbar, Iraq

Nasreen M Abdulrahman
Department of Animal Production, Faculty of Agricultural sciences, University of Sulaimani, Iraq

Ikenna Kelvin Obiyor, Christopher Didigwu Nwani, Gregory Ejikeme Odo, Josephine Chinenye Madu, Doris Ulumma Ndudim and Ifeanyi Oscar Ndimkaoha Aguzie
Department of Zoology and Environmental Biology, University of Nigeria, Nsukka, Enugu State, Nigeria

Thomas P Simon and Nicholas J Cooper
School of Public and Environmental Affairs, 1315 E. Tenth Street, Indiana University, Bloomington, Indiana 47405, USA

Imtiaz Ahmed and Amir Maqbool
Fish Nutrition Research Laboratory, Department of Zoology, University of Kashmir, Hazratbal, Srinagar

Frank J Zadlock IV, Satshil B Rana, Zain A Alvi and Wyatt Murphy
Department of Biological Science, Seton Hall University, South Orange, New Jersey, USA

Ziping Zhang
College of Animal Science, Fujian Agriculture and Forestry University, Fuzhou, China

Carolyn S Bentivegna
Department of Chemistry and Biochemistry, Seton Hall University, South Orange, New Jersey, USA

Erkie Asmare and Dereje Tewabe
Bahir-Dar Fisheries and Other Aquatic Life Research Center, P.O. Box: 794, Bahir-Dar, Ethiopia

Sewmehon Demissie
Amhara Regional Agricultural Research Institute, P.O. Box: 527, Bahir Dar, Ethiopia

Edwin Pei Yong Chow, Kah Heng Liong and Elke Schoeters
Kemin Industries (Asia) Pte Limited, 12 Senoko Drive, Singapore 758200, Singapore

Richard Ffrench-Constant and Matthew J Witt
Centre for Ecology and Conservation, University of Exeter, Penryn Campus, Cornwall, TR10 9FE, UK

Stephen Long
Centre for Ecology and Conservation, University of Exeter, Penryn Campus, Cornwall, TR10 9FE, UK
Environment and Sustainability Institute, University of Exeter, Penryn Campus, Cornwall, TR10 9FE, UK
Department of Geography, University College London, Pearson Building, Gower Street, London, WC1E 6BT, UK

Kristian Metcalfe
Environment and Sustainability Institute, University of Exeter, Penryn Campus, Cornwall, TR10 9FE, UK

Thompson OA and Mafimisebi TE
Department of Agricultural and Resource Economics, the Federal University of Technology, Akure, Nigeria

Kebede B
Wacale District Livestock and Fisheries Development Office, Oromia, Ethiopia

Habtamu T
Veterinary Drug and Animal Feed Administration and Control Authority, Ethiopia

Ren DL, Chen M, Ge SC and Bing Hu
Chinese Academy of Sciences Key Laboratory of Brain Function and Disease, China

Yajuan Li
Chinese Academy of Sciences Key Laboratory of Brain Function and Disease, China
Laboratory of Structural Immunology and School of Life Sciences, University of Science and Technology of China, P R China

Austin Saye Wehye
Bureau of National Fisheries, Ministry of Agriculture, Liberia

Patrick K Ofori-Danson and Angela Manekuor Lamptey
Department of Marine and Fisheries Sciences, University of Ghana

Ayissi I
University of Abdelmalek Essaâdi, Department of Biology, Faculty of Science, Tetouan 2121, Morocco
Cameroon Marine Biology Association, Morocco
Specialized Research Center for Marine Ecosystems in Kribi-Cameroon, Cameroon
Institute of Fisheries and Aquatic Sciences (ISH) at Yabassi, University of Douala, PO Box 2701, Douala, Cameroon

Jiofack TJE
Sub-Regional School and Postdoctoral Water Development and Integrated Management of Forests and Tropical Territories, Kinshasa, RDC, Congo

Muhammad Talib Kalhoro, Mu Yongtong, Shah Syed Babar Hussain, Memon Aamir Mahmood and Mohsin Muhammad
Ocean University of China, College of fisheries, Qingdao, Shandong, China

Kalhoro Muhsan Ali
Faculty of marine Sciences, Lasbela University of Agriculture, Water and marine Sciences, 90150, Balochistan, Pakistan

Pavase Tushar Ramesh
College of Food Science and Engineering, Seafood Safety Lab, Ocean University of China, Qingdao, 266003, China

Huicab-Pech ZG, Castaneda-Chavez MR and Lango-Reynoso F
National Technological Institute of Mexico/ Technological Institute of Boca del Rio Veracruz, Mexico

Perumal Rajakumaran and Baskralingam Vaseeharan
Department of Animal Health and Management, Alagappa University, Karaikudi 630003, Tamil Nadu, India

Kidanie Misganaw and Addis Getu
Department of Animal Production and Extension, University of Gondar, P.O. Box: 196, Gondar, Ethiopia

Debraj Kole and Apurba Ratan Ghosh
Ecotoxicology Lab, Department of Environmental Science, The University of Burdwan, Golapbag, Burdwan 713104, West Bengal, India

Palas Samanta
Ecotoxicology Lab, Department of Environmental Science, The University of Burdwan, Golapbag, Burdwan 713104, West Bengal, India
Division of Environmental Science and Ecological Engineering, Korea University, Anam-dong, Sungbuk-gu, Seoul 02841, Republic of Korea

Sandipan Pal
Department of Environmental Science, Aghorekamini Prakashchandra Mahavidyalaya, Subhasnagar, Bengai, Hooghly 712611, West Bengal, India

Aloke Kumar Mukherjee
Department of Conservation Biology, Durgapur Government College, Durgapur 713214, West Bengal, India

Nawwar Zawani Mamat, Mohd Idrus Shaari and Nur Amirul Anas Abdul Wahab
Marine Technology Programme, Universiti Teknologi MARA (Perlis), 02600 Arau, Perlis, Malaysia

Adamu Yimer, Minwyelet Mingist
Department of Fisheries, Wetlands and Wildlife Management, College of Agriculture and Environmental Sciences, Bahir Dar University, P.O. Box 5501, Bahir Dar, Ethiopia

Behailu Bekele
School of Food and Chemical Engineering, Bahir Dar Institute of Technolgy, Bahir Dar University, P.O. Box 79, Bahir Dar, Ethiopia

Nisreen E Mahmoud and MM Fahmy
Department of Parasitology Faculty Of Veterinary Medicine, Cairo University, Egypt

Mohga FM Badawy
Department of veterinary hygiene and management, Faculty Of Veterinary Medicine, Cairo University, Giza 11221, Egypt

Ayeloja AA and Adebisi GL
Fisheries Technology Department, Federal College of Animal Health and Production Technology Moor Plantation, PMB 5029, Ibadan, Nigeria

George F
Department of Aquaculture and Fisheries Management, Federal University of Agriculture, Abeokuta (FUNAAB) PO Box 2240, Abeokuta, Nigeria

Sodeeq E
Department of Agric. Extension and Management, Federal College of Animal Health and Production Technology Moor Plantation, PMB 5029 Ibadan, Nigeria

Harlioglu MM and Farhadi A
Department of Fisheries, Fırat University, Elazig, Turkey

Saeed Ganjoor M
Genetic and Breeding Research Centre for Cold Water Fishes (Shahid-motahari Centre), Iranian Fisheries Science Research Institute, agricultural Research education and e tension Organisation (areeo)Yasuj, Iran

Index

A

Aflp Marker, 1

Allometry, 36, 38

Aquaculture, 1, 5, 11, 16, 23, 26, 36, 40, 45, 52, 55, 72, 83, 89, 95, 98, 101, 104, 110, 115, 119, 122, 138, 153, 156, 162, 174, 180-183, 195-196, 200, 203, 206, 209, 212

Artisanal, 7, 10, 31, 35, 37, 93, 95, 103, 110-111, 116, 131-132, 137, 140, 142, 144

Artisanal Fisheries, 10, 116, 140, 144

Assembly Validation, 83

B

Bacteria, 87, 97, 118, 121, 126, 155-161, 181, 183, 186-188, 195

Barley, 52-55

Barley Sprout Powder, 52-55

Benthic, 56-57, 60-64, 69, 108, 164, 183, 194

Biochemical Parameters, 71, 178

Blood, 15, 45-46, 55, 71, 76, 80, 89, 99-102, 121, 126-127, 129-130, 178-179, 195

By-catch, 49-50, 140-144, 164, 167

C

Cameroon, 9-10, 134-135, 137, 140-144, 188

Catfish Aquaculture, 97, 110

Circadian Rhythm, 22

Clarias Gariepinus, 26, 30, 38-39, 45, 47-50, 56, 60, 81, 91, 93, 97, 117, 119, 170, 181, 183-184, 187, 194-195, 199

Classical Fisheries Management, 28-29, 31

Coastal Waters, 1, 4, 6, 8-10, 63, 131, 136-138, 162

Common Carp, 16, 23, 26, 52-55, 71, 77, 90, 101-102, 194, 202

Condition Factor, 36-39, 48, 154

Crayfish, 64-70, 79, 200, 202-203, 207

D

Danio Rerio, 11, 15-16, 84, 129

De Novo Assembly, 83-84, 88-90

Disease, 11, 15-16, 50, 55, 94, 99, 101-104, 117-118, 121-123, 128, 155-157, 159-161, 194-195, 206, 208-209, 211

Dolphin, 140-141

E

Earthen Ponds, 45-48, 113, 161, 206-207

Economic Management, 145

Efficiency Ratios, 110-111

Eleyele Lake, 36-38

Encapsulated Butyric Acid, 97-99, 101

Exploitation Rate, 6-9, 131, 133-134, 136-138, 152

F

Fingerlings, 45-46, 49-50, 52-53, 72, 74, 76, 79, 81-82, 94, 110-115, 180

Fish, 5-12, 14-23, 40, 43, 56, 60-63, 67, 69-72, 74-86, 88-102, 108-122, 125, 129-132, 134-138, 142, 146, 150-156, 159-162, 168-176, 192, 200, 206, 209, 211-212

Fish Species, 6-7, 17, 25, 31, 36, 39, 49, 56, 61-62, 72, 77-79, 86, 89, 91, 95, 119, 122, 132, 138, 146, 154, 170, 173, 176, 178, 182, 188, 190, 196-197, 200, 203, 208

Fishery, 1, 6-7, 9-10, 26, 29, 31-36, 43, 72, 79, 83, 91-95, 103-110, 118, 121, 132, 138, 145-146, 150, 153-154, 162-163, 172, 174, 200, 204, 206

Fishmeal Replacement, 45

Flood Pulse, 28-31, 33, 35

Floodplain Fisheries, 28-30, 32, 34

Full Fish, 110-115

G

Genetic Variability, 1, 4, 25, 109

Growth Performance, 6-7, 9, 45, 47, 49-50, 52, 55, 74, 77, 79-81, 99-101, 131, 133, 137, 145, 147, 150-151, 182

Gynogenesis, 11-12, 14-16

H

Habitat Suitability, 64

Heat-shock, 11, 14

Hepsetus Odoe, 36-38

Hybrid Catfish, 97-100

Hydroponic Germination, 52, 54

I

Immune Parameters, 97-98, 101

Indian Scad, 145-146, 150, 152-154

Inflammation, 123-130, 155, 157, 159-160

Intestine, 51-52, 54-55, 101, 117-121, 155-157, 159, 161, 175-180, 191

Investigation Production, 17

J

Jebel Aulia Dam, 17, 20-21

Juveniles, 6, 26, 50-51, 67, 79-80, 82, 110-115, 153, 162, 168-169, 181, 192

K

Kidney, 117-121, 156, 159, 180

L

Langanoo, 117-121

Length-weight Relationship, 36-39, 145-147, 149-150, 154

Leukocyte, 76, 123-124, 129

Liberia, 6-9, 131-138

Livelihood Diversification, 91

Liver, 72, 77, 80-81, 117-121, 156-157, 159-160, 179-180

M

Mantee, 140

Marine Fish Viscera, 45-49

Menhaden Fish, 83

Microemulsified Carotenoids, 97-98, 101

Microhabitat Scale, 64-65

Migration, 4, 123-130, 154

Mortality, 6-10, 29, 33, 74, 92, 98-99, 107, 114, 131-132, 134-135, 137-138, 144-147, 149, 151-154, 156, 189-191, 194, 208-209, 211

N

Natural Planting, 52-54

O

Okavango Delta, 28-35

Otamiri River, 56-62

Oysters, 103-109, 155

P

Physicochemical, 27, 56-61, 63, 156-157, 159, 191

Pigmentation, 97-98, 101, 208

Portunus Trituberculatus, 1-2, 4

Profitability, 9, 52, 108, 110-111, 114-115, 138, 196-199

R

Reach-scale, 64, 66, 68

Reservoir, 17, 21, 36, 38, 61-62

River Fisheries, 91, 93-95

S

Salmon, 5, 16, 40, 42-44, 50-51, 78, 81, 90, 101, 121, 203, 207-208

Sardinella Maderensis, 6, 9-10

Sea Turtles, 140-141, 144

Seasons, 17, 20, 29, 36, 59-60, 91-93, 106-108, 153, 168

Shellfish, 102-103, 108-109, 122, 130, 160, 162

Small Scale Fisheries, 33, 103

Sperm Motility, 12, 22-27

Stock Assessment, 7, 9-10, 32-35, 138, 145-146, 150, 153-154

Swimming Crab, 1-4

Synergistic, 97-98, 101

T

Tilapia, 16-18, 27, 31-32, 39, 49-51, 78, 80-81, 84, 95, 100, 102, 115, 117-121, 155-161, 170-173, 179-181, 184, 187-189, 194-195, 198

Transcriptome Analysis, 83, 89-90

Trout, 4, 11, 16, 22, 26-27, 40, 42-44, 50, 78-82, 101-102, 155-156, 161, 179, 200, 202-205, 208, 211-212

V

Variation, 1, 4-5, 9, 17, 20, 26, 35, 40, 56, 58-59, 61, 107, 109, 119-121, 137, 143, 159-161, 164

Vitamin D, 40, 42-44

W

Watershed-scale, 64-68

Whale, 140

Z

Zebrafish, 11-12, 14-16, 23, 26, 84-88, 123-130

Zeway, 117-122